Power Systems

Electrical power has been the technological foundation of industrial societies for many years. Although the systems designed to provide and apply electrical energy have reached a high degree of maturity, unforeseen problems are constantly encountered, necessitating the design of more efficient and reliable systems based on novel technologies. The book series Power Systems is aimed at providing detailed, accurate and sound technical information about these new developments in electrical power engineering. It includes topics on power generation, storage and transmission as well as electrical machines. The monographs and advanced textbooks in this series address researchers, lecturers, industrial engineers and senior students in electrical engineering.

Power Systems is indexed in Scopus

Taulant Kërçi · Federico Milano

Frequency Quality in Power Systems

Definition, Metrics, Control and Operation

Springer

Taulant Kërçi
Future Operations Design Authority
EirGrid Plc
Dublin, Ireland

Federico Milano
School of Electrical and Electronic Engineering
University College Dublin
Dublin, Ireland

ISSN 1612-1287 ISSN 1860-4676 (electronic)
Power Systems
ISBN 978-3-032-04720-5 ISBN 978-3-032-04721-2 (eBook)
https://doi.org/10.1007/978-3-032-04721-2

This Springer imprint is published by the registered company Springer Nature Switzerland AG
The registered company address is: Gewerbestrasse 11, 6330 Cham, Switzerland

Quality takes you out of yourself, makes you aware of the world around you.
Quality is opposed to subjectivity… Quality is not a thing. It is an event… Quality is the event at which awareness of both subjects and objects is made possible.

—Zen and the Art of Motorcycle Maintenance, Robert Pirsig

The frequency is the heartbeat of a power system.

—ENTSO-E

We power engineers have demonstrated an extraordinary ability to mess around with names and definitions.

—Harold Kirkham

To my parents, Sulejman and Sheqere

To Yolanda and Alessandro

Foreword by Eoin Kennedy

This is an excellent and timely book that addresses several key technical challenges related to frequency quality. This topic is of huge interest to EirGrid as we work collaboratively with SONI towards our ambition to be able to safely, securely, and efficiently operate the Ireland and Northern Ireland power system with system non-synchronous penetration levels up to 100% and zero conventional units online by 2035. I want to express my thanks and congratulate the authors for their tireless efforts in producing such a valuable resource for the power system industry.

June 2025

Eoin Kennedy
Director of Innovation and Planning
EirGrid
Dublin, Ireland

Foreword by Marta Val Escudero

Maintaining the frequency within defined limits has become a challenging task in inverter-based low-inertia power systems. This state-of-the-art book, written by two noted researchers on the topic, is a welcome addition to helping system operators address frequency-related risks. It defines, analyzes, and compares the frequency quality of different real-world power systems and provides various operational and structural means and countermeasures that can help prevent the occurrence of poor frequency quality. I take this opportunity to congratulate the authors of this book which will have a wide appeal.

June 2025

Marta Val Escudero
Chair of Study Committee C4—Power System Technical Performance
CIGRE
Dublin, Ireland

Preface

Frequency quality has become a key issue of power systems with high shares of inverter-based generation and converter-based demand. Maintaining a good level of frequency quality during normal and abnormal operations is crucial to achieving a secure operation of power systems.

The authors work in Ireland, a leading country in the study of the dynamic performance and operation of low-inertia systems. The aips is currently the largest in the world that can operate with instantaneous inverter-based generation of up to 75%. At the time of writing, the first author is also part of a recently established task force by the entsoe to address the poor frequency quality in the different European power systems (i.e., ce power system, Nordic, and aips). The material of the book is the result of the authors' research with University College Dublin and practical experience with EirGrid, the Irish tso, for a period of more than 10 years.

This is the first book entirely dedicated to frequency quality and blends power system expertise from both academia and industry. The book considers real-world power systems and describes state-of-the-art mathematical approaches and *ad hoc* software tools to evaluate the various causes of frequency quality problems. The book also provides recommendations on how to maintain acceptable levels of frequency quality.

The book aims to provide power system engineers and researchers with theoretical tools and practical insights into the frequency quality concept, including the challenges and opportunities in keeping it within predefined limits. We hope the book will be a valuable reference for the design of the control, the operation and the planning of current and future inverter-based power systems.

Dublin, Ireland
June 2025

Taulant Kërçi
Federico Milano

Acknowledgments We wish to thank the following people for the fruitful collaborations throughout the recent years, for the interesting discussions and suggestions:

- Eoin Kennedy, EirGrid, Ireland.
- Simon Tweed, EirGrid, Ireland.
- Marta Val Escudero, EirGrid, Ireland.
- David McGowan, SONI, Northern Ireland.
- Manuel Hurtado, EirGrid, Ireland.
- Connor Duggan, EirGrid, Ireland.
- Dusko Nedic, EirGrid, Ireland.
- Bart Franken, EirGrid, Ireland.
- Mo Cloonan, EirGrid, Ireland.
- Muhammad Adeen, EirGrid, Ireland.
- Juan Giraldo Chavarriaga, TNO, Netherlands.
- Mohammed Ahsan Adib Murad, DIgSILENT, Germany.
- Mariglen Gjergji, OST, Albania.
- Emanuele Crisostomi, University of Pisa, Italy.
- Michela Moschella, University of Pisa, Italy.

We also acknowledge the support of EirGrid and SONI, especially with regard to the provision of data and figures.

Last but not least, we thank our families for their unwavering support while writing the book.

Competing Interests The authors have no competing interests to declare that are relevant to the content of this manuscript.

Contents

Notations

General Notation

v	Variable
V	Parameter
$\boldsymbol{v}$	Vector
$\mathbf{V}$	Matrix
$\bar{v}$	Complex variable
$v_d,\ v_q$	Components of the dq-axis reference frame of the Park transform
v_i	Element i of a vector
V_n	Nominal/rated quantity
V_{hk}	Element (h, k) of a matrix
$v_\alpha,\ v_\beta$	Components of the Clarke $\alpha\beta$ transform
$\bar{v}_{dq}$	Park vector, $\bar{v}_{dq} = v_d + \jmath v_q$
v'	Time derivative of v
$v^{(\alpha)}$	Fractional order derivative of v of order α
$\hat{v}$	Hilbert transform of v
$\mathfrak{v}$	Fourier transform of v

Common Symbols

Due to the variety of topics discussed in the book, and in order to use the usual notation found in the literature, some symbols are overloaded. We carefully avoided notation clashes and made sure that the context always makes the meaning of the symbols unambiguous.

0_n	Null vector of order n
1_n	Unitary vector of order n
$a,\ b,\ c$	Coefficients

d	Demand power (UC problem)
B	Susceptance
$\mathbf{B}$	Susceptance matrix
C	Cost
D	Damping
e_i	Unit vector along i-th coordinate
f	Frequency
$\boldsymbol{f}$	Vector of differential equations
$\boldsymbol{g}$	Vector of algebraic equations
G	Conductance
$G(s)$	Transfer function
H	Inertia constant
K	Control gain
i, h, k	Indexes
$\imath$	Current magnitude
$\jmath$	Imaginary unit
ℓ	Length
p	Active power
q	Reactive power
r	Reserve power (UC problem)
R	Resistance, ramp rate
$\mathcal{R}$	Droop coefficient
s	Variable of the Laplace transform
S	Apparent power
t	Time
T	Time period, time constant
u	Input variable
v	Voltage magnitude
w	Wind power (UC problem)
x	State variable
X	Reactance
y	Algebraic variable
$\bar{Y}$	Admittance
$\boldsymbol{Y}$	Admittance matrix
z	Discrete variable
$\bar{Z}$	Impedance
α, β, γ	Factors
γ_f	Skewness
δ	Rotor angle
ε	Error
ζ	White noise
η	Stochastic process
θ	Phase angle of voltage phasor
θ_p	Pitch angle of the blades of a wind turbine
ϑ	Phase angle of instantaneous voltage

κ	Curvature of a space curve
λ	Electricity price
μ	Mean value
ν	Wind speed
ϖ	Probability
ρ	Air density
ϱ	Radial frequency/instantaneous bandwidth ($\varrho = \varsigma'$)
σ	Standard deviation
ς	Natural logarithm of the voltage magnitude ($\varsigma = \log(\upsilon)$)
φ	Phase angle of analytic signal
φ'	Instantaneous frequency
ω	Angular frequency, rotor angular speed
$\mathit{\Omega}$	Set
Ω	Geometric frequency
ω_o	Synchronous reference angular frequency
ω_κ	Azimuthal frequency

Acronyms

ACE	Area Control Error
ACER	Agency for the Cooperation of Energy Regulators
AEC	Australian Energy Council
AEMO	Australian Energy Market Operator
aFRR	Automatic Frequency Restoration Reserve
AGC	Automatic Generation Control
AI	Artificial Intelligence
AIMD	Additive-Increase-Multiplicative-Decrease
AIPS	All-Island Power System
APC	Active Power Control
AUS	Australia
AVR	Automatic Voltage Regulator
AW	Anti-Windup
BA	Balancing Authority
BESS	Battery Energy Storage Systems
BRP	Balance Responsibility Party
CAES	Compressed Air Energy Storage
CAISO	California Independent System Operator
CE	Continental Europe
CL	Consequential Loss
CLI	Consequential Loss for Infeed
CLO	Consequential Loss for Outfeed
CoI	Center of Inertia
CPS	Control Performance Standard
CSIG	Constant-Speed Induction Generator
DAE	Differential-Algebraic Equation
DASSA	Day-Ahead System Services Auction
DC	Dynamic Containment
DER	Distributed Energy Resource
DFD	Deterministic Frequency Deviation
DFIG	Doubly-Fed Induction Generator

DFT	Discrete Fourier Transform
DM	Dynamic Moderation
DR	Disturbance Recorder
DR	Dynamic Regulation
DS3	Delivering a Secure, Sustainable Electricity System
DSM	Demand Side Management
DSO	Distribution System Operator
DSU	Demand Side Unit
d-UC	Deterministic Unit Commitment
EDIL	Electronic Dispatch Instruction Logger
EMS	Energy Management System
EMT	Electro-Magnetic Transient
ENTSO-E	European Network of TSOs for Electricity
ERCOT	Electric Reliability Council of Texas
ESIG	Energy Systems Integration Group
EU	European Union
EV	Electric Vehicle
FAT	Full Activation Time
FCR	Frequency Containment Reserve
FCR-D	Frequency Containment Reserve for Disturbances
FCR-N	Frequency Containment Reserve for Normal Operation
FDR	Frequency Disturbance Recorder
FFR	Fast Frequency Response
FFT	Fast Fourier Transform
FIR	Fast Instantaneous Reserve
FoR	Facilitation of Renewables
FOS	Frequency Operating Standard
FPD	Frequency Probability Distribution
FRCE	Frequency Restoration Control Error
FRCR	Frequency Risk and Control Report
FRR	Frequency Restoration Reserve
FRT	Fault Ride Through
GB	Great Britain
GFL	Grid-Following Converter
GFM	Grid-Forming Converter
HVDC	High Voltage Direct Current
IBR	Inverter-Based Resource
IC	Interconnector
IE	Ireland
ISP	Imbalance Settlement Period
IT	Information Technology
KPI	Key Performance Indicator
LCIS	Low Carbon Inertia Services
LEU	Large Energy User
LF	Loop Filter

LFC	Load Frequency Control
LFSM	Limited Frequency Sensitive Mode
LLTF	Large Loads Task Force
LPF	Low-Pass Filter
LSAT	Look-ahead Security Assessment Tool
LSI	Largest Single Infeed
LSO	Largest Single Outfeed
LTS	Long-Term Schedule
MAGC	Market-based Automatic Generation Control
MC	Monte Carlo Method
mFRR	Manual Frequency Restoration Reserve
MG	Microgrid
MILP	Mixed-Integer Linear Programming
MPC	Model Predictive Control
MPPT	Maximum Power Point Tracking
MUON	Minimum Number of Conventional Units Online
NEM	National Electricity Market
NERC	North American Electric Reliability Corporation
NESO	National Energy System Operator
NI	Northern Ireland
N-S	North-South
OEM	Optimal Energy Management
OF	Over-frequency
OFGS	Over Frequency Generation Shedding
OPRC	Operational Policy Review Committee
ORA	Oustaloup's Recursive Approximation
PCC	Point of Common Coupling
PD	Phase Detector
PEV	Plug-in Electric Vehicle
PFC	Primary Frequency Control
PFR	Primary Frequency Regulation
PI	Proportional Integral
PICASSO	Platform for the International Coordination of Automated Frequency Restoration and Stable System Operation
PJM	Pennsylvania-New Jersey-Maryland
PLL	Phase-Locked Loop
PMU	Phasor Measurement Unit
PN	Physical Notification
POR	Primary Operating Reserve
PSS	Power System Stabilizer
PtG	Power to Gas Demand
PV	Photovoltaic
RA	Regulatory Authority
RES	Renewable Energy Sources
RG	Synchronous Region

RI	Reference Incident
RII	Reference Incident for Infeed
RIO	Reference Incident for Outfeed
RMS	Root Mean Square
RoCoF	Rate of Change of Frequency
RR	Replacement Reserve
RRD	Replacement Reserve Desynchronized
RRS	Replacement Reserve Synchronized
RTC	Real-Time Commitment
RTD	Real-Time Dispatch
SCADA	Supervisory Control and Data Acquisition
s-DFT	Short-time Discrete Fourier Transform
SFC	Secondary Frequency Control
SFR	Secondary Frequency Regulation
SIR	Synchronous Inertial Response
SNSP	System Non-Synchronous Penetration
SOC	State of Charge
SOGL	System Operations Guideline
SONI	System Operator for Northern Ireland
SOR	Secondary Operating Reserve
SRF	Synchronous Reference Frame
STATCOM	Static Synchronous Compensator
s-UC	Stochastic Unit Commitment
SVC	Static VAR Compensator
TAS	Tasmania
TDS	Time Domain Simulation
TEC	Time Error Control
TFC	Tertiary Frequency Control
TG	Turbine Governor
TOR1	Tertiary Operating Reserve 1
TOR2	Tertiary Operating Reserve 2
TREx	Transient analysis Results Extractor
TSAT	Transient Security Assessment Tool
TSO	Transmission System Operator
UC	Unit Commitment
UF	Under-frequency
UFLS	Under Frequency Load Shedding
ULTC	Under-Load Tap Changer
UPS	Uninterruptible Power Supply
US	United States
V2G	Vehicle-To-Grid
VCO	Voltage Controlled Oscillator
VDIFD	Voltage Dip Induced Frequency Dip
VFM	Volume Forecasting Methodology
VPP	Virtual Power Plant

VSAT	Voltage Security Assessment Tool
VSC	Voltage Sourced Converter
VSG	Virtual Synchronous Generator
WDT	Wind Dispatch Tool
WSCC	Western Systems Coordinating Council
ZCD	Zero-Crossing Detection

List of Figures

List of Tables

Part I
Basic Concepts and Definitions

Chapter 1
Instantaneous Frequency

This chapter provides an overview of the definitions and estimation techniques of the fundamental frequency of ac grids. Section 1.1 explains the link between frequency variations and power imbalances and explains why frequency, similarly to bus voltage magnitude, is different from bus to bus in transient conditions. This section also discusses the conceptual challenges of defining the frequency of non-stationary ac voltages. Section 1.2 describes relevant frequency estimation methods and discusses their features and limitations. Finally, Sect. 1.3 defines the Rate of Change of Frequency (RoCoF) and its role in power system operation and protection.

1.1 Frequency in Electric Power Systems

The vast majority of electric power systems in the world is ac and operates at a single frequency, typically, 50 or 60 Hz or, as they were used to be called [1], 50 or 60 *cycles*. Figure 1.1 shows the synchronous regions of the Pan-European grid. For each region, in ideal steady-state conditions, the frequency is constant and has the same value everywhere. So, for example, the frequency is the same in Poland and in Portugal. However the frequency in Synchronous Region (RG) Continental Europe is different, in general, from the frequency in RG Ireland (IE), which is connected through High Voltage Direct Current (HVDC) links and thus decoupled from a frequency perspective.

As it has to be expected, ideal steady-state conditions are never satisfied and the frequency of the system is never really perfectly constant. Due to synchronous machines, frequency is intertwined with the mechanism the generators keep the power balance with load consumption and system losses. The well known equation that regulates this balance is:

$$2H_{\text{agg}}(t)\,\omega'(t) = p_{\text{supply}}(t) - p_{\text{load}}(t) - p_{\text{loss}}(t), \tag{1.1}$$

T. Kërçi and F. Milano, *Frequency Quality in Power Systems*, Power Systems,
https://doi.org/10.1007/978-3-032-04721-2_1

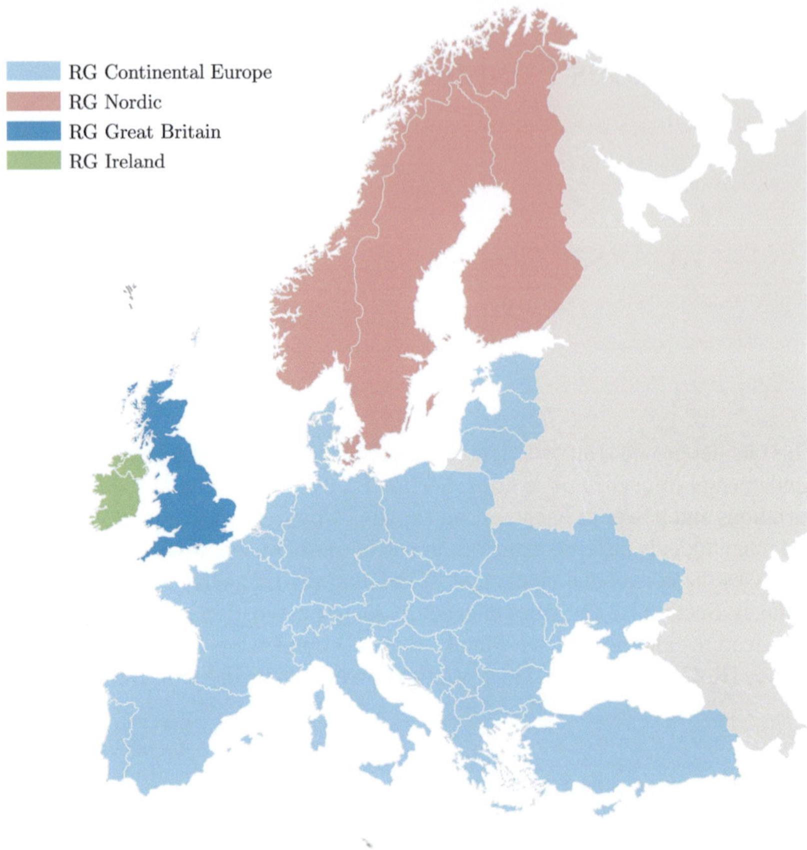

Fig. 1.1 Synchronous Region (RG) of the Pan-European grid

where H is the total aggregated inertia constant of the synchronous machines connected to the system, and p_{supply}, p_{load} and p_{loss} are the powers of the generation, demand and losses, respectively. ω is the angular frequency of the system and ω' is the time derivative of ω, typically called RoCoF. In (1.1), the starting time, which is proportional to the inertia of the machines, is not assumed to be constant as this can change, even abruptly, if part of the generation changes from conventional synchronous machines to Inverter-Based Resource (IBR)-based generators, as shown in Fig. 1.2.

From (1.1), the frequency is given by:

$$f(t) = \frac{\omega(t)}{2\pi}. \tag{1.2}$$

In the remainder of this book, we will often use the term "frequency" to actually mean "angular frequency", as this is the quantity that arises from the models of the

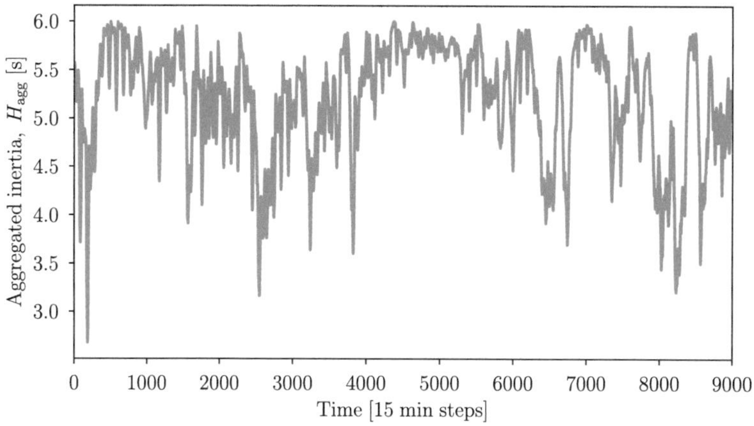

Fig. 1.2 Time-varying aggregated inertia in Germany (courtesy of A. Ulbig, see also reference [2])

machines and other devices that compose the power system. As most of the times we will use per unit quantities, the 2π proportionality coefficient is immaterial.

Equation (1.1) states that any unbalance of the power leads to an acceleration of the rotor angular speeds of the synchronous machines and, hence, a variation of the system frequency. This is why system operators require the frequency to be as constant as possible, as this indirectly implies that the power balance is achieved accurately. We note since now that for most devices, with very few notable exceptions as synchronous clocks and some other instrumentation, the fact that the frequency of the system is precisely equal to the nominal synchronous reference, is not overly important. In other words, ensuring $\omega' = 0$, and hence the power balance, does not necessarily implies that ω is equal to the nominal angular frequency. As a matter of fact, Primary Frequency Control (PFC) is generally not perfect tracking the synchronous reference frequency. We further elaborate on this point in Chap. 2.

Equation (1.1) suggests a possible way to estimate the frequency in a power system as:

$$f(t) = \frac{1}{2\pi} \int_t \frac{1}{H_{\text{agg}}(t)} (p_{\text{supply}}(t) - p_{\text{load}}(t) - p_{\text{loss}}(t))dt. \tag{1.3}$$

Calculating the frequency through (1.3) is highly unpractical. One should have to estimate p_{supply}, p_{load}, p_{loss} and H_{agg}, and this estimation should be very precise as the deviations of the frequency with respect to the nominal value are generally small. Even if one were able to solve these technical issues, the question would arise of what f represents.

Equation (1.3) does not model a specific node of the grid. Rather, it is an *equivalent* system frequency which is likely different from the frequency that can be measured at any physical node of the grid. A more precise representation of the "frequencies" of the grid can be given looking at every machine terminal bus [3], as follows.

Let us consider the augmented admittance matrix of a network with n buses and m synchronous machines. This matrix accounts for network branches, that is, overhead transmission lines, cables and transformers, as well as the m nodes of the electromotive forces behind the internal reactances of the synchronous machines [4]. The resulting matrix can be structured as follows:

$$\bar{\mathbf{Y}} = \begin{bmatrix} \bar{\mathbf{Y}}_{\mathrm{GG}} & \bar{\mathbf{Y}}_{\mathrm{GB}} \\ \bar{\mathbf{Y}}_{\mathrm{BG}} & \bar{\mathbf{Y}}_{\mathrm{BB}} \end{bmatrix}, \tag{1.4}$$

where G and B indicate generator buses, and load and transition buses, respectively, and $\bar{\mathbf{Y}}_{\mathrm{GG}} \in \mathbb{C}^{m\times m}$, $\bar{\mathbf{Y}}_{\mathrm{BB}} \in \mathbb{C}^{n\times n}$; $\bar{\mathbf{Y}}_{\mathrm{GB}} \in \mathbb{C}^{m\times n}$; and $\bar{\mathbf{Y}}_{\mathrm{BG}} \in \mathbb{C}^{n\times m}$. Matrix $\bar{\mathbf{Y}}_{\mathrm{BB}}$ is:

$$\bar{\mathbf{Y}}_{\mathrm{BB}} = \bar{\mathbf{Y}}_{\mathrm{bus}} + \bar{\mathbf{Y}}_{\mathrm{G}}, \tag{1.5}$$

where $\bar{\mathbf{Y}}_{\mathrm{bus}}$ is the well-known network admittance matrix, and $\bar{\mathbf{Y}}_{\mathrm{G}}$ is a diagonal matrix whose i-th diagonal element is 0 if no machine is connected to bus i, and the inverse of a machine internal transient reactance if such a machine is connected at bus i. Load buses are all buses where there is no generation. Transition buses are thus assumed to be load buses with zero consumption. If there are no phase shifting transformers in the grid, then $\bar{\mathbf{Y}}_{\mathrm{bus}}$ is symmetrical and, hence, $\bar{\mathbf{Y}}_{GL} = \bar{\mathbf{Y}}_{LG}^{T}$. It can also be shown that $\bar{\mathbf{Y}}_{\mathrm{BB}}$ is full rank [5].

Based on (1.4), one can find an expression that links the frequencies at the network buses with the rotor angular speeds of the synchronous machines, as follows:

$$\mathbf{0}_n = \mathbf{B}_{\mathrm{BB}}(\boldsymbol{\omega}_{\mathrm{B}}(t) - \omega_o \mathbf{1}_n) + \mathbf{B}_{\mathrm{BG}}(\boldsymbol{\omega}_{\mathrm{G}}(t) - \omega_o \mathbf{1}_m), \tag{1.6}$$

where $\boldsymbol{\omega}_{\mathrm{G}} \in \mathbb{R}^m$, is the vector of machine angular rotor speeds; $\boldsymbol{\omega}_{\mathrm{B}} \in \mathbb{R}^n$, are the frequencies at the system buses; ω_o is the nominal system frequency; and

$$\mathbf{B}_{\mathrm{BB}} = \mathrm{Im}\{\bar{\mathbf{Y}}_{\mathrm{BB}}\}, \qquad \mathbf{B}_{\mathrm{BG}} = \mathrm{Im}\{\bar{\mathbf{Y}}_{\mathrm{BG}}\},$$

where $\mathbf{B}_{\mathrm{BB}}$ has same rank and symmetry properties as $\bar{\mathbf{Y}}_{\mathrm{BB}}$.

To simplify the notation, we rewrite (1.6) in terms of frequency deviations with respect to the nominal frequency ω_o, as follows:

$$\mathbf{0}_n = \mathbf{B}_{\mathrm{BB}} \Delta\boldsymbol{\omega}_{\mathrm{B}}(t) + \mathbf{B}_{\mathrm{BG}} \Delta\boldsymbol{\omega}_{\mathrm{G}}(t), \tag{1.7}$$

where $\Delta\boldsymbol{\omega}_{\mathrm{B}} = \boldsymbol{\omega}_{\mathrm{B}} - \omega_0 \mathbf{1}_n$ and $\Delta\boldsymbol{\omega}_{\mathrm{G}} = \boldsymbol{\omega}_{\mathrm{G}} - \omega_0 \mathbf{1}_m$. The expression in (1.7) is known as *frequency divider formula*. Its formulation was first proposed in this form in [3] although expressions obtained with similar hypotheses can be found also in earlier works, for example, [6]. The name "divider" is justified by the fact that for a two bus grid, (1.7) shows that the frequency is linear interpolation between the sending and receiving buses, similarly to a resistive voltage divider.

Equation (1.7) can be rewritten as

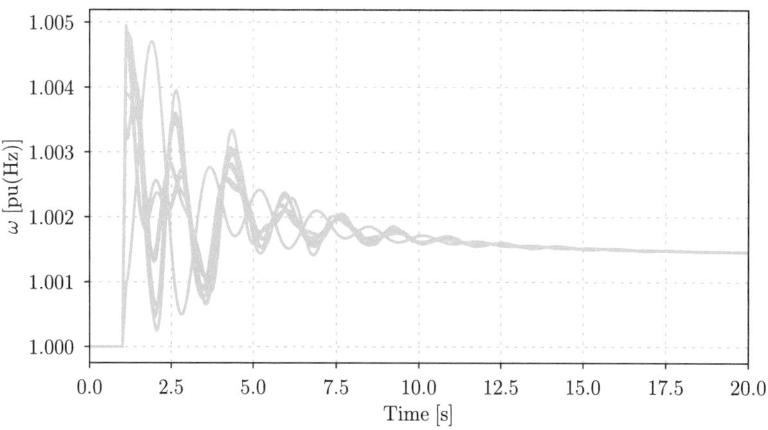

Fig. 1.3 Transient behavior of bus frequencies following the clearing of a short circuit as obtained with the frequency divider formula for a grid with exclusively conventional synchronous generation

$$\Delta\omega_{\mathrm{B}}(t) = -\mathbf{B}_{\mathrm{BB}}^{-1}\mathbf{B}_{\mathrm{BG}}\,\Delta\omega_{\mathrm{G}}(t) = \mathbf{D}\Delta\omega_{\mathrm{G}}(t), \tag{1.8}$$

which indicates that knowing *all* rotor angular speeds and the topology of the grid, one can obtain the frequencies at all buses. The inertia of the machine does not appear explicitly in (1.7) and, hence, in (1.8). However, since the internal admittance of each machine is, in per unit, proportional to the capacity of the machine itself, bigger machines weight more than smaller ones in the system.

Matrix **D** is, in general, not full rank because in a power system there are way less generators than nodes, $m \ll n$. However, it can be shown that the Moore-Penrose pseudo-inverse of **D** is an extremely sparse matrix and, in fact, one can estimate $\Delta\omega_{\mathrm{G}}$ with a very small subset of $\Delta\omega_{\mathrm{B}}$. In [7], it is shown that one needs only the frequency at terminal bus of the machine and the frequencies at every bus directly connected to terminal bus to estimate the rotor speed of the machine. This also leads to the possibility to estimate the frequency of the center of inertia of an entire system using a number of bus frequency measurements approximately double than the number of machines in the system [8].

Figure 1.3 shows a typical transient behavior for a system with conventional power plants of the bus frequencies following the occurrence and then the clearing of a short circuit. The bus frequencies are calculated using (1.8). The frequencies differ from bus to bus in the first seconds, and mostly follow the rotor speed of the closest machine(s). After a few seconds, the electromechanical oscillations are fully damped and the system becomes synchronous, namely, all buses have the same frequency. This is the effect of the PFC of the synchronous machines. However, as explained in Chap. 2, this frequency is not exactly equal to the reference speed, which is 1 pu in this case.

Taking to the extreme consequences the concept of the frequency divider, one can argue that, during a transient, frequency has to be different not only from bus

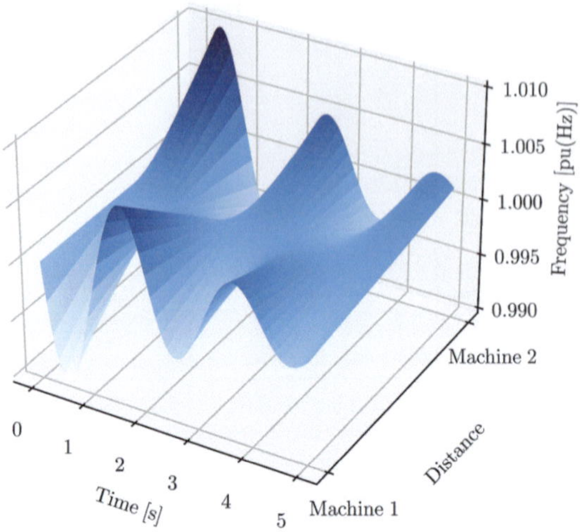

Fig. 1.4 Representation of the time-space variation of frequency as a continuum in the electric grid during transients as obtained from the frequency divider formulation. The rotor speeds of the synchronous machines define the boundary conditions for the frequency distribution

to bus, but also from point to point of every transmission line. Frequency is, thus, a time-space continuum where the transmission lines constitute the medium and the rotor speeds of synchronous machines represent the boundary conditions [9, 10]. This behavior is illustrated in Fig. 1.4.

We see, thus, that the single system frequency that appears in (1.1) is substantially an approximation of the model of the system which is valid a few seconds after the occurrence of a contingency. The single bus model cannot, of course, capture the local dynamics, which are due to the inertial response of the synchronous machines, in the interval from 0 to 5-10 seconds after the occurrence of a contingency. This period, once dominated exclusively by the synchronous machine inertial response and their electromechanical oscillations, is impacted also by the primary controllers of IBRs which can significantly change the behavior of the frequency at the point of connection of the IBR. But the other way around is also true: local frequency variations can impact the way the controllers of the IBRs respond to system contingencies. The interest in frequency quality has risen in recent years precisely because the relevance of frequency transients has moved from global to local. In other words, short-term bus frequency variations are becoming more and more relevant for the control and the overall dynamic performance of power grids.

The frequency divider formula, while relatively accurate as shown in [11], is obviously a simplified expression. The main assumption is that only synchronous machines are able to impose/modify the frequency at buses. This can be shown not to be true [12]. And, in fact, only linear admittance loads are pure *frequency followers*. Any other device will modify, even if possibly very slightly, the frequency at its point of connection with the grid, thus being a *frequency influencer*. Of course, only very few devices are able to impact significantly on the frequency. Among these devices, synchronous machines and Grid-Forming Converters (GFMs) are the most relevant ones. The frequency divider formula can be modified to account for other devices, such as induction machines [13] or frequency Grid-Following Converters (GFLs) with droop control [14]. But this exercise is not particularly efficient as one should derive an expression for every possible device and controller included in the grid. This has been done in [15], at the price of extending the definition of frequency to the complex domain, which we discuss more in detail later in this section. However, the main issue remains, that is, the need for the knowledge of the model of every device connected to the grid.

We need, thus, to find some other approach to estimate frequency. This problem has been widely studied in the literature and there exist, as we show in the following section, a significant number of different approaches. And, as it is often the case, when different approaches to achieve same results coexist, it means that no one manages to solve the problem in a satisfactory way.

The first issue to solve is what "frequency" actually is. In [16], frequency is assumed to be a property of a signal rather than a physical quantity like time, distance or mass. However, Eq. (1.1) and the frequency divider formula (1.7) seem to state the opposite. In (1.1), the frequency is a state variable of the model of a physical, although admittedly highly simplified, system. This is again due to synchronous machines, which not only link frequency variations to the power balance but also link their rotor angular speed to the frequency of the ac voltage at their terminal bus. The vast majority of other electrical devices connected to the grid, notably power electronic converters, cannot impose this link, among other reasons, because most devices do not have a rotor. One can also argue that the equivalence between rotor angular speed and the frequency of the voltage is just a metrological trick, the same to say that the current is equivalent to the deviation of the needle of an electromechanical ammeter.

Let us accept the fact that frequency is the *property of a signal*. This shifts the problem to define such a property and then how to calculate (or, better, measure) it. According to the most basic definition, the frequency f of a signal is the inverse of the period T of the signal itself, namely:

$$f = \frac{1}{T}. \tag{1.9}$$

This definition has two fundamental issues. The first is methodological, the problem of defining f is not solved, but simply "moved" to the problem of defining T. The second issue is conceptual, as it assumes that the signal is stationary and the period is constant. In practice, a real-world signal, and in particular the voltage at a node of a power grid, is rarely (arguably never) stationary. And it is only when the signal is *not* stationary that one really needs to estimate the frequency, as its deviation with respect to the synchronous reference speed is a good metric of the dynamic of the power unbalance of the system as stated by (1.3). Equation (1.9) leads thus to an impasse: one can use it to define frequency only if the signal is stationary but it turns out that the stationary condition is precisely the only situation when one does not really need to estimate the frequency.

The IEEE Std C37.118.1-2011 defines the frequency as the time derivative of the phase angle of a sinusoidal signal [17]. Let us assume that the voltage measured at a node of an ac grid is:

$$v(t) = V \cos \vartheta(t), \tag{1.10}$$

then the frequency is given by:

$$f(t) = \frac{1}{2\pi} \vartheta'(t). \tag{1.11}$$

This definition makes a substantial progress with respect to (1.9) as it allows for the frequency to be a function of time. However, as for (1.9), the problem is just shifted, not solved, as one has to define and measure ϑ in the first place. There is, however, a crucial conceptual (methodological) issue: definition (1.11) works only if the signal is effectively a sinusoidal function. In [16], this approach is called *representationalist*, which consists in defining a model first and then fitting measurements to that model. In the ancient Greek mythology, the same approach is resembled by the pictoresque character of Procrustes, a criminal who had an iron bed and was used to adapt to this bed, by stretching or cutting, the unfortunate people that he kidnapped. In this context, the iron bed is represented by Eq. (1.10).

As there is no guarantee that the voltage or any other quantity measured at given node of an ac system is actually sinusoidal, the approach above needs to be improved. There are two ways: one is to persists in the Procrustes' approach but using different models (functions), with more or less parameters and more or less complicated expressions. Reference [18] provides some of the several hundreds models that have been proposed in the literature. A recent development of the representationalist approach can be found in [19], where the name *model* is called *functional dictionary*.

The main issue with all these models is that they cannot be *falsified*, as parameter fitting techniques will always return some number to adjust whatever model to the measurements. But at the same time, there is no way to decide which model is *correct*. The reader familiar with theoretical physics will recognize that this situation is very similar to the issues posed by string theory and its intrinsically unfalsifiable nature.

A conceptually different methodology, which is called *operationalist* in [16], is to defined a process or algorithm, rather than the model. This category is probably the most interesting from the technical and mathematical point of view as it requires to inventing a procedural technique rather than fitting a signal to a given function. Well-known examples of the operationalist approach are the Fourier transform and the Hilbert transform. The former decomposes the original signal into a set of sinusoids with constant magnitude, frequency and phase; the latter finds the time-varying magnitude and frequency that make a sinusoidal function equal to the original signal. We discuss more on the Fourier approach in the following section. In the context of a formal definition of frequency, the Hilbert transform is more relevant.

As anticipated above, the Hilbert transform is capable of transforming "any" signal, say v, into an equivalent, time-varying sinusoidal function [20]:

$$v(t) = A(t) \cos \varphi(t). \tag{1.12}$$

The first step is to calculate the Hilbert transform of the original signal, say $\hat{v}$, then the parameters of the equivalent sinusoidal signal are given by:

$$A(t) = \sqrt{v^2(t) + \hat{v}^2(t)},$$
$$\varphi(t) = \operatorname{atan}\left(\frac{\hat{v}(t)}{v(t)}\right).$$

Then, the frequency of $v(t)$ is defined as:

$$f(t) = \frac{1}{2\pi}\varphi'(t) = \frac{1}{2\pi}\,\frac{v(t)\hat{v}'(t) - v'(t)\hat{v}(t)}{v^2(t) + \hat{v}^2(t)}. \tag{1.13}$$

In signal processing, the quantity φ' is commonly called *instantaneous frequency*.

It is important to note that, while (1.12) has the same shape as (1.10), they are coming from conceptually different approaches. Equation (1.10) is a *model* of the signal, that is, the signal has to fit a sinusoidal function regardless its actual shape. Equation (1.12) is the result of a *procedure*, namely the Hilbert transform, that leads to obtain the values of A and φ. Only if the original signal is actually a sinusoid, then $\varphi = \vartheta$. In all other cases, φ is a fictitious phase angle that does not resemble any specific parameter of the original signal $v(t)$. We also note that the procedure based on the Hilbert transform does not require the magnitude to be constant and, in fact, the value of the magnitude can be also determined as a byproduct of the Hilbert transform.

The instantaneous frequency (1.13) and the definition (1.11) given by the IEEE Std C37.118.1-2011 both define the frequency as the time derivative of a phase angle. This correspondence between frequency and a time derivative can be generalized. First, we note that if one defines the following analytic signal:

$$\bar{v}(t) = v(t) + \jmath\hat{v}(t) = A(t)\exp(\jmath\varphi(t)) = \exp(\varsigma(t) + \jmath\varphi(t)), \tag{1.14}$$

where $\jmath$ is the imaginary unit and $\varsigma(t) = \log(A(t))$, then:

$$\bar{v}'(t) = [\varsigma'(t) + \jmath\varphi'(t)]\,\bar{v}(t), \tag{1.15}$$

where the term $\varsigma' + \jmath\varphi'$ is a *complex frequency*. This definition of frequency appeared related to electric circuits as early as in [21], which dates 1964, although its definition in the context of signal processing dates at least five years earlier in [22]. The term $\varsigma'(t)$ is also called *instantaneous bandwidth*:

$$\varsigma'(t) = \frac{d}{dt}\log(A(t)) = \frac{A'(t)}{A(t)} = \frac{v(t)v'(t) + \hat{v}(t)\hat{v}'(t)}{v^2(t) + \hat{v}^2(t)}. \tag{1.16}$$

The study of the instantaneous bandwidth has had a hype in the 1990s, with the works by Cohen on time-frequency analysis [23], but later faded away as the problem of finding a suitable time-frequency distribution was eventually deemed unsolvable. The applications of (1.16) to power systems are quite recent, see for example [15, 24].

The concept of complex frequency has been recently extended to multi-dimensional spaces giving rise to the definition of *geometric frequency*, which can be formally written as a multivector [25]:

$$\varrho(t) + \boldsymbol{\Omega}(t) = \frac{\boldsymbol{v}(t) \cdot \boldsymbol{v}'(t)}{|\boldsymbol{v}(t)|^2} + \frac{\boldsymbol{v}(t) \wedge \boldsymbol{v}'(t)}{|\boldsymbol{v}(t)|^2}, \tag{1.17}$$

where $\boldsymbol{v} = (v_1, v_2, \dots, v_n)$ is a vector of arbitrary order n; and $\cdot$ and $\wedge$ represent the inner and outer products, respectively, as defined in geometric algebra; and the voltage magnitude is calculated as the Euclidean norm:

$$|\boldsymbol{v}(t)| = \sqrt{\sum_{i=1}^{n} v_i^2(t)}. \tag{1.18}$$

The first component of the multivector, ϱ, is scalar in any dimensions and represents a *radial frequency*. The bivector $\boldsymbol{\Omega}$ is a generalization of φ' and accounts for the possible angular rotations that can happen in a n-dimensional space.

For $n = 2$, the ϱ coincides with the instantaneous bandwidth, in fact:

$$\varrho = \frac{v_1(t)v_1'(t) + v_2(t)v_2'(t)}{|\boldsymbol{v}(t)|^2} = \frac{|\boldsymbol{v}(t)|'}{|\boldsymbol{v}(t)|} = \varsigma'(t). \tag{1.19}$$

Then, in two dimensions, there is only one possible rotation axis, which is perpendicular to the plane of the coordinates. Assuming that the voltage vector is $\boldsymbol{v} = (v_1, v_2)$, then one has:

$$\boldsymbol{\Omega}(t) = \frac{v_1(t)v_2'(t) - v_1'(t)v_2(t)}{|\boldsymbol{v}(t)|^2}\, \boldsymbol{e}_1 \wedge \boldsymbol{e}_2 = \omega_\kappa(t)\, \boldsymbol{e}_1 \wedge \boldsymbol{e}_2, \tag{1.20}$$

where ω_κ is called *azimuthal frequency* and the absolute value of the wedge product $\wedge$ is calculated, in two dimensions, as the determinant of the following matrix:

$$|\boldsymbol{v}(t) \wedge \boldsymbol{v}'(t)| = \begin{vmatrix} v_1(t) & v_2(t) \\ v_1'(t) & v_2'(t) \end{vmatrix}.$$

Formally, the imaginary unit can be defined as:

$$\jmath = (\boldsymbol{e}_1 \wedge \boldsymbol{e}_2). \tag{1.21}$$

Hence, (1.20) can be rewritten as:

$$\boldsymbol{\Omega}(t) = \jmath\, \omega_\kappa(t). \tag{1.22}$$

In three dimensions, $\boldsymbol{\Omega}(t)$ is equivalent to a vector with three independent elements:

$$\star\,\boldsymbol{\Omega}(t) = \boldsymbol{\omega}(t) = \frac{\boldsymbol{v}(t) \times \boldsymbol{v}'(t)}{|\boldsymbol{v}(t)|^2}, \tag{1.23}$$

where $\star$ is the Hodge star operator that maps bivector into vectors and *vice versa*. where $\times$ is the conventional cross product, which is defined only up to three dimensions. For example, consider $\boldsymbol{v} = v_1\boldsymbol{e}_1 + v_2\boldsymbol{e}_2 + v_3\boldsymbol{e}_3$, where $(\boldsymbol{e}_1, \boldsymbol{e}_2, \boldsymbol{e}_3)$ is a orthonormal basis, that is, a basis composed of orthogonal unit vectors. Then, one has:

$$\boldsymbol{v}(t) \cdot \boldsymbol{v}'(t) = v_1 v_1' + v_2 v_2' + v_3 v_3', \tag{1.24}$$

and

$$\boldsymbol{v}(t) \times \boldsymbol{v}'(t) = (v_2 v_3' - v_2' v_3)\boldsymbol{e}_1 + (v_3 v_1' - v_3' v_1)\boldsymbol{e}_2 + (v_1 v_2' - v_1' v_2)\boldsymbol{e}_3, \tag{1.25}$$

and

$$\begin{aligned}\boldsymbol{\Omega}(t) &= \star\boldsymbol{\omega}(t) \\ &= \frac{1}{|\boldsymbol{v}(t)|^2}\begin{bmatrix} 0 & (v_1 v_2' - v_1' v_2) & -(v_3 v_1' - v_3' v_1) \\ -(v_1 v_2' - v_1' v_2) & 0 & (v_2 v_3' - v_2' v_3) \\ (v_3 v_1' - v_3' v_1) & -(v_2 v_3' - v_2' v_3) & 0 \end{bmatrix}.\end{aligned} \tag{1.26}$$

For $n = 4$, the independent elements are 6 and so on. Discussions on the properties and applications of the geometric frequency can be found in [26, 27].

So far, we have presented the definitions of mathematical objects, of various degrees of complexity and generality, to define the frequency. However we have not discussed whether these objects have an actual physical meaning. And we have not discussed if these definitions are consistent with each other. Let us consider a few examples.

First, assume an ideal sinusoidal signal, with constant magnitude and angular frequency:

$$v(t) = V\cos(\omega_o t). \tag{1.27}$$

As the angular frequency is constant, one can calculate it directly from the period T, namely, $\omega_o = 2\pi/T$, where the period can be obtained for example using a zero-crossing detection technique (see Sect. 1.2.1).

The IEEE Std C37.118.1-2011 definition gives immediately that the angular frequency is ω_o, as expected.

The approach based on the analytic signal gives:

$$\bar{v}(t) = v(t) + \jmath\,\hat{v}(t) = V\cos(\omega_o t) + \jmath\,V\sin(\omega_o t), \tag{1.28}$$

where we have utilized the fact that the Hilbert transform of the cosine is the sine. Then, we obtain $A = V$ and $\varphi(t) = \omega_o t$, again as expected, and the complex frequency of the the signal is:

$$\varsigma'(t) + \jmath\,\varphi'(t) = \jmath\,\omega_o,$$

as it is straightforward to deduce that $v(t)v'(t) + \hat{v}(t)\hat{v}'(t) = 0$ and hence $\varsigma'(t) = 0$. To apply the geometric approach, let us define the vector:

$$\begin{aligned}\boldsymbol{v}(t) &= [V\cos(\omega_o t),\, V\sin(\omega_o t)]\\ &= v(t)\boldsymbol{e}_1 + \hat{v}(t)\boldsymbol{e}_2,\end{aligned}$$

where $\boldsymbol{e}_1$ and $\boldsymbol{e}_2$ are unit vectors that define an orthonormal basis in a space of two dimensions. Then we obtain:

$$\begin{aligned}\boldsymbol{v}'(t) &= v'(t)\boldsymbol{e}_1 + \hat{v}'(t)\boldsymbol{e}_2\\ &= -\omega_o\hat{v}(t)\boldsymbol{e}_1 + \omega_o v(t)\boldsymbol{e}_2\end{aligned}$$

and:

$$\begin{aligned}\boldsymbol{v}(t)\cdot\boldsymbol{v}'(t) &= -\omega_o v(t)\hat{v}(t) + \omega_o\hat{v}(t)v(t),\\ \boldsymbol{v}(t)\wedge\boldsymbol{v}'(t) &= \omega_o\big(v^2(t) + \hat{v}^2(t)\big)\,\boldsymbol{e}_1\wedge\boldsymbol{e}_2\\ &= \omega_o V^2\,\boldsymbol{e}_1\wedge\boldsymbol{e}_2,\end{aligned} \tag{1.29}$$

which gives the following expression for the geometric frequency:

$$\varrho(t) + \jmath\,\omega_\kappa(t) = 0 + \jmath\,\omega_o,$$

where we have used the identity $\jmath = \boldsymbol{e}_1 \wedge \boldsymbol{e}_2$. This case is thus equivalent to a curve that describes a circle, which is the only curve with constant positive curvature and constant radius (see Fig. 1.5).

So far, all definitions are consistent and return same results, which is indeed reassuring. Let us consider a second case for which the magnitude and the phase angle of the signal v are time varying:

$$v(t) = V(t)\cos(\vartheta(t)). \tag{1.30}$$

One can still obtain an estimation of the period through a zero-crossing of the signal but, as ϑ' is not constant, the instantaneous frequency will be different, in general and except at most a finite number of instants, from the value obtained with the zero-crossing technique. The definition of the frequency given by the IEEE Std C37.118.1-2011, is still valid, but one has to assume the knowledge of $\vartheta(t)$, separately from $V(t)$, which is not a trivial operation. With this aim, one can use the representationalist approach, which fixes *a priori* a function for $\vartheta(t)$ and $V(t)$ with given parameters, which can then be fit with some numerical technique such as least mean square.

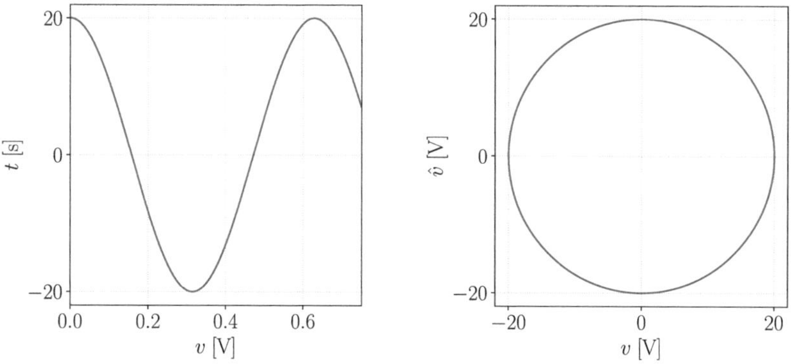

(a) Time domain (b) Space $(v(t), \hat{v}(t))$

Fig. 1.5 Representation of signal (1.27) in time domain and in the state space $(v(t), \hat{v}(t))$ for $\omega_o = 10$ rad/s and $V = 20$ V

With this regard, the IEEE Std C37.118.1-2011 provides three "models" of the (1.30), as follows:

$$\begin{aligned}
v_1(t) &= V\big(1 + k_x \cos(\omega t)\big) \cos\big(\omega_o t + k_a \cos(\omega t - \pi)\big), \\
v_2(t) &= V \cos(\omega_o t + \pi R_f t^2), \\
v_3(t) &= V\big(1 + k_x \mathcal{H}(t - t_1)\big) \cos(\omega_o t + k_a \int_t \mathcal{H}(t - t_1) dt),
\end{aligned}$$

where ω is the modulation frequency, R_f is the frequency ramp rate, and $\mathcal{H}(t_1)$ is the Heavyside step function. The three models are utilized in the standard for dynamic compliance, that is, they allow testing under same conditions, the performance of synchrophasors after three specific events: amplitude and frequency sweep, ramp of system frequency, and step changes in phase and magnitude. Consistently with (1.11), the angular frequencies of the three models above are defined in the standard as[1]:

$$\begin{aligned}
\omega_1(t) &= \omega_o - k_a \omega \sin(\omega t - \pi), \\
\omega_2(t) &= \omega_o + R_f t, \\
\omega_3(t) &= \omega_o + k_a \mathcal{H}(t - t_1).
\end{aligned} \tag{1.31}$$

The purpose of the standard is to define a meaningful set of tests that allows comparing in well-defined reproducible conditions synchrophasors. In other words, the standard does not assume that the three models above are "actual" representations of the voltage signal. It is up to the make of the synchrophasor to make sure that if the signal is one of the three signals provided in the standard, then the frequency estimated by

[1] The standard only defines the frequency at multiple of the period $T = 2\pi/\omega_o$. The fundamental frequency, thus, is assumed to be constant and always equal to the synchronous reference speed.

the device resembles the expressions given in (1.31). Moreover, while the models considered in the standard certainly reflect common events in power systems, they do not account for *all* possible events, as this is not the objective of the standard.

It remains, however, the problem of how to estimate the instantaneous frequency given a generic signal. And for this, a synchrophasor has necessarily to follow some sort of recipe, which implies using an operationalist approach. Considering again the signal (1.30), one can proceed using the Hilbert transform and creating an analytic signal. However, we encounter immediately a major issue. In general in fact one has that, in general the Hilbert transform of the signal (1.30) is:

$$\hat{v}(t) \neq V(t)\sin(\vartheta(t)). \tag{1.32}$$

If this inequality holds, formulas (1.13) and (1.16) will return some number but these are not what we expect for the instantaneous frequency and bandwidth. To come to rescue, there exists a theorem of the Hilbert transform, namely the Bedrosian's theorem [28], which states that if the spectrum of $V(t)$ and $\theta(t)$ do not overlap, then:

$$\hat{v}(t) \approx V(t)\sin(\vartheta(t)). \tag{1.33}$$

Interestingly enough, the hypothesis of the Bedrosian's theorem does apply, in most cases, to power systems, as the frequency spectrum of $\vartheta(t)$ is generally in the region equal or above 50 (or 60) Hz, whereas the spectrum of $V(t)$ is generally in a region of the order of few Hz. It is important to note that only thanks to the fortunate combination of the Bedrosian's theorem and common operating conditions of power systems, the Hilbert transform makes sense for the calculation of the instantaneous frequency.

There are however two relevant caveats. First, analytic signals are defined only for positive frequencies. This might not be the case in power systems, where the negative sequence appears whenever loads are unbalanced. Second, and more important, we have assumed so far that one is able to actually calculate the Hilbert transform of a signal. This transform is defined as follows:

$$\hat{v}(t) = \frac{1}{\pi}\,\text{p.v.}\int_{-\infty}^{+\infty} \frac{v(\tau)}{t-\tau}\,\mathrm{d}\tau, \tag{1.34}$$

where p.v. indicates the Cauchy principal value. Neglecting technicalities, the main issue is that one needs to know the signal $v(t)$ for *all* times to be able to calculate its Hilbert transform. This is certainly possible for measured signals, such as the voltage of an electric circuit, as measured signals are characterized by finite-energy (square-integrable) and can be thus represented with a finite spectrum. However, it appears inconsistent that to calculate the *instantaneous* frequency $\varphi'(t)$, one needs to have the full knowledge of a signal for a finite—that is, noninfinitesimal—time interval. This is recognized as one of the paradoxes of the definition of the instantaneous frequency based on analytic signals [23].

Let us further delve into the apparent inconsistencies of the instantaneous frequencies by means of a simple example. Consider a signal composed of the sum of two cosines:

$$v(t) = V_a \cos(\omega_a t) + V_b \cos(\omega_b t), \tag{1.35}$$

where $\omega_a > \omega_b$ and $V_a > V_b$, so that, in the language of signal processing, $V_a \cos(\omega_a t)$ is the *carrier*, and $V_b \cos(\omega_b t)$ is the *upper side-band*. This is also a relevant case for power systems, as we can interpret $V_a \cos(\omega_a t)$ as the fundamental frequency and $V_b \cos(\omega_b t)$ and harmonic component, hence $\omega_b = h\omega_a$, where h is an integer, typically greater than 3. Then, calling $m = V_b/V_a$, $V_a = V$ and $\omega_a = \omega_o$, $\omega_h = h\omega_o$, the signal $v(t)$ can be rewritten as:

$$v(t) = V\big(\cos(\omega_o t) + m\cos(\omega_h t)\big). \tag{1.36}$$

The Fourier transform (or *spectrum*) of this signal is obtained as:

$$\mathfrak{v}(\omega) = \mathcal{F}[v](\omega) = \frac{1}{\pi}\int_{\infty}^{\infty} v(t)e^{-\jmath\omega t}dt. \tag{1.37}$$

The function $\mathfrak{v}(\omega)$ provides a representation of the signal $v(t)$ in the frequency domain, that is, transform is a collection (integral) of the projections (convolution product) of the signal $v(t)$ on a new coordinate system ($e^{-\jmath\omega t} = \cos(\omega t) - \jmath\sin(\omega t)$). In turn, the signal is parametrized with respect to the parameter ω of the new coordinates. The signal in (1.35) consists of the sum of two cosines, then, for $\omega > 0$, its Fourier transform has only two non-null components precisely at ω_o and ω_h:

$$\mathfrak{v}(\omega) = \mathcal{F}[v](\omega) = V\Delta(\omega - \omega_o) + mV\Delta(\omega - \omega_h), \quad \omega > 0, \tag{1.38}$$

where $\Delta(\cdot)$ is the Dirac delta function, which is zero everywhere except in zero, where $\Delta(0) = 1$.

The Hilbert transform retains time for the parametrization of the signal. In particular, the first coordinate remains the signal itself. Then, the second coordinate is built using the projection (convolution) of the signal onto a fictitious time axis given by the function $1/(t-\tau)$—see (1.34). As the Hilbert transform of the cosine is the sine, the analytic signal associated with (1.35) is:

$$\begin{aligned}\bar{v}(t) &= A(t)\exp(\jmath\varphi(t)) \\ &= V\big(\cos(\omega_o t) + m\cos(\omega_h t)\big) + \jmath V\big(\sin(\omega_o t) + m\sin(\omega_h t)\big),\end{aligned} \tag{1.39}$$

where:

$$\begin{aligned}A(t) &= V\sqrt{1 + m^2 + 2m\cos(\omega_{h-1}t)}, \\ \varphi(t) &= \omega_o t + \operatorname{atan}\left(\frac{m\sin(\omega_{h-1}t)}{1 + m\cos(\omega_{h-1}t)}\right),\end{aligned} \tag{1.40}$$

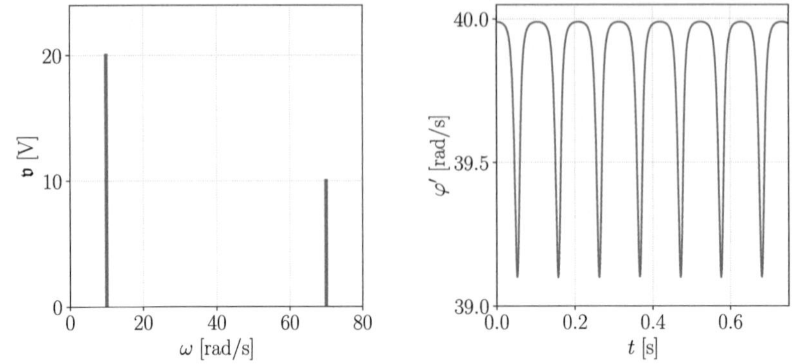

(a) Fourier transform ($\omega > 0$) (b) Instantaneous frequency

Fig. 1.6 Representation of signal (1.36) through its Fourier transform and its instantaneous frequency for $\omega_o = 10$ rad/s, $h = 7$, $V = 20$, and $m = 0.5$

where $\omega_{h-1} = \omega_h - \omega_o = (h-1)\omega_o$. Then, the instantaneous bandwidth and frequency are:

$$\begin{aligned} \varsigma'(t) &= \frac{m\omega_{h-1}\sin(\omega_{h-1}t)}{1+m^2+2m\cos(\omega_{h-1}t)}, \\ \varphi'(t) &= \omega_o + \frac{\omega_{h-1}}{2}\left(1 - \frac{1-m^2}{1+m^2+2m\cos(\omega_{h-1}t)}\right). \end{aligned} \tag{1.41}$$

The expressions above give rise to a variety of concerns. Depending on the values of h and m, in fact, $\varphi'(t)$ can be negative, never equal to ω_o, or bigger or smaller than ω_o and ω_h. Figure 1.6 illustrates some of these inconsistencies for a possible choice of the parameters of the signal (1.39). In this example, $\varphi'(t)$ is not constant, despite the signal being periodic, and is never equal to ω_o and ω_h.

The fact that the information obtained with Fourier and the information obtained with the Hilbert transform seem to be inconsistent led to consider these results as paradoxes of the instantaneous frequency [23]. However, we beg to differ from this interpretation and we boldly state that these paradoxes are such only if one assumes the Fourier transform as the *ground truth* of the signal.

It is worth noting at this point that the radial and azimuthal frequency obtained with the geometric approach are *invariants*, that is, quantities that have always the same value independently from the coordinates used to represent the signal. In particular, the azimuthal frequency can be written as:

$$\omega_\kappa(t) = \kappa(t)\,|\boldsymbol{v}(t)|, \tag{1.42}$$

where κ is the *curvature* of an hypothetical curve $\boldsymbol{r}$, the velocity of which is the vector $\boldsymbol{v} = \boldsymbol{r}'$. As we have seen, in two dimensions, the instantaneous frequency

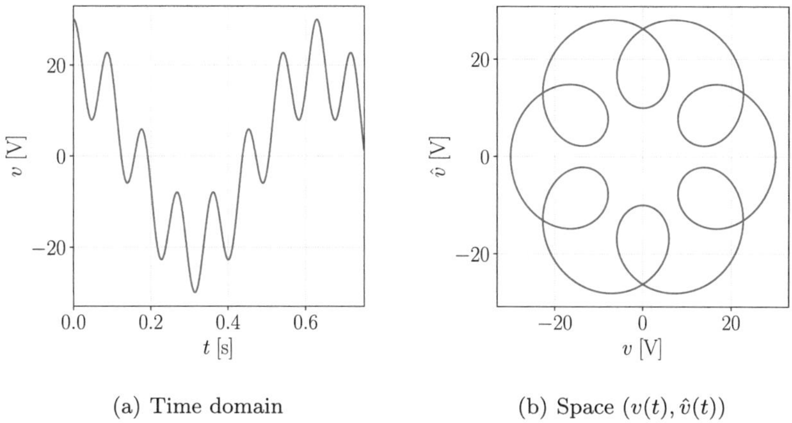

Fig. 1.7 Representation of the signal (1.36) in time domain and in the space $(v(t), \hat{v}(t))$ for $\omega_o = 10$ rad/s, $h = 7$, $V = 20$, and $m = 0.5$

and the azimuthal frequency coincide. Hence, also the instantaneous frequency is an invariant [26].[2]

Interpreting $\varphi'(t)$ as the ratio of the curvature by the magnitude of the velocity of a curve justifies why it can be negative or take values that are never equal, bigger or smaller than ω_o and ω_h: φ' represents how the curve $\boldsymbol{r}$ bends in the plane $(\boldsymbol{e}_1, \boldsymbol{e}_2)$. This is illustrated in Fig. 1.7. Using this interpretation, then, it appears irrelevant whether the spectrum of the Fourier transform of the signal is crisp or a continuum. The components of the signal in the frequency domain are not geometric invariants but, rather, projections on arbitrary coordinates $(\cos(\omega t) - \jmath \sin(\omega t))$ and arbitrary parameter (ω). They are constant only for a "fortunate" choice of these coordinates and parameter as well as for the peculiar structure of the signal [30].

The main conclusion of this discussion is that the instantaneous frequency has a geometric meaning and is thus always "correct." However, undoubtedly, if the signal has the form of (1.35), the instantaneous frequency might not provide the most useful information. In stationary conditions, in fact, one would expect to obtain a constant value and, specifically, the frequency of the carrier. The only way to obtain this result using a geometric approach is to filter the signal and remove any harmonic content, which corresponds to reduce the curve $\boldsymbol{r}$ of which the analytic signal is the velocity to a circle.

We have still to solve the deepest and most critical of the paradoxes, namely the fact that the signal has to be known forever to be able to calculate the Hilbert transform and construct the analytic signal. This issue cannot be solved recurring to the invariants. It is in effect a structural issue of analytic signals. There are practical ways to solve the issue which, in turn, utilize the typical approach that is also adopted for the Fourier transform, namely, using a window of the signal and then assume that

---

[2] In [29], it is shown that the instantaneous frequency is unique but being invariant is a stronger condition. Uniqueness is, in fact, implicit in the concept of invariance.

this window repeats forever. The trick is to assume that the signal is stationary and periodic even if it is not. This approach, however, does not solve the fundamental point, namely, that the Hilbert transform substantially creates an extra dimension, which is, in effect, totally arbitrary.

Reference [31] suggests that all signals are complex, that is, are composed of a real and an imaginary part, called *latent*. However, all we can actually measure is the real part, and we do not have any information on how the latent part looks like. Utilizing the Hilbert transform to define the latent component of a signal is arbitrary and might lead, as in fact it does, to inconsistencies. The fact that we need to know the signal for $t \in (-\infty, \infty)$ is one of such inconsistencies. Again, the inconsistency arises only for the choice of the Hilbert transform as the tool to calculate the latent signal, not to the nature of the signal itself.

The discussion on what is the best choice for the latent signal does not seem to be solvable in conventional signal processing. One may also raise reasonable doubts on the existence at all of the latent component, such as, for example, the Occam razor.

Fortunately, the situation is much simpler in power systems. In fact, substantially all transmission and distribution systems are three-phase. This means that, for every node of the grid, we do not have just a single (scalar) signal for the voltage but, in effect, at least three values, one for each phase of the voltage. One can thus consider that each phase voltage is the component of a vector and assume the each phase is a coordinate in a three-dimensional space [25, 26]. Moreover, if one filters the zero-sequence, only two coordinates are needed, for example, the α and β components of the Clarke transform [30].

Assume that the phase voltages at a node of the grid is represented as the following vector:

$$\boldsymbol{v}_{\mathrm{abc}}(t) = v_{\mathrm{a}}(t)\boldsymbol{e}_{\mathrm{a}} + v_{\mathrm{b}}(t)\boldsymbol{e}_{\mathrm{b}} + v_{\mathrm{c}}(t)\boldsymbol{e}_{\mathrm{c}}, \tag{1.43}$$

then, applying the Clarke transform to obtain the α and β components gives:

$$\begin{bmatrix} v_\alpha(t) \\ v_\beta(t) \end{bmatrix} = \sqrt{\frac{2}{3}} \begin{bmatrix} 1 & -\frac{1}{2} & -\frac{1}{2} \\ 0 & \frac{\sqrt{3}}{2} & -\frac{\sqrt{3}}{2} \end{bmatrix} \begin{bmatrix} v_{\mathrm{a}}(t) \\ v_{\mathrm{b}}(t) \\ v_{\mathrm{c}}(t) \end{bmatrix}. \tag{1.44}$$

At this point, one can define the following complex signal:

$$\bar{v}_{\alpha\beta}(t) = v_\alpha(t) + \jmath v_\beta(t). \tag{1.45}$$

In a balanced, sinusoidal system, the calculation above leads to:

$$\bar{v}_{\alpha\beta}(t) = V\cos(\omega_o t + \theta_o) + \jmath V\sin(\omega_o t + \theta_o), \tag{1.46}$$

which is, in effect, an analytic signal. However, in general, (1.45) is not an analytic signal as it can contain a negative sequence. As we have discussed above, this is not an issue, as the definition of the α and β components of the Clarke transform are ultimately, simply a way to define two possible coordinates. As the calculation of the geometric invariants does not depend on the choice of the coordinates, results are guaranteed to be meaningful. However, in general, the instantaneous frequency that can be obtained by calculating the Hilbert transform of each phase voltage does not coincide with the frequency obtained using (1.45). That is, in general:

$$\frac{v_{\rm i}(t)\hat{v}_{\rm i}'(t) - v_{\rm i}'(t)\hat{v}_{\rm i}(t)}{v_{\rm i}(t)^2 + \hat{v}_{\rm i}^2(t)} \neq \frac{v_\alpha(t)\hat{v}_\beta'(t) - v_\alpha'(t)\hat{v}_\beta(t)}{v_\alpha^2(t) + v_\beta^2(t)}, \quad \text{for } i = \{\text{a, b, c}\}. \tag{1.47}$$

We illustrate this with an example. Consider the following unbalanced stationary voltage:

$$\boldsymbol{v}(t) = V_\alpha \cos\theta(t)\boldsymbol{e}_\alpha + V_\beta \sin\theta(t)\boldsymbol{e}_\beta, \tag{1.48}$$

where $V_\alpha \neq V_\beta$ are constant and $\theta(t) = \omega_o t + \theta_o$; ω_o is the fundamental synchronous reference frequency; θ_o is constant and its value depends on the chosen phase angle reference. As the signal is stationary and does not contain harmonics, one would expect that the instantaneous frequency is constant. This result is in fact obtained by processing each component of $\boldsymbol{v}(t)$ in (1.48) separately and applying the Hilbert transform. However, since the curve described by (1.48) is an ellipse, its curvature and radius are not constant and hence we obtain that the complex frequency is a time-varying quantity with:

$$\varsigma'(t) + \jmath\,\varphi'(t) = \frac{\omega_o(V_\beta^2 - V_\alpha^2)\sin 2\theta(t)}{2|\boldsymbol{v}(t)|^2} + \jmath\frac{\omega_o V_\alpha V_\beta}{|\boldsymbol{v}(t)|^2}, \tag{1.49}$$

where

$$|\boldsymbol{v}(t)|^2 = V_\alpha^2 \cos^2\theta(t) + V_\beta^2 \sin^2\theta(t).$$

In this specific case, one can of course utilize other formulas that exploits the properties of the ellipse to obtain the value of the fundamental frequency, see for example [32]. The main issue however remains: the curvature and hence φ' does not represent, in general, the fundamental frequency of the system. It does not "have to," as the fundamental frequency is just a parameter, whereas the curvature is a geometric invariant.

Whenever (1.47) is an identity, the Hilbert transform is the *correct* choice for the definition of the latent component of a signal. On the other hand, the calculation of the geometric frequency based on the Clarke components is *always* consistent, assuming that one accepts that the phase voltages of an ac circuit are equivalent to coordinates of a voltage vector. The geometric approach also solves the issue that arises with the Hilbert transform, that is, the need of knowing the signal for all times. In fact, the geometric approach is intrinsically *local* and only requires the instantaneous values of v_α and v_β and their derivatives at thc same instant t.

Table 1.1 summarizes the various definitions of frequency given in this section. We note that, in general, these definitions all leads to different values of the frequency and coincide only in the trivial case of stationary balanced sinusoidal voltages.

Table 1.1 Frequency definitions

Type	Model	Formula [rad/s]
Stationary	$T = \text{const.}$	$\dfrac{2\pi}{T}$
IEEE standard	$v(t) = V \cos \vartheta(t)$	$\vartheta'(t)$
Analytic signal	$\bar{v}(t) = v(t) + \jmath \hat{v}(t)$	$\varphi'(t) = \dfrac{v(t)\hat{v}'(t) - v'(t)\hat{v}(t)}{v^2(t) + \hat{v}^2(t)}$
Complex frequency	$\bar{v}(t) = v_\alpha(t) + \jmath v_\beta(t)$	$\omega(t) = \dfrac{v_\alpha(t)v'_\beta(t) - v'_\alpha(t)v_\beta(t)}{v^2_\alpha(t) + v^2_\beta(t)}$
Geometric frequency	$\boldsymbol{v}(t) = [v_a(t), v_b(t), v_c(t)]$	$\omega_\kappa(t) = \left\lvert \dfrac{\boldsymbol{v}(t) \times \boldsymbol{v}'(t)}{\lvert \boldsymbol{v}(t) \rvert^2} \right\rvert$

As a final remark, it is important to note that, in practice, and with proper filtering, the difference between the instantaneous and geometric frequency is quite small during power system transients. Also, transient deviations of the frequency with respect to the synchronous reference are small and slow. Thus, independently on the method utilized to calculate the frequency, results tend to be similar. We illustrate this point in the following section.

1.2 Estimation Methods

Accurate measurement of grid frequency is a critical component of reliable grid control. For synchronous generators, the rotor angular speed coincides with the electrical frequency up to the number of pole pairs. It is thus just enough to measure a mechanical quantity.[3] For all other devices, in particular, inverters and Phasor Measurement Units (PMUs), the estimation of the frequency is obtained from measurements of electrical quantities. The most traditional methods are Phase-Locked Loops (PLLs) and Discrete Fourier Transform (DFT). However, these methods are not always fully reliable, especially when utilized for the regulation of IBRs. The biggest challenge is, as it can be expected from the discussions given in the previous section, a reliable estimation of the variation of the frequency in fast transient conditions. Next we illustrate advantages and drawbacks of some common methods to estimate the instantaneous frequency.

[3] Although, in some cases, the electrical frequency at the terminal bus of the machines is measured. This is the case, for example, of the Itaipu Binacional 14 GW power plant located on the Paraná River between Brazil and Paraguay.

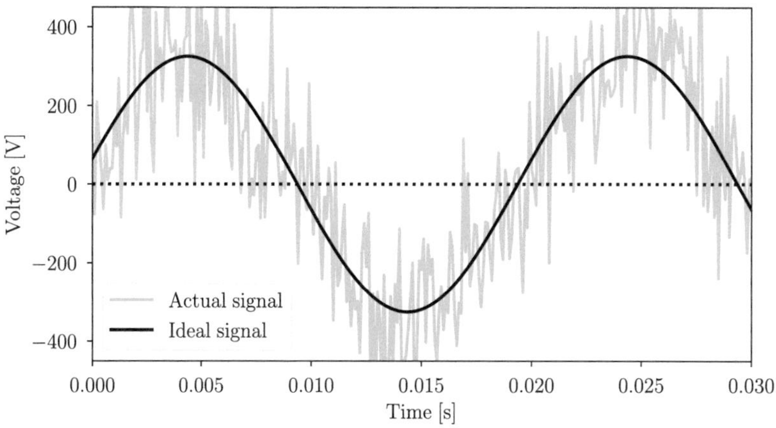

Fig. 1.8 Effect of noise on a sinusoidal signal. The noise makes difficult to detect precisely the zero crossing of the signal

1.2.1 Zero-Crossing Detection

Zero-Crossing Detection (ZCD) is a common grid frequency measurement method. It consists in the measurement of time elapsed between two consecutive zero-crossings of the voltage waveforms. An ideal sine wave is zero twice per period, so at 50 Hz, this technique returns an estimation every 10 ms. Noise and harmonics can bias the identification of the correct instant at which the signal is null, as illustrated in Fig. 1.8. Of course, noise and harmonics can be filtered. It remains the fact, however, that with a sample frequency of 10 Hz, the zero-crossing method can be too coarse for applications such as the fast frequency control of IBRs or the estimation of the RoCoF.

In simulations, ZCD can be correctly emulated only through fully-fledges Electro-Magnetic Transient (EMT) models. In conventional transient stability analysis software tools, bus voltages are represented using slowly time-varying phasors. Thus ZCD cannot be simulated unless the time-domain phase voltages are "reconstructed" and noise, harmonics and negative sequence are added to the signal obtained from the phasors.

1.2.2 Fourier Transform

As anticipated in the previous section, the main idea of the Fourier transform is that, if a signal can be approximated with a series of sines or cosines, then one obtains a sharp spectrum with only values corresponding to the frequencies of the sinusoids that compose the time-domain signal.

The application of the Fourier transform to power system analysis has found its natural application in harmonic analysis, which is the study of the effect of frequencies multiple of the fundamental one in stationary conditions [33]. In a harmonic analysis, a signal can be represented as a sum of sinusoids and, possibly, a non-null constant term:

$$v(t) = \sum_h V_h \cos(h\omega_o t + \theta_h), \quad h \in \{0, 1, \dots, n\}, \tag{1.50}$$

which can be conveniently studied in the frequency domain rather than in the time domain.

If the signal is not periodic, Fourier can still be used but the spectrum becomes a continuum and the integral of (1.37) has to be calculated for $t \in (-\infty, \infty)$. Obviously, this is impractical in real-world applications, specifically for on-line estimations of the frequency of a voltage during an electromechanical transient.

To overcome this issue, the most common technique is to calculate the short–time Discrete Fourier Transform (s-DFT) of a window of samples. This approach works because, even if the fundamental frequency is not perfectly constant, it never varies too much, and a window of 20 ms or multiples of it provides a reasonable representation of the signal. However, as it has to be expected, this approach has also several issues and several techniques, more or less sophisticated, have been proposed to compensate the approximation introduced by taking a relatively small number of samples, for example, aliasing and spectral leakage.

The discrete implementation of the Fourier transform in practical applications, namely, the Fast Fourier Transform (FFT), it also makes difficult to identify to estimate in a timely manner small deviations of the fundamental frequency. For example, with sampling frequency of $f_s = 15$ kHz, and a signal composed of $L = 3000$ samples, which at 50 Hz represents 10 cycles or 0.2 s, the frequency resolution obtained with the FFT is $f_s/L = 5$ Hz. As frequency deviations during transients are of the order of a hundred mHz. One has thus to take a window of hundreds of cycles, or several seconds, to obtain a precision comparable to the system frequency deviations. But in this way, the window becomes comparable with the duration of the transient itself. This makes the utilization of the Fourier transform inadequate for control and on-line applications. It is commonly utilized instead in PMUs, which are typically utilized for off-line monitoring and aftermath analysis.

Among the techniques to reduce the impact of these issues on the estimation, we mention the appropriate choice of the window profile, length and sampling rate, as well as spectral interpolation. There is indeed a vast literature on this topic. For a gentle introduction, the reader is referred to Chap. 3 of [34].

All studies lead to the conclusion that the Fourier transform can be adapted to transient signals and can provide a relatively good estimation of the frequency variations. However, the Fourier approach is intrinsically inadequate to capture transient variations of a signal. One may thus ask why the FFT is still utilized in practice. There are mainly three important reasons.

First, the FFT is an extremely robust numerical technique that applies particularly well to discrete signals with constant sampling rate. A signal transformed in frequency domain can be transformed back to time time without any loss of information except for the frequencies that were lost in the original sampling. The same cannot be said for other continuous transforms such as the Park transform, where even a small noise makes the inverse transformation return very poor results. The FFT transform works so well that is in fact often utilized also to determine the Hilbert transform, to which it is linked by the following relationship:

$$\mathcal{F}\left[\hat{v}\right](\omega) = -\jmath\, \mathcal{F}\left[v\right](\omega)\,, \qquad \omega > 0, \tag{1.51}$$

which also explains why the Hilbert transform "creates" a second time coordinate: it is substantially equivalent to a 90° rotation of the original signal. The second reason is that there exists an immense literature on signal processing based on the Fourier transform. All this experience can be proficiently exploited. Last but not least, despite its intrinsic limitations in transient conditions, the FFT is very robust as it does not require to calculate derivatives or integrals of the signal.

1.2.3 Phase-Locked Loop

A PLL is an electronic device that attempts to *track* the signal while it is evolving in time [35, 36]. PLLs are commonly used by IBRs to synchronize the converter with the grid. The estimation of the frequency based on PLLs is a byproduct of the synchronization, and is conceptually close to the definition of the instantaneous frequency given in [17].

PLLs consist of a Phase Detector (PD), a Loop Filter (LF) and a Voltage Controlled Oscillator (VCO), which is substantially an integrator. A basic scheme of the PLL is shown Fig. 1.9. The sophistication of the LF can vary, but, ultimately, is prone to the tradeoff common to all controllers: the faster is the tracking the more sensitive to noise is the estimation. In Fig. 1.9, the LF is represented with a Proportional Integral

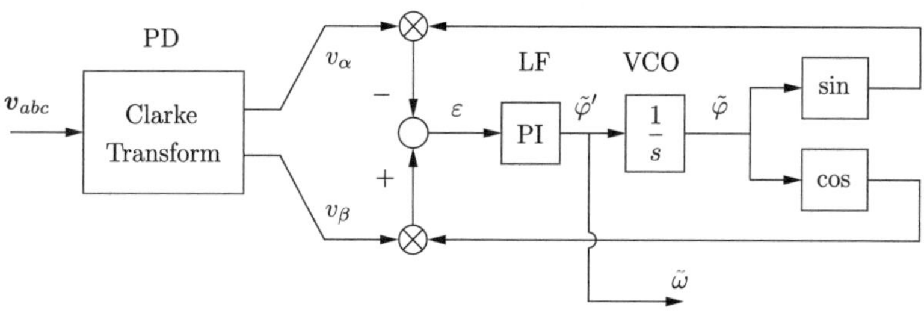

Fig. 1.9 Scheme of a three-phase synchronous reference-frame PLL. PD: phase detector; LF: loop filter; VCO: voltage controlled oscillator

(PI) controller, which leads to the Synchronous Reference Frame (SRF)-PLL, which is arguably the simplest possible PLL implementation.

The α and β components of the Clarke transform are obtained starting from the the measurements of the phase voltages. Assuming that the complex signal $v_\alpha + \jmath\, v_\beta$ is analytic and the system is balanced, from (1.46) one obtains:

$$\frac{v_\beta(t)}{v_\alpha(t)} = \frac{V \sin(\varphi(t))}{V \cos(\varphi(t))}, \tag{1.52}$$

or, equivalently:

$$v_\beta(t)\cos(\varphi(t)) - v_\alpha(t)\sin(\varphi(t)) = 0, \tag{1.53}$$

and the instantaneous frequency is $\omega = \varphi'$.

The PLL can only provide an estimation of the phase angle. In transient conditions, thus, the estimated angle, say $\tilde{\varphi}$, is not equal to φ. The estimation error is obtained as:

$$\varepsilon(t) = v_\beta(t)\cos(\tilde{\varphi}(t)) - v_\alpha(t)\sin(\tilde{\varphi}(t)). \tag{1.54}$$

Then, this error is tracked through a PI controller, as shown in Fig. 1.9. If the gains of the PI are K_p and K_i, respectively, one has:

$$\tilde{\varphi}'(t) = K_p\varepsilon(t) + x_\varepsilon(t), \tag{1.55}$$

where

$$x_\varepsilon'(t) = K_i\varepsilon(t). \tag{1.56}$$

The signal $\tilde{\omega}(t) = \tilde{\varphi}'(t)$ is the sought estimation of the instantaneous frequency.

In practice, the α and β components of the Clarke transform are projected on a frame rotating at nominal synchronous angular speed, say ω_o. This, in turn, leads to the Park transform, as shown in Fig. 1.10. In this case, the angle to estimate is:

$$\theta(t) = \arctan\left(\frac{v_q(t)}{v_d(t)}\right). \tag{1.57}$$

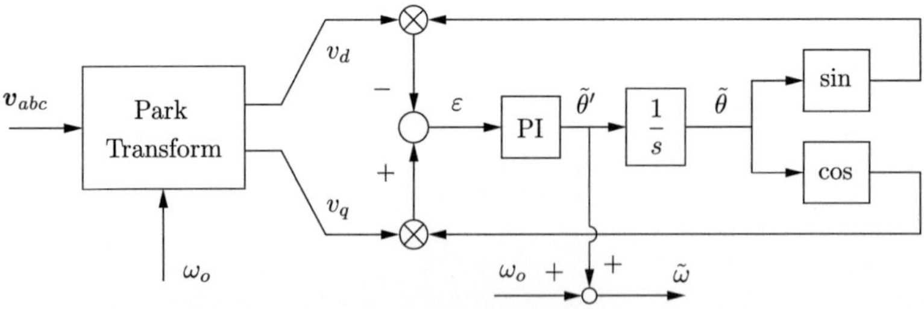

Fig. 1.10 Scheme of a SRF-PLL based on the Park transform

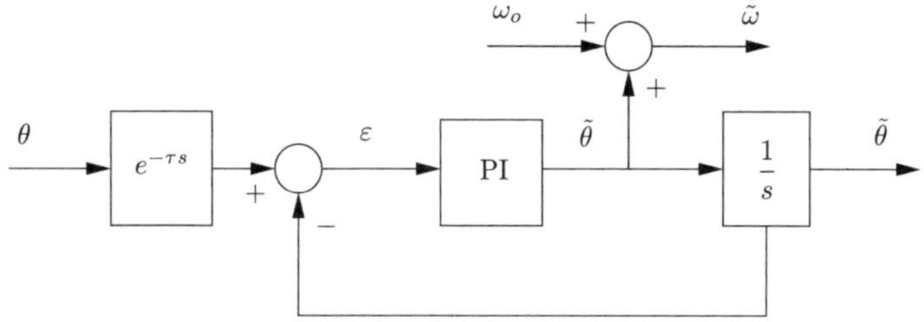

Fig. 1.11 Polar representation of a SRF-PLL

Calling the estimated angle is $\tilde{\theta}$, its derivative, namely θ' represents the estimated deviation of the frequency of the voltage with respect to the angular speed of the Park reference frame, which is conveniently set equal to the nominal angular frequency of the system ω_o. Hence the estimated instantaneous frequency is $\tilde{\omega}(t) = \tilde{\theta}'(t) + \omega_o$.

The accuracy of the estimation of the PLL depends on the implementation and tuning of the LF, and, as discussed in the previous section, by the negative sequence and by harmonics. To solve these issues several types of PLLs have been proposed. The SRF-PLL shown in Fig. 1.9. A comprehensive study on a variety of more sophisticated solutions commonly utilized in practice is provided in [37]. However, regardless of the complexity of the PLL, one has, in turn, to remove noise, harmonics and the negative sequence. What remains after that is an analytic signal whose instantaneous frequency reflects the time-varying fundamental frequency of the measured voltage.

PLLs can be easily simulated both in EMT and transient stability models. For the latter, in rectangular coordinates, one can use directly the dq-axis coordinates of the bus voltages and the scheme shown in Fig. 1.9. The PD, namely the voltage measurement and Park transform, can be modeled as a delay. In polar coordinates, the scheme becomes even simpler as the voltage is already represented as time-varying magnitude and phase angle. The resulting scheme of the SRF-PLL in polar coordinates is shown in Fig. 1.11.

Table 1.2 summarizes the main features of the estimation methods described in this section. Note that the PLL can estimate either the instantaneous frequency or the imaginary part of the complex frequency, depending whether the negative sequence is filtered or not. On the other hand, the DFT does not estimate the frequency of any specific model discussed in Sect. 1.2 and Table 1.1, but assuming that the voltage can be written as:

$$v(t) \approx \sum_{i=0}^{n} V_i \cos(\omega_i t + \theta_i), \tag{1.58}$$

then, all techniques proposed in the literature aim at estimating the value of the fundamental frequency ω_1 for every window of samples. This can be considered an approximation of the expression (1.10) considered in IEEE Std C37.118.1-2011, that is:

$$V_1 \cos(\omega_1 t + \theta_1) \approx V \cos \vartheta(t). \tag{1.59}$$

Table 1.2 Frequency estimation methods

Method	Estimation	Advantages	Drawbacks
ZCD	T	Simplicity	Error prone
DFT	$\omega_1 \approx \vartheta'(t)$	Robust, reliable	Only off-line applications
PLL	$\varphi'(t), \omega(t)$	For control applications	Need filtering

Next, we compare the dynamic performance of the PLL considering two different models of the grid. In the first case, we use a standard positive-sequence transient-stability model. In the second case study, we consider a fully-fledged EMT model.

1.2.3.1 Example 1

We have mentioned that, as any control loop, the accuracy of the PLL during transients depends on the tuning of its parameters. Figure 1.12 shows a comparison of the estimation obtained with an SRF-PLL and the "exact" instantaneous frequency obtained calculating analytically the complex frequency at a bus of a system during a transient following the occurrence and then clearing of a fault. The analytical expression for the complex frequency in a transient stability model can be written as [15]:

$$\bar{\imath}'(t) = [\bar{\mathbf{Y}}_{\text{bus}}\, \text{diag}(\bar{\boldsymbol{v}}(t))](\varsigma'(t) + \jmath\, \varphi'(t)), \tag{1.60}$$

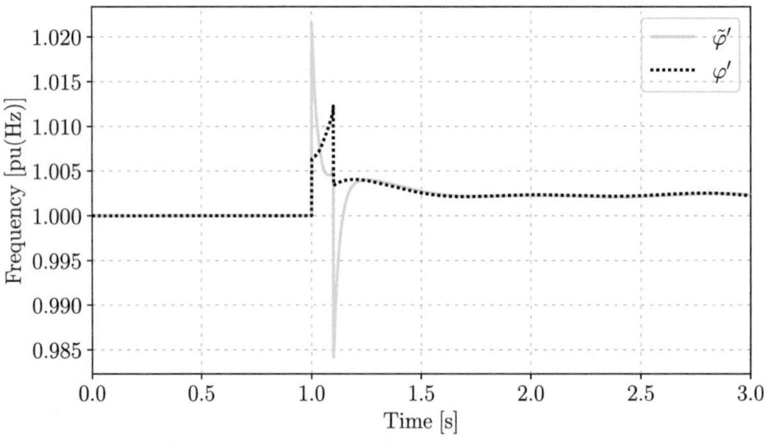

Fig. 1.12 Comparison of the frequency estimated with a SRF-PLL ($\tilde{\varphi}'$) and the complex frequency calculated analytically (φ')

where $\varsigma' + \jmath\,\varphi'$ is the complex frequency vector, $\bar{\boldsymbol{v}}$ is the vector of bus voltages and $\bar{\boldsymbol{\imath}}$ is the vector of currents injected at the buses. The expression of the time derivatives of injected currents depends on the model of the devices connected at the bus. In general, the currents can be written as:

$$\bar{\boldsymbol{\imath}} = \boldsymbol{\psi}(\boldsymbol{x}, \bar{\boldsymbol{v}}, t), \tag{1.61}$$

where $\boldsymbol{x}$ is the vector of the state variables of the devices. Hence, one has:

$$\bar{\boldsymbol{\imath}}'(t) = [\boldsymbol{\psi}_x]\boldsymbol{x}'(t) + [\boldsymbol{\psi}_{\bar{v}}\mathrm{diag}(\bar{\boldsymbol{v}}(t))](\varsigma'(t) + \jmath\,\varphi'(t)), \tag{1.62}$$

where $\boldsymbol{\psi}_x$ and $\boldsymbol{\psi}_{\bar{v}}$ are the Jacobian matrices of the functions $\boldsymbol{\psi}$ with respect to state variables and bus voltages, respectively. Merging (1.60) and (1.62), it is possible to find the exact expressions of φ' at every step of the integration of the differential algebraic equation that model the system:

$$\varphi'(t) = \mathrm{Im}\{[[\bar{\mathbf{Y}}_{\mathrm{bus}} - \boldsymbol{\psi}_{\bar{v}}]\,\mathrm{diag}(\bar{\boldsymbol{v}}(t)]^{-1}[\boldsymbol{\psi}_x]\boldsymbol{x}'(t)\}. \tag{1.63}$$

The imaginary part of the complex frequency jumps when the fault occurs and then again when it is cleared. This also happens to the estimation obtained with the PLL. However, due to the dynamic of the PLL LF, the spikes of the PLL are higher and take a relatively long time to recover. Thus, the estimation of the PLL is fully incorrect during the fault and for some instants after the fault is cleared. For this reason, PLLs are not adequate for relays and protections [38]. Outside this region, the complex frequency and the PLL estimation are in good agreement.

1.2.3.2 Example 2

In high-voltage transmission systems, harmonics and imbalances are minimized by design and proper filtering, in order to comply with network codes. This makes easier to estimate the instantaneous frequency. In this example, we compare the instantaneous frequency $\tilde{\varphi}'$ obtained with an SRF-PLL with the azimuthal frequency ω_κ using a fully fledged EMT model. In both cases, the starting point is the measurement of the voltage vector $\boldsymbol{v}_{abc}(t)$ at a high-voltage bus of the grid. The transient is caused by a three phase fault that is cleared by the protections after 100 ms. Then we consider two scenarios: (i) balanced sinusoidal system; and (ii) unbalanced sinusoidal system with load imbalances ranging from 5 to 10% on one of the phases. Results are shown in Fig. 1.13.

For the estimation of the azimuthal frequency, we have adopted a simple numerical procedure. The time derivatives of the voltage sample v_i at time t_i is obtained using a numerical central difference:

$$v_i' \approx \frac{v_{i+1} - v_{i-1}}{2\Delta t},$$

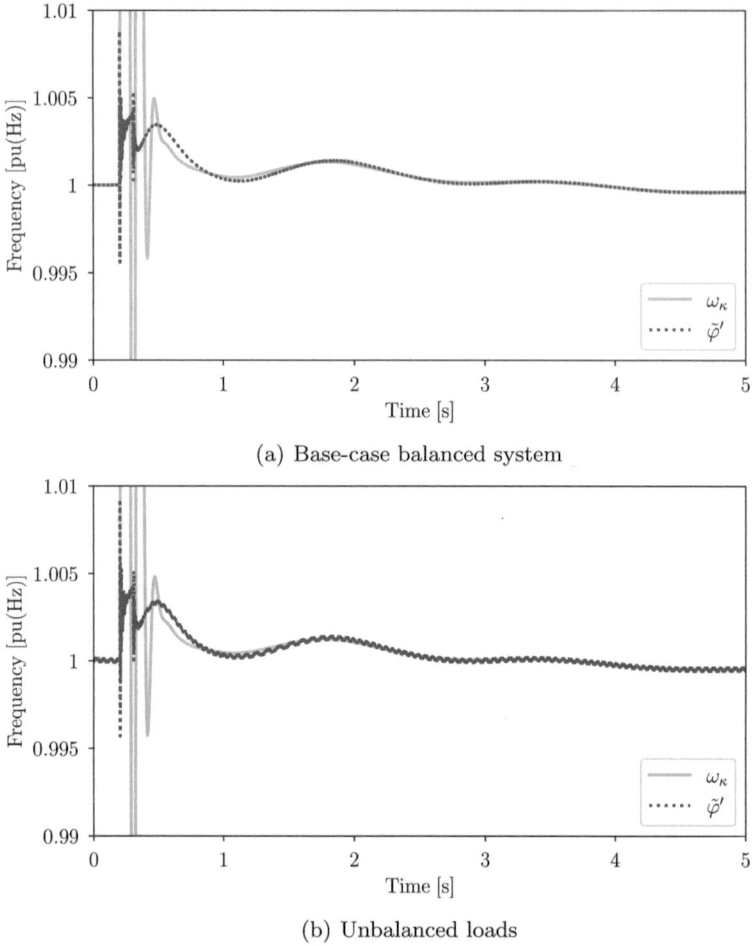

Fig. 1.13 Frequency estimation for an ac transmission system

where Δt is the time interval of the sampling rate and v_{i+1} and v_{i-1} are the voltage samples at $t_i + \Delta t$ and $t_i - \Delta t$, respectively. Then we have calculated the azimuthal frequency with (1.17) and finally used a properly tuned lead-lag filter to remove high frequency noise. The best results have been obtained using $\Delta t = 0.1$ ms. For bigger Δt the estimation becomes inaccurate and for smaller Δt the effect of noise becomes dominant. In turn, the estimation of the azimuthal frequency presents similar challenges and same tradeoffs as the estimation of the instantaneous frequency with PLLs.

Overall, there is a very good match between ω_κ and $\varphi'(t)$, also for the scenarios with unbalanced voltages and harmonic contents. Both the time derivatives needed for the calculation of the azimuthal frequency and the PLL LF can be implemented in more sophisticated ways, which can lead to better results. However, even the simple implementations considered in this example show that a reasonable estimation of the frequency is feasible in practice.

1.3 Rate of Change of Frequency

We conclude this chapter with a brief discussion on another important metric in power system operation and control, namely, the RoCoF. This is, in turn, the time derivative of the frequency:

$$\mathrm{RoCoF}(t) = \frac{1}{2\pi}\omega'(t). \tag{1.64}$$

This quantity is important as it is strictly related to the power imbalance. From (1.1), one has:

$$\mathrm{RoCoF}(t) = \frac{1}{2\pi}\frac{p_{\mathrm{gen}}(t) - p_{\mathrm{dem}}(t) - p_{\mathrm{loss}}(t)}{2H_{\mathrm{agg}}(t)}. \tag{1.65}$$

If H_{agg} is known and can be assumed to be constant in a given period, the RoCoF is in effect a direct consequence of any power imbalance occurring in the system.

Grid codes impose specific conditions on the RoCoF. For example, the current Grid Code in IE requires generators to be able to ride through RoCoF of 1 Hz/s [39]. There exist also protections that are utilized to identify islanding scenarios, in particular, for medium voltage grids with distributed generation.

As it can be expected from the discussion on the estimation of the frequency in the previous sections, a precise estimation of the RoCoF is not an easy task. In first place, one has to define the instantaneous frequency. Then one has to define a robust way to calculate its derivative. To this difficulty, one should also add the fact that if the inertia of the system vary significantly, then the threshold value of the RoCoF should also be updated accordingly, which leads to the even more complex problem of estimating the effective inertia of a system. This can differ from the physical one if inverter-based generation provide Fast Frequency Response (FFR) or synthetic inertia.

A robust, although rough, approach to estimate the RoCoF is with a finite difference:

$$\mathrm{RoCoF}(t) = \frac{1}{2\pi}\frac{\omega(t) - \omega(t - \Delta t)}{\Delta t}, \tag{1.66}$$

where the value of Δt has to be chosen carefully to obtain a good tradeoff between accuracy and quick response. A rolling window of $\Delta t = 500$ ms is a frequent choice [40].

Another definition is the initial value of the RoCoF, that is the time derivative of the frequency right after the occurrence of an imbalance. This is defined as [40]:

$$\mathrm{RoCoF}|_{t=0+} = \frac{\Delta p_{\mathrm{imbalance}}}{p_{\mathrm{load}}} \frac{\omega_o}{2\pi(2H_{\mathrm{agg}})}, \tag{1.67}$$

which descends directly from (1.65) and requires the knowledge not only of the aggregated inertia of the system but also of the current loading condition and of the power imbalance. Equation (1.67) is usually used by Transmission System Operators (TSOs) to define minimum inertia floors in their systems to keep RoCoF within defined limits (see Chap. 5).

Apart from "practical" definitions, also the calculation of the RoCoF for simulation tools poses some challenges. In standard positive-sequence transient stability models, the calculation RoCoF based on the complex frequency obtained with Eq. (1.63) is the most accurate approach as it is based on an analytical model. However, no commercial software tool provides this feature.[4] Another option is to utilize the frequency divider formula where instead of the frequency deviations, one can use the time derivatives of the rotor speeds of the synchronous machines:

$$\mathbf{0}_{n,1} = \mathbf{B}_{\mathrm{BB}}\boldsymbol{\omega}'_{\mathrm{B}}(t) + \mathbf{B}_{\mathrm{BG}}\boldsymbol{\omega}'_{\mathrm{G}}(t), \tag{1.68}$$

where $\boldsymbol{\omega}'_{\mathrm{G}}$ are available during the simulation as they are defined by the differential equations of the machine. But, again, the calculation of the frequency divider formula is not included in most commercial software tools that utilize a positive sequence transient stability model.

These tools generally provide an estimation of bus frequencies based on PLL models as described in Sect. 1.2.3, which have difficulties to properly estimate the frequency, especially in the first instants after the occurrence of a large disturbance, which is precisely when the estimation of the RoCoF is most needed.

Reference [42] provides a pragmatic approach to overcome the limitations of the PLL to estimate the frequency during transients, TREx, and proposes a sequence of remediation actions to improve the estimation of the RoCoF, namely, moving average function; blocking function; switching filter; and jump shaver (see Fig. 1.14). All these strategies are substantially aimed at reducing the effects of the "jumps" in the voltage phasors that are inevitable in conventional transient stability models, as bus voltage magnitudes and phase angles are algebraic variables and, thus, can show large abrupt variations during the simulation.

In Brief

- There exist several definitions of frequency, which may lead to different values for the same signal.

[4] Simulation results shown in the examples of this chapter where obtained with the software tool Dome, which has been developed by the second author of the book and his research group [41].

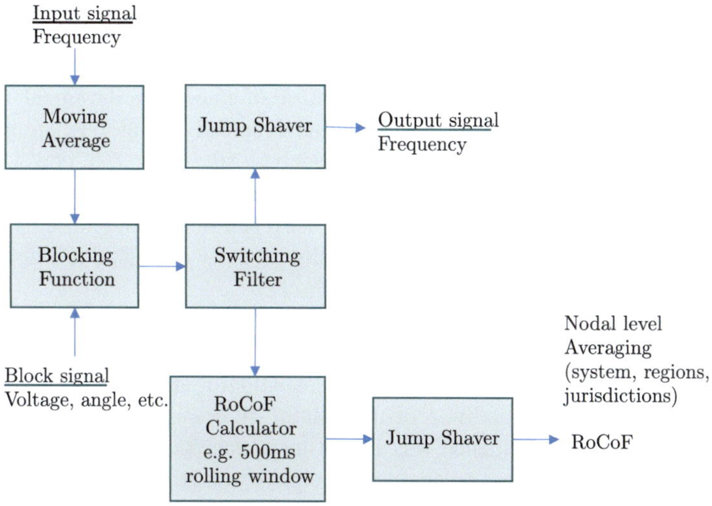

Fig. 1.14 Scheme of the TREx implemented by EirGrid [42]

- The instantaneous frequency is substantially a time derivative of the phase angle of the time-varying bus voltage phasor.
- Differential geometry provides "invariants," which are well-defined and unambiguous mathematical quantities. Invariants also have the property that are independent from the choice of coordinates. However, the geometric invariants do not always coincide with the sought instantaneous frequency.
- As a corollary of the above, the instantaneous frequency is not an invariant and its value does depend on the method utilized to estimate it.
- Despite being designed for stationary signals, Fourier transform is possibly still the best approach to estimate the variations of the fundamental frequency of power systems. However the Fourier transform is not suitable for controls.
- The complex frequency can be calculated analytically in simulation tools.
- PLLs are well-suited for control and are the most utilized synchronization mean for converters. However, they are not suited for protections as their accuracy is low in the first instants after a large disturbance.
- The RoCoF is the time derivative of the frequency.
- Grid codes provide simple formulas to evaluate RoCoF but these formulas are far from the theoretical definition.
- The calculation of the actual theoretical value of the RoCoF is a challenge both in practice and in simulation tools.

Open Questions

Before being able to define frequency and RoCoF of a non-periodic signal, one has to first state clearly what "property" of that signal is actually being sought (curvature, harmonic spectrum, phase of the associated analytic signal, etc.). In power system transients, the most relevant quantity is arguably the instantaneous frequency of the positive sequence of the voltage at network buses. Harmonics and noise have to be removed. Whether this quantity can be actually properly estimated is still an open question as all existing techniques have drawbacks.

Even if the instantaneous frequency could be estimated with any given accuracy, it remains a deeper issue to be solved, namely, whether the information provided by the frequency and RoCoF are actually a meaningful and unambiguous metric of the current operating condition of the system. In conventional power systems where the dynamic performance was driven by synchronous machines and their primary controllers, there was no doubt that frequency and RoCoF are key quantities to determine the power unbalance of the system. This is not so obvious for power systems with high share of IBRs or, in the extreme case, 100% IBR-based generation. The relevance of frequency and RoCoF will highly depend on the control strategies that become standard for IBRs.

References

1. Furfari F (2000) The evolution of power-line frequencies 133 1/3 to 25 Hz. IEEE Ind Appl Mag 6(5):12–14
2. Ulbig A, Borsche TS, Andersson G (2014) Impact of low rotational inertia on power system stability and operation. IFAC Proc Vol 47(3):7290–7297. 19th IFAC world congress
3. Milano F, Ortega Á (2017) Frequency divider. IEEE Trans Power Syst 32(2):1493–1501
4. Milano F (2010) Power system modelling and scripting. Springer, London
5. Kettner AM, Paolone M (2018) On the properties of the power systems nodal admittance matrix. IEEE Trans Power Syst 33(1):1130–1131
6. Ilić MD, Liu S (2012) Hierarchical power systems control. Springer, London, UK
7. Milano F, Ortega Á, Conejo AJ (2018) Model-agnostic linear estimation of generator rotor speeds based on phasor measurement units. IEEE Trans Power Syst 33(6):7258–7268
8. Milano F (2018) Rotor speed-free estimation of the frequency of the center of inertia. IEEE Trans Power Syst 33(1):1153–1155
9. Tzounas G, Dassios I, Milano F (2022) Frequency divider as a continuum. IEEE Trans Power Syst 37(6):4970–4973
10. Ponce I, Milano F (2024) Complex frequency propagation along transmission lines. In: IEEE PES general meeting, pp 1–5
11. Ortega Á, Musa A, Monti A et al (2018) Hardware-in-the-loop validation of the frequency divider formula. In: IEEE PES general meeting, pp 1–5
12. Milano F, Ortega Á (2021) A method for evaluating frequency regulation in an electrical grid–part i: theory. IEEE Trans Power Syst 36(1):183–193
13. Tan B, Zhao J, Milano F et al (2021) Extended frequency divider for bus frequency estimation considering virtual inertia from DFIGs. In: IEEE PES innovative smart grid technologies conference—Latin America (ISGT Latin America), pp 1–5

14. Aghahassani M, Castronuovo ED, Ledesma P et al (2023) Extended frequency divider formula with inclusion of DER control dynamics. In: IEEE PES general meeting, pp 1–5
15. Milano F (2022) Complex frequency. IEEE Trans Power Syst 37(2):1230–1240
16. Kirkham H, Dickerson W, Phadke A (2018) Defining power system frequency. In: IEEE PES general meeting, pp 1–5
17. IEEE/IEC International Standard—Measuring relays and protection equipment—Part 118-1: Synchrophasor for power systems–measurements. IEEE, pp 1–78 (2018)
18. Boashash B (1992) Estimating and interpreting the instantaneous frequency of a signal. I. Fundamentals. Proc IEEE 80(4):520–538
19. Karpilow A, Derviškadić A, Frigo G et al (2022) Characterization of non-stationary signals in electric grids: a functional dictionary approach. IEEE Trans Power Syst 37(2):1126–1138
20. Hahn SL (1996) Hilbert transforms in signal processing. Artech House, Boston, MA
21. Hahn S (1964) Complex variable frequency electric circuit theory. Proc IEEE 52(6):736–737
22. Hahn SL (1959) The instantaneous complex frequency concept and its application to the analysis of building-up of oscillations in oscillators. In: proceedings of vibration problems, vol 1. Państwowe Wydawn. Naukowe, pp 24–46
23. Cohen L (1995) Time frequency analysis: theory and applications. In: Prentice-hall signal processing. Upper Saddle River, NJ
24. Milano F, Alhanjari B, Tzounas G (2024) Enhancing frequency control through rate of change of voltage feedback. IEEE Trans Power Syst 39(1):2385–2388
25. Milano F (2022) A geometrical interpretation of frequency. IEEE Trans Power Syst 37(1):816–819
26. Milano F, Tzounas G, Dassios I et al (2022) Applications of the Frenet frame to electric circuits. IEEE Trans Circuits Syst I Regul Pap 69(4):1668–1680
27. Milano F (2023) The Frenet frame as a generalization of the Park transform. IEEE Trans Circuits Syst I Regul Pap 70(2):966–976
28. Bedrosian E (1963) A product theorem for Hilbert transforms. Proc IEEE 51(5):868–869
29. Hahn SL (2003) On the uniqueness of the definition of the amplitude and phase of the analytic signal. Signal Process 83(8):1815–1820
30. Milano F, Tzounas G, Dassios I et al (2022) Using differential geometry to revisit the paradoxes of the instantaneous frequency. IEEE Open Access J Power Energy 9:501–513
31. Sandoval S, De Leon PL (2018) The instantaneous spectrum: a general framework for time-frequency analysis. IEEE Trans Signal Process 66(21):5679–5693
32. Alshawabkeh A, Tzounas G, Molina-García Á et al (2025) Instantaneous frequency estimation in unbalanced systems using affine differential geometry. IEEE Trans Circuits Syst II Express Briefs 72(1):343–347
33. Das JC (2015) Power system harmonics and passive filter designs. Wiley IEEE Press, New York
34. Milano F (ed) (2022) Advances in power system modelling, control and stability analysis, 2nd edn. The IET, London, UK
35. Karimi H, Karimi-Ghartemani M, Iravani MR (2004) Estimation of frequency and its rate of change for applications in power systems. IEEE Trans Power Delivery 19(2):472–480
36. Nicastri A, Nagliero A (2010) Comparison and evaluation of the PLL techniques for the design of the grid-connected inverter systems. In: IEEE international symposium on industrial electronics. IEEE, pp 3865–3870
37. Teodorescu R, Liserre M, Rodríguez P (2011) Grid converters for photovoltaic and wind power systems. Wiley, Chichester, UK
38. Pourbeick P, Weber J, Majumdar R et al (2018) WECC white paper on understanding frequency calculation in positive sequence stability programs, pp 1–4
39. Eirgrid & SONI, RoCoF alternative & complementary solutions project. [Online]. https://www.eirgrid.ie/
40. ENTSO-E (2020) Inertia and rate of change of frequency (RoCoF). [Online]. https://www.eepublicdownloads.entsoe.eu

41. Milano F (2013) A Python-based software tool for power system analysis. In: IEEE PES general meeting, pp 1–5
42. Bakhtvar M, Nedic DP, Jafarian M et al (2024) Enhancing the evaluation of rate of change of frequency during fault contingencies simulated in phasor-domain tools. In: CIGRE Paris session. CIGRE, pp 1–13

Chapter 2
Frequency Control

Frequency is currently the best measurable quantity that contains information about the overall balance between generation and consumption. For this reason, an efficient and tight tracking of the frequency allows a secure operation of power systems. The chapter revisits the concept of frequency control and provide a comprehensive overview of it. In particular, we present various terms and concepts related to frequency control that are often utilized with different meanings by different TSOs. Then, we discuss how to navigate through the current plethora of terms surrounding the topic of frequency control. With this aim, we provide a classification that emphasizes the difference between *frequency response*, *frequency control*, and *frequency regulation* (Sect. 2.2). Section 2.3 presents the different stages of frequency control and includes a variety of real-world cases that illustrate the latest trends in the industry on how to improve frequency control of converter-based power systems. Finally, Sects. 2.4 and 2.5 discuss the role of TSO control centers and provide several examples based on Time Domain Simulation (TDS).

2.1 Background

The concept of "frequency control" has existed since the early days of ac power systems, the history of which was initiated more than a century ago. Initially, very basic control systems were used to perform frequency control, and this task was typically handled by a single power plant [1]. As the provision of high-quality electrical supply became more important for customers, automatic and sophisticated control systems were designed, and more resources were dedicated to performing frequency control to the highest standards as is the case nowadays. Moreover, as power systems span a range of time scales starting from milliseconds to hours and years [2], frequency control had to be designed to be performed across different time scales.

T. Kërçi and F. Milano, *Frequency Quality in Power Systems*, Power Systems,
https://doi.org/10.1007/978-3-032-04721-2_2

To manage power unbalances and, as a byproduct, transient frequency variations, for example, those following the loss of a large generating unit, TSOs utilize various frequency response stages that can be broadly classified: (i) electromagnetic energy; (ii) inertial response; (iii) PFC; (iv) Secondary Frequency Control (SFC); (v) Tertiary Frequency Control (TFC); and (vi) Time Error Control (TEC). The remainder of the chapter describes the specific features and purposes of each of these stages.

2.2 Frequency Control Versus Frequency Regulation

Before delving into the details of frequency control, we revisit the meaning of two commonly used but often confused terms, namely *frequency control* and *frequency regulation*. Indeed, one may find hundreds of papers published by academia in the literature where such terms are used interchangeably. For example, we cite [3–5]. The industry is no different from this confusion. However, we show below that while these two terms are, in effect, synonyms, in industry, they generally have different meanings. For most system operators, in fact, *frequency regulation* is a subset of *frequency control*. Quoting Harold Kirkham, discussing the definition of "smart grid" in the PowerGlobe forum in 2024: *We power engineers have demonstrated an extraordinary ability to mess around with names and definitions*. We agree with this statement and have recently tried to raise awareness to address these issues [6]. This book represents our attempt to reduce "the mess" with the definitions of frequency control, response, regulation, quality, and FFR. One might argue with Jeffrey Wischkaemper in the same PowerGlobe thread above that the power system community has much bigger problems to solve than "fixing" definitions. However, there has to be a general consensus on most basic concepts. Otherwise, it will be difficult to understand each other and there is little hope for solving complex power system problems.

To do so, we take a few definitions available in industry and academia and highlight existing issues/differences/similarities between these terms. Based on that, we then propose a comprehensive definition of *frequency control* to address the misunderstanding and confusion that exists today. With this aim, Table 2.1 presents all the relevant definitions. It can be seen that NERC and the selected reference from the literature do not use a definition for *frequency control*, whereas EirGrid does so even though it is a simple and vague definition. This is surprising as NERC defines the different stages of *frequency control*, that is, PFC, SFC, TFC, and TEC (see below). It should be noted though that there are other TSOs such as the National Energy System Operator (NESO) in Great Britain (GB) that very rarely use the terms *frequency control* or *frequency regulation*, if at all.

With regard to *frequency regulation*, all three references provide a detailed definition yet quite different. The main similarity is that both NERC and EirGrid recognize the governor response as contributing toward frequency regulation. However, there are some important differences. NERC mentions AGC as a type of control assisting with frequency regulation while EirGrid does not. This makes sense considering

Table 2.1 Definitions of frequency control and regulation

Term	North American electric reliability corporation (NERC)	EirGrid	Literature [7]
Frequency control	–	The control of the frequency on the power system	–
Frequency regulation	The ability of a balancing authority to help the interconnection maintain scheduled frequency. This assistance can include both turbine governor response and automatic generation control	The automatic adjustment of active power output by a generation unit, initiated by free governor action in response to continuous minor fluctuations of frequency on the power system	Centralized increase or decrease in active power provided by resources with AGC to match the nominal frequency at all times under normal operation
Regulating reserve	An amount of reserve responsive to automatic generation control, which is sufficient to provide normal regulating margin	Response by dynamic or spinning resources, usually conventional generation	–

that AGC is widely implemented in NERC but not in IE. Indeed, the vast majority of power system engineers use AGC as a synonym to frequency regulation. However, frequency regulation is much more than just AGC as we explain below and throughout the book.

Then there is another significant difference between the three definitions, that is, EirGrid and [7] specifically relate *frequency regulation* to the ability to respond to continuous minor fluctuations of the frequency, or, in other words, relate *frequency regulation* to normal operating conditions and exclude abnormal/contingency conditions. For example, in practice, in IE, this is achieved by setting a narrow governor frequency dead-band such as ± 15 mHz on conventional units and wind power plants, or manual actions taken by the dispatcher. However, NERC recognizes that frequency regulation (through SFC) includes reserves to restore frequency to its scheduled value following disturbances as well and not just normal conditions [8]. This aspect is confirmed in the illustrative examples at the end of this chapter. In addition, all three definitions suggest automatic actions as the means to provide frequency regulation but, as mentioned above, manual actions are also a key source of frequency regulation for TSOs that, for example, do not implement an AGC as is the case in IE, Northern Ireland (NI) or GB. The confusion continues if one looks at the *regulating reserve* definition with NERC explicitly mentioning regulating units that receive and respond to AGC signals and EirGrid referring to dynamic/spinning resources such as conventional units. Therefore, it is clear that there is no consistent definition and use of the frequency regulation and frequency control terminology in the power system community.

Motivated by this gap in the literature we propose the following definitions of *frequency response* and *frequency control* in power systems:

> ***Frequency response*** *refers to the natural, automatic, and manual increase or decrease of power production or consumption by power system components and resources in response to frequency fluctuations and time error. Power adjustments are performed as a result of the inherent dynamic behavior of power system components as well as any controller, in all time scales.*

> ***Frequency control*** *refers to the automatic or manual increase or decrease of power production or consumption by power system resources in response to frequency fluctuations and time error. Power adjustments are performed as a result of PFC, SFC, TFC and TEC in all time scales.*

Frequency response includes all active and passive devices, as long as they can store/release energy. Examples of passive devices are transmission lines and the inertia of synchronous machines. Frequency control, on the other hand, includes only controllers in all time scales. On the other hand, there is no need to provide a specific definition of frequency regulation, as "regulation" and "control" are synonyms. We recognize, however, that most system operators consider frequency regulation a subset of frequency control, which in turn is a subset of frequency response, as illustrated in Fig. 2.1.

In particular, Fig. 2.1 shows that "frequency regulation" is often used in industry to indicate: (i) a sub-set of frequency control; (ii) both automatic and manual active power adjustments; (iii) a continuous process managing frequency fluctuations during normal and abnormal operating conditions; (iv) three processes namely automatic PFC with narrow frequency dead-band, and manual and automatic SFC; and (v) a specific time scale, usually from 1 second (e.g., fast PFC or FFR) to 15 min (e.g., manual AGC).[1] Moreover, if PFC is used with a narrow dead-band, then it can provide both Primary Frequency Regulation (PFR) within the standard frequency range and contingency response.

The definitions above refer to "power," without specifying whether it is active or reactive one. Indeed, in high-voltage transmission systems, the link between active power and frequency is much stronger than that between reactive power and frequency. However, as the control move from transmission to distribution systems, the coupling between active and reactive powers and frequency becomes fuzzier. This can be deduced from power flow equations.

With this aim, let us consider the active and reactive power injections in the branch connecting buses h and k:

[1] Interestingly, AGC, which includes the word "control," is often considered by system operators as part of frequency regulation. This semantic inconsistency is removed with the proposed definition of frequency control.

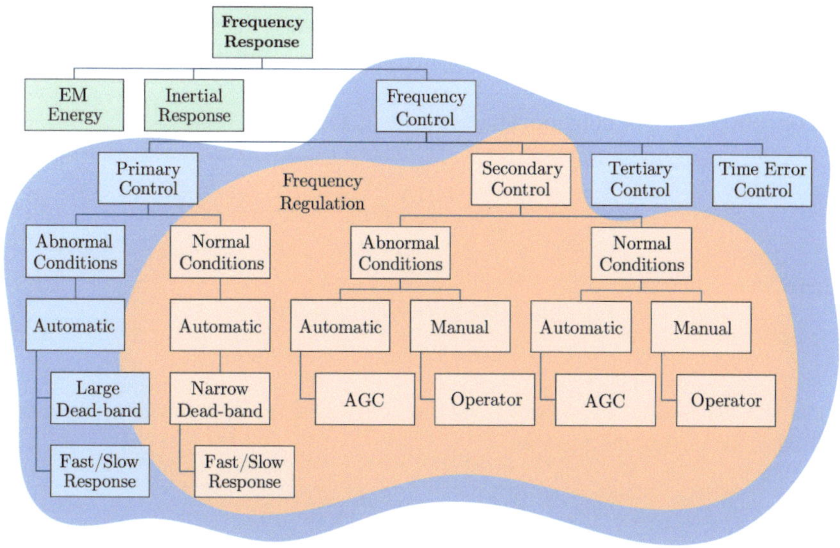

Fig. 2.1 Frequency response and control in power systems. The pink-shaded area within frequency control represents common meanings that system operators assign to the term "frequency regulation"

$$
\begin{aligned}
p_{hk} &= v_h^2 G_{hh} + v_h v_k \left(G_{hk} \cos\theta_{hk} + B_{hk} \sin\theta_{hk} \right), \\
q_{hk} &= -v_h^2 B_{hh} + v_h v_k \left(G_{hk} \sin\theta_{hk} - B_{hk} \cos\theta_{hk} \right),
\end{aligned}
\tag{2.1}
$$

where p_{hk} and q_{hk} are the active and reactive powers, respectively, injected in bus h flowing toward bus k, respectively; $G_{hh} + jB_{hh}$ and $G_{hk} + jB_{hk}$ are the diagonal h-th and (h, k)-th elements, respectively, of the grid admittance matrix; v_h and v_k represent the magnitude of the voltages at buses h and k, respectively; and $\theta_{hk} = \theta_h - \theta_k$ indicates the difference between voltage phase angles.

Equation (2.1) is valid for both steady-state and transient conditions. In the latter case, which is the relevant situation for control purposes and thus the focus of this chapter, voltage magnitudes and angles are functions of time. Differentiation of (2.1) with respect to the angles and voltages gives:

$$
\begin{aligned}
dp_{hk} &= \underbrace{\frac{\partial p_{hk}}{\partial \theta_h} d\theta_h + \frac{\partial p_{hk}}{\partial \theta_k} d\theta_k}_{dp_{hk}^{\theta}} + \underbrace{\frac{\partial p_{hk}}{\partial v_h} dv_h + \frac{\partial p_{hk}}{\partial v_k} dv_k}_{dp_{hk}^{v}}, \\
dq_{hk} &= \underbrace{\frac{\partial q_{hk}}{\partial \theta_h} d\theta_h + \frac{\partial q_{hk}}{\partial \theta_k} d\theta_k}_{dq_{hk}^{\theta}} + \underbrace{\frac{\partial q_{hk}}{\partial v_h} dv_h + \frac{\partial q_{hk}}{\partial v_k} dv_k}_{dq_{hk}^{v}},
\end{aligned}
\tag{2.2}
$$

where dp^{θ}_{hk} and dq^{θ}_{hk} are the sensitivities of dp_{hk} and dq_{hk} that depends on the angle. Accordingly, dp^{v}_{hk} and dq^{v}_{hk} are the sensitivities of dp_{hk} and dq_{hk} that depends only on the voltage.

For convenience of notation, let us rewrite (2.1) as:

$$\begin{aligned} p_{hk} &= \phi_{hh} + \phi_{hk} \\ q_{hk} &= \psi_{hh} + \psi_{hk} \end{aligned} \tag{2.3}$$

where

$$\begin{aligned} \phi_{sr} &= v_s v_r \left(G_{sr} \cos\theta_{sr} + B_{sr} \sin\theta_{sr}\right), \\ \psi_{sr} &= v_s v_r \left(G_{sr} \sin\theta_{sr} - B_{sr} \cos\theta_{sr}\right). \end{aligned} \tag{2.4}$$

Then, the differential expressions for the active and reactive power components related to the angles are given by:

$$dp^{\theta}_{hk} = -\psi_{hk}\, d\theta_{hk}\,, \tag{2.5}$$

$$dq^{\theta}_{hk} = \phi_{hk}\, d\theta_{hk}\,, \tag{2.6}$$

and, for the active and reactive power components related to the voltages, the differential expressions are given by:

$$dp^{v}_{hk} = 2\,\phi_{hh}\,\frac{d(v_h)}{v_h} + \phi_{hk}\,\frac{d(v_h v_k)}{v_h v_k}\,, \tag{2.7}$$

$$dq^{v}_{hk} = 2\,\psi_{hh}\,\frac{d(v_h)}{v_h} + \psi_{hk}\,\frac{d(v_h v_k)}{v_h v_k}\,. \tag{2.8}$$

In high-voltage transmission systems, $B_{sr} \geq G_{sr}$ and, hence, the terms dp^{θ}_{hk} and dq^{v}_{hk} are dominant. However, in distribution systems, $B_{sr} \approx G_{sr}$, which makes the terms dp^{v}_{hk} and dq^{θ}_{hk} non-negligible. The interested reader can find in [9, 10] some examples of how the coupling between reactive power and frequency can be exploited for the control of Distributed Energy Resources (DERs).

In the literature, other interpretations of the difference between *control* and *regulation* have been proposed. For example, an interesting one is given in [11] that reads as follows: *Regulation involves a continuous adjustment based on measured values. Imagine you want to maintain a constant temperature in a room. A thermostat continually measures the temperature and adjusts it to ensure it stays close to the desired value. This is regulation–a dynamic adjustment to achieve a specific goal. Control is more like a fixed setting without constant adjustments. Suppose you turn on a light. You decide it should be on but not necessarily how bright it needs to be. That's control–a one-time setting of a state or value without ongoing adjustments. In summary, […] regulation is continuous and actively responsive to changes, while control is more of a fixed setting without continuous adjustments.*

While the above may make a perfect analogy if ones considers frequency as the control parameter instead of the temperature, and the PFC and SFC as the "thermostat", we see two main problems to apply it in power systems. Specifically, while PFC with narrow dead-band is indeed continuous and actively responsive to (small) frequency changes providing frequency regulation, it does so based on a "fixed setting", for example, ±15 mHz dead-band. In addition, both PFC (with narrow dead-band) and SFC have included, for a long time now in power systems, the word "control" despite being called *frequency regulation*. While changing these terms (for example, removing the word "control" and changing it with the word "regulation") may be doable, it will need to be done in a consistent way. Another interesting use of the term is provided in [12] where both the words control and regulation are utilized, as in *automatic generation control regulation*. However, while technically correct, the expression "control regulation" is redundant.

Most importantly, the classification in Fig. 2.1 shows that frequency regulation is much more than just AGC [7] and, indeed, much more than even SFC [13]. High performance of frequency regulation can also be achieved by PFC with narrow dead-band or manual dispatch operating in similar time scales to AGC, as is the case in AIPS. Indeed, as will be illustrated in the examples in this chapter and discussed in Chap. 5, PFC with a narrow dead-band is quite effective in providing frequency regulation as compared to AGC. Australian Energy Market Operator (AEMO) states that it will not meet its frequency operating standard obligation without PFC continuously active under normal operating conditions (or without PFR) [14].

When the first author tells power system engineers not familiar with the AIPS that there is no AGC in place in the AIPS, they invariably ask: but how do you manage frequency then? Well, one of the objectives of this book is to show that it is possible to operate power systems, including low-inertia ones such as the AIPS, without an AGC and just PFC with ±15 mHz dead-band and manual (re-)dispatch.

Another important feature to highlight is that due to its time scales frequency regulation includes both normal and abnormal operating conditions and not just normal conditions. For example, AGC or "manual AGC" will contribute to restore frequency (assuming there are regulation reserves in place) toward the nominal value following a large contingency. This will be illustrated in the examples below.

2.3 Time Scales of Frequency Control

Despite TSOs using improved dispatching processes and sophisticated load and generation forecasting techniques, still it is not possible to perfectly maintain the balance between generation and consumption, in particular, following suddenly occurring events, such as the tripping of a generator (momentary imbalances). Such events require fast and automatic responses from the available resources such as the online generating units. On the other hand, continues changes between demand and generation require slower acting response which could take the form of automatic or manual responses. For this reason, frequency control takes the form of a hierarchical control based on response time and duration. This is illustrated in Fig. 2.2. While each stage acts over a different time scale, there is still some overlap between them. In fact, such

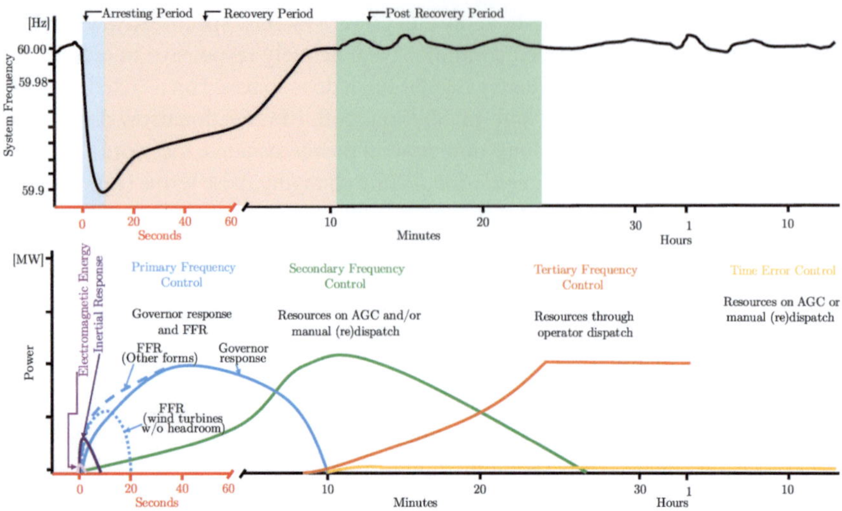

Fig. 2.2 Time scales of frequency control (adapted from [16])

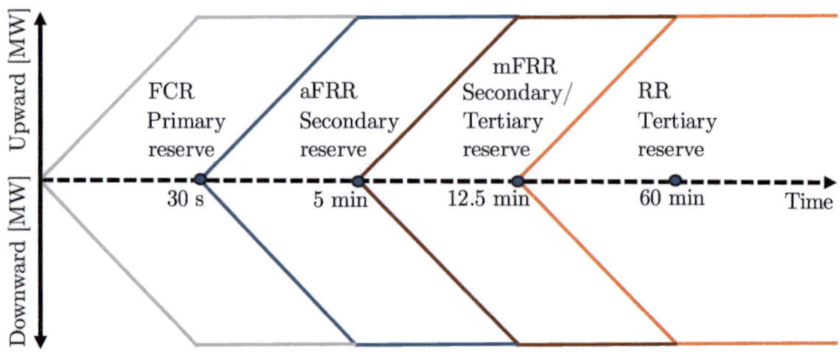

Fig. 2.3 Active power reserve products envisioned by ENTSO-E (adapted from [17])

a temporal overlap between the different frequency control services, which have substantively different characteristics, presents opportunities for mis-coordination that could lead to oscillations [15].

It is relevant to note that different power systems and jurisdictions may use different terminology for the different frequency control stages even though often they refer to the same thing. For example, ENTSO-E use the term Frequency Containment Reserve (FCR) for PFC, Frequency Restoration Reserve (FRR) including both Automatic Frequency Restoration Reserve (aFRR) and Manual Frequency Restoration Reserve (mFRR) with the former representing automatic SFC (AGC) and the latter part of both manual SFC and TFC, and Replacement Reserve (RR) representing TFC, see Fig. 2.3.

To make things even more complicated there are TSOs within ENTSO-E that use different products, such as the Nordic TSOs that split the FCR product into two

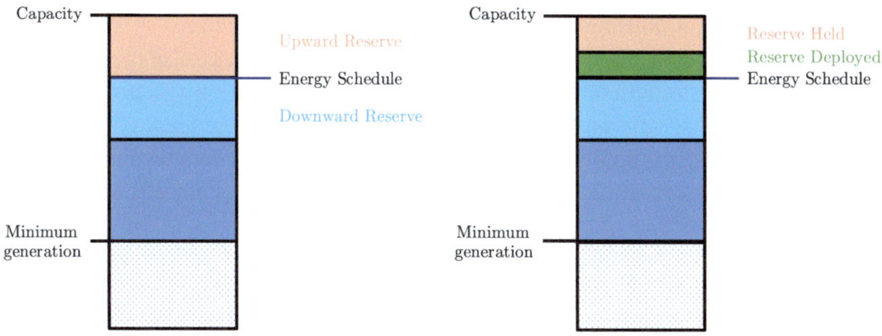

Fig. 2.4 Upward and downward reserves provision (adapted from [21])

different products namely FCR for normal operation (Frequency Containment Reserve for Normal Operation (FCR-N)) and FCR for disturbances (Frequency Containment Reserve for Disturbances (FCR-D))[2] [18]. FCR-N is basically slow PFC with FAT of 3 min and narrow dead-band that activates continuously due to small variations in generation and demand that occur during normal operation, whereas FCR-D has a FAT of 10 s and aims at containing frequency during disturbances. The only term that is used more consistently across NERC and ENTSO-E, but also in other parts of the world, is FFR. However, FFR is not currently defined in the European Union (EU) legislation as a standard frequency reserve product [19]. The use of different terminology is not ideal and generates confusion among practitioners of different regions.

The frequency control products described above are usually provided both in an upward and downward direction, that is, to deal with Under-frequency (UF) (upward reserves) and Over-frequency (OF) (downward reserves) events, respectively. These reserves and their deployments are illustrated in Fig. 2.4. TSOs monitor the availability of these reserves in real-time and generally recognize the need for all categories of reserves to be distributed in the system to avoid potential congestion in the grid [20].

2.3.1 *Electromagnetic Energy*

This stage of frequency response represents the energy stored in the magnetic fields of synchronous generators and the energy stored in the capacitance of transmission lines. It is inherent in the way power systems are designed and work. The generators that are electrically close to, for example, the loss of generation will supply energy from their magnetic fields and provide the largest active power response almost

[2] We use the word "product," commonly used by TSOs to refer to "market products" for reserve services, which are generally defined based on response times or Full Activation Time (FAT). For instance, while Battery Energy Storage Systems (BESS) can provide both FFR and FCR (market products), it does so based on one PFC (physical controller) and not two different PFCs.

immediately and proportionally to their electrical distance. Such a response may be sustained for approximately $\frac{1}{3}$ of a second [1]. How long this response is sustained is dependent on the strength and speed of the generator's excitation system. However, given the very small amount of energy stored, this stage of frequency response is neglected by TSOs in the design of frequency control strategies and dimensioning of reserves. For this reason, apart from the example below, it is not discussed in the rest of this book.

2.3.2 *Inertial Response*

This is another inherent response of power systems that does not involve a control design and follows the electromagnetic energy stage. From the physics we know that inertia represents the resistance of an object to a change in its position. Power systems are composed of large and small "moving/rotating" parts or masses such as the rotors of the synchronous machines, including both synchronous generators and motors. Inertial response to frequency deviations means the natural resistance of such masses to changes in their stored kinetic energy. In fact, the inertial response is immediate and represents the first relevant measure against frequency changes or, differently, it can also be considered as fastest relevant reserve or "instantaneous reserve" available to TSOs sustained for a few seconds. As such, it is a key parameter helping to maintain the power imbalance, that is, the power difference between production and consumption. Indeed, the low-inertia issue is considered as one of great challenges that power electronics introduce to the conventional power system [22].

For example, following a frequency deviation, kinetic energy that is intrinsically stored in the rotating masses of the generator system is released, rendering power system frequency dynamics slower and, hence, easier to regulate. The rotational energy is given as [23]:

$$E_{kin} = \frac{1}{2} J (2\pi f_{\mathrm{m}})^2 , \tag{2.9}$$

with J (kg m^2) as the moment of inertia of the synchronous machine and f_{m} the rotating frequency of the machine. The inertia constant H for a synchronous machine is defined by[3]:

$$H = \frac{E_{kin}}{S_n} = \frac{J(2\pi f_m)^2}{2S_n} , \tag{2.10}$$

[3] The unit of the inertia constant is seconds. However, unlike conventional time constants, the inertia constant is referred to the machine rate power and needs to be scaled to the system power base when interconnecting several machines with a grid in numerical simulations. For this reason, it is customary to indicate the units of the inertia constant as MWs/MVA. This also remembers that H represents the number of seconds that a synchronous machine at nominal speed can generate the rated power before stopping completely if the TG is suddenly removed.

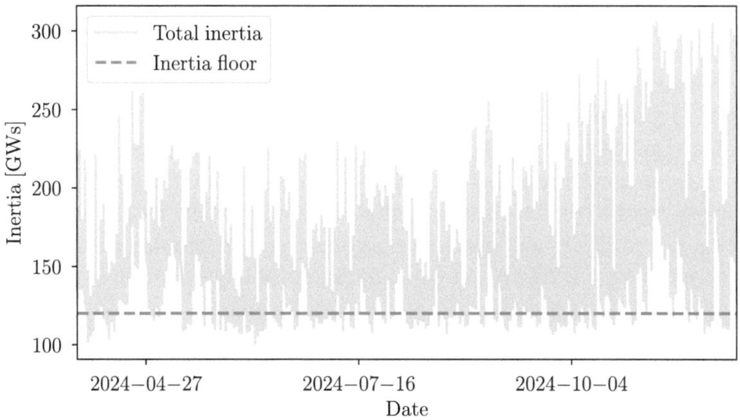

Fig. 2.5 Time-variance of inertia based on data from the GB power system [25]

with S_n as the rated power of the generator and H denoting the time duration during which the machine can supply its rated power solely with its stored kinetic energy. Typical values for H are in the range from 2 to 10 s. Total aggregated inertia from the synchronous machines is then calculated as:

$$H_{\text{agg}} = \sum_{i \in \Omega_{\text{G}}} H_i \, S_{n,i} \,, \tag{2.11}$$

where Ω_{G} is the set of online synchronous machines, $\text{S}_{\text{n,i}}$ is the MVA power rating of the online machines i.

In converter-dominated power systems, inertia becomes a highly time-dependent parameter as Renewable Energy Sources (RES) such as wind and solar power generation (weather-dependent) vary significantly throughout the day/month/year [24]. This variability of total inertia is illustrated in Fig. 2.5 based on data from the GB power system. Unlike synchronous generators, IBRs such as solar and wind generation have very distinct characteristics, including a lack of physically coupled inertial response, except for the Doubly-Fed Induction Generator (DFIG) due to its stator being synchronously connected to the power network.

Naturally, the displacement of large conventional generators for IBRs lead to a decrease in inertia. This reduction then alters the sensitivity of the system frequency to changes in operating conditions. In fact, inertia is recently becoming a local issue rather than just a system-wide issue as there could be significant inertia distribution differences between regions of a power system. These differences can then lead to increasing the system's local RoCoF [26]. Given that inertia is such an important parameter, TSOs monitor and maintain minimum inertia levels to ensure the overall power balance and, in particular, to maintain RoCoF within predefined limits during large frequency excursions [27, 28]. For example, this is the case of the Irish, British and Australian TSOs (see Chap. 5). These TSOs are currently deploying synchronous condensers to replace the inertia lost by conventional synchronous generators. Note that a synchronous condenser is a type of dc excited synchronous motor that is unloaded, that is, its shaft is free spinning [29]. An example of synchronous condensers procurement from the AIPS grid is provided below.

The fact that TSOs are installing or planning to install synchronous machines such as synchronous condensers (but also nuclear generation in many countries) means that power systems will experience inertia for, at least, a few more decades. Hydro units, as a type of RES, will always be a good option to provide clean energy and induction motors are certainly not going to disappear anytime soon.

2.3.2.1 Regional Inertia

IE and NI have set ambitious decarbonization targets. For example, the ambition of both countries is to have at least 80% of electricity demand met by RES by 2030 and ultimately transition to net zero by 2050 [30]. This means that the TSOs need to operate the future AIPS with zero carbon-intensive conventional generator units online. To do so, first, the TSOs need to replace some of the services that these conventional units provide today such as inertia, reactive power, and short-circuit contribution. Here is where an old conventional technology such as synchronous condensers comes in.

Back in 2022, EirGrid and SONI launched a consultation to procure LCIS namely 10 GWs of inertia with a requirement of 4 GWs in NI and of 6 GWs in IE for 2026 [31]. What is interesting though is that the TSOs identified and incentivized certain zones in the AIPS (see Fig. 2.6) where the location of LCIS devices such as synchronous condensers would bring more benefits. The main driver for this locational requirement were weak areas of the AIPS. In other words, the LCIS is one single service comprising the provision of synchronous inertia, reactive power and short-circuit contribution. However, the distribution of inertia will also help address potential differences in the post-contingency RoCoF values [32]. Regional inertia distribution is also critical with respect to damping of power system oscillations [33].

In 2024, EirGrid and SONI completed the procurement of a total of 10.963 GWs inertia, with 6.963 GWs in IE and 4 GWs in NI, see Table 2.2 (awarding fixed term contracts with a six-year duration) [34]. 5 out of 6 of the procured synchronous condenser devices are located within the three incentivized zones in Fig. 2.6. At the time of writing, these inertia volumes equate to approximately 45% of the system's current inertia floor requirement. The TSOs are currently considering another phase of procuring LCIS. GFM could be another potential technology providing LCIS in the future. This is considered crucial given GFM could provide inertia support at a lower cost than synchronous condensers, making it a cost-effective option for grid stabilization.

It is worthwhile to mention that other TSOs operating converter-based low-inertia power grids such as AEMO have regional inertia requirements as well [35]. Finally, real-world spot (day-ahead) inertia markets are not yet available despite the literature

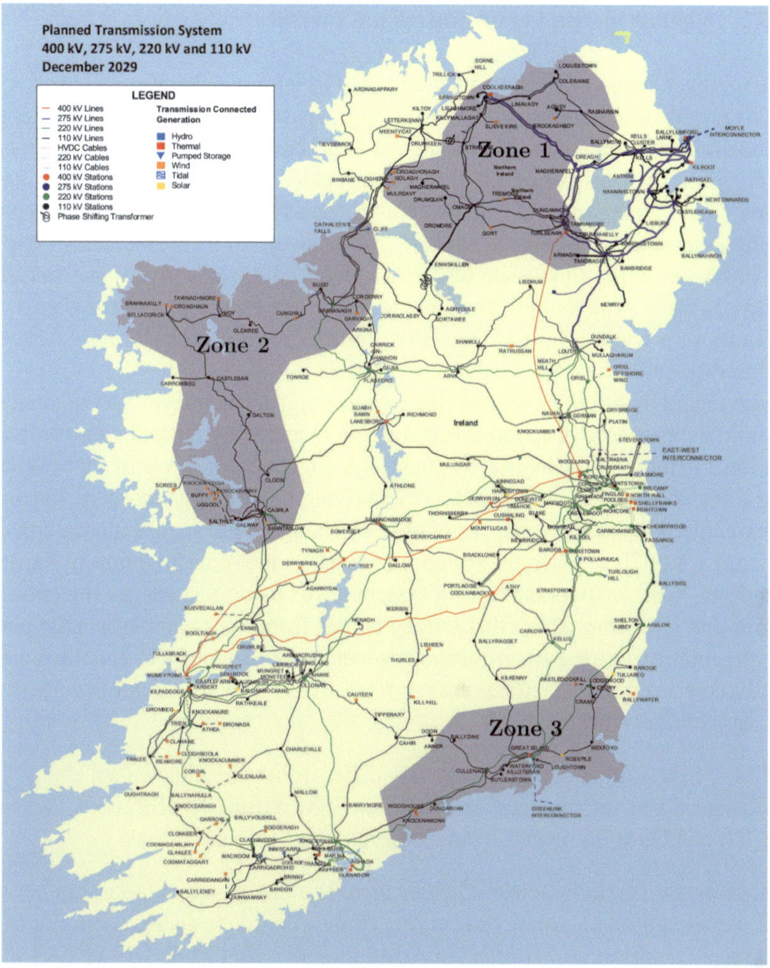

Fig. 2.6 Zones incentivized for placement of LCIS devices [31]

Table 2.2 Summary of LCIS phase 1 procurement

Jurisdiction	Requirement [GWs]	Procured [GWs]	Synchronous condensers	Target go-live year
IE	6	6.963	4	2027
NI	4	4	2	2027

advocating these as key to the cost-effective operation of the future low-carbon system [36, 37].

2.3.2.2 Virtual Inertia Versus Rotational Inertia

Since inertia is such a critical parameter for TSOs, it is considered useful to provide a brief overview of other types of it. For instance, apart from synchronous condensers, advanced control strategies of IBRs such as GFM control are able to provide inertia-like response (or "virtual inertia"), that can replace the inertia lost by phasing out conventional generators. There exist another term for inertia-like response namely "synthetic inertia" or "emulated inertia" [38]. However, this is just another form of FFR provision in the time scale of 2 to 6 s that could serve to mitigate frequency stability at times of high wind availability where, for example, the stored energy in the blades of a wind turbine is converted into electrical energy during a frequency dip [39, 40], and so will not be discussed further.

With regard to different types of GFM available, it can be said that all of them seek to mimic the response of a traditional rotating machine by reproducing the "swing equation." In terms of the inertial response, it is created by the inverter controls which bring a lot of flexibility compared to "fixed" inertia from synchronous machines. For example, converters make it possible to change their virtual inertia constant, say, from 1 s to 20 s, based on system needs and under consideration of hardware limitations. What matters is meeting system needs using fast energy injection into the grid and less from where that energy comes from, for example, kinetic energy from the rotating masses of synchronous generators and motors vs chemical energy directly converted into electrical energy from BESS. It is important to remark that from an inertia perspective [41]:

- GFM inertia estimation varies and may differ from nameplate value; and
- Different GFM implementations such as GFM Virtual Synchronous Generator (VSG) and GFM droop could have different inertia contributions.

Moreover, it is still not clear whether there is a one-to-one relationship between GFM inertia and rotational inertia coming from synchronous generation. For example, a recent demonstration project in Australia shows that while GFM BESS shows faster response than standard GFL it is slower in providing effective active power response compared to synchronous generators over the first 2 seconds [42]. More work is needed to understand such differences. This can then increase the trust of TSOs on GFM being a mature technology providing inertia services. However, even without GFM, real-world experiences have shown that the combination of some synchronous inertia with FFR from GFL IBRs such as BESS injecting active power in the inertial time frame could be enough to safely and securely operate power systems [43]. In this case, particular attention should be paid to mandatory frequency requirements vs market-based approaches to procure reserve services and balance between cost and reliability.

Table 2.3 Inertia monitoring

Monitoring method	Principle	Relevant Jurisdiction
Unit Commitment (UC)	Sum of all inertial contributions from online synchronous units	Electric Reliability Council of Texas (ERCOT), EirGrid/SONI, AEMO
UC + Adjustment factor	Same as above plus inertia from load and DERs	NESO
Continuous frequency measurement ("ambient" or with stimulus)	Real-time inertia based on "ambient" frequency changes or using known stimulus	NESO
Historic disturbance events	Based on historic analysis of large disturbance events	Not used for inertia monitoring but rather for calibration of other methods

2.3.2.3 Inertia Monitoring

TSOs monitor inertia levels in real-time to make sure enough inertia is available at all times. Doing so brings the following main benefits in converter-based power systems [44, 45].

- Help maintain system stability through situational awareness in real-time and ensure inertia stays above a minimum threshold.
- Help avoid RES curtailment.
- Optimal generation dispatch.
- Allows TSOs to take preventive measures and activate relevant controls to prevent, for example, potential grid outages.
- Allows to optimize inertia sources and frequency reserves, in particular, PFC reserves.
- Can help in terms of future planning and placement of GFM sources.

Table 2.3 provides a summary of the main inertia monitoring methods in use with TSOs [45]. The table shows that most TSOs still use simple methods to enhance the visibility of inertia across the power system. A common approach consists of simply summing up the inertia of synchronous machines connected to the grid. An exception is NESO, GB, where recently it is using pioneering new approaches such as utilizing an ultra-capacitor to inject a few MW oscillatory power and then using the swing equation with known Δp and RoCoF to determine the inertia level [46].

2.3.3 *Primary Control*

As shown in Fig. 2.2 PFC is the third stage of frequency response and is defined by NERC as [47]: *Actions provided by the Interconnection to arrest and stabilize frequency in response to frequency deviations. Primary Control comes from automatic generator governor response, load response (typically from motors), and other devices that provide an immediate response based on local (device-level) control systems.* Note that *a governor can be either an electronic or mechanical device that increases or decreases a generator's output based on frequency changes in the system* [13]. PFC is mandatory in many jurisdictions around the world.

PFC has a ramping capability of the order of hundreds of MW/s [48]. Historically, only synchronous generating units provided PFC. For this reason, it can be said that PFC is a well-established art for synchronous generation [15]. On the other hand, converter-based resources such as BESS have recently become the dominant providers of PFC and are capable of extremely fast PFC. Such a capability is beneficial but can also give rise to system-wide frequency oscillations [15].

PFC is also commonly known as frequency response. This is the case of US and GB [8, 49].[4] It is reported in [12] that the US Eastern Interconnection has had a significant decline in its frequency response over the past 30 years with the main reason being attributed to the lack of incentives for generators to provide PFC. PFC can be divided into proportional/dynamic/continuous and discrete/static response depending on how it responds to frequency deviations. For example, the vast majority of conventional and new RES generating units and BESS provide a proportional response to frequency deviations (also called a droop response) while demand side units are typical providers of static response based on predefined frequency thresholds. See Chap. 12 on examples of such responses from the AIPS. Since PFC is mainly implemented through droop control we will provide a brief overview of this control below.

In many cases, PFC is assumed to only help deal with large disturbance. For example, reference [51] states that: *Primary frequency control is critical for maintaining the reliability of the interconnection after a disturbance, by restoring the generation-load balance, and it is implemented through governor control and automatically assisted by the response of frequency dependent loads*. Although this is the main goal of the PFC, note that if a narrow PFC dead-band is used such as, for example, ± 15 mHz, then PFC hugely contributes to frequency regulation in normal operating conditions (see below). This key aspect is often overlooked. A power system can operate without SFC but it cannot operate at all without the continuous exertion of effective PFC, especially following contingencies such as large generator trips [52].

2.3.3.1 Primary Frequency Regulation

For TSOs that do not employ and AGC, PFR is the key service that helps maintain frequency within limits during normal operating conditions. There are generally two

[4] We believe it is better to separate "control" from "response" and consider *primary frequency control as an objective and primary frequency response as the means by which resources contribute to this objective* [50].

Table 2.4 PFR products in various jurisdictions

Jurisdiction	Service	Direction (Up/Down)	FAT [s]	Dead-band [±mHz]	JType
AIPS	Primary operating reserve (POR)	Up	5	15	Pre/Post-fault continu-ous/contingency
AIPS	Secondary operating reserve (SOR)	Up	15	15	Pre/Post-fault continu-ous/contingency
GB	Dynamic moderation (DM)	Up/Down	1	15	Pre-fault continuous
GB	Disturbance recorder (DR)	Up/Down	10	15	Pre-fault continuous
CE	FCR	Up/Down	30	10	Pre/Post-fault continu-ous/contingency
Nordic	FCR-N	Up/Down	180	10	Pre-fault continuous
Australia	Fast reserve	Up/Down	6	15	Pre/Post-fault continu-ous/contingency

ways/options that PFR can be implemented in practice (see Table 2.4), that is, by introducing an "explicit" PFR product or by embedding PFR in the PFC product ("implicit PFR") designed for contingency conditions such as the trip of a large generator. The latter can be achieved by applying a narrow dead-band, for example, ±10-15 mHz and relatively slow PFC response. For example, this is the case in the CE[5] and AIPS (see Fig. 12.16). The former PFR products are designed solely to keep frequency within normal limits (standard frequency range).

GB is an interesting example as it has recently (2022) introduced two new explicit PFR products namely DM and DR to improve pre-fault frequency performance. These two PFR products are shown in Fig. 2.7 and are currently being provided by BESS.

In particular, apart from the different FAT (1 and 10 seconds), DM utilizes a variable droop/trajectory, that is, a different droop/trajectory is used when frequency is within the ±100 mHz range and then a different droop/trajectory when frequency is between ±100 mHz range and ±200 mHz range. The impact of these new products and frequency performance/strength in GB will be covered in detail in Chap. 5.

[5] Full activation of FCR occurs when the frequency deviation reaches ±200 mHz.

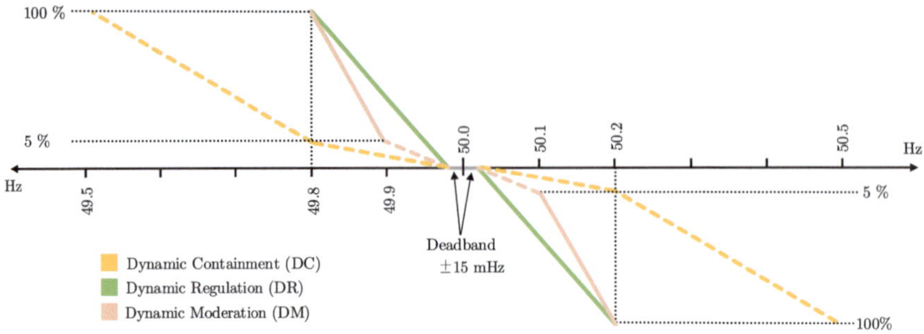

Fig. 2.7 New PFC/PFR reserve products introduced in GB (adapted from [53])

2.3.3.2 Fast Frequency Response

A term related to PFC but sometimes used in a different context is FFR. As shown in Fig. 2.2, FFR is basically a class of PFC, only faster than the traditional PFC provided by conventional units. This is sometimes called "fast governor." It was first introduced as an ancillary service[6] by EirGrid and SONI [54].

The main issue is the definition of FFR itself. For instance, reference [55] defines FFR as: *FFR corresponds to the frequency response of all types of generation technologies not responding to Eq.* (2.2)[7] *immediately after major power imbalances. The FFR is determined by the additional active power injected by these generation units responding to an additional control loop. The time frame involved can last until several seconds after the contingency depending on the control parameters*. A similar definition is provided in the IEEE Std 2800-2022 [16]: *Active power injected to the grid in response to changes in measured frequency during the arresting phase of a frequency excursion event to improve the frequency nadir or initial rate-of-change of frequency*.

The above definitions and many others similar in the literature [56] have three main problems: (i) only assume FFR from generation technologies and not, for example, from demand side units that are key FFR service providers in practice such as in the AIPS; (ii) assume that only active power is injected but active power may also be withdrawn from the system for OF events such as, for example, in GB, AIPS and Australia (see Chap. 5); and (iii) assume that FFR is only activated during abnormal conditions. However, the latter is not always the case in practice. For example, NESO has recently introduced three different FFR products (see Fig. 2.7) with two of them (DR and DM) designed to provide frequency regulation and one to deal with contingency events. In fact, even the DC reserve product which is designed

[6] According to the European legislation on common rules for the internal market for electricity ancillary service means: *a service necessary for the operation of a transmission or distribution system, including balancing and non-frequency ancillary services, but not including congestion management*.

[7] Swing Equation (1.1).

for contingency purposes, provides frequency regulation under normal operating conditions as it uses a narrow dead-band namely ±15 mHz (even though small volumes are used). Therefore, we note (again) that frequency regulation is much more than AGC in practice.

Motivated by this confusion and gap in the literature we propose the following definition of FFR.

> ***Fast frequency response*** *refers to the automatic increase or decrease of power production or consumption by power system resources in response to frequency fluctuations. Power adjustments are performed through automatic PFC with narrow (e.g., ±15 mHz) or wide dead-bands (e.g., ±200 mHz) in the time frame from tens of milliseconds to tens of seconds.*

Similar to the frequency control definition above, we have excluded the word "active" from the FFR definition as one can, for example, use combined voltage-frequency control to improve the stability and transient frequency performance of the overall power system [9]. Further, essentially the three definitions are similar with key differences being time scales and key frequency control characteristics such as dead-bands and the form of the control implemented, that is, manual or automatic (see Fig. 2.2). For instance, because FFR time scales range from tens of ms to tens of seconds it means FFR helps to improve the frequency nadir, zenith, and initial RoCoF and, if a narrow dead-band is used with relatively slow response (for example, 1 s instead of tens of milliseconds), then it will also improve the long-term frequency quality.

The amount of energy injected to or withdrawn from the grid is another key factor but we have excluded it from the definitions as it may be confusing and, depending on the power system, it can vary a lot. For example, FFR needs in low-inertia systems may be quite high whereas in high-inertia grids there may not be a need for FFR at all.

Tables 2.5 and 2.6 provide information on common types of FFR response and list a number of few relevant jurisdictions that have recently introduced an FFR product, respectively. All technology providers in Table 2.5 are currently providing all types of FFR in the AIPS. In particular, it is worth noting here that while the vast majority of people link FFR exclusively with power electronic converters, including both GFL and GFM, FFR provision capabilities of newer conventional generators with a response time down to a couple of seconds are also successfully demonstrated. This point is further illustrated in Chap. 12. Additionally, due to its nature, static-based FFR in the form of low-cost static reserves tend to be used with larger dead-bands as it is the one that arrest the frequency excursion rapidly following a contingency, namely, injects large amount of energy almost instantaneously by tripping demand.

Table 2.6 shows the characteristics of FFR in the different selected jurisdictions.[8] While the minimum response time is below 2 s for the contingency products, the dead-band varies a lot and, in the Nordic case, the FFR type includes just static and not dynamic/proportional response. Because of procuring such an FFR type, the

[8] In the table, FIR stands for Fast Instantaneous Reserve.

Table 2.5 Common types of FFR response and relevant technologies

FFR response	Main technology providers
Inertia-like/"synthetic/emulated inertia"	Wind turbines
Droop/proportional-based	BESS, solar PV, HVDC interconnectors (ICs), conventional
Discrete/static/step-based	Demand side units (DSUs)

Table 2.6 FFR products in different jurisdictions

Jurisdiction	Service	Direction (Up/Down)	Response time [s]	Duration [s]	Dead-band [±mHz]	Type
AIPS	FFR	Up	0.15–2	10	<100	Post-fault contingency
GB	DC	Up/Down	1	900	15	Post-fault contingency
GB	DM	Up/Down	1	1800	15	Pre-fault continuous
GB	DR	Up/Down	10	3600	15	Pre-fault continuous
Nordic	FFR (static)	Up	0.7–1.3	5/30	300–500	Post-fault contingency
Australia	FFR	Up/Down	0.1–1	900	15	Post-fault contingency
ERCOT, United States (US)	FFR	Up	0.25	900	150	Post-fault contingency
Ontario	FFR	Up	1.5	9	100	Post-fault contingency
New Zealand	FIR (static)	Up	1	60	800	Post-fault contingency

Nordic TSOs use quite a wide range of dead-bands to avoid continuously activating FFR resources such as demand units through switched controllers. Recall that the frequency response dead-band is a key parameter because if narrow values are applied, FFR can provide fast frequency regulation like in the GB power system. However, NESO uses a variable droop to avoid activating large volumes of FFR during normal system conditions, that is, if the frequency is within the ±200 mHz range (see Fig. 2.7 and Chap. 5).

In practice, FFR is predominantly provided by converter-based resources, primarily BESS units with a response time or FAT down to hundreds of ms (see Chapters 5 and 12). FFR provision is subject to measurement, detection, processing, filtering, and activation and hence characterizes an inherent delay. For instance, conventional PFC from IBRs is implemented at the power plant controller rather than at the inverter level and thus subject to latency and delays associated with communications networks and measurements loops between the power plant controller and controlled inverters [57]. This inherent delay, in fact, means that even when using aggressive droop in the FFR of GFL inverters, they cannot fully replace the inertial response provided by GFM inverters (immediate frequency response to disturbances) [58]. In contrast, frequency controls are generally implemented at the inverter level in GFM control topologies which result in a direct and near-instantaneous response to changes in system frequency due to constant voltage angle at the inverter terminals [57].

The fast response can lead to potential FFR-induced oscillations (see the example in Sect. 3.7.2) and rotor angle instability due to overlapping with the generator angle swing dynamics [15, 46]. Due to this overlapping, FFR also provides some characteristics of inertia but cannot fully replace it. FFR can, however, support power system operation with low inertia and, allows for larger contingency sizes at lower inertia levels [59]. For instance, reference [60] suggests that 20% FFR from the total frequency control reserve would reduce 6.5% of total inertia requirements for a system of 20 GWs.

It is important to keep in mind that inertia and FFR are intrinsically different services. The inertial response of synchronous machines compensates the internal machine imbalance between mechanical and electrical power. In other words, synchronous machines "relocate" the imbalance from the system to their air gaps. This happens almost instantaneously, that is, in the time scale of stator flux dynamics. Then, relatively slowly, the kinetic energy stored in the rotor of the machines is exchanged in order to compensate these internal power imbalances. In the process, the machines swing and force the entire system to oscillate. FFR, on the other hand, lacks the initial "relocation" of the power imbalance. FFR can only exchange power with the grid. Hence, a very fast FFR can be an issue as a large sudden power exchange will have substantially the same effect as an outage and will be instantaneously compensated by the synchronous machines connected to the grid. This can later cause electromechanical oscillations. In summary, FFR and inertia are not interchangeable as they deliver different benefits to the system.

2.3.3.3 Active Power Control

Active Power Control (APC) is another term often misunderstood in the power system community. For example, according to [14]: *AEMO notes that some stakeholders have expressed that the term primary frequency control is confusing or at least unnecessary and that the term "active power control" should be used instead.* Indeed, APC is often used to refer to the active power control of IBRs only and, in some case, it goes even further such as in the Irish context where it refers to APC as PFC with narrow dead-band (±15 mHz) from RES (wind) [61].

In [62], APC is defined as: *The control of active power is basically the curtailment of the available energy production by operating the system be low the maximum power point.* Whereas [63] reads: *Active power control (APC) refers to the power regulation of individual units, relieving network constraints, or increasing system stability and*

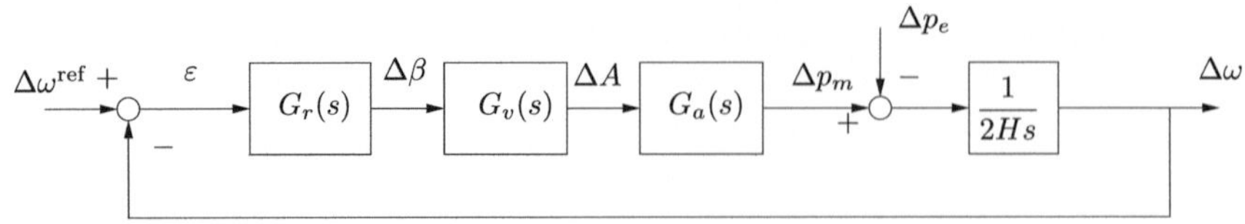

Fig. 2.8 High-level linearized model of the machine, turbine, valve and governor

efficiency. The first definition is quite specific and relates APC to RES curtailment to be able to provide frequency support, whereas the second definition is more general and relates to the ability to provide demand-supply and congestion management. We believe, however, that APC is a general term that refers to the control of active power output of a resource, rather than active power output being solely dependent on the energy source availability. This understanding is used in [12] where three forms of APC are considered, namely, "synthetic inertial control", PFC, and AGC.

2.3.3.4 Droop Control

The droop control is the most common control for the TGs of conventional power plants as well as IBRs. In this section, we provide a qualitative justification of the reason why this controller, despite not perfect tracking the reference frequency is, in general, to be preferred. For this qualitative analysis, let us consider first a conventional power plant with a synchronous generator. A high-level linearized model of the machine, turbine, valve and governor is shown in Fig. 2.8.

The quantities in the scheme have the following meaning: s is the Laplace variable; H is inertia constant of the machine; ω is the rotor speed; p_m is the mechanical power; p_e is the electrical power; $\omega^{\rm ref}$ is the reference speed; ε is the frequency error; β is angular position for the valve regulation; and A is the cross section of the pipe that feeds with steam, water or gas the turbine. Δ indicate variations of the variables with respect to the current operating point. Note that $\Delta\omega^{\rm ref}$, which represents the variations for the reference speed with respect to the synchronous reference frequency of the system is typically zero, except when modified by the TEC, see Sect. 2.3.6. Finally, the transfer functions have the following meaning:

$G_r(s) = \Delta\beta/\Delta\varepsilon$ is the TG transfer function.
$G_v(s) = \Delta A/\Delta\beta$ is the actuator transfer function of the valve.
$G_a(s) = \Delta p_m/\Delta A$ is the turbine transfer function.

The expressions of these transfer functions depend on the technology and the specific implementation of the turbine. There are in effect hundreds of different models, with various degrees of complexity. However, for the purposes of the design of the transient response of the PFC of the vast majority of conventional power plants, it suffices to focus on the overall transfer function of governor, valve and turbine (see Fig. 2.9):

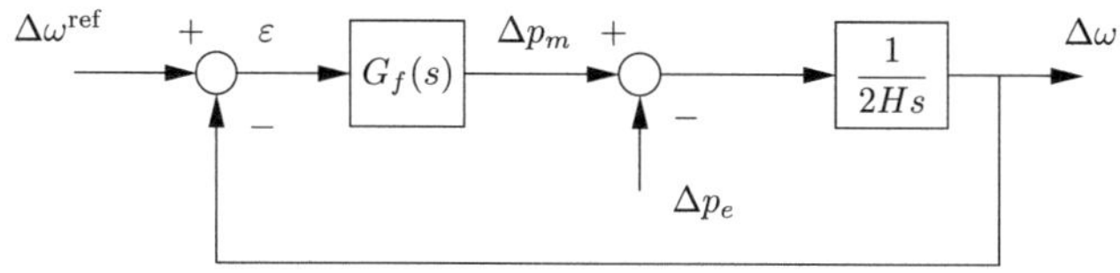

Fig. 2.9 Simplified frequency control scheme of a conventional power plant

$$G_f(s) = G_r(s)G_v(s)G_a(s)\,. \tag{2.12}$$

Once the sought structure of $G_f(s)$ is determined, one can determine the implementation of $G_r(s)$ based on the specific behavior of the turbine and valve. In other words, $G_r(s)$ is determined *a posteriori* as:

$$G_r(s) = \frac{G_f(s)}{G_v(s)G_a(s)}\,. \tag{2.13}$$

Interestingly enough, while the structure of $G_a(s)$ can vary considerably depending on the type of turbine, the overall, high-level dynamic behavior of $G_f(s)$ is quite consistent and similar across most power plants, especially hydro and steam ones. Gas or combined cycle power plants may differ slightly and we consider these at the end of this section.

The requirements for $G_f(s)$ are substantially three:

- the frequency has to be restored as soon as possible after a contingency;
- the power plant dynamic response has to be stable; and
- the TG should be perfect tracking, namely, $\varepsilon = 0$ or, equivalently, $\omega = \omega^{\rm ref}$ in steady state.

The first requirement is constrained by the rate of change of power of the turbine itself. The second requirement imposes that the TG has to be slower (at least a decade) than the slowest dynamic of the synchronous machine, in this case, its electromechanical oscillations. This, in turn, implies that one cannot increase the gain of the TG indefinitely as, at a certain point, the overall system machine+TG will be unstable. The third requirement implies the utilization of a controller that includes an integrator.

The ubiquitous conventional PI control can be tuned in such a way to be both fast and stable and, due to the integrator channel, is perfect tracking. Let us consider the following transfer function for $G_f(s)$:

$$G_f(s) = K\frac{1 + sT_2}{s}\,, \tag{2.14}$$

which represents a conventional PI control. Figure 2.10 shows the open loop transfer function for the simplified control scheme of Fig. 2.9, namely:

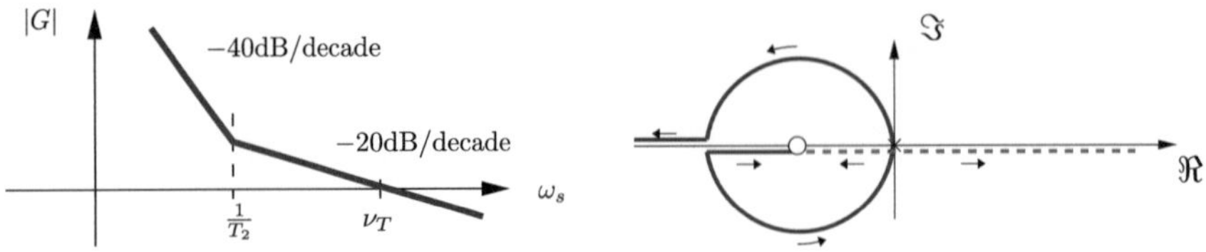

Fig. 2.10 Root loci and Bode's diagram of a conventional power plant with PI control

$$G(s) = G_f(s)\,\frac{1}{2Hs} = K\frac{1+sT_2}{2Hs^2}\,. \tag{2.15}$$

In the left panel of Fig. 2.10, the crosses and the circle represent the poles and the zeros, respectively, of the open loop transfer function. The continuous line represents the root loci for $K > 0$, whereas the dashed line represents the root loci for $K < 0$. The gain K can be chosen to properly tune the damping and the velocity of the response of the control. In general, the cut frequency ν_T satisfies the condition:

$$\nu_T \gg \frac{1}{T_2}\,. \tag{2.16}$$

Of course, the model is very simplified and does not take into account faster dynamics. These constitute the upper limit for ν_T.

We note that a very similar dynamic response can be obtained with a lead-lag. Let us assume the following transfer function:

$$G(s) = KT_1\frac{1+sT_2}{(1+sT_1)}\,, \tag{2.17}$$

which leads to the total open-loop transfer function:

$$G(s) = G_f(s)\,\frac{1}{2Hs} = KT_1\frac{1+sT_2}{2Hs(1+sT_1)}\,. \tag{2.18}$$

Figure 2.11 shows the root loci and the Bode's diagram for the transfer function in (2.18) assuming $T_1 > T_2$ and $K > 0$. The only difference with respect to the PI control, is that the two zeros do not coincide in the origin. Apart form that, with sufficiently high gain K, the cut frequency ν_T and thus the dynamic response of the controller can be tuned in the same way as the PI.

The other difference, which we will see has substantial consequences, is that fact that the lead-lag is not perfect tracking the reference frequency. The third requirement above is thus not satisfied.

Let us define two relevant quantities. First, note that for $\Delta\omega^{\text{ref}} = 0$:

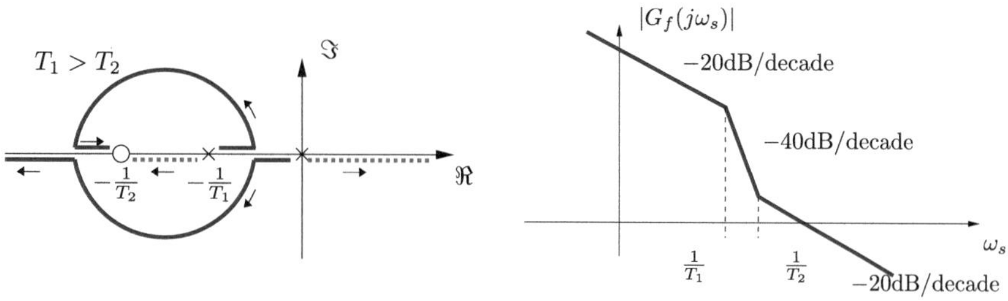

Fig. 2.11 Root loci and Bode's diagram of a conventional power plant with lead-lag control

$$\frac{\Delta\omega}{\Delta p_m}(0) = -\frac{1}{K\,T_1} \tag{2.19}$$

$$\frac{\Delta\omega}{\Delta p_m}(\infty) = -\frac{1}{K\,T_2} \tag{2.20}$$

The *static droop* $\mathcal{R}$ on machine bases is defined as:

$$\mathcal{R} = -\frac{\Delta\omega/\omega_n}{\Delta p_m/S_n}(0) = \frac{S_n}{\omega_n}\,\frac{1}{K\,T_1}\,, \tag{2.21}$$

where ω_n and S_n are the rated angular frequency and capacity of the generator, respectively. The transient droop $\mathcal{R}_t$ on machine bases is defined as:

$$\mathcal{R}_t = -\frac{\Delta\omega/\omega_n}{\Delta p_m/S_n}(\infty) = \frac{S_n}{\omega_n}\,\frac{1}{K\,T_2} = \frac{T_1}{T_2}\mathcal{R}\,. \tag{2.22}$$

The closed loop transfer function of the rotor speed with respect to variations of the electrical power is:

$$\frac{\Delta\omega(s)}{\Delta p_e(s)} = \frac{-\dfrac{1}{2Hs}}{1+\dfrac{1}{2Hs}K T_1\dfrac{1+sT_2}{1+sT_1}}\,. \tag{2.23}$$

It is well known that $\lim_{t\to\infty}\Delta\omega(t) = \lim_{s\to 0} s\,\Delta\omega(s)$. Then, assuming again $\Delta\omega^{\rm ref} = 0$ and using a step variation for the electrical power, $\Delta p_e(s) = \Delta p/s$, one has:

$$\lim_{t\to\infty}\Delta\omega(t) = \lim_{s\to 0} s\,\frac{-\frac{1}{2Hs}}{1+\frac{1}{2Hs}K T_1\frac{1+sT_2}{1+sT_1}}\,\frac{\Delta p}{s} = -\frac{\Delta p}{K\,T_1} \neq 0\,. \tag{2.24}$$

and, using the definition of static droop:

$$\frac{\Delta\omega}{\omega_n} = -\frac{1}{\mathcal{R}}\,\frac{\Delta p}{S_n}\,. \tag{2.25}$$

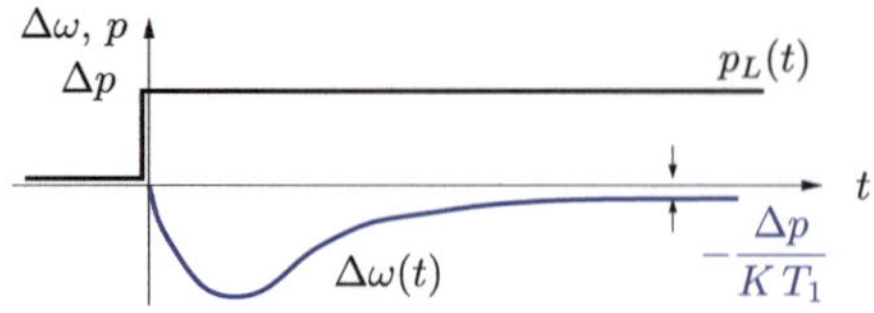

Fig. 2.12 Typical transient of the frequency following a power unbalance

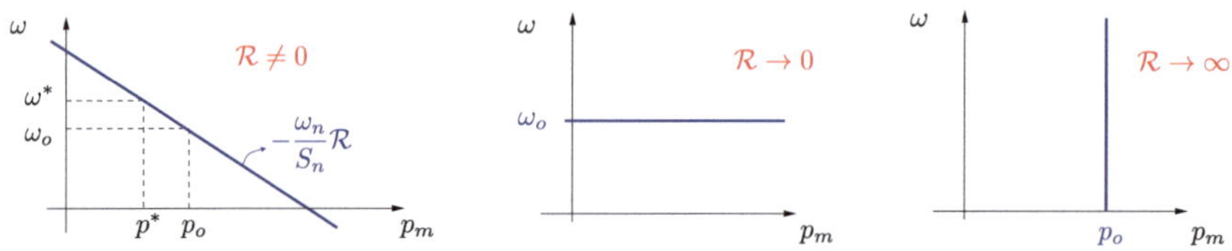

Fig. 2.13 Steady-state frequency/power characteristic of the PFC of a conventional power plant for different values of the droop $\mathcal{R}$

As expected, thus, the static error of the frequency for the lead-lag control is not zero. The transient following the power unbalance is illustrated in Fig. 2.12 A similar calculation, which is left for exercise, for the PI control leads to a zero static error of the frequency.

Considering the droop $\mathcal{R}$ as a parameter, on can identify two limit cases. The first limit case is $\mathcal{R} \to \infty$, which is equivalent to "opening" the loop, or equivalently, not providing any frequency control at all. The case for $\mathcal{R} \to \infty$ is equivalent to a "constant" power, independently from the value of the rotor speed or of the frequency of the system. Small power plants which are not required to provide PFC, can be operated with $\mathcal{R} \to \infty$. The second limit case for $\mathcal{R} = 0$, is also called *isochronous control* as, in steady-state the frequency is always equal to the reference. As the PI is perfect tracking, one can consider the PI control as the limit case of the lead-lag control for $\mathcal{R} \to 0$. The three steady-state power/frequency characteristic for various $\mathcal{R} > 0$, $\mathcal{R} = 0$, and $\mathcal{R} \to \infty$ are shown in Fig. 2.13.

It remain to justify why the lead-lag control is considered at all, given that it is not perfect tracking the frequency, whereas the PI, which has substantially same transient behavior, is intrinsically perfect tracking. We explain this point through an example. Let us consider a grid with three power plants providing PFC but with different values of the droop $\mathcal{R}$ as shown in Fig. 2.14.

At the initial operating point, the power balance gives:

$$p_e^o = p_{m,1}^o + p_{m2}^o + p_{m,3}^o , \tag{2.26}$$

where p_e^o is the total power consumption of the grid, i.e., load and losses. Let us assume now that there is a variation of the load consumption leading to a total power variation $\Delta p_e > 0$. As $\mathcal{R}_1 = 0$, any variation Δp_e is provided by power plant 1. In practice, this is an undesirable situation in most power systems for various

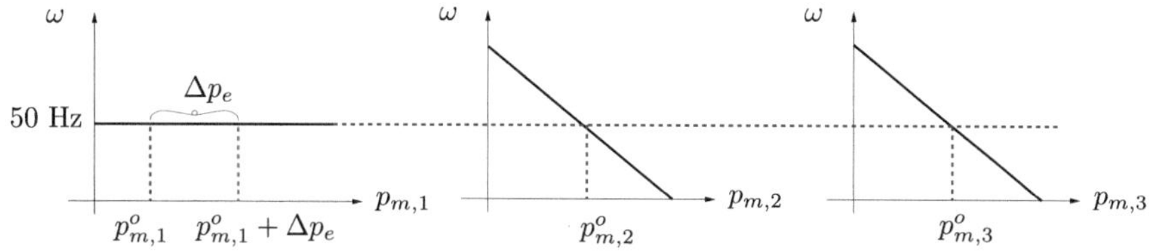

Fig. 2.14 System with three power plants providing PFC, one of which providing perfect tracking regulation ($\mathcal{R}_1 = 0$) and two with conventional droop control ($\mathcal{R}_2 \neq 0$, $\mathcal{R}_3 \neq 0$)

reasons. First, Δp_e might be bigger than the power reserve of power plant 1, making the contingency unfeasible. Then, it is expected that all power plants bigger than a certain size participate to the frequency regulation proportionally to their capacity.

One could then set the droops of all power plants to zero. This is however another undesirable situation, as in this case, the final operating point is undetermined. In mathematical terms, one has one equation:

$$\Delta p_e = \Delta p_{m,1} + \Delta p_{m2} + \Delta p_{m,3} \,, \tag{2.27}$$

and three unknowns, namely $\Delta p_{m,1}$, Δp_{m2}, and $\Delta p_{m,3}$. This impasse can be resolved using lead-lag controllers. In fact, if the droop of all power plants if $\mathcal{R} \neq 0$, one obtains a well determined system in four unknowns, namely, $\Delta\omega$, $\Delta p_{m,1}$, Δp_{m2}, and $Delta p_{m,3}$, and four equations, which in per unit can be written as:

$$\begin{aligned}
\Delta p_e &= \Delta p_{m,1} + \Delta p_{m2} + \Delta p_{m,3} \,, \\
\Delta p_{m,1} &= -\frac{1}{\mathcal{R}_1}\Delta\omega \,, \\
\Delta p_{m,2} &= -\frac{1}{\mathcal{R}_2}\Delta\omega \,, \\
\Delta p_{m,3} &= -\frac{1}{\mathcal{R}_3}\Delta\omega \,.
\end{aligned} \tag{2.28}$$

Figure 2.15 illustrates the solution for the case $\mathcal{R}_1 \neq 0$, $\mathcal{R}_2 \neq 0$, $\mathcal{R}_3 \neq 0$. The operating point post contingency is well defined and unique but $\Delta\omega \neq 0$, in fact:

$$\Delta\omega = \frac{-\Delta p_e}{\frac{1}{\mathcal{R}_1} + \frac{1}{\mathcal{R}_2} + \frac{1}{\mathcal{R}_3}} \,. \tag{2.29}$$

This example also shows that only if all the droops have same value, the power plants will contribute equally in proportion to their capacity. On the other hand, if some droop is significantly smaller than others, the power plant with smallest droop will contribute more (proportionally to its capacity) than the others. The limit case is, again, for $\mathcal{R} = 0$. As a matter of fact, in interconnected power systems, the droops of power plants have similar values (see Table 2.7). The value of the droop

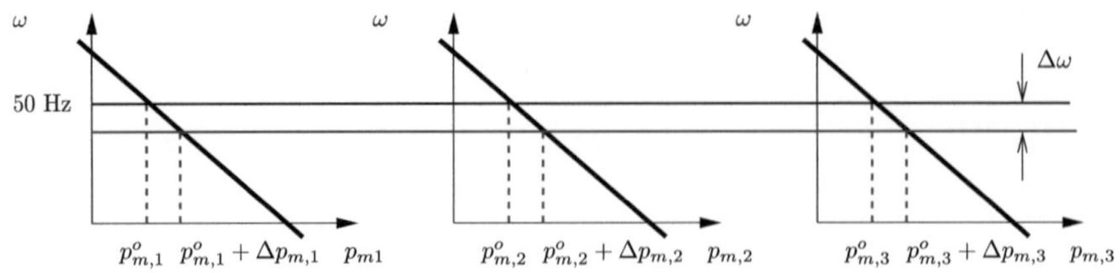

Fig. 2.15 System with three power plants providing PFC, all of which with conventional droop control ($\mathcal{R}_1 \neq 0, \mathcal{R}_2 \neq 0, \mathcal{R}_3 \neq 0$)

Table 2.7 PFC droop for different jurisdictions

Jurisdiction	Droop (%)
Pennsylvania-New Jersey-Maryland (PJM), US	≤5
ERCOT, US	4–5
Ontario	3–7
CE	2–12
Australia	≤5
GB	3–5
AIPS	3–5
New Zealand	0–7

is *small*, as the steady state error is proportional to the value of the droop. In fact, if we assume $\mathcal{R}_1 \approx \mathcal{R}_2 \approx \mathcal{R}_3 = \tilde{\mathcal{R}}$, i.e., the three machines have about same capacity, Eq. (2.29) leads to:

$$\Delta\omega = -\frac{1}{3}\,\tilde{\mathcal{R}}\,\Delta p_e\,. \tag{2.30}$$

In general, if we assume that in a system there are n power plants, the frequency variation after a contingency that leads to a power unbalance Δp can be roughly estimated as:

$$\Delta\omega = -\frac{1}{n}\,\hat{\mathcal{R}}\,\Delta p\,, \tag{2.31}$$

where $\hat{\mathcal{R}}$ is the weighted sum of the droops of the power plants:

$$\hat{\mathcal{R}} = \frac{\sum_{i=1}^{n} S_{n,i}\mathcal{R}_i}{\sum_{i=1}^{n} S_{n,i}}\,. \tag{2.32}$$

It remains the question whether PI controllers are utilized at all for PFC. The answer is, surprisingly enough, yes. The obvious case is that of small power systems where there is only one power plant providing frequency support. However, PI con-

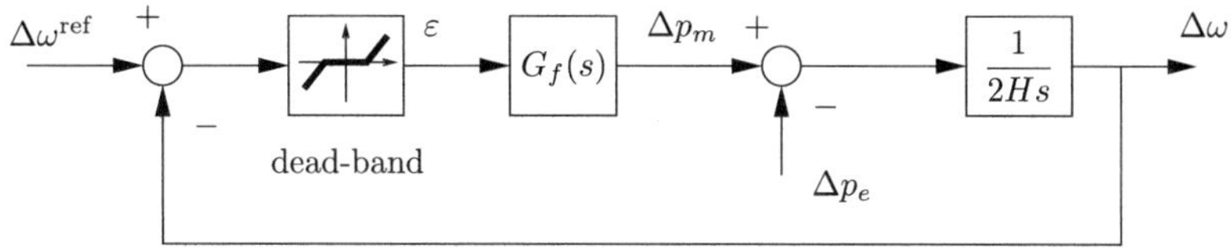

Fig. 2.16 Frequency control scheme with dead-band on the frequency error

Table 2.8 PFC dead-band for different jurisdictions

Jurisdiction	Dead-band s [±mHz]
PJM, US	≤ 36
ERCOT, US	≤ 17
Ontario	≤ 36
CE	≤ 10
Australia	≤ 15
GB	≤ 15
AIPS	≤ 15
Nordic	≤ 10
New Zealand	≤ 20
Malaysia	≤ 25

trollers are utilized also in large interconnected systems. For example, most TGs of gas turbines and combined cycles are based on PIs. This may appear inconsistent with the discussion above but it is simply a modeling issue. So far, we have assumed that the error tracking is perfectly continuous. However, in practice, the governor is made insensible to small values of the frequency error through a dead-band as shown in Fig. 2.16.[9]

The main purpose of the dead-band is that of reducing the operations of the mechanical parts, such as the actuator of the valve, and thus reducing maintenance costs. Moreover, mechanical systems always have a backlash that results in an involuntary dead-band. Table 2.8 shows some typical values of the dead-band utilized in real-world power systems.

As it stops the action of the governor for small frequency errors, the dead-band has an effect similar to the droop, that is, it leads to a non-null steady state error of the frequency. This error is independent from the transfer function $G_f(s)$, which can thus be either a lead-lag or a PI, or any other perfect tracking control. There is, however, a substantial difference between the effect of the droop and that of the dead-band, namely the distribution of the frequency. If there were no dead-band in

[9] The dead-band definition provided in System Operations Guideline (SOGL) [19] states: *frequency response dead-band means an interval used intentionally to make the frequency control unresponsive.*

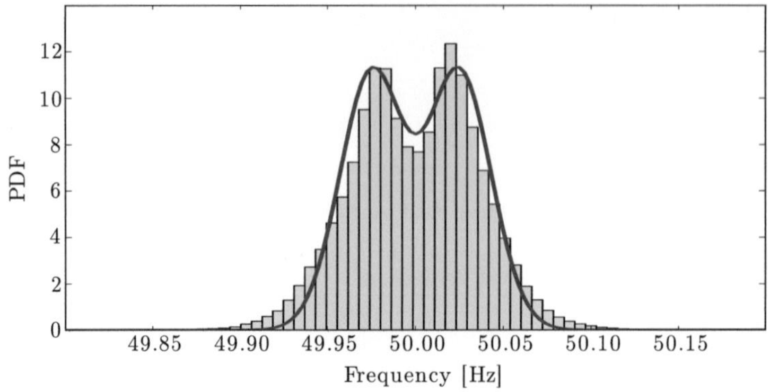

Fig. 2.17 Bimodal distribution of the frequency in the Irish system. Measurements were taken for a week with a sampling rate of 0.01 s by means of a FDR (courtesy of Prof. Yilu Liu, University of Tennessee)

the TGs of the power plants, the distribution of the frequency would be approximately normal. The dead-band, on the other hand, leads to a bimodal distribution, as shown in Fig. 2.17. We just mention this point here without further details as the whole Chap. 4 is dedicated to the discussion of the effect of control on the frequency distribution in real-world power systems.

The need for a non-perfect tracking PFC implies that also power electronic converters, when regulating the frequency, are equipped with droop controllers. In the simplest implementation, this controller can be written as:

$$T \Delta p'(t) = K(\Delta\omega^{\text{ref}} - \Delta\omega(t)) - \Delta p(t)\,, \tag{2.33}$$

and, as in general these controllers do not participate to the time regulation, thus leading to $\Delta\omega^{\text{ref}} = 0$, one as:

$$T \Delta p'(t) = -K\,\Delta\omega(t) - \Delta p(t)\,, \tag{2.34}$$

which shows the same steady-state characteristic of the PFC of conventional power plants. Also in case of IBRs, the droop control can include a dead-band. However, in this case, rather than avoiding unnecessary stress of the mechanical parts of the turbine, the dead-band of the droop controllers of IBRs is aimed at limiting the frequency support only in cases where the frequency deviation is greater than average, thus indicating the need for extra power reserve. Thus, if present, the frequency dead-band of the droop control of IBRs is larger than that of conventional power plants.

We conclude that PFC cannot be perfect tracking the frequency, either by design of the control or because of practical requirements of the TGs. It is desirable, however, that a controller, aimed at reducing the frequency error in steady-state is in place in power systems. This is again for the sake of frequency quality. This objective is achieved with the SFC, which is described in the next section.

2.3.4 Secondary Control

As described above, PFC arrests frequency in response to large deviations (from further decline/excursion) by bringing it to a new steady-state level. It also provides frequency regulation if narrow dead-band is used. However, it does not correct frequency deviations to zero. Here is when SFC comes in. SFC is the fourth stage of frequency response and is defined, based on NERC, as [47]: *Actions provided by an individual Balancing Authority (BA) or its Reserve Sharing Group to correct the resource–load unbalance that created the original frequency deviation, which will restore both Scheduled Frequency and Primary Frequency Response. Secondary Control comes from either manual or automated dispatch from a centralized control system.*

SFC is usually used as a synonym to AGC, Load Frequency Control (LFC) and aFRR. However, these terms represent the automatic version of SFC but not its manual version. Not all TSOs, in fact, use automatic SFC. SFC has a ramping capability of the order of hundreds of MW/min [48]. The goal of SFC is to restore both the scheduled frequency or bring back frequency to nominal value and restore PFC reserves. It does so on a time frame from tens of seconds up to 15 min based on a combination of automatic and manual dispatch actions, for example, electronic dispatch logger or phone calls to generators. The time scale decoupling between PFC and SFC ensures overall power system stability [64]. However, it should be noted that since SFC, in particular, manual SFC, take minutes to restore frequency to the nominal value following a contingency event, it is critical that the PFC delivery is sustained as much as possible (see Fig. 12.11 as a real-world example of sustained delivery of PFC).

The most common means of exercising automatic SFC is through AGC (or aFRR in Fig. 2.3). AGC/aFRR is widely used by TSOs. For example, reference [65] suggests that there are approximately 60 AGCs implemented in the US that perform the power-balancing function. AGC implementation is also very common in Europe with few exceptions such as GB and AIPS. NESO, the GB TSO, has identified two main barriers to the implementation of AGC [66]:

- As GB is no longer a member of EU, it cannot participate in the common EU platform Platform for the International Coordination of Automated Frequency Restoration and Stable System Operation (PICASSO) for cross-border balancing energy exchange.
- There are technical challenges to implement cross-border exchange of aFRR as GB is connected to the rest of the world through HVDC links which require the exchange to be explicitly defined and controlled. Other reasons include handling of HVDC losses by PICASSO platform and issues such as maximum ramp rates and dead zones.

NESO suggests that these challenges and cost of implementing AGC estimated to be around £25m would exceed the benefits of an AGC installed such as having fewer and smaller frequency deviations and reduction of the number of manual actions taken by the TSO. Therefore, a key and open question that NESO but, in general,

Table 2.9 AGC FAT for different jurisdictions

Jurisdiction	FAT [minutes]
ERCOT, US	5
PJM, US	5
Ontario	5
South Africa	10
CE	2–15
China	< 1
Australia	5
India	15

all TSOs not utilizing an AGC need to address is whether future 100% RES power systems could be operated without an AGC.

AGC is a simple yet effective central feedback controller that usually involves a PI response. It is designed and intended to work in parallel with PFC as the AGC cannot by itself deliver adequate frequency control. There are different implementations of AGC in practice such as "constant frequency control," "constant interchange control" and "tie-line bias control". As the names suggest, the first two control modes look to only control frequency and the tie-line interchanges, respectively. The third AGC mode, on the other hand, is the most common implementation worldwide and controls both frequency and tie-line interchanges, that is, the first two modes are reduced versions of the third AGC mode. Additionally, despite AGC being a simple controller, there is no harmonization of it, both from a market and a technical perspective. For example, Table 2.9 shows different FATs.

ENTSO-E reports an AGC FAT[10] between circa 2 min and 15 min for most European TSOs [67]. In practice, having a faster AGC FAT (say, 2 min) might mean that very few resources can meet the requirements while a slow FAT (say, 15 min) value might be desirable from a market point of view as it will result in higher liquidity, that is, more resources are technically able to offer AGC capacity. But faster AGC FAT are desirable from a frequency control performance point of view as it will result in better frequency quality. For instance, China has recently introduced stringent requirements for hydropower and thermal power plants to have an AGC FAT of less than 2 s and 1 min, respectively [68]. As mentioned above, high-quality AGC services may have an impact in terms of the number of conventional power plants meeting the requirements and, thus, impact their revenue streams. Therefore, there should always exist a tradeoff between market/cost implications and system needs.

AGC is used during both normal and abnormal operating conditions, that is, after PFC has recovered frequency following major disturbances [12]. The contribution of AGC during contingency conditions is often overlooked by power system engi-

[10] FAT can be divided into a preparation period during which no energy is delivered and a ramping period.

neers/researchers. But as it is be demonstrated in the illustrative examples below, AGC can be a very effective tool to recover frequency quickly to the nominal value following large disturbances. What it is common, though, is that TSOs usually suspend AGC during emergency conditions, such as partial blackouts or when multiple islands are formed, to avoid uncoordinated actions with the control room operators until the system is restored. Another important aspect often overlooked is the interaction of SFC reserves, for example, AGC reserves, with PFC reserves during intact system conditions. That is, if AGC reserves are inaccurately forecasted or under-procured, AGC cannot manage frequency within normal limits. In this case, PFC reserves act automatically to oppose changes in frequency. If this were to happen, there would be two main concerns [69]: (i) PFC reserves are depleted and may not be enough to contain and stabilize the frequency following a sudden imbalance between load and generation; and (ii) PFC and SFC reserves are fully utilized during normal operating conditions and thus leave the system short in reserves if further power imbalances occur. This situation can lead to cascading outages and a widespread blackout. Therefore, the adequacy of PFC reserves depends on the adequacy of SFC reserves [69]. It is imperative, thus, that adequate SFC reserves are procured or available to TSOs at all times.

AGC operates in conjunction with Supervisory Control and Data Acquisition (SCADA) systems, that is, SCADA gathers information about, for example, frequency, generator outputs, and actual tie-line interchange between different control areas. Using such inputs, it is then possible to determine the overall balance of the system and issue automatic dispatch targets or set-points remotely to all enabled plant every couple/few seconds. The aim of these set-points is to correct the Area Control Error (ACE) which is an input to the the AGC/LFC controller. ACE represents the active power imbalance between generation and demand, and is a linear combination of net tie-line interchange and frequency deviations, such as, for example, AGC tie-line bias control [12, 70, 71]. Its calculation by NERC ($\mathrm{ACE}_{\mathrm{NERC}}$), ENTSO-E ($\mathrm{ACE}_{\mathrm{ENTSO\text{-}E}}$) and AEMO ($\mathrm{ACE}_{\mathrm{AEMO}}$) is given by:

$$\mathrm{ACE}_{\mathrm{NERC}} = \Delta p_{\text{tie-line}} - 10\, K_{\mathrm{bias}}\, \Delta f \,, \tag{2.35}$$

$$\mathrm{ACE}_{\mathrm{ENTSO\text{-}E}} = \Delta p_{\text{tie-line}} + K_f\, \Delta f \,, \tag{2.36}$$

$$\mathrm{ACE}_{\mathrm{AEMO}} = 10\, K_{\mathrm{bias}}\, \Delta f \,, \tag{2.37}$$

where $\Delta p_{\text{tie}-\text{line}}$ represents the difference between the total actual and scheduled net interchange with neighboring control areas; K_{bias} (MW/0.1 Hz) is the balancing authority's bias factor or frequency bias, which represents the amount of frequency response that the control area has, or should have; Δf represents the frequency deviation, namely, actual vs scheduled frequency and, if TEC is used, it can include a frequency offset representing accumulated frequency deviation; and K_f represents the power-frequency characteristic of the control area in MW/Hz and positive. K_f shows the relationship between active power imbalance and frequency deviation. For instance, ENTSO-E report an approximate value of $K_f = 27,000$ MW/Hz including FCR/PFC, self-regulation of load, etc., for the CE power system [70]. This means

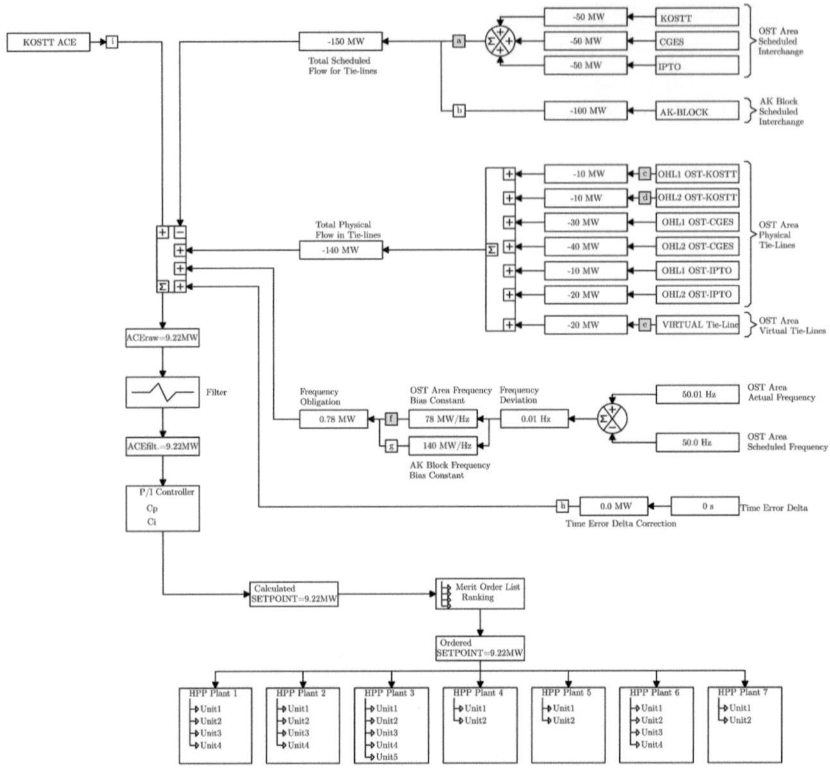

Fig. 2.18 Control diagram of AGC implemented by the Albanian TSO, OST (*Source* OST)

that, for example, a 500 MW change/deviation would lead to approximately 20 mHz frequency deviations. For comparison, the GB power system estimates K_f= 2,250 MW/Hz [72]. In practice K_f varies continuously with changing system conditions. Additionally, the second part of the equations are introduced to ensure that systems are providing PFC, and that the AGC does not counter its PFC obligation [12]. Note that AEMO does not currently use $\Delta p_{\text{tie-line}}$ in the ACE calculation. Moreover, ACE control is becoming increasingly difficult due to tie-line flows between different control areas becoming more volatile due to variations of solar and wind in relatively short time scales. Oscillations in ACE have also been observed recently due to market parties switching large volumes of power on and off very quickly (e.g., less than 10 s) in response to price signals.

Figure 2.18 shows a real-world implementation of AGC, namely that implemented by the Albanian TSO, OST. Note that numbers are not exact except for the values of K_f. Additionally, the small square boxes represent different switches that, if filled, are closed and, if empty, are opened. OST uses a PI controller, and ACE is composed of both tie-line and frequency deviations similar to Eq. (2.36). The Albanian transmission network is connected to neighboring countries through six overhead tie-lines (OHLs), that is, two with Kosovo (KOSTT TSO), two with Greece (IPTO TSO), and two with Montenegro (CGES TSO).

Although traditionally conventional generators have provided the AGC functionality, recently IBR-based generation and energy storage systems such as BESS are becoming the main providers of AGC. For instance, references [73, 74] study the participation of a 30 and 10 MW wind farms in AGC and show their potential by testing the behavior against field measurements and experimental results, respectively. On the other hand, the potential of solar PV providing AGC services under different conditions (solar resource intensity) is shown in [75] through a successful 300 MW power plant test. BESS is shown to respond well to AGC commands/set-points and provide SFC in [76]. As a matter of fact, BESS are on a daily basis the main or even the only assets under AGC providing both up and down regulation for California Independent System Operator (CAISO), whereas conventional generation is hardly under AGC anymore [77]. However, based on control room experience, CAISO reports one significant challenge, that is, how to join together such conventional cycle and hydro units and new resources such as BESS to control ACE, due to extremely fast ramp rates of BESS [78]. The solution that worked for CAISO was to put a logic in their AGC to reduce the ramp rate in certain bands of ACE. It is also interesting to mention that, counterintuitively, AGC reserve requirements may not always increase with the increasing installed variable RES such as wind and solar PV. For example, ERCOT, reports that AGC reserves are trending down due to mainly increased geographical dispersion of wind generation resources and improved methodology for determining the requirements [79].

Not all TSOs have an AGC and therefore manual dispatch actions to maintain energy balance and scheduled frequency have also the same importance to AGC. Sometimes TSOs, in particular, those that do not implement an AGC may deliberately run the frequency higher or lower than nominal sometimes as well in order to have a better margin. For instance, before the peak demand TSOs may prefer to keep frequency slightly above 50 Hz as they know that in the next hour or so demand will increase and frequency thus decrease assuming generation is more or less the same. Then, they may need to deliberately run the frequency low during the middle of the night to ensure the cumulative change in frequency over the day is zero, or simply just because they know demand will decrease. This is a common operational practice in the Nordics where the *Balance Operators*[11] keep the frequency higher than 50 Hz before hour shift to have a better margin toward expected low frequencies shortly before and after hour shift due to ramp up of HVDC ICs and increase in consumption [20].

In summary, the main differences between PFC and SFC are as follows:

- PFC is a fast-acting control while SFC is a slow-acting control complementary to PFC, that is, it allows resources providing PFC to return to their set-points and thus be ready to provide further PFC, if required.

[11] Swedish and Norwegian TSOs responsible for balancing the Nordic synchronous area.

- PFC arrests and stabilize frequency but does not return frequency to the nominal value and, if narrow dead-band is used, it also provides PFR, while SFC acts to eliminate the residual frequency error remained after PFC through Secondary Frequency Regulation (SFR).
- PFC provides a response to locally-measured frequency through the controls of individual generators while SFC is dependent on a centralized reference frequency and involves both an automatic (AGC) and manual centralized response (operator dispatch instructions) that simultaneously controls multiple generators.
- As such, PFC has no major communications issues/delays while SFC is subject to response delays up to few seconds.
- The input of PFC involves only the locally measured frequency error while the objective of SFC is to look after both the frequency error and the net tie-line flow or interchange error–see Eq. (2.35).
- In interconnected power systems, PFC in one control area will react to a frequency deviation in another area and thus cause generation changes, while the aim of SFC is to keep overall balance of its own area and not respond to normal changes in neighboring control areas.

2.3.4.1 AGC Model

The first AGC system was installed by an US East Coast utility almost a century ago namely in 1927 [1]. Since then, it has contributed in maintaining a good level of frequency quality by adjusting the MW set-points of controllable generating units every few seconds, acting over tens of seconds to minutes. For example, AGC was recently installed in India to enhance the frequency control efficiency and stability of the system [80].

In practice, AGC is implemented in a discrete form. That is, the AGC sends power signals (also known as AGC pulses) to participating conventional generation units at fixed intervals, for example, every 1-10 seconds [81, 82]. This time interval or cycle time has shown to be adequate in managing real-time power mismatches and has not caused any power system instability (see Chap. 9 for more details). Figure 2.19 shows the standard control scheme of the AGC employed in this book.

The input of the controller is the difference between the reference frequency ω^{ref} and the measured frequency $\omega_{\text{CoI}}(t)$, that is, the Center of Inertia (CoI). This represents a "constant frequency control" AGC mode and is similar to the one used in Australia, where only deviation of the frequency is considered and tie-line power flows are neglected [83]. In practice, TSOs use the frequency of a pilot bus of the system as an input for the AGC [1]. Next, the main controller of the AGC is an integral control with gain K_o, as follows:

$$\Delta p'(t) = K_o(\omega^{\text{ref}} - \omega_{\text{CoI}}(t)) , \tag{2.38}$$

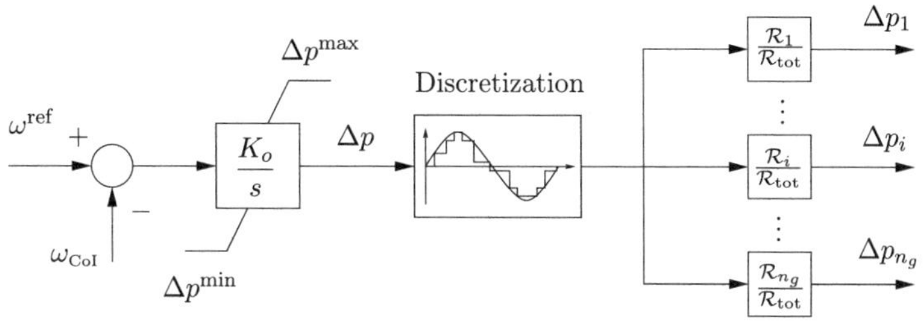

Fig. 2.19 AGC control diagram. The discretized output signals Δp_i are sent to the PFC of synchronous generators and wind power plants

where Δp is the output of the integrator. The integral term is needed to perfect tracking the reference frequency and nullify the steady-state frequency error introduced by the PFC (FCR in ENTSO-E terminology). In general, TSOs use AGC based on PI controller that includes other functionalities, among others, filtering and heuristics in order to reduce the area control error [84]. However, the fact that $\omega_{\text{CoI}}(t)$ is used as an input for the controller represents actually a sort of filter as the frequency of the center of inertia is a weighted average quantity. The output of the continuous integrator is discretized at given fixed-time intervals and sent to each TG.

These signals (Δp_i) are proportional to the capacity of the machines and the TG droops ($\mathcal{R}_i$) and normalized with respect to the total droop of the system:

$$\mathcal{R}_{\text{tot}} = \sum_{i=1}^{n_g} \mathcal{R}_i \,. \tag{2.39}$$

The model of the TG considered in this book is depicted in Fig. B.1 in Appendix B. In particular, $p_{\text{ord},i}$ represent the power order set-point as obtained by the electricity market. This model is suitable for transient stability analysis [85].

In practice, the output of the AGC is limited by the upward/downward secondary regulation reserve [82]. Therefore, the active power output of the integrator block of the AGC (see Δp in Fig. 2.19) is limited [86], in this case, through an Anti-Windup (AW) limiter, as follows:

$$\begin{aligned}
\text{if } \Delta p(t) \geq \Delta p^{\max} \;\&\; \Delta p'(t) \geq 0 : \Delta p(t) = \Delta p^{\max} \;\&\; \Delta p'(t) = 0 \,, \\
\text{if } \Delta p(t) \leq \Delta p^{\min} \;\&\; \Delta p'(t) \leq 0 : \Delta p(t) = \Delta p^{\min} \;\&\; \Delta p'(t) = 0 \,, \\
\text{otherwise : } \; \Delta p'(t) = K_o(\omega^{\text{ref}} - \omega_{\text{CoI}}(t)) \,.
\end{aligned} \tag{2.40}$$

The AW limiter is needed to limit the windup phenomenon of the integrator state variable Δp.

2.3.5 Tertiary Control

Tertiary Frequency Control (TFC) is similar to manual SFC but takes place on an even longer time scales. More specifically, NERC defines TFC as [87]: *Actions provided by Balancing Authorities on a balanced basis that are coordinated so there is a net zero effect on ACE. Examples of Tertiary Control include dispatching generation to serve native load; economic dispatch; dispatching generation to affect Interchange; and re-dispatching generation. Tertiary Control actions are intended to replace Secondary Control Response by reconfiguring reserves.*

As it can be inferred from the definition, different to PFC and SFC that have an automatic response, TFC is all about manual control and centrally coordinated actions to restore PFC and SFC reserves. Indeed, while the FAT of TFC may vary from system to system it is generally in the range of tens of minutes and so much slower than PFC and SFC. For example, AEMO achieves TFC through the central energy re-dispatch process that runs every five minutes [88]. TFC reserves are unique as they are the only reserve category that is not used for energy imbalance, but instead used for reserve imbalance [89]. In fact, they can be provided by both synchronized and desynchronized units as it is currently the case in the AIPS.

Common types of TFC reserve are RRs after a disturbance (that is, there are enough reserves to respond to a second event). Chapter 12 will provide a real-world example of RR from the AIPS including its definition and volume forecasting methodology.

2.3.6 Time Error Control

The TEC is the final frequency control process. Its origin dates back to 1916 when Henry Ellis Warren invented the self-starting synchronous motor used in the Telechron Clock in 1919 [90]. The Telechron Clock accuracy was heavily dependent on the system frequency, that is, it would only provide the correct time if the correct average frequency over the long-term. Because of this, the Warren Clock Company gave clocks to electric operators that could regulate system frequency and thus maintain the correct time on these clocks. Since then, the time error correction service was incorporated in general operating practice by power system operators. Therefore, the current form of TEC is a carryover of these practices.

In terms of TEC definition, SOGL defines it as [19]: *a process for time control, where time control is a control action carried out to return the electrical time deviation between synchronous time and UTC time to zero.* In mathematical terms, time error is calculated as the integral of frequency deviations over a given time t as follows [91]:

$$\epsilon(t) = \int_0^t \frac{\omega(t) - \omega_n}{\omega_n} dt \,, \tag{2.41}$$

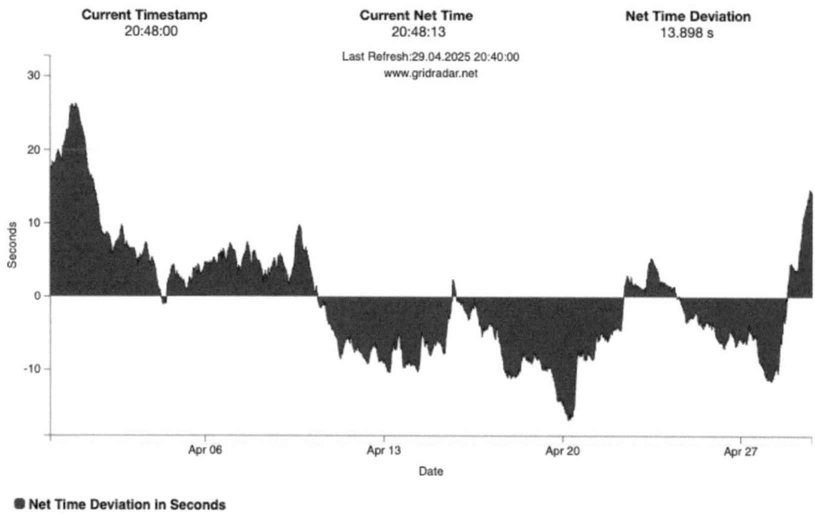

Fig. 2.20 Example of TEC in the CE power system (*Source* Gridradar [92])

where ω is the instantaneous frequency and ω_n is the nominal synchronous reference frequency.

TEC is a process whereby TSOs "manipulate" frequency up or down to reduce any accumulated time error. A positive time error can be corrected by targeting/scheduling frequency slightly below the nominal value for a certain time period. SOGL sets a clear methodology on how TSOs in Europe should correct the electrical time deviation, for example, through the actions to increase or decrease the average frequency of the system. For example, if the daily error exceeds ± 20 s, the frequency setpoint is adjusted by ± 10 mHz for the whole next day [91]. As a real-world example, Fig. 2.20 shows the evolution and control of time error in the CE power system in April 2025. It can be seen from the figure that around early April 2025, the error exceeded its limit of 20 s. Then, the TSOs of the CE system brought it within the limit after several hours.

In practice, TEC is performed through manual control by TSOs not implementing an AGC, or through AGC by TSOs that have an AGC installed and use the time error as an input to calculation of ACE [93].

The use of synchronous electric clocks has decreased over time. However, there are still many industrial processes and devices that use the frequency of the power grid as timing source. For example, we mention the following [94, 95]:

- Meters of electrical energy which calculate different tariff periods in a precise time measurement based on frequency as input value;
- Power plants control energy;
- Power quality devices;
- Old industries processes;
- Customers in textile industries, and synchronous motors;

Table 2.10 TEC limits for different jurisdictions

Jurisdiction	Limit [± s]
AIPS	±10
GB	±10
PJM, US	±10
CE	±20
Nordic	±30

- Traffic lights;
- Programmable household appliances;
- Clocks of railway stations;
- Vinyl disc players.

Table 2.10 provides a list of relevant TEC limits for different jurisdictions. Complying with these limits might be challenging, at times, particularly at high RES conditions such as windy nights where the TSOs may prefer to run frequency slightly above 50 Hz to reduce wind curtailment. In the US, each interconnection has a TEC process to maintain the long-term average frequency at 60 Hz [8]. The calculation and monitoring of ϵ in real-time has been questioned in recent years. For instance, NERC has recently reviewed the purpose of TEC and has concluded that: *there is not a reliability reason to continue Time Error Corrections* [90]. AEMO has taken a similar path and has decided: *to report time error, but there are no longer formal limits on accumulated time error in the Frequency Operating Standard (FOS) from 9 October 2023* [96]. Therefore, it can be said that the purpose of TEC in power systems is diminishing and so this stage of frequency control may be completely removed in the future.

One obvious question would be how has time error evolved in light of the integration of massive intermittent generation such as wind and solar power. Reference [94] suggest that time error for ERCOT, has not shown any correlation with the annual wind power generation from 2008 to 2017 although wind had tripled from 2008 (16,225 GWh) to 2016 (57,551 GWh). This is explained by the fact that ERCOT had provided TEC services during this period. The same reference reports acceptable levels of time error in most jurisdictions worldwide except Brazil and China that have time error greater than one minute due to not employing TEC services. Both countries also appear to have a significant deviation of their long-term average frequency from the nominal frequency.

Fig. 2.21 Albanian system operator control room (*Source* OST)

2.4 Control Centers

While not a frequency control stage *per se*, TSO control rooms are the heart of power system operation, control, and management. We provide a brief overview below of the main tasks performed by control rooms.

Figure 2.21 shows an image of the control room of the system operator, the Albanian TSO, where the first author worked at the beginning of his career. It is from these rooms that expert control engineers make important decisions in ensuring "lights stay on" at all times. From a frequency control point of view, control centers generally show on a big video wall real-time frequency measurements and ACE. Future control centers will continue the trend of automation of manual processes that do not increase operators' situational awareness but are time-consuming, tedious, and repetitive tasks [97]. This increased automation will help operate an increasingly complex power system.

Depending on the size of the power system, control centers could include from up to a couple of operators to tens of operators in large interconnected power systems working on 24/7 shifts. There are at least two roles that need to be performed in the control centers, that is, power system balancing which usually includes scheduling and market dispatch that help keep active power balance close to real-time and SFC and, depending on the arrangements in each jurisdiction, it can also include dealing

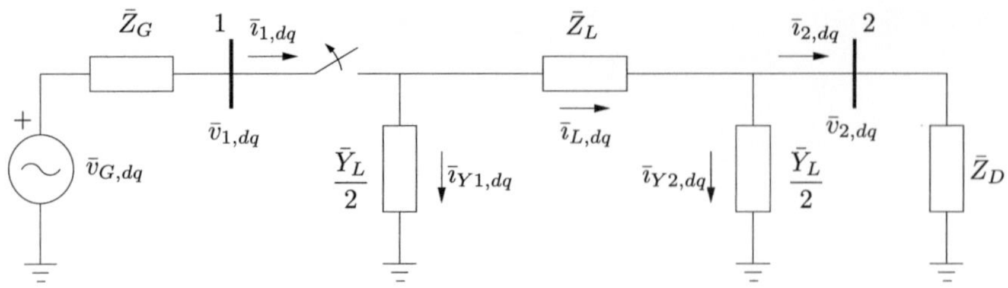

Fig. 2.22 π-model of an over-head transmission line connecting a voltage source and a load through

with cross-border energy exchange. The other critical role is focused on transmission network management in terms of managing scheduled and unscheduled outages and congestion management, including voltage control.

2.5 Examples

The illustrative examples included below are aimed at supporting the discussion of previous sections. All examples, except for the first one, utilize the Western Systems Coordinating Council (WSCC) system described in Appendix and illustrate the response and effectiveness of different controls such as AGC and different parameters such as governor dead-bands and droops ($\mathcal{R}$). To make the comparison as relevant as possible, we replace a synchronous generator with a wind power plant modeled as DFIG with PFC and dead-band of ±200 mHz as is normally the case in the AIPS. In addition, we introduce load and wind noise and ramps similar to Chap. 10 and run different short- and long-term TDSs. Finally, unless stated otherwise, the AGC gain K_o is set to 0.5 and the AGC set-point update happens every 2 s.

2.5.1 *Energy of Transmission Lines*

This example discusses the ability of transmission lines to store energy and release it to the load during a voltage dip. We consider the simplified lumped π-model of the transmission line shown in Fig. 2.22.

Three-phase ac transmission systems are huge RLC circuits where resistances, inductances and capacitances are due to, respectively, transmission losses, conductor inductive effects and transformer flux leakages and magnetization, and overhead line and cable parasite capacitive charging.

Of course, transmission lines are not designed to be energy storage systems and, in effect the energy stored in the conductors is small compared to that the inertia of synchronous machines and the power reserve of the prime movers of turbines.

Table 2.11 Equivalent π-circuit parameters per phase for long transmission lines at 60 Hz [98]

Length, ℓ [km]	R_L [Ω]	X_L [Ω]	$G_L/2$ [μS]	$B_L/2$ [μS]
200	3.64573	73.74022	0.1260473	449.89134
400	6.80630	142.59986	1.0503017	915.21820

However, the stored energy of transmission lines is not null and has an effect, even if tiny, in the very first instants after a fault.

The parameters of the line are indicated in Table 2.11 and represent the equivalent π-model parameters for a standard long transmission line [98]. The load connected to bus 2 is purely resistive, namely, $\bar{Z}_D = R_D$, and consumes the rated current at the nominal voltage. The source is a constant 550 kV, 60 Hz voltage generator and is assumed to be the phase reference:

$$\bar{\upsilon}_{G,dq} = \frac{\sqrt{2}}{\sqrt{3}} 550 + j0 \text{ kV} , \tag{2.42}$$

where the dq-axis magnitude is assumed to be the RMS phase-to-neutral voltage and the equivalent Thévenin equivalent impedance per phase of the source is set as:

$$\bar{Z}_G = 10 \, \Omega/\text{phase} . \tag{2.43}$$

We use Park vectors to represent voltages and currents, namely:

$$\begin{aligned} \bar{\upsilon}_{dq}(t) &= \upsilon_d(t) + \jmath \, \upsilon_q(t) , \\ \bar{\imath}_{dq}(t) &= \imath_d(t) + \jmath \, \imath_q(t) . \end{aligned}$$

Then, the equations of the circuit in dq-frame coordinates rotating at ω_o are:

$$\begin{aligned}
0 &= \bar{\upsilon}_{G,dq} - R_G \, \bar{\imath}_{1,dq}(t) - \bar{\upsilon}_{1,dq}(t) \\
0 &= \bar{\imath}_{1,dq}(t) - \bar{\imath}_{Y1,dq}(t) - \bar{\imath}_{L,dq}(t) \\
0 &= 0.5 \, B_L \left(\frac{\mathrm{d}}{\mathrm{d}t} + j\omega_o \right) \bar{\upsilon}_{1,dq}(t) + 0.5 \, G_L \, \bar{\upsilon}_{1,dq}(t) - \bar{\imath}_{Y1,dq}(t) \\
0 &= \bar{\upsilon}_{1,dq}(t) - \bar{\upsilon}_{2,dq}(t) - R_L \, \bar{\imath}_{L,dq}(t) - X_L \left(\frac{\mathrm{d}}{\mathrm{d}t} + j\omega_o \right) \bar{\imath}_{L,dq}(t) \\
0 &= \bar{\imath}_{L,dq}(t) - \bar{\imath}_{Y2,dq}(t) - \bar{\upsilon}_{2,dq}(t) \\
0 &= 0.5 \, B_L \left(\frac{\mathrm{d}}{\mathrm{d}t} + j\omega_o \right) \bar{\upsilon}_{2,dq}(t) + 0.5 \, G_L \, \bar{\upsilon}_{2,dq}(t) - \bar{\imath}_{Y2,dq}(t) \\
0 &= \bar{\upsilon}_{2,dq}(t) - R_D \, \bar{\imath}_{2,dq}(t) ,
\end{aligned} \tag{2.44}$$

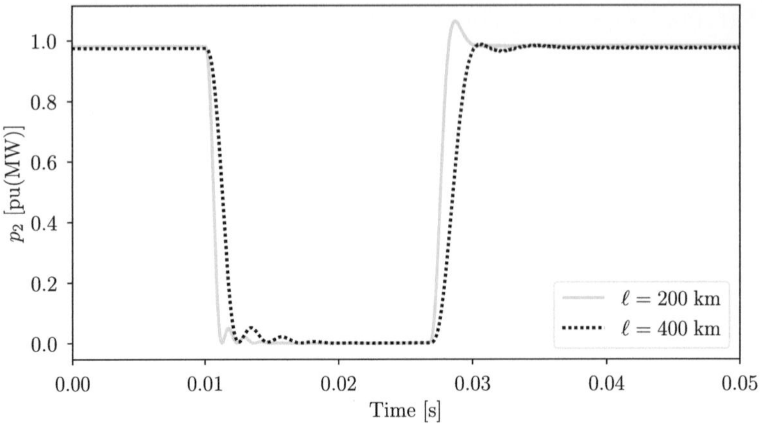

Fig. 2.23 Effect of the line length on the power consumed by the load connected to bus 2 following the disconnection of the source at $t = 0.01$ s during 1 cycle. ℓ is the length of the transmission line. The power base is 500 MVA

where

$$\begin{aligned} \bar{Z}_L &= R_L + jX_L \\ \bar{Y}_L &= G_L + jB_L \, , \end{aligned} \tag{2.45}$$

are the series impedance and the shunt admittance, respectively, of the transmission line, and $\omega_o = 2\pi\,60 = 377$ rad/s is the fundamental angular speed of the system.

Figure 2.23 shows the dynamic behavior of the power consumed by the load connected to bus 2 after the disconnection of the source at $t = 0.01$ s during 1 cycle, namely 16.7 ms at 60 Hz. While the energy stored in the transmission line cannot sustain the power of the load and decays in a few milliseconds but, notably, does not drop instantaneously. This energy is useful to reduce the effects of micro imbalances in the load consumption.

2.5.2 *Impact of Reduced Inertia*

Modern power systems operate with less and less inertia following the phase out of large synchronous machines. Figure 2.24 shows an example where the base case is modified such as the inertia constants H of the two synchronous machines is reduced by four.

The effect of this inertia reduction is illustrated assuming the loss of the load connected at bus 6 at $t = 1$ s. This represents a significant incident for the WSCC as almost one third of the total load is lost in the system. Because of this, the frequency goes well above 51 Hz which is usually the maximum over-frequency deviation in some jurisdictions such as in the AIPS (see Chap. 3). However, any power system

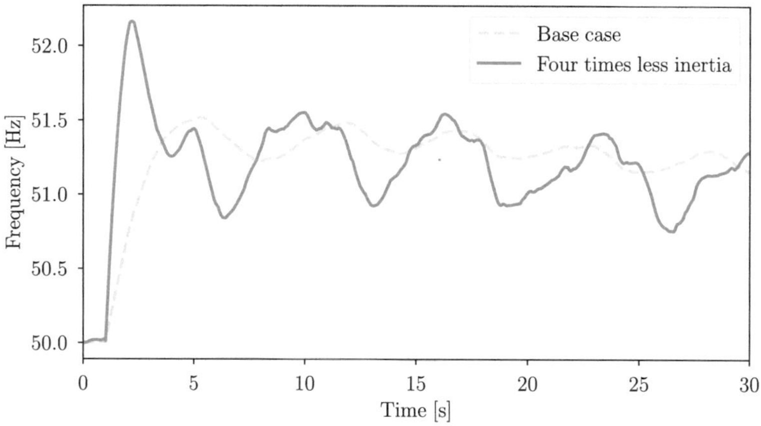

Fig. 2.24 Effect of reduced inertia levels on frequency response

that suddenly loses more than 30% of its demand or generation is at serious risk of blackout.

In terms of the main impacts of reduced inertia levels on the transient frequency response it can be seen that: (i) RoCoF gets significantly higher; (ii) frequency zenith is larger and occurs earlier; and (iii) the transient is followed by higher sustained oscillations that indicate reduced damping. Needless to say, inertia is a critical parameter of power systems that need to be maintained at certain minimum levels to keep frequency/RoCoF within limits following disturbances.

We simulate the same scenario again but without any contingency. We run the simulation for a long time to visualize the statistical properties of frequency variations.

Figure 2.25 suggests that reducing inertia significantly, in this case by four times, has an impact on normal operation as it makes frequency more volatile. This is to be expected considering the system is "lighter." However, the differences are not as dramatic as in Fig. 2.24. Thus, the deterioration of frequency behavior in normal operations due to less inertia should be less of a concern for TSOs.

2.5.3 Impact of Droop Coefficient

The droop coefficient ($\mathcal{R}$) is key frequency control parameter. As shown in Table 2.7, TSOs normally apply a value of $\mathcal{R}$ between 4-5%. To illustrate its effect, Fig. 2.26 depicts the frequency traces using various droop coefficients of both synchronous machines and the wind power plant. Notice that since the focus is on PFC, we assume the two synchronous machines and the wind power plant are not under AGC. Also, the dead-bands of TGs of conventional power plants are ±15 mHz in this case.

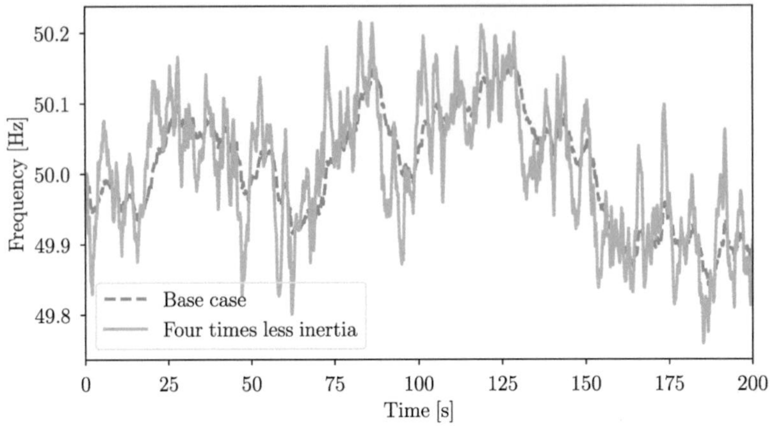

Fig. 2.25 Effect of reduced inertia levels on frequency variations

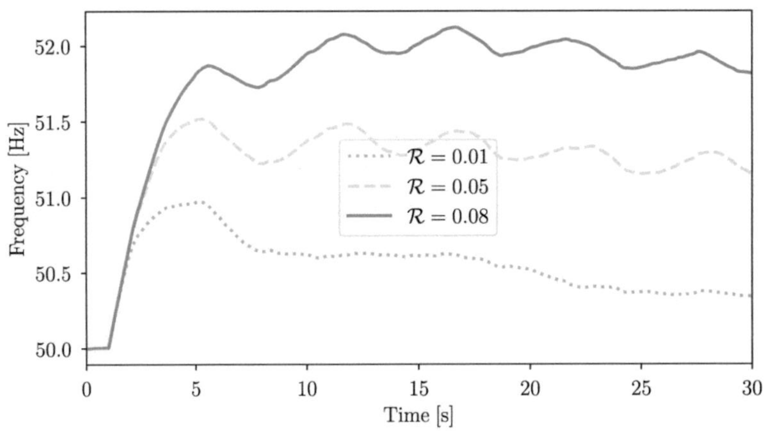

Fig. 2.26 Effect of PFC droop $\mathcal{R}$ on transient conditions

As expected, a small value of $\mathcal{R}$, in this case 1%, leads to a much better frequency performance. However, "aggressive droops" with very small values of $\mathcal{R}$ may be impractical for most technologies.[12] and, most importantly, may lead to unnecessary oscillations as it will be discussed throughout the book. Therefore, TSOs need to carefully choose the value of $\mathcal{R}$.

[12] This is not the case of BESS, which, for example, can inject into the grid from zero to 300 MW in 150 ms.

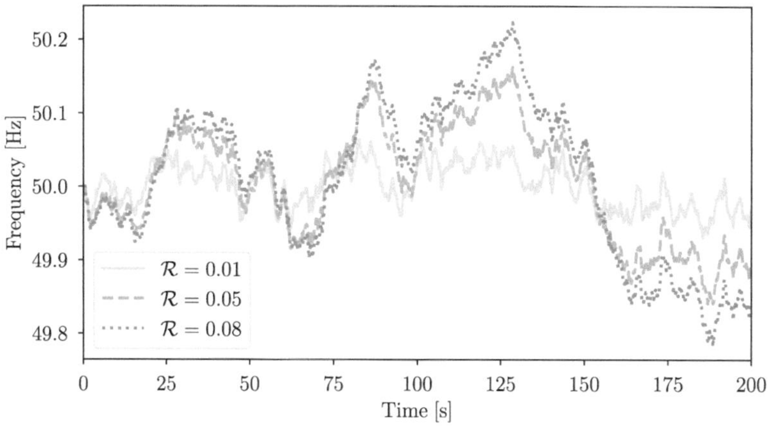

Fig. 2.27 Effect of PFC droop $\mathcal{R}$ on normal operating conditions

When discussing droop control, the focus is generally only on transient frequency response. The impact of droop control on long-term frequency variations and frequency regulation performance is often neglected. To illustrate this impact, Fig. 2.27 shows the results of a scenario similar to the case above except for the contingency and for longer time.

The combination of narrow dead-band (± 15 mHz) and small droop value namely $\mathcal{R} = 0.01$ leads to an excellent frequency regulation performance with frequency staying within the ± 50 mHz range for all the time. Therefore, TSOs need to keep in mind these results when designing their frequency control products, in this case, those designed for frequency regulation purposes.

2.5.4 Impact of Frequency Error Dead-Band

In this scenario, we test the effectiveness of different dead-band values on the transient frequency response following the loss of the load connected at bus 6 at $t = 1$ s. Figure 2.28 shows the relevant results.

The dead-band does not play a significant role to mitigate RoCoF. This is due to the fact that, for the three case, the injected active power is similar. In other words, frequency deviations are so fast during a transient that there is no much difference between using a narrow or wider dead-band, for example, ± 15 mHz and ± 200 mHz. On the other hand, the dead-band slightly helps to reduce the frequency zenith and the steady-state error in the first seconds following the contingency.

Given PFC with narrow dead-bands plays a crucial role in providing PFR, it is a good idea to illustrate its impact on normal frequency variations. To do so, we implement three different governor dead-bands for the two synchronous machines namely ± 15 mHz, ± 100 mHz, and ± 200 mHz. Similar to the droop example above,

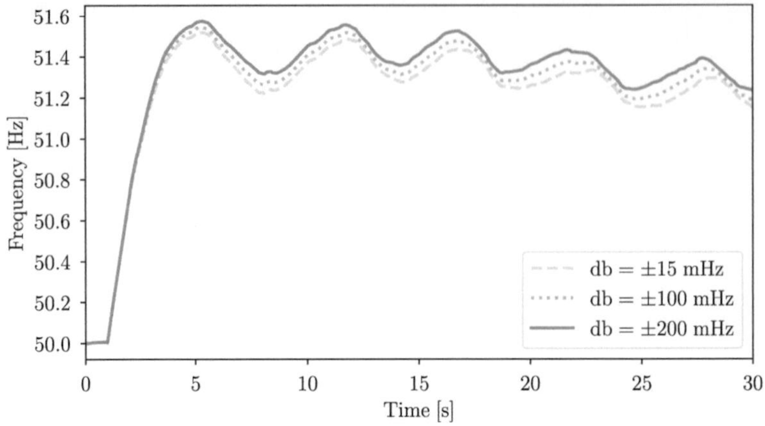

Fig. 2.28 Effect of dead-bands on abnormal operating conditions

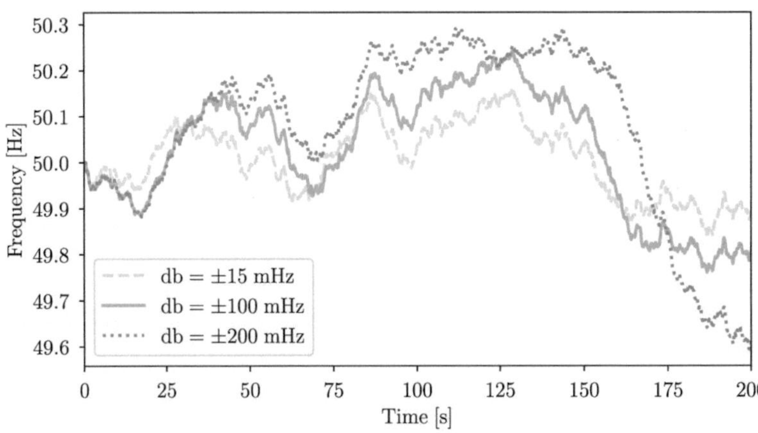

Fig. 2.29 Effect of governor dead-bands on normal operating conditions

we assume AGC is not installed as the focus is on the effectiveness of PFC with different dead-bands, and the droop values are all the same namely $\mathcal{R} = 0.05$. Figure 2.29 shows relevant results.

Narrow dead-bands such as ±15 mHz lead to a much better frequency regulation performance, so much so that is able to keep frequency within the ±200 mHz range which is the standard frequency range used in, for example, AIPS and GB power systems (see Chap. 3). Most importantly, it obtains this result without an AGC in place. In other words and surprisingly enough, an AGC may not be necessary if narrow dead-bands are utilized. This is the reason why Irish TSOs normally use narrow dead-bands in practice as those shown in Table 2.8. Additionally, when the penetration of IBRs is high, for example, greater than 50%, these resources should

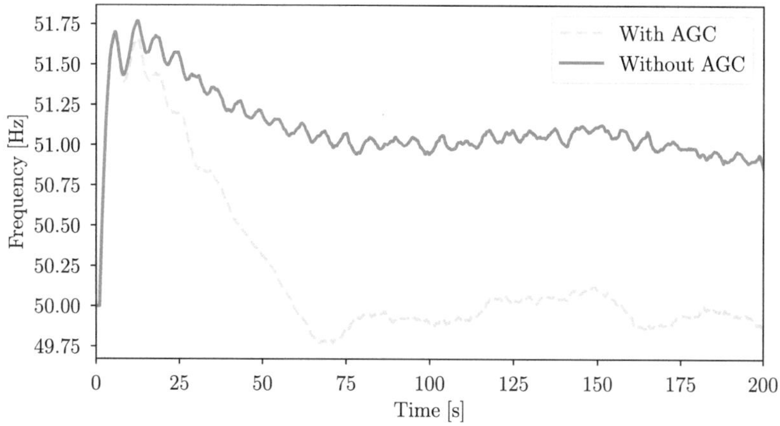

Fig. 2.30 Effect of AGC on abnormal operating conditions

implement a narrow dead-band similar to conventional generators to help with PFR during normal operating conditions.

While narrow PFC dead-bands may be preferred by TSOs from a system need point of view given they help ensure tight distribution of frequency around the nominal value, the costs and benefits to consumers associated with mandatory requirements for such dead-bands should be carefully assessed. Requiring all (capable) resources to provide narrow PFC dead-band could mean that the necessary service could not be provided by those that are best able to or at least cost [99]. Instead, it could risk over-consumption that can result in higher costs to market participants which, in turn, can result in higher costs to consumers. For instance, different power system industry actors in Australia argue that the recent mandatory PFC with ±15 mHz dead-band (which excludes BESS if dispatched at 0 MW) has led to an oversupply of the service, and believe that market-based approaches are more efficient and should replace the mandatory requirement and deliver lower costs to consumers [99]. Therefore, careful consideration should be given when introducing new ancillary service arrangements and mandatory requirements in order to always strike a balance between reliability and cost.

2.5.5 Impact of AGC

The short-term contribution of AGC is often neglected or overlooked. We show in this example that, on the contrary, the effect AGC is crucial for the frequency control of power systems. Figure 2.30 compares two cases namely with AGC and without AGC following the loss of the load at bus 6.

Although the AGC does not help to address RoCoF and frequency zenith (as expected considering it is a slow controller), it provides excellent response and helps

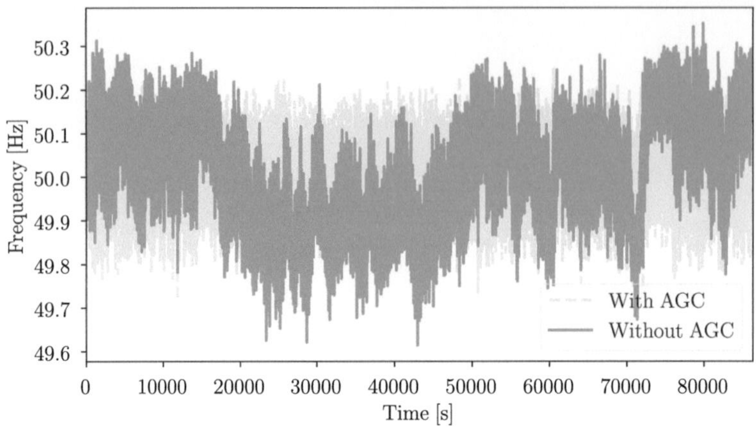

Fig. 2.31 Effect of AGC on frequency variations

recover frequency to nominal after around 1 min following the contingency. The case without AGC, instead, leads to an oscillatory frequency with a significant steady-state error. As discussed in Chap. 5, power systems that do not employ an AGC but rely on manual actions suffer from poor frequency recovery. On the other hand, TSOs that employ an AGC such as AEMO demonstrate similar frequency recovery as shown in Fig. 2.30. To conclude, it is fair to say that AGC is a viable solution if TSOs face frequency recovery problems.

With regard to the effectiveness of AGC on long-term frequency deviations, Fig. 2.31 shows the results of a 24 h simulation. The long simulation is selected on purpose because the AGC is a slow controller.

Results in Fig. 2.31 suggest that having an AGC in place leads, overall, to a better long-term dynamic frequency performance as, for example, the frequency spends less time outside the ±200 mHz range. Hence, the AGC improves frequency quality.

2.5.6 Comparison Between PFR and AGC

We have seen so far that both PFC with narrow dead-band and AGC are good candidates for frequency regulation. In this last example, we compare these two control strategies under intact conditions.

Figure 2.32 shows a comparison between PFC with narrow dead-band of ±15 mHz and without AGC, and the case with dead-band of ±200 mHz and with AGC. The second scenario leaves frequency ineffectively controlled in the ±200 mHz range. The case of PFC with narrow governor dead-band of ±15 mHz outperforms the case of PFC with wider dead-band (±200 mHz) combined with an AGC. This is due to the fact that PFC is faster than AGC and thus more effective in compensating frequency deviations.

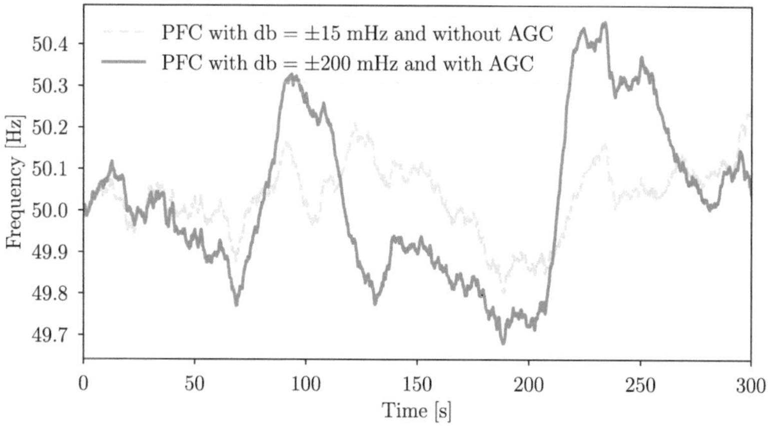

Fig. 2.32 Comparison of the effect of PFC with narrow dead-band and AGC on frequency variations

This example confirms that frequency regulation is not limited to AGC. AIPS and GB power systems are two excellent examples of systems still being operated without an AGC and just PFR and manual AGC. However, both systems are concerned by current challenges posed by frequency regulation and are considering solutions to address this problem. An obvious one is to install an AGC, which is also an EU requirement.

In Brief

- There is currently no consistent notation for the frequency control across system operators and practitioners.
- The chapter provides new definitions of frequency response, frequency control, and FFR.
- Frequency regulation is a sub-set of frequency control and is not limited to AGC.
- Dead-band of PFC is a key parameter that significantly impacts the overall frequency regulation performance.
- Thinking in terms of time scales and energy helps navigate the inconsistencies of the many confusing terms utilized in industry.
- TSOs are looking for solutions on how to make "slow" conventional synchronous machines and "fast" IBRs work together while ensuring the safe and secure operation of power systems.
- BESS have become the main providers of frequency control and frequency regulation.
- Time error is a robust long-term frequency performance indicator but it is becoming less and less important in power systems.

Open Questions

The control of low-inertia and zero-inertia power systems poses a large number of questions that currently neither academia nor industry can answer. Some of these questions are reported below.

- Can large-scale power systems operate without synchronous machines?
- Can future 100% IBR-based power systems operate without an AGC?
- Does it make sense to use FFR under normal conditions in future low-inertia power systems?
- What are the recommended FFR settings in terms of FAT, dead-band and droop/trajectory?
- How to control power system frequency when demand is entirely met by distributed PV?
- Can TSOs replace synchronous inertia with GFM inertia assuming a one-to-one relationship between them?
- How to aid control-room operators automatically damp power system oscillations?
- Will TEC service be necessary in the future?
- What is the impact of topology, granularity, and geographical distribution on the frequency regulation performance?

References

1. EPRI (2009) EPRI power system dynamics tutorial. Electric Power Research Institute, United States, Tech. Rep
2. Kërçi T (2021) Modelling and simulation of long-term dynamics in power systems. PhD dissertation, University College Dublin. School of Electrical and Electronic Engineering
3. Cao X, Engelhardt J, Ziras C et al (2024) Battery energy storage systems providing dynamic containment frequency response service. Electr Power & Energy Syst 162:110288
4. Astero P, Evens C (2020) Optimum operation of battery storage system in frequency containment reserves markets. IEEE Trans Smart Grid 11(6):4906–4915
5. Zhao T, Parisio A, Milanović JV (2020) Distributed control of battery energy storage systems for improved frequency regulation. IEEE Trans Power Syst 35(5):3729–3738
6. Kërçi T (2023) On the definition and classification of power system flexibility. In: 2023 IEEE PES innovative smart grid technologies Europe (ISGT EUROPE), pp 1–5
7. Viola L, Mohammadi S, Dotta D et al (2024) Ancillary services in power system transition toward a 100% non-fossil future: market design challenges in the United States and Europe. Electr Power Syst Res 236:110885
8. NERC Rs Committee (2011) Balancing and frequency control: a technical document prepared by the NERC resources subcommittee. Tech. Rep., NERC Technical Report
9. Zhong W, Tzounas G, Milano F (2022) Improving the power system dynamic response through a combined voltage-frequency control of distributed energy resources. IEEE Trans Power Syst 37(6):4375–4384
10. Bernal R, Milano F (2024) Improving voltage and frequency control of DERs through dynamic power compensation. Electr Power Syst Res 235:110768

11. cold.world GmbH (2023) Difference between control and regulation. [Online]. www.cold.world
12. Ela E, Gevorgian V, Fleming P et al (2014) Active power controls from wind power: bridging the gaps. National Renewable Energy Lab. (NREL), Golden, CO (United States), Tech. Rep
13. PJM (2024) State of the market report for PJM 2024. January through June. [Online]. www.monitoringanalytics.com
14. AEMO (2018) Advice from AEMO—Primary frequency control. [Online]. www.aemo.com.au
15. ESIG (2024) Diagnosis and mitigation of observed oscillations in IBR-dominant power systems. a practical guide. ESIG, Tech. Rep
16. IEEE standard for interconnection and interoperability of inverter-based resources (IBRs) interconnecting with associated transmission electric power systems. IEEE, pp 1–180 (2022)
17. Kraftwerke N (2025) What is frequency containment reserve (FCR)? [Online]. www.next-kraftwerke.com/knowledge/frequency-containment-reserve-fcr
18. ENTSO-E (2024) Overview of frequency control in the nordic power system. [Online]. https://eepublicdownloads.blob.core.windows.net/
19. European Commission (2017) Establishing a guideline on electricity transmission system operation. [Online]. www.eur-lex.europa.eu
20. ENTSO-E (2024) Nordic balancing philosophy. [Online]. www.eepublicdownloads.entsoe.eu
21. EPRI (2021) Video tutorial: illustrating operating reserve needs and methods. [Online]. www.epri.com
22. Peng Q, Jiang Q, Yang Y et al (2019) On the stability of power electronics-dominated systems: challenges and potential solutions. IEEE Trans Ind Appl 55(6):7657–7670
23. Ulbig A, Borsche TS, Andersson G (2014) Impact of low rotational inertia on power system stability and operation. IFAC Proc Vol 47(3):7290–7297. 19th IFAC world congress
24. Milano F, Dörfler F, Hug G et al (2018) Foundations and challenges of low-inertia systems (invited paper). In: Power systems computation conference (PSCC), pp 1–25
25. NESO (2024) System inertia. [Online]. www.neso.energy/
26. Osbouei BA, Taylor GA, Bronckart O et al (2019) Impact of inertia distribution on power system stability and operation. IEEE Milan PowerTech 2019:1–6
27. Fresia M, Minetti M, Rosini A et al (2024) A techno-economic assessment to define inertia needs of the Italian transmission network in the 2030 energy scenario. IEEE Trans Power Syst 39(3):5050–5061
28. Du P (2023) System inertia trend and critical inertia. In: Renewable energy integration for bulk power systems: ERCOT and the Texas interconnection. Springer, pp 199–222
29. Gordon S (2024) Dynamic interactions between voltage and frequency events in future power systems. PhD dissertation, University of Strathclyde
30. EirGrid & SONI (2023) Shaping our electricity future roadmap. [Online]. https://www.eirgrid.ie/
31. EirGrid & SONI (2022) Consultation on low carbon inertia service (LCIS) competitive procurement. [Online]. https://www.eirgrid.ie/
32. Misyris G, Graham B, Mitra P et al (2023) Methodology for identifying regional inertia issues in future power grids. In: IEEE PES general meeting, pp 1–5
33. Borsche TS, Liu T, Hill DJ (2015) Effects of rotational inertia on power system damping and frequency transients. In: 2015 54th IEEE conference on decision and control (CDC), pp 5940–5946
34. SEM Committee. Low carbon inertia services phase 1 procurement. Information note. [Online]. www.semcommittee.com
35. AEMO (2024) 2024 inertia report. [Online]. www.aemo.com.au
36. Baig AM, Badesa L, Strbac G (2021) Importance of linking inertia and frequency response procurement: the great britain case. IEEE Madrid PowerTech 2021:1–6
37. Liang Z, Mieth R, Dvorkin Y (2023) Inertia pricing in stochastic electricity markets. IEEE Trans Power Syst 38(3):2071–2084

38. Bonfiglio A, Invernizzi M, Labella A et al (2019) Design and implementation of a variable synthetic inertia controller for wind turbine generators. IEEE Trans Power Syst 34(1):754–764
39. Al Kez D, Foley AM, Ahmed F et al (2023) Overview of frequency control techniques in power systems with high inverter-based resources: challenges and mitigation measures. IET Smart Grid 6(5):447–469
40. EirGrid & SONI (2017) DS3 system services qualification trials process outcomes and learnings 2017. [Online]. www.cms.eirgrid.ie
41. Tan J (2025) Inertia monitoring in Hawaii: insight and lessons learned. ESIG 2025 Spring Technical Workshop, Tech. Rep., [Online]. www.esig.energy
42. Transgrid, (2024) Flagship report. Large-scale battery storage as an inertia substitute Wallgrove grid battery. Transgrid, Tech. Rep
43. Kërçi T, Hurtado M, Gjergji M et al (2023) Frequency quality in low-inertia power systems. In: IEEE PES general meeting, pp 1–5
44. Lavanya L, Swarup KS (2024) Inertia monitoring in power systems: critical features, challenges, and framework. Renew Sustain Energy Rev 190:114076
45. Matevosyan J (2022) Inertia monitoring. [Online]. www.esig.energy
46. Zhang Z (2024) Impact of fast frequency response on the angular stability of low inertia power systems. PhD dissertation, The University of Manchester (United Kingdom)
47. NERC (2012) Frequency response initiative report-the reliability role of frequency response
48. Chávez H, Baldick R, Matevosyan J (2015) The joint adequacy of AGC and primary frequency response in single balancing authority systems. IEEE Trans Sustain Energy 6(3):959–966
49. Nedd M, Browell J, Egea-Alvarez A et al (2020) Operating a zero-carbon GB power system: implications for Scotland
50. Eto JH, Undrill J, Roberts C et al (2018) Frequency control requirements for reliable interconnection frequency response. Tech. Rep., Lawrence Berkeley National Laboratory (United States)
51. Noela Sofia Guzman E, Caizares CA, Bhattacharya K et al (2022) Frequency regulation model of bulk power systems with energy storage. IEEE Trans Power Syst 37(2):913–926
52. Undrill J (2018) Primary frequency response and control of power system frequency. Tech. Rep., Lawrence Berkeley National Laboratory (United States)
53. NESO (2025) Webinar: pre-fault frequency control modelling. [Online]. www.neso.energy/document/356951/download
54. AEMO (2016) International review of frequency control adaptation. Tech. Rep. 2016
55. Rahmann C, Castillo A (2014) Fast frequency response capability of photovoltaic power plants: the necessity of new grid requirements and definitions. Energies 7(10):6306–6322
56. Ela E, Hytowitz R, Helman U (2019) Ancillary services in the United States: technical requirements, market designs and price trends. Electric Power Research Institute
57. Matevosyan J, Halley A, Ramasubramanian D et al (2024) Considerations, challenges and evolution in dynamic modeling of synchronous machines and inverter-based resources. In: Power system dynamic modelling and analysis in evolving networks. Springer, pp 351–435
58. Jalali A, Farahani E, Delac M et al (2023) Impact of grid-forming inverters on frequency control of a grid with high share of inverter-based resources. Cigre Sci Eng
59. Modi N, Jalali A, Commerford I et al (2022) Fast frequency response from transmission-connected solar farms: Australian experience. Cigre Sci Eng :187–199
60. Badrzadeh B, Emin Z (2024) Power system dynamic modelling and analysis in evolving networks. Springer Nature
61. Kërçi T, Milano F (2025) Asymmetry of frequency distribution in power systems: sources, estimation, impact and control. IEEE Open Access J Power Energy 12:135–145
62. Zepter JM, Engelhardt J, Ledro M et al (2023) Experimental assessment of active power control of distributed generation units. Sustain Energy Technol Assess 60:103500
63. Khan A, Hosseinzadehtaher M, Shadmand MB et al (2020) On the stability of the power electronics-dominated grid: a new energy paradigm. IEEE Ind Electron Mag 14(4):65–78
64. Milano F, Ortega Á (2020) Frequency variations in power systems: modeling, state estimation, and control. John Wiley & Sons

65. Bose A, Overbye TJ (2021) Electricity transmission system research and development: grid operations. Office of Electricity U.S. Department of Energy, Tech. Rep
66. Ofgem (2021) Decision to grant an exemption to NGESO from requirement to implement an automatic frequency restoration process in the GB synchronous area. [Online]. www.ofgem.gov.uk
67. ENTSO-E (2016) Impact of merit order activation of automatic frequency restoration reserves and harmonised full activation times. [Online]. www.eepublicdownloads.entsoe.eu
68. Liang Z, Lu G, Li L et al (2023) Coordinated optimization control strategy of hydropower and thermal power AGC units. Energy Rep 9:1292–1303. Selected papers from 2022 international conference on frontiers of energy and environment engineering
69. Eto JH (2011) Use of frequency response metrics to assess the planning and operating requirements for reliable integration of variable renewable generation
70. ENTSO-E (2019) Continental Europe significant frequency deviations–January 2019. [Online]. www.eepublicdownloads.entsoe.eu
71. Bevrani H (2009) Robust power system frequency control. Springer
72. NESO (2023) ENTSO-E draft network code on load-frequency control and reserves. [Online]. www.neso.energy/document/39246/download
73. Doenges K, Sigrist L, Egido I et al (2022) Wind farms in AGC: modelling, simulation and validation. IET Renew Power Gener 16(1):139–147
74. Rebello E, Watson D, Rodgers M (2019) Performance analysis of a 10 MW wind farm in providing secondary frequency regulation: experimental aspects. IEEE Trans Power Syst 34(4):3090–3097
75. NREL (2017) Demonstration of essential reliability services by a 300-MW solar photovoltaic power plant. [Online]. www.nrel.gov/docs/fy17osti/67799.pdf
76. Doenges K, Egido I, Sigrist L et al (2020) Improving AGC performance in power systems with regulation response accuracy margins using battery energy storage system (BESS). IEEE Trans Power Syst 35(4):2816–2825
77. CAISO (2024) 2023 special report on battery storage. [Online]. www.caiso.com
78. ERCOT (2024) ERCOT innovation summit 2024: essential reliability services panel. [Online]. www.youtube.com
79. Du P (2023) Renewable energy integration for bulk power systems: ERCOT and the Texas interconnection. Springer Nature
80. Sharma A, Chilukuri P, Dora MH et al (2023) Operational challenges and their mitigation strategies in grid level implementation of automatic generation control in Indian power system. In: IEEE 20th India council international conference, pp 1116–1121
81. Apostolopoulou D, Sauer PW, Domínguez-García AD (2014) Automatic generation control and its implementation in real time. In: 2014 47th Hawaii international conference on system sciences, pp 2444–2452
82. Pandurangan V, Zareipour H, Malik O (2012) Frequency regulation services: a comparative study of select North American and European reserve markets. In: 2012 North American power symposium (NAPS), pp 1–8
83. Alhelou HH, Bahrani B, Ma J et al (2024) Australia's power system frequency: current situation, industrial challenges, efforts, and future research directions. IEEE Trans Power Syst 39(3):5204–5218
84. Rebours YG, Kirschen DS, Trotignon M et al (2007) A survey of frequency and voltage control ancillary services–Part I: technical features. IEEE Trans Power Syst 22(1):350–357
85. Milano F (2010) Power system modelling and scripting. Springer, London
86. Pavlovsky V, Steliuk A (2014) Modeling of automatic generation control in power systems. Springer International Publishing, Cham, pp 157–173
87. NERC Rs Committee (2021) Balancing and frequency control: reference document. Prepared by the NERC resources subcommittee. Tech. Rep, NERC Technical Report
88. AEMO (2022) AEMO advice: reliability panel review of frequency operating standard. [Online]. www.aemo.com.au

89. Ela E, Milligan M, Kirby B (2011) Operating reserves and variable generation. National Renewable Energy Lab. (NREL), Golden, CO (United States), Tech. Rep
90. NERC (2020) Time error correction. [Online]. www.nerc.com
91. Scherer M, Andersson G (2015) How future-proof is the continental European frequency control structure? IEEE Eindh PowerTech 2015:1–6
92. Gridradar (2025) Network time deviation of the UCTE power grid. [Online]. www.gridradar.net/en/net-time
93. Meng Y (2021) Economic analysis of centralized energy storage and impact on NEM. In: IOP conference series: earth and environmental science, vol 680, no 1. IOP Publishing, p 012004
94. Zhang Y, Yao W, You S et al (2017) Impacts of power grid frequency deviation on time error of synchronous electric clock and worldwide power system practices on time error correction. Energies 10(9)
95. ENTSO-E (2018) 'All CE TSOs' agreement on frequency restoration control error target parameters in accordance with article 128 of the commission regulation (EU) 2017/1485 of 2 August 2017 establishing a guideline on electricity transmission system operation. [Online]. www.eepublicdownloads.entsoe.eu
96. AEMO (2024) Frequency monitoring–Quarter 1 2024. [Online]. www.aemo.com.au
97. Marot A, Kelly A, Naglic M et al (2022) Perspectives on future power system control centers for energy transition. J Mod Power Syst Clean Energy 10(2):328–344
98. Anderson PM, Farmer RG (1996) Series compensation of power systems. PBLSH! Inc, Encinitas, California
99. Iberdrola Australia Limited (2022) Iberdrola Australia submission to primary frequency response incentive arrangements. [Online]. www.iberdrola.com.au

Chapter 3
Frequency Quality

Frequency quality is an emerging topic in power system analysis [1]. In this chapter, we aim at (re)defining frequency quality, describe the various causes of frequency quality problems, and presenting different metrics, particularly those utilized by TSOs, to evaluate and quantify frequency quality in power systems. The chapter also discusses how TSOs monitor frequency quality and provides examples of its recent deterioration.

The chapter is organized as follows. Section 3.1 provides a general overview of frequency quality, discusses its current definitions and proposes a new definition and classification. Section 3.2 presents a list of metrics utilized by TSOs to quantify the quality of system frequency. Section 3.3 focuses on the importance of maintaining a good level of frequency quality. Section 3.4 describes at a high-level the main causes of frequency quality problems. Section 3.5 discusses the main reasons for the recent frequency quality degradation worldwide. Section 3.6 describes how TSOs monitor and report on both short- and long-term frequency quality. Section 3.7 provides two illustrative examples. The first example discusses the long-term frequency quality challenges the CE power system faces. The second example illustrates a recent IBR-driven oscillation which is another type of frequency quality issue for TSOs. Section 3.8 provides concluding remarks on the importance of engagements with industry and the Regulatory Authoritys (RAs) when developing technical solutions to any power system problems.

3.1 Overview, Definition and Classification

Frequency quality has been an issue since the early days of power systems due to the absence of interconnection and simple manual frequency control from single power plants [2]. Indeed, the choice of today's single standard frequency, that is, 50 or 60 Hz, was based on practical quality issues such as impact on lighting flicker and

T. Kërçi and F. Milano, *Frequency Quality in Power Systems*, Power Systems,
https://doi.org/10.1007/978-3-032-04721-2_3

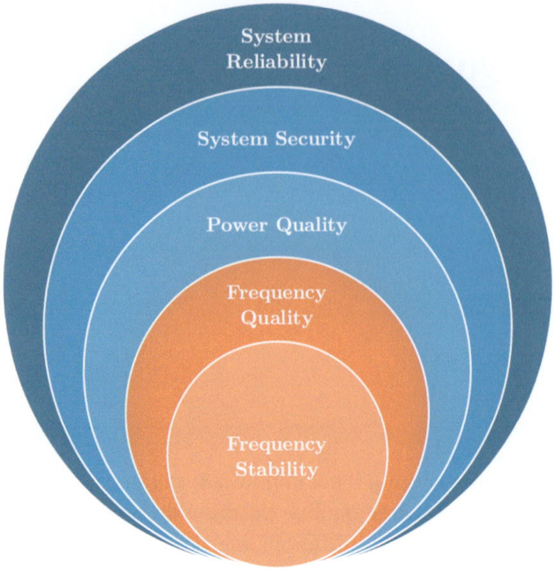

Fig. 3.1 System reliability and its subsets

transformers [3]. Frequency quality became less of a concern since power systems were interconnected, that is, from mid-1920s to 1990s [4], so much so that anecdotal experience says that system frequency used to be "flat" some 30–40 years ago compared to frequency variations experienced nowadays.

Frequency quality is a subset of the more general concept of *power quality*. This has been defined as *any power problem manifested in voltage, current, or frequency deviations that results in failure or mis-operation of customer equipment* [5]. Traditionally, however, the focus of power quality has been on voltage fluctuations, in particular on waveform distortions such as harmonics, sags, flickers and unbalances [6]. Reference [7] dedicates some space to frequency variations but considers that the frequency in a power system is substantially the same everywhere. Note that conventional power quality focuses on customer equipment, not on the system. Power quality is a subset of system security and reliability. Figure 3.1 shows a taxonomy of system reliability and its subsets, which include frequency quality.

Power quality has been one of the most popular topics in the power system community in the period from the late 1980s to the early 2000s. Then the interest eventually shifted toward other topics, in particular, RES, and distributed inverter-based generation. By the mid-2010s, mainstream research and industry considered power quality as a substantially "solved" topic. However, as power systems have moved toward extremely high shares of distributed, inverter-based renewable generation that reduce the inertia of synchronous machines to unprecedented low levels, power quality has gradually made its way back to the scene. The main focus is now on frequency variations, both local and system-wide, which have become a critical issue for the control, operation, and planning of power systems.

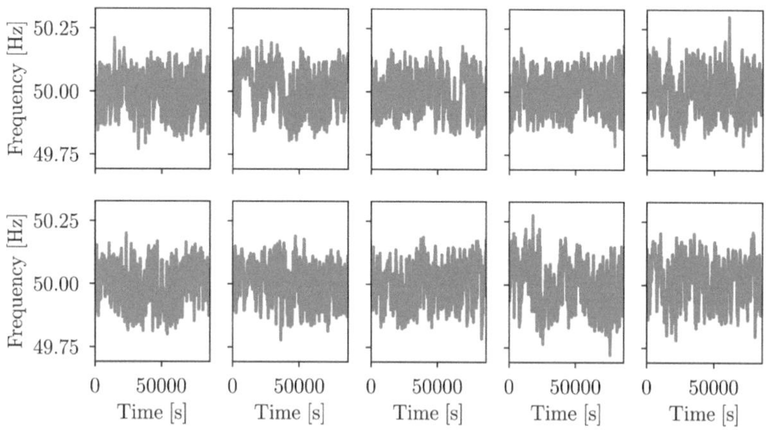

Fig. 3.2 Frequency in the GB power system in the period from April 1 to 10, 2024

Except after large contingencies, frequency variations are small, usually in the range of a few mHz in large interconnected power systems and tens of mHz in small island power systems. This is illustrated in Fig. 3.2 which shows the frequency in the GB power system for the first ten days of April 2024. While no day looks exactly the same, frequency is fairly predictable as it shows a stochastic behavior around 50 Hz.

Nevertheless, it is not always straightforward to interpret the severity of frequency variations. The same frequency variation, in fact, can be considered small or large depending on the size of the power system and the time scale at which the variation occurs (see also the discussion in Chap. 1) as well as on the operational limits that the system operator imposes on the frequency. Figure 3.3 shows a 5-minute intervals of real-world frequency measurements data from the AIPS that are plotted at four different y-axis scales. The bottom-right panel uses as y-axis limits the frequency nadir and zenith standards of the AIPS. Looking exclusively at the bottom-right panel, one may argue that frequency is "constant" in the considered period. However, if one looks at the top-left figure, the frequency appears to "vary significantly."

A similar illustration of different frequency variations is shown in Fig. 3.4 where four consecutive five-minute frequency data are plotted using similar y-axis limits. Frequency variations have more or less the same magnitude in all periods. However, the figure also shows quite different dynamic behaviors, such as ramps and fluctuations around the average. These behaviors are driven by different events and stochastic processes such as those due from data centers (see Chap. 7) or the RES (see Chap. 10).

Before further discussing why frequency quality is a relevant issue in modern power systems, it is useful to provide a good definition, the need of which has been recognized in recent works on the topic [8].

The Oxford English Dictionary definition of the word "quality" is [9]: *the standard of something when it is compared to other things like it; how good or bad something is.* Quality is a relative concept as it involves the comparison of something with a

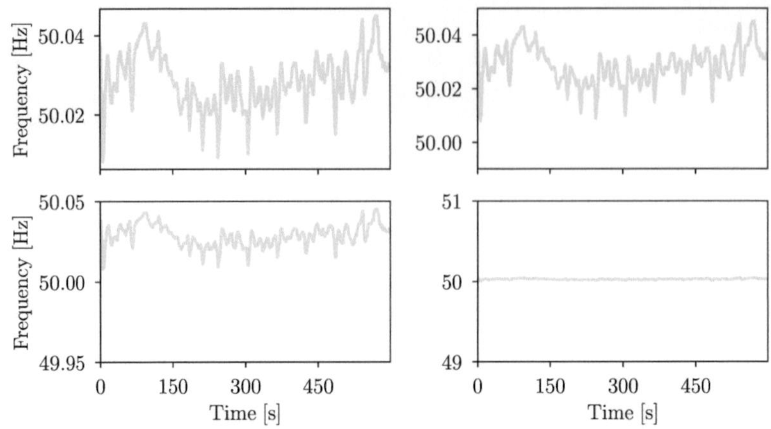

Fig. 3.3 Five-minutes frequency data plotted with different min and max y-axis limits (data provided by AIPS)

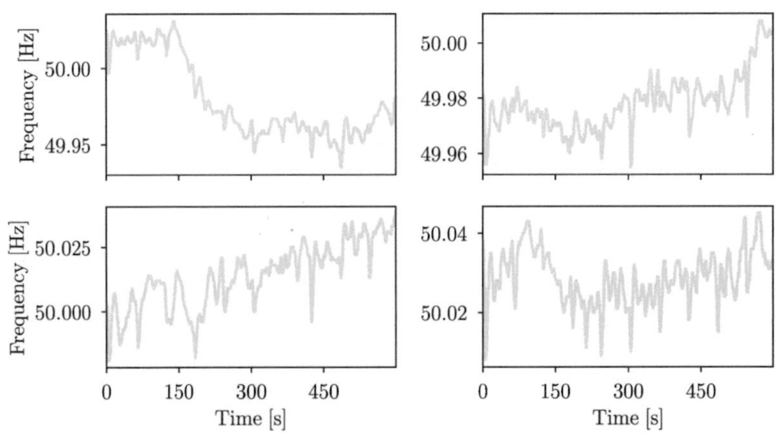

Fig. 3.4 Four consecutive five-minute frequency data (data provided by AIPS)

reference, which is assumed to be "good." The quality of a power system frequency, thus, measures how good or bad frequency is compared to, for example, the nominal one. However, similar to the frequency control and frequency regulation terminology, there is no common definition and use of the term "frequency quality." Moreover, while it is used extensively in Europe, the term "frequency quality" is not commonly utilized in the rest of the world.

The European legislation on electricity transmission system operation, the SOGL, defines many frequency quality targets (see below in the chapter) that allow the evaluation of the quality of the system frequency but it "forgets" to define the term itself in the same document [10]. Rather, it defines it differently in different documents. Specifically, in [1] ENTSO-E defines frequency quality as *a measure of the power systems ability to maintain a stable operation during the changes of consumption and production that are creating imbalances; and to handle disturbances.* This definition explains frequency quality as the capability of a power system to maintain the power balance rather then the deviations of the frequency from the nominal one [11]. The same report states that *frequency quality is mainly caused by insufficient balancing between production and consumption, and measured as "minutes outside the normal operating band.* In a similar report a couple of years later ENTSO-E mentions that *frequency quality is a measure on how well the TSOs maintain the balance in the synchronous transmission system.*

The above definitions are based on the understanding that frequency quality measures the ability of power systems to maintain active power imbalances during normal and abnormal operating conditions. This interpretation provides useful insight into the meaning of frequency quality. However, the focus is on active power imbalances rather than on frequency itself and does not include an explicit comparison as expected from the Oxford English Dictionary definition.

Once an unambiguous definition of frequency is accepted (see Chap. 1), then there is no need to mention active power imbalances to define its quality. This is the approach used, for example, in the definition of power quality and voltage/current quality in [6]: *power quality is concerned with deviations of the voltage from its ideal waveform (voltage quality) and deviations of the current from its ideal waveform (current quality).* In the same vein, we give the following definition of frequency quality:

> **Frequency quality** is concerned with deviations of the frequency from predefined limits/standards during normal and abnormal operating conditions. It is ensured by having adequate frequency response in place at all times.

The main features of this definition are: (i) it captures the standard and how frequency quality is measured; (ii) it defines the period during which the frequency quality is evaluated; and (iii) it indicates how frequency quality is maintained within limits. Moreover, this definition explicitly mentions normal and abnormal operating conditions. This point is important as several power system engineers and researchers relate frequency quality only to normal operating conditions.

Figure 3.5 illustrates the proposed definition for frequency variations considering normal and abnormal system conditions and the various stages of frequency control that ensure that frequency quality stays within predefined limits. It is worth highlighting that frequency stability is a subset of frequency quality, or, in other words, frequency stability refers only to frequency deviations during abnormal operating conditions (contingency events) [12].

In terms of time scales, frequency quality can be broadly split into *short-term* that spans from 0_+ seconds to minutes, and *long-term* that spans from minutes to years, see Fig. 3.6. The minutes limit could, for example, represent the maximum time to restore frequency to the frequency restoration range (see Table 3.1) following the occurrence of an instantaneous power imbalance that is smaller than or equal to the Reference Incident (RI) (or the largest single contingency). In other words, it is assumed that after 15 min (maximum limit) following a contingency the frequency

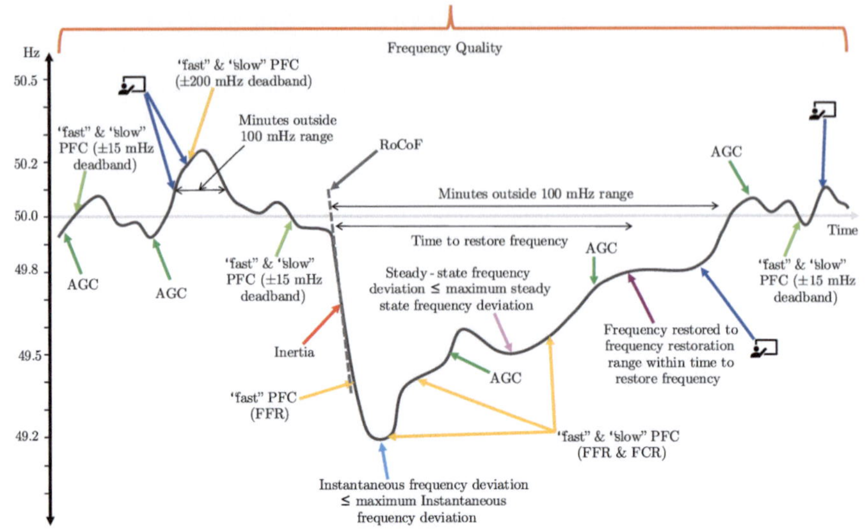

Fig. 3.5 Illustration of frequency quality concept

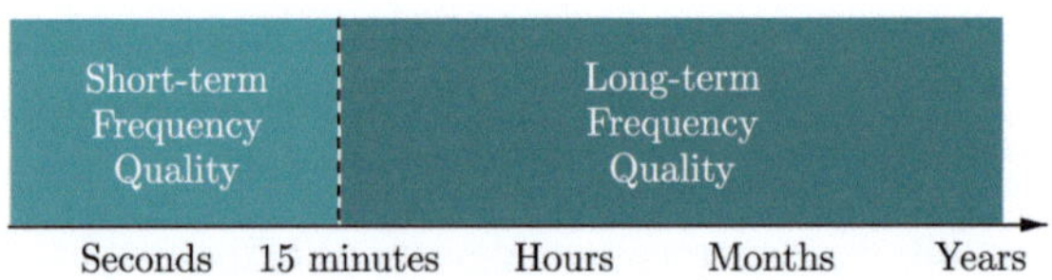

Fig. 3.6 Time scales of frequency quality

will return to normal frequency quality limits and, thus, can be classified as long-term. It is worthwhile to mention that in practice, and especially in power systems that employ FFR and AGC, frequency will tend to recover much quicker to normal limits (see Chap. 5). This time scale classification is considered useful as based on our experience, and similar to the definition above, power system engineers/researchers tend to relate it as a long-term phenomena only.

3.2 Quantifying Frequency Quality

Different TSOs use different metrics (and a different number of metrics) to quantify the quality of system frequency. Table 3.1 presents a variety of metrics utilized by different jurisdictions. Some metrics not only include ranges but also the time durations in which the respective ranges should be reached. Table 3.1 suggests that while the TSOs in Europe utilize a comprehensive range of indices to measure quality, that is not the case in the US. Specifically, in the US, the TSOs, known as independent system operators, focus more on long-term frequency quality performance rather than short-term ones. That is, NERC uses long-term CPS as an indicator of suffi-

Table 3.1 Key frequency quality metrics for the CE, GB, IE/NI, Nordic, US and Australian jurisdictions [10, 18–20]

Metric	CE	GB	IE/NI	Nordics	US	Australia
Std. frequency range [mHz]	±50	±200	±200	±100	–	±150
Max. instantaneous freq. deviation [mHz]	800	800	1000	1000	–	1000
RoCoF [Hz/s]	–	0.125–1	1	–	–	1
Max. steady-state freq. deviation [mHz]	200	500	500	500	–	500
Time to recover frequency [min]	Not used	1 min	1 min	Not used	–	–
Frequency recovery range [mHz]	Not used	±500	±500	Not used	–	–
Time to restore frequency [min]	15	15	15	15	–	–
Frequency restoration range [mHz]	Not used	±200	±200	±100	–	–
FRCE level 1	30	3	3	30	–	–
FRCE level 2 [%]	5	1	1	5	–	–
Alert state trigger time [min]	5	10	10	5	–	–
No. of threshold crossings	–	–	–	✓	–	✓
Amount of frequency oscillation	–	–	–	✓	–	–
UFLS events in 40 years	–	–	–	< 1	–	–
Frequency area	–	–	–	✓	–	–
Frequency step around the hour shift	–	–	–	✓	–	–
Length of frequency path	–	–	–	✓	–	–
Max. no. of minutes during a year outside the standard frequency range	15,000	15,000	15,000	15,000	–	–
Total time outside the ±100 mHz frequency range during a year [%]	–	≤98%	≤98%	–	–	–
Total time outside the ±150 mHz frequency range during a year [%]	–	–	–	–	–	≤99
Mean [mHz]	✓	✓	✓	✓	–	–
σ_f [mHz]	✓	✓	✓	✓	✓	✓
Δf [mHz]	–	–	–	–	–	–
Target σ_f of one-minute average [mHz]	–	–	–	–	≤18–30	–
Asymmetry of freq. distr. $\Delta\sigma_f$ [mHz]	–	–	–	–	–	–
Time to nadir/zenith [s]	–	–	–	–	–	–
Electrical time deviation [s]	20	✓	10	30	–	–
Percentile-based (the 1-, 5-, 10-, 90-, 95- and 99-percentile)	✓	✓	✓	✓	–	–
Total time instantaneous freq. deviation > Std. frequency range [min]	✓	✓	✓	–	–	–
No. of events freq. deviation > 200% Std. below 50% restoration range [min]	✓	✓	✓	–	–	–
Minutes outside normal operating freq. band during normal (abnormal) system conditions	–	–	–	–	–	≤ 5 (5)
Min. control performance standard 1 (CPS1) during a year [%]	–	–	–	–	≥ 100	–
Min. Balancing Authority ACE limit [MW]	–	–	–	–	✓	–
Interconn. freq. response obligation [$\frac{\text{MW}}{0.1\,\text{Hz}}$]	–	–	–	–	(−1042, −395)	–

ciency SFC [13]. The CPS is also used in other large interconnected power systems such as China [14, 15]. The definition of the CPS1 is [16]:

$$\mathrm{CPS1} = (2 - \mathrm{CF})\,100\%\,, \tag{3.1}$$

where CF is defined as:

$$\mathrm{CF} = \frac{(\mathrm{CF}_{1\mathrm{min}})_{12\mathrm{months}}}{(\sigma_f)^2}\,, \tag{3.2}$$

where $\mathrm{CF}_{1\mathrm{min}}$ is defined as:

$$\mathrm{CF}_{1\mathrm{min}} = \frac{\mathrm{ACE}_{1\mathrm{min}}}{-K_{\mathrm{bias}}}\,\Delta f_{1\mathrm{min}}\,, \tag{3.3}$$

where CF is the frequency-related compliance factor, $\mathrm{CF}_{1\mathrm{min}}$ represents the average CF within a minute, $(\mathrm{CF}_{1\mathrm{min}})_{12\mathrm{months}}$ is the $\mathrm{CF}_{1\mathrm{min}}$ obtained over 12 months, σ_f is the target standard deviation of frequency of one-minute average, $\mathrm{ACE}_{1\mathrm{min}}$ is the average ACE within a minute, K_{bias} is the balancing authority's bias factor (MW/0.1 Hz) and $\Delta f_{1\mathrm{min}}$ is the average Δf within a minute.

The use of long-term frequency quality indices can be explained by the fact that being large interconnected power systems with huge inertia, short-term frequency quality issues such as RoCoF are not a problem. On the other hand, AEMO in Australia uses similar indices to TSOs in Europe. Most of these frequency quality indices are implemented in legislation, for example, in the SOGL in Europe. As such, there may be legal consequences for TSOs if standards are not met.

Most importantly, frequency quality cannot be fully captured with a single parameter. Indeed, there are at least six critical indices, as follows:

- RoCoF;
- Nadir;
- Zenith;
- Minutes outside the standard frequency range, say, during a year;
- Standard deviation of the frequency, σ_f; and
- An oscillation-based metric similar to what the Nordic TSOs use given oscillations are becoming an emerging concern for TSOs.

Note that nadir, zenith and RoCoF together define frequency stability. Moreover, the minutes outside the standard frequency and σ_f define long-term frequency quality.

For this reason, any mention of "frequency quality" should also indicate which metric(s) is(are) used to quantify it and what type of frequency quality is evaluated, that is, short-term or long-term. A valid question is whether one needs all these parameters or others used in the literature [11] to quantify frequency quality. Another compelling question is what is the origin of different frequency quality limits. For instance, based on historical research the Nordic TSOs have found out that the standard frequency range limit of ± 100 mHz was set due to: (i) it is a reasonable band of operation for units equipped with electro-hydraulic governors; and (ii) it results in an increased margin for activation of PFC reserves for disturbances (FCR-D reserves). Similarly, ENTSO-E suggests the following reasons for setting the 15,000 min threshold as the maximum number of minutes outside the standard frequency range per year per synchronous area [17]:

- CE: Standard set based on a probabilistic risk calculation for exhaustion of FCR reserves due to a combination of a persisting frequency deviation and a sudden disturbance due to a generator or HVDC tripping.
- Nordic: Standard set based on historical data and an evaluation of increased deterministic and stochastic imbalances resulting from further energy market developments.
- GB: Standard set based on the number of occasions the frequency could deviate outside the standard frequency range per year and that the frequency should return to this range within 10 min.
- AIPS: Standard set based on system performance incentives related to system frequency quality at the time.

Despite setting the above standard based on careful considerations of the behavior of the different synchronous areas, ENTSO-E recognizes a need to keep the possibility of using modified values in the future, especially as Europe is transitioning towards a more sustainable energy market, and the integration of a large amount of RES.

It is interesting to observe the wide range of most of the metrics of IE/NI compared to the metrics defined for the CE control area. For example, the standard frequency range for the CE and IE/NI control areas is ± 50 mHz and ± 200 mHz, respectively. These metrics make sense considering the CE synchronous area accounts for around 435 GW of peak demand (the largest synchronous electrical grid in the world) compared to approximately 7.5 GW of the AIPS. That is, it is harder to control frequency and RoCoF in a small power system. For example, due to the relatively big size of the Nordic power system, the Nordic TSOs do not see RoCoF as a critical aspect following the RI because it is never big enough to cause generators to disconnect [21]. On the other hand, a relatively small power system such as AIPS, which is not meshed with another synchronous system, has RoCoF as the main stability issue [22]. Moreover, the Largest Single Infeed (LSI)/Largest Single Outfeed (LSO) and the RI represent a significant percentage $(20 - 30\%)$ of total demand in AIPS. To better illustrate these differences, Fig. 3.8 shows the frequency trace of the Albanian TSO (part of CE) and AIPS (IE/NI) on January 1, 2021. As expected, frequency fluctuates much more in AIPS than in CE power system.

In general, RoCoF is not explicitly considered as a frequency quality metric by TSOs [10]. This is somehow surprising given that RoCoF:

- is an important indicator of the robustness of an electrical grid;
- is used as key operational stability constraint/metric in practice;

- involves significant costs to keep it within limits, for example, the cost related to bringing inertia online, the cost to reduce the LSI/LSO and the costs of resultant RES curtailment [23].
- is employed as an anti-islanding protection technique [24];
- is usually the main binding frequency stability constraint in low-inertia systems (see Chap. 6);
- presents a significant risk to different power system equipments such as gas-fired synchronous machines, if too high [25].

For these reasons, we believe that RoCoF is a key indicator of frequency quality.

The other metric that is not used in practice by TSOs is $\Delta\sigma_f$. However, this is included in Table 3.1 as it is an important parameter to measure the asymmetry of frequency distribution and the strength of frequency control, both of which can be seen as frequency quality aspects (see Chaps. 4 and 5).

Then, we note that only the Nordic TSOs recognize oscillations as a frequency quality indicator and use an explicit parameter to quantify the impact. For example, apart from causing wear and tear on synchronous machines, oscillations have a direct impact on time outside the standard frequency range as well [20]. It appears sensible, thus, to use an oscillation-based metric for frequency quality. This metric can be, for example, the amount of frequency oscillation used by the Nordic TSOs.

Some of the parameters included in Table 3.1 quantify the level of frequency quality during disturbances while others during normal operating conditions. These metrics are summarized in Fig. 3.7 into short-term (up to 15 min) and long-term ones (a year).

While TSOs utilize both short and long-term parameters to measure and monitor frequency quality, the literature has, so far, mostly focused on short-term aspects of frequency quality and stability, and very little attention has been given to how to maintain long-term frequency quality parameters within limits. This is understandable given the importance of dealing with contingencies. But, in recent years, it has become increasingly difficult for TSOs operating both small and large-scale grids to keep frequency within limits during normal operating conditions. For example, despite being one of the biggest power grids in the world and using both PFR (FCR with $\pm$ 10 mHz dead-band) and AGC, the CE system frequency spent more than 15,000 min outside $\pm$50 mHz range in 2023. Similarly, due to increased intra-interval fluctuations and limited ramping from generators, the frequency deviations in a provincial power system in China increased from 0.019 to 0.032 Hz from 2014 to 2020, which is a whopping 68.4% increase [26].

The RA and TSO in New Zealand also predict challenges with keeping frequency within normal limits [27]: *An increasing amount of variable and intermittent resources, primarily in the form of wind and solar photovoltaic generation, is likely to cause more variability in frequency within the "normal band" of 49.8–50.2 Hz, which is likely to be exacerbated over time by decreasing system inertia.* Three main options are being investigated to address the above challenge in New Zealand's power system as follows:

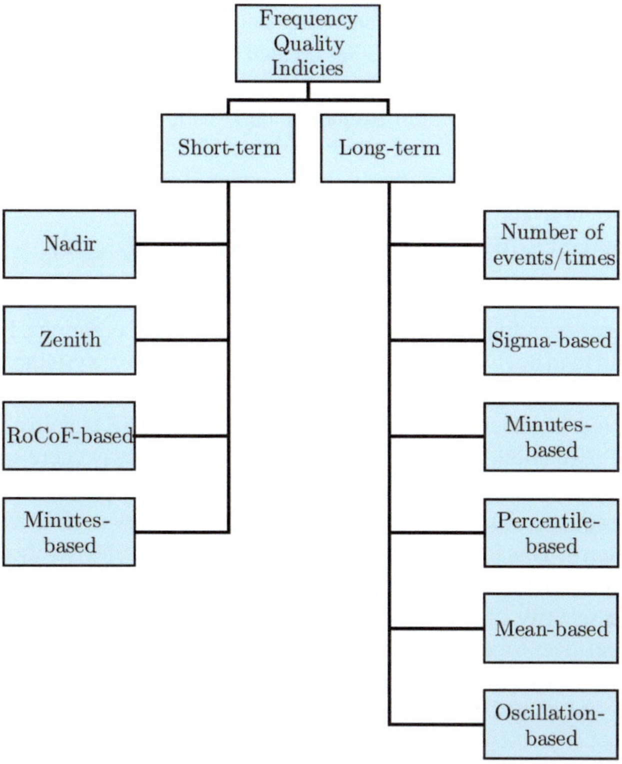

Fig. 3.7 Common metrics used to quantify frequency quality

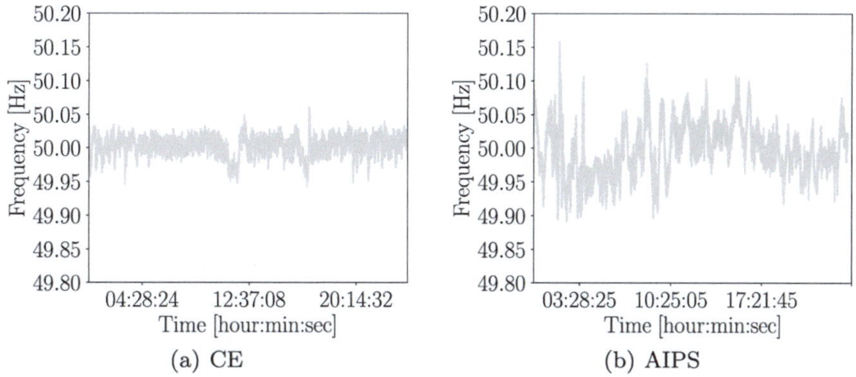

Fig. 3.8 Frequency in the CE and AIPS power systems on January 1, 2021

- Lower the 30 MW threshold for generating stations to be excluded by default from providing frequency support. This has created an incentive for generators to build generating units that export less than 30 MW.
- Stetting a permitted maximum frequency dead-band that generating units must contribute to PFR. Recall that while a narrow dead-band reduces the costs of reserve procurement, it leads to higher equipment life cycle costs for generating units due to more active frequency response.
- Procure more SFR/AGC reserves.

Finally, it is important to remember that the parameters that affect frequency quality change depending on the network and the operating conditions. For example, frequency will tend not to move much away from nominal if the system inertia is high and PFR and SFR during peak demand and low RES generation. On the other hand, if there is little inertia and most generation is provided with RES that do not contribute to frequency regulation, the frequency will tend to drift away more quickly from nominal. For this reason, night hours might show larger frequency variations than day hours. Frequency quality, thus, can be viewed as a "time-varying" characteristic of power systems. This feature of frequency quality is comprehensively illustrated in the remainder of the book. If one then considers that frequency quality is an essential aspect of the security of supply level, it could be argued that also the security of supply changes over time.

3.3 Importance of Frequency Quality

Electric power systems are unique in that production and consumption must be matched instantaneously and at every instant. Because frequency reflects this balance, it corresponds to the "heartbeat" of a power system. Power systems are thus designed to operate with frequency being close to the nominal. TSOs strive to maintain high frequency quality in order to reduce the risk of supply interruptions caused by power imbalances (primary objective). However, there are quite a number of other more specific reasons why it is a good idea to ensure good frequency quality level at all times. The consequences of lax frequency quality include but are not limited to [25, 28–32]:

- Blades on synchronous turbines may be damaged.
- Wear and tear on generator governing equipment. Large frequency deviations can have a negative effect on generating units participating in primary control, due to the required fast changes in output power, especially if they happen often.
- Inefficient operation of generators resulting in increased emissions.
- Stable control of synchronous machines may be compromised, for example, gas-fired generation tripping due to lean blow out on high positive RoCoF).
- System stability may be compromised and could result in the collapse of the power system itself (by triggering protective system actions, for example).

- Potential OF tripping of IBRs and DERs and converter-based loads.
- Active and reactive injection errors in utility-scale asynchronous generation such as wind generation. These errors can be due to, for example, high RoCoF.
- ICs may deviate from their dispatch targets and during a contingency may lead to a cascading failure.
- Frequency controlled clocks will not keep time correctly.
- Measurements of quantities that require a certain level of accuracy may be compromised.
- Generation units moving away from their dispatched MW levels due to frequency drift leading to unscheduled power system operation.
- Dispatch of energy and inter-regional loads may no longer be optimized owing to measurement error.
- Load measurement and its billing may be affected.
- Impedance measurement may affect stable control of voltage.
- Feedback through stabilizers may affect voltage.
- Harmonic dependent measurement devices may be affected.
- Harmonic filter efficiency.
- Power factor increase/decrease.
- Locked rotor torque, increase/decrease.
- Friction and wind losses increases.
- The tripping on over-speed of synchronous generators and motors.
- High magnetizing currents in induction motors and transformers.
- The loss of auxiliary motor.
- Relevance of power system analysis that is generally performed at the nominal frequency is questionable.
- The stalling of generators and motors.
- The increase in the reactances of circuit elements.
- The loss of blocks of demand through OF.
- Decrease the trust of TSO customers and market participants on supply reliability and quality.
- Misuse of PFC or FCR, such as, for example, contingency reserves being used under normal operation.
- System damping issues in situations of low availability of PFC reserves.
- It could degrade the quality of the product being delivered to end users (too low and lights would flicker unacceptably, for example).
- It could result in overloading transmission lines as various generators try to restore system frequency impacting markets efficiency.
- Or failure to meet the applicable frequency quality limits.

The list above makes clear how important is for TSOs to ensure that frequency quality is maintained at all times. However, as opposed to synchronous generators, *frequency quality usually does not have a substantial impact on IBR-based generation, as these generation types often cover a large band of frequency* [33]. In practice, TSOs maintain frequency quality in two ways, as follows.

- By setting mandatory minimum frequency control requirements in network/grid codes. That is, for a user to be connected to the network, it has to show that it can provide, for example, PFC and SFC if required by the TSO. For instance, the EU Network Code on requirements for generators requires an FCR-like frequency response called Limited Frequency Sensitive Mode (LFSM). However, LFSM has a large dead-band requirement (in the range of $\pm(200, 500)$ mHz) and, thus, serves as an emergency power measure. Additionally, while LFSM is a mandatory function and not a market product, generators are not required to reserve any capacity to ensure the availability of the function. Instead, LFSM utilizes any free capacity that is available at any point in time [34]. Note that this approach may not be prudent from a system security perspective. For instance, if there is a significant UF event and RES are being operated at Maximum Power Point Tracking (MPPT), then these RES would not have any margin to increase generation through the LFSM function. For this reason, grid codes have traditionally required conventional generators to reserve a minimum capacity, for example, not less than 5% and 10% of registered capacity in IE and Sardinia, Italy, respectively [35]. We believe that the whole concept of operating RES at MPPT needs to be revisited by TSOs and the respective RAs.
- By procuring enhanced frequency control services through the TSOs ancillary/system services market. For example, TSOs operating converter-based renewable-dominated grids have recently introduced the FFR product to mainly deal with faster happening frequency nadir and zenith and, to a certain extent, arrest RoCoF as well.

The two options above are illustrated in Fig. 3.9. The approach in Fig. 3.9 is also applied to voltage management in practice.

TSOs have other emergency frequency control actions to deal with extreme frequency deviations such as automatic schemes such as UFLS or Over Frequency Generation Shedding (OFGS). However, these actions are generally referred to as *system defense measures* rather than options to maintain frequency quality [36]. TSOs do not account for these defense measures when dimensioning their frequency control reserve requirements. Figure 3.10 provides an example of the measures taken at different frequency thresholds in the Nordic power system [34].

Another factor that can impact frequency quality, especially following significant imbalances, is the self-regulation (voltage/frequency dependency) of loads or K_f in Eq. (2.36). Due to the uncertainty of the self-regulating effect of the loads [37], this term is often neglected by TSOs [36]. Other TSOs explicitly consider it when dimensioning their frequency disturbance reserve requirements. For example, Nordic TSOs assume a value of 200 MW when determining the FCR-D needs [38]. However, Nordic TSOs recognize that the self-regulating effect has a locational aspect. For instance, a 1000 MW loss of production in the south of Sweden will lead to a different system frequency response compared with a trip of a 1000 MW production in the south of Finland [1]. This is due the different power flow and resulting voltage dip. Concretely, "moving power" to the south of Sweden results in a larger voltage reduction which, in turn, leads to a larger reduction in load (due to voltage dependent

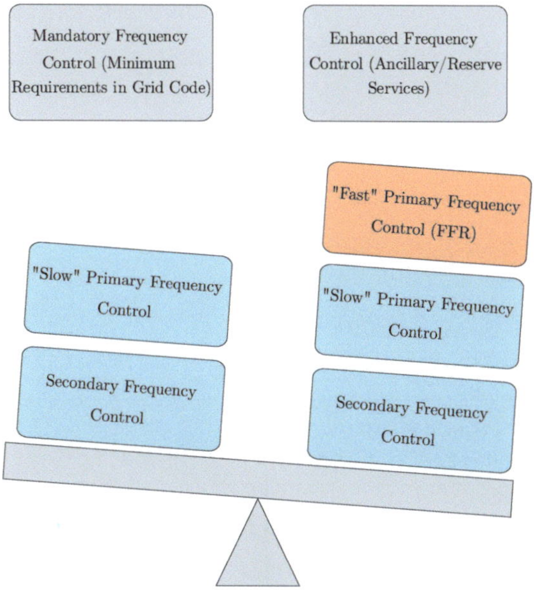

Fig. 3.9 Frequency quality management in power systems

loads). The different power imbalances in the two scenarios will affect both frequency nadir and steady-state value.

Network codes generally require that demand and generation units remain connected within specific frequency ranges which are lower than relevant frequency quality parameters. For instance, while the maximum frequency deviations (or nadir and zenith) in IE are 49 Hz and 51 Hz, demand and generation units are required to remain connected if frequency is within 47–47.5 Hz for 20 s (see Table 3.2). European Standard EN 50160 requires that – under normal operating conditions – the system frequency shall be within 49.5–50.5 Hz range for 95% of the time [38]. This standard is outperformed in practice in terms of frequency quality in Nordic and the AIPS. Indeed, excursions outside this frequency band are very rare and would represent a significant contingency event such as loss of two generating units, that is, $N-2$ rather than $N-1$.

3.4 Causes of Frequency Quality Problems

So far, we have discussed what we mean by frequency quality, how we can quantify it, and why it is important to maintain a good level of it. It remains to discuss the causes of frequency quality issues. On a high-level, these are summarized and illustrated in Fig. 3.11. Frequency quality is dependent on how these imbalances with different causes can be handled. In this section, we provide a brief overview of each of these different types of imbalances below and discuss their impact on frequency quality

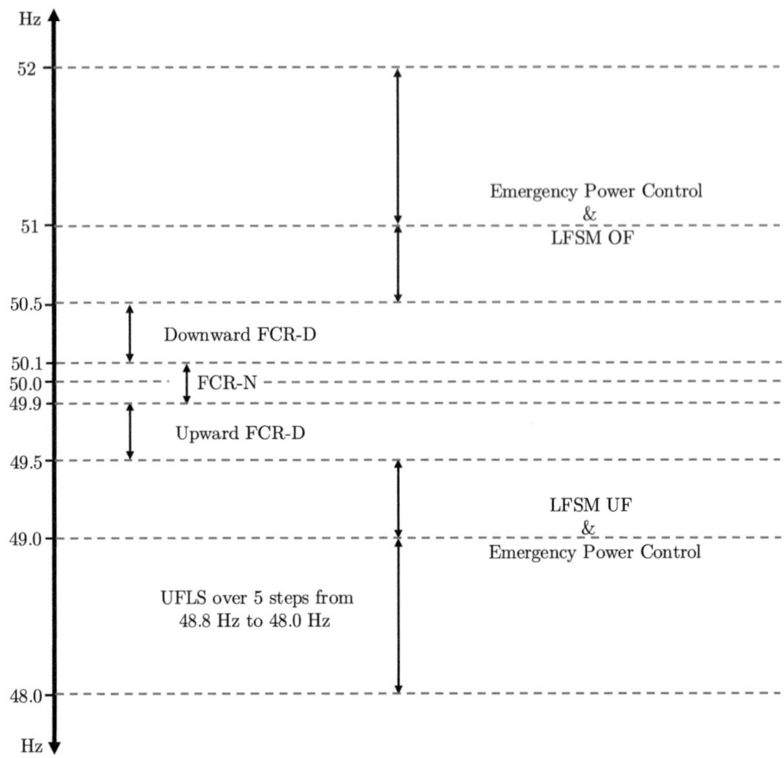

Fig. 3.10 Different measures taken at different frequency thresholds

Table 3.2 Minimum time periods for generation units to remain operational without disconnecting in IE

Frequency range [Hz]	Time period
47–47.5	20 s
47.5–48.5	90 min
48.5–49	90 min
49–51	Unlimited
51–51.5	90 min
51.5–52	60 min

throughout the book. We also provide detailed reasons why an ongoing frequency quality degradation has been recently observed in power systems worldwide. The focus is, in particular, on the CE power system.

Different TSOs may employ different "balancing philosophies", that is, *proactive* or *reactive* balancing [39], to deal with the different imbalances in Fig. 3.11. The former approach means the TSO uses forecasts and tries to solve the imbalances by

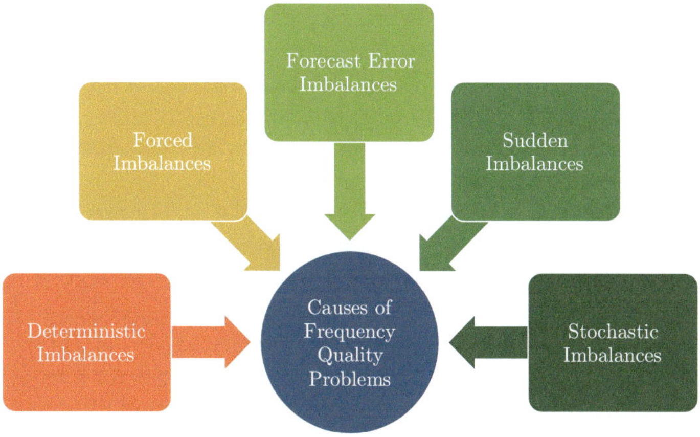

Fig. 3.11 Main causes of frequency quality problems

taking actions in advance such as using FRR to prevent future imbalances that are known with a high probability. This approach is used by some European TSOs with the main drivers being economic optimization/efficiency, or maintaining operating margins in systems with limited availability of fast resources, in particular automatic reserves [40]. In contrast, other TSOs may employ a reactive approach whereby they wait until imbalances arise in real-time and then take appropriate actions. This is done to provide strong incentives for market participants to reduce imbalances, thereby also reducing the need for balancing actions by the TSO [39].

3.4.1 Sudden Imbalances

Sudden imbalances are one of the less frequent imbalances but have the highest impact on frequency quality. The most well-known examples include the abrupt unexpected loss of a large production unit that leads to an UF event. Recently, sudden large demand disconnections/reductions are also increasing (see Chap. 7). TSOs deal with these types of imbalances by ensuring enough inertia and power reserve all of the time, for example, by procuring sufficient primary and secondary reserves. For example, a static method is used by TSOs when determining reserve requirements to deal with the LSI/LSO loss [41]. If these measures are not enough, then TSOs have designed defense measures such as UFLS and OFGS, to avoid complete blackouts.

Sudden imbalances may not always necessarily mean the disconnection of a resource unit. The introduction of massive amounts of RES can lead to significant sudden imbalances, up to hundreds of MW, in time scales of several minutes (see Chap. 11). Such imbalances lead to significant frequency deviations similar to that of a sudden loss of a generator/demand unit. In [42], it is reported that due to cloud passing in Western Australia in 2016, roof-top PV generation was reduced by some 400 MW in less than 30 min leading to system frequency falling to 49.5 Hz. These frequency deviations are easily comparable to a large generator trip. Note that these types of imbalances are usually referred to as (short-term) forecast error imbalances, but given their size and impact on system frequency we think they may be better grouped under the umbrella of "sudden imbalances."

Therefore, TSOs need to rethink the classification of "contingencies" and the fact that they may need to hold reserves and ramping margin to accommodate sudden ramps down and up of RES. This is what CAISO did in 2016 when it introduced a short-term (10-minute ramping capacity) flexible ramping product to help manage uncertainty from the load, wind, and solar production [43, 44]. In this context, the determination of reserves and ramping requirements calls for a more dynamic approach (compared to the static methods used for sudden loss of LSI/LSO) whereby the amount of reserve/ramping capacity is adapted according to observable conditions [41].

3.4.2 Stochastic Imbalances

Power systems involve many different stochastic processes such as continuous fluctuations of consumer behavior and RES generation. Stochastic imbalances (or volatility) can be considered as "fast variations" time-continuous stochastic processes compared to, for example, "slow variations" coming from forecast errors (uncertainty imbalances). In mathematical terms, stochastic processes are usually modeled as stochastic differential equations. These are combined with traditional deterministic models, thus leading to time-continuous stochastic Differential-Algebraic Equations (DAEs). This model is then utilized to study the dynamic behavior of power systems subjected to volatility.

Traditionally, due to stochastic imbalances involving small active power magnitude changes, TSOs were not worried so much about studying their impact on stability [45]. However, the ever-growing penetration of RES leads to significant volatility, the effect of which needs to be considered carefully. Despite stochastic imbalances involving small magnitude changes, they can have an impact on both short-term and long-term frequency quality. In [46], it is shown that stochastic imbalances can lead to frequency instability, in certain system conditions. Therefore, considering that such imbalances will continue to increase with the increase of IBR-based RES penetration and flexible loads, TSOs must have the necessary tools to study their impact on frequency quality. In this book, we provide a comprehensive analysis of the effect of stochastic imbalances on frequency quality (see Chaps. 8 and 10).

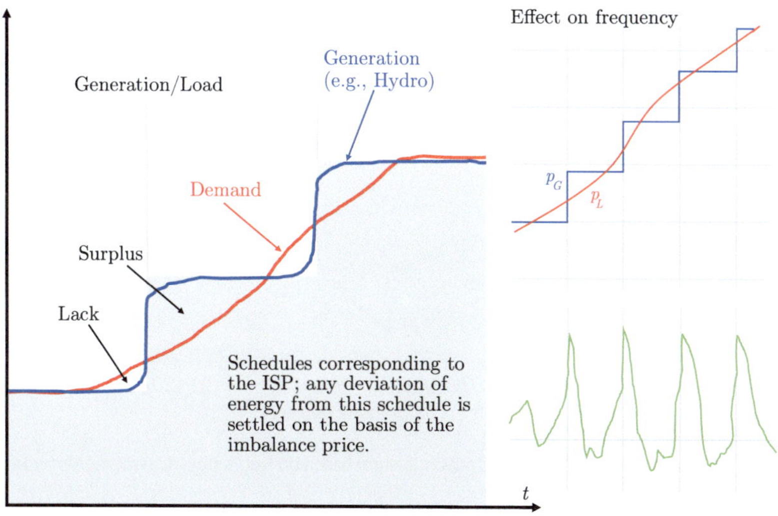

Fig. 3.12 DFDs phenomenon in the CE power system (adapted from [47])

3.4.3 Deterministic Imbalances

Deterministic imbalances happen at known given times. They can be considered as "foreseen" disturbances. In Europe, these are usually known as Deterministic Frequency Deviations (DFDs) and occur regularly at the beginning of hourly or sub-hourly intervals. The phenomenon of DFDs is mainly caused by the development and design of electricity markets, in particular, in European self-dispatch systems, where the scheduling and dispatch of generation and large consumption units are determined by the scheduling agents of those units rather than the TSOs. The latter is referred to as "central-dispatch" (see Chap. 10). Specifically, generation units tend to follow a stepwise behavior during a short period of time around the change of the hour rather than following the real-time physical conditions of the system. On the other hand, as it is well-known, the demand curve follows a more linear change which gives rise then to significant imbalances between generation and demand at the beginning of hourly or sub-hourly intervals of electricity trading. This behavior and its impact on system frequency is illustrated in Fig. 3.12.

While DFDs are caused by market actors, the latter do not violate any market rules. An illustration of actual DFDs in Europe is given in Fig. 3.13 which shows a few hours of the original CE frequency depicted in Fig. 3.8. The behavior of the frequency indicates that DFDs happen at the change of the hour and represent the largest frequency deviations.

Figure 3.14 provides a comparison of the magnitude of the DFDs between the CE (mostly self-dispatch systems) and AIPS (central-dispatch) power systems. The DFD magnitude is calculated using the concept of the *frequency step around the hour shift* utilized by the Nordic TSOs [48]. Specifically, it is evaluated as the absolute

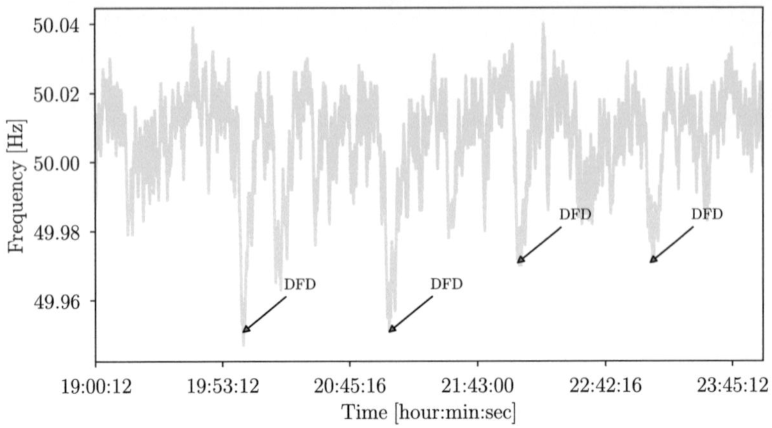

Fig. 3.13 DFDs phenomenon in the CE power system based on frequency measurements on January 1, 2021

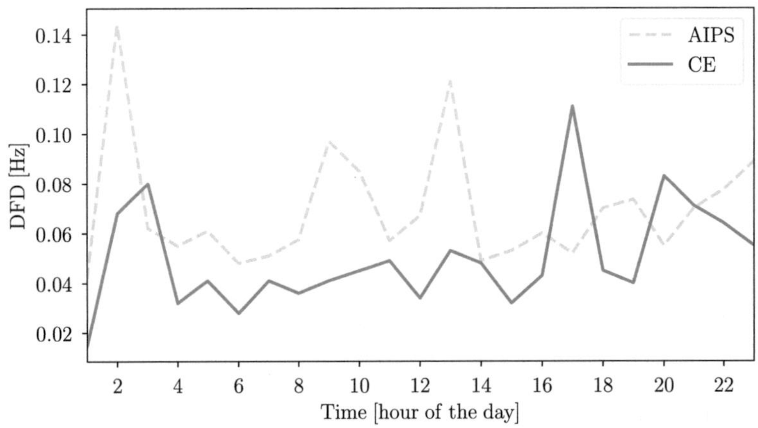

Fig. 3.14 DFDs phenomenon in the CE and AIPS power systems based on frequency measurements on January 1, 2021

difference between the highest and lowest frequency during the period from 5 min before to 5 min after the hour shift. Figure 3.14 suggests that the DFD magnitudes are clearly visible and comparable despite the CE power system being way bigger than the AIPS power system. While the magnitude of frequency deviations is much lower throughout the day in CE (see Fig. 3.8), around the hour shift they are comparable with those in the AIPS. This indicates the much larger impact of the market in CE power system compared to AIPS.

Surprisingly enough, market-induced load-generation imbalances or DFDs, not generation trips, are the main source of frequency deviations in Nordic and CE power systems [49]. ENTSO-E reports variations with peak-to-peak values up to 150 mHz within a time window of ten minutes centered on the change of the hour, which corresponds with the standard time interval for cross-border schedule changes [50]. For this reason, in [51], it is recognized that DFDs critically affect power grid frequency quality performance. As mentioned above, they are likely to be more evident in self-dispatch systems such as GB, CE, and Nordic systems [51] rather than central-dispatch systems such as AIPS and US power systems where the TSOs have full control from a scheduling and dispatch perspective from day-ahead to real-time (see Chap. 10).

In summary, the following are the main causes of DFDs [33]:

- The link between generation and consumption dynamics becomes weak as the market rules are based on the exchange of energy blocks of fixed time periods rather than physical conditions of the system.
- The physical ramping of generation and load is not aligned between all market participants. This imbalance is then reflected in the DFDs.
- Market rules also enforce Balance Responsibility Partys (BRPs) to be balanced over the whole market time unit. This rule makes BRPs to adjust their generation at the latest point in time by the fastest gradient possible.
- Last but not least, different national legislation such as noise emission constraints, night tariff changes, etc. can cause rapid changes in generation or load at specific hours.

DFDs are a real concern for TSOs as they can lead to the depletion of frequency control reserves. If this happens, it can make the power system more vulnerable to additional, unforeseen disturbances or failures. DFDs can also have a direct impact on the correct operation of synchronous generators (as developed in the turbine and gas engine industry) that have their drive-train coupled with grid frequency [33]. DFDs have been an already known phenomenon for quite some time now and, yet, no effective solution has been found to eliminate or, at least, reduce their impact. To make things even more difficult for TSOs, it is understood that increasing control reserves does not seem to substantially improve the situation; rather, and obviously, it increases system operation costs considerably.

Proposed solutions to reduce the impact of DFDs include, among others [33, 51, 52]:

- Impose ramping on generation and load market units on a local (national) level.
- Increase temporarily FCR volumes while a more efficient solution is found.
- Additional FFR reserves from BESS.
- Additional aFRRs;
- Introduce a dynamic frequency setpoint in control areas/blocks.
- Increase market resolution, that is, increase number of dispatch intervals from 1 h to 15 min.
- Balancing authority's bias factor (K_{bias}) adaptation.
- Forward activation of active power based on power imbalance forecasts.
- Impose a limit of change in net position of a market area and bidding zone between two successive market time periods;

- Introduce a spot power balancing (or residual central dispatch) such as in GB.
- Introduce a central dispatch in the CE similar to the AIPS.
- Use changes between Imbalance Settlement Period (ISP) for BRPs as market-based aFRRs bids;

When trying to address DFDs, but in general any power system frequency quality issue, the cost of the remedial action needs to be carefully considered as, ultimately, this cost will inevitably inflate consumer invoices. Note also that, while increasing the market resolution has the potential to reduce the magnitude of DFDs (see Chap. 10), it does not guarantee their removal. In the worst-case scenario, it may just shift the problem from happening every hour to every 15 min. Moreover, spot power balancing and central dispatch involve significant market changes and thus will be a major change politically in many control blocks in CE.

Finally, there exist different types of DFDs, not only market-driven ones. For example, Chap. 7 discusses DFDs caused by large demands such as data centers. These involve regular, cyclical fluctuations in data center demand due to potential digital processes run by data center customers in their sites. These fluctuations cause ongoing frequency quality concerns for TSOs.

3.4.4 Forced Imbalances

Forced imbalances are driven by an external forcing function acting on the system. Relevant examples include, among others:

- Unexpected equipment failures that require (automatic/manual) forced disconnection/outage;
- Control interactions that lead to significant frequency oscillations;
- Wrong functioning of AGC or wrong AGC parameters;
- Incorrect or poorly tuned generator controller;
- Operator-driven imbalances, for example, intentionally run frequency below/above nominal for TEC purposes;
- Telecommunication line failures such as in one or more tie-lines, between substations or between TSOs central controllers and substations leading to, for example, long-lasting steady-state frequency deviations [53]; and
- TSOs doing a poor job in controlling system frequency proactively.

Most non-intentional forced imbalances are considered dangerous from a TSO perspective as they do not explicitly procure reserves to deal with such imbalances. TSOs implement operational measures and policies when these types of imbalances appear in the system. For example, the Irish TSOs suggest reducing the dead-band of wind power plants to ±15 mHz, thus turning On APC [54], to help damp forced (low-frequency) oscillations appearing on the system. However, these actions might not be efficient in all cases (see Fig. 3.21), and therefore TSOs will have to learn and adapt to new scenarios as they arise.

3.4.5 *Forecast Error Imbalances*

As weather-dependent resources, RES, such as solar and wind generation, are both variable and uncertain at multiple time scales. For this reason, TSOs use regular forecasts of RES but also demand and the availability of dispatchable generation. As power systems are operated in real-time, there will always be uncertainty or forecast errors about the future. These errors or energy imbalances tend to increase as more RES are integrated and can lead, in general, to increased real-time balancing needs. In particular, high ACE and frequency excursions are expected.

TSOs address forecast error imbalances by procuring reserves and ramping products, for example, as is the case in the AIPS. Forecast errors lead to increased reserves/ramping volumes required. An interesting example is from Denmark, where Energinet, the Danish TSO, procures additional SFC/FRR reserves (both aFRR and mFRR) to tackle imbalances resulting from forecast errors. The following equation is used by Energinet [55]:

$$\mathrm{FRR} = \mathrm{SR} + \mathrm{RI} + \mathrm{IMB} - \mathrm{SH} - \mathrm{VB}\,, \tag{3.4}$$

where FRR represents the total aFRR and mFRR procured, SR represents FRR needs due to potential local congestion issues during reserves delivery, RI is the RI which for Energinet includes the trip of the largest resource unit, SH represents the sharing of reserves with other areas (that is to be subtracted), and VB represents voluntary reserve bids available in the balancing reserve market.

3.5 Reasons for Frequency Quality Degradation

In recent years, there has been a worldwide gradual deterioration of frequency quality. For instance, deterioration of frequency quality, including system-wide frequency and local frequency variations, has been reported for real-world power electronic-dominated power grids such as the European, Nordic, British, Irish, Australian, US, and Brazilian power systems, among others [56–60]. As mentioned above, the CE power system exceeded, for the first time, the maximum number of minutes outside the standard frequency range in 2023. Specifically, frequency spent 15,389 min outside the ±50 mHz range in 2023 [61]. In 2024, the total minutes outside the ±50 mHz range ended close to the limit, and the trend in 2025 suggests that the threshold will be violated again. Then, for example, in the AIPS, it is more difficult for TSOs to maintain frequency within the standard frequency range under intact conditions due to, for example, significant solar PV ramps in the minute time scales [57].

Projections suggest that this deterioration will continue during the shift toward RES production. For example, in the Nordic power system, it is expected that the events outside the normal frequency band will increase from 0.4% in 2019 to 1.1-

Table 3.3 Summary of key factors impacting on short-term (ST) and long-term (LT) frequency quality

Factor	ST impact	LT impact
RES variability and uncertainty	Low	High
Demand variability and uncertainty	Low	High
Reduced generators synchronous inertia	High	Low
Reduced load relief	High	Low
Oscillations	High	Low
Lack of frequency control and regulation from IBRs/DERs	High	High
Lack of FRT from IBRs/DERs and converter-based demand	High	Low
Lack of observability, controllability and forecasting of DERs	High	High
Lack of dynamic models for DERs and converter-based demand	High	Low
Delayed active power from IBRs/DERs and converter-based demand	High	Low
System splits	High	Low
Delayed network investments	High	Low
Aging of conventional generating portfolio	High	High
Increase of LSI/LSO	High	Low
Electricity markets	Low	High
Insufficient reserve volumes	High	High
Poor frequency regulation	Low	High
Extreme weather events	High	Low
Frequency/RoCoF definition/estimation/ measurement	High	High
Poor coordination, cooperation and information sharing	High	High

5.0% in 2040, with an average of 2.8% [62]. New phenomena have also appeared such as low/high-frequency oscillations and asymmetry in the frequency distribution [63].

Challenges related to frequency quality have already been widely discussed in the literature [64–74]. However, these works mostly focus on the short-term aspects of frequency quality such as inertia/RoCoF and frequency nadir/zenith and neglect long-term frequency performance.

While the literature mainly attributes the reduction of inertia from conventional synchronous generators as one of the main causes of (short-term) frequency quality degradation [67, 75], there are many other factors greatly influencing it as summarized in Table 3.3 and discussed in detail below.

Integrating variable and uncertain large-scale RES such as solar and wind generation is one of the most obvious and understood reasons for poor frequency quality.

However, RES represent an opportunity as well given their higher level of controllability and superior frequency control performance compared to conventional plants. Hence, TSOs, in collaboration with industry and academia, should further explore and trial the use of RES to support higher frequency quality standards. Meanwhile, frequency control and, in particular, PFC provision by the remaining conventional generators becomes more important than ever under high RES penetrations.

While IBRs such as BESS play a critical role in maintaining high-frequency quality standards, in particular, frequency stability, they can also be a source for IBR-driven oscillation. These oscillations can be considered as difficult "storms" for TSOs to manage given their frequency of occurrences, lack of experience to deal with them including the coordination with slow conventional resources, and the potential impact on stability. Therefore, there is an urgent need for TSOs to "weather these storms." For instance, fast-responding IBRs may lead to unintended oscillations when very high transient gains are used [59, 76]. In the same vein, the TSO in the Netherlands reports a significant number of oscillations (3.5 Hz) since Spring 2023 during high RES penetration (above 20 GW) and a reduced number of conventional generators online [77]. It is reported based on analysis performed so far that the oscillations are likely initiated by the voltage control of solar PV power plants. Key questions raised by the Dutch TSO include:

- How to locate the source(s) of oscillations especially if no PMU data are available.
- How dangerous are oscillations from a grid security perspective and what amplitudes can be tolerated.
- How to mitigate oscillations in real-time in terms of, for example, minimum number of conventional units online and testing of IBR units such as HVDC P/Q and V-/Q-control modes.

An excellent technical report [76] prepared by ESIG can help address the above questions and, in general, serve as a practical guide to identify the causes and determine the mitigation of oscillations observed in either field measurements or simulations.

Delayed network investments, that is, failing to build key network assets such as interconnection between different countries or control areas may result in serious frequency deviations. A relevant example here is from the AIPS where the lack of a second North-South (N-S) IC could lead to significant RoCoF values in IE and NI following a potential system split scenario [22].

As more countries are building interconnections such as HVDC cables in the order of GW to help with increased security of supply and better market integration also means increases in the LSI/LSO and the RI that the TSOs used to operate the system before. For this reason, inertia and reserves held in the system should increase accordingly (for example, % of LSI/LSO) to maintain a similar frequency quality level. However, it is noteworthy that HVDC links provide excellent frequency control and regulation. For instance, the Sardinia-Peninsula 500 kV HVDC link between Sardinia and mainland Italy is the main source of PFC, including PFR, in the Sardinian power system and ensures high-frequency quality standards. Additionally, future HVDC links will be equipped with GFM capability and thus provide an inertial-like response.

In general, there is a lack of frequency support from IBRs under normal and abnormal operating conditions. For instance, wind and solar PV are designed to operate at MPPT and, in practice, implement wide frequency dead-bands. This could be attributed to a lack of financial incentives for IBRs [78]. However, it should be noted that IBRs such as solar PV can provide frequency control through proper control techniques [79]: (i) solar PV deloading where the PV plant is operated at an operational point below the MPPT; and (ii) dc link capacitor response providing virtual inertia.

It is well-known that reducing PFC within the standard frequency range means that for any given supply-demand imbalance, the size and duration of the frequency deviation increases. In this case, it is likely that a given imbalance may result in frequency moving outside the standard frequency range and therefore increasing the risk of breaching the standard. For this reason, there needs to exist some balance between minimizing movement for IBRs and supporting grid frequency for normal, steady-state changes in frequency [80]. For instance, in most jurisdictions, IBRs are rarely used for frequency regulation on a daily basis. While one might argue that BESS can be the sole providers of frequency regulation, as discussed in the previous chapter, it should be noted that it is always better from a TSO point of view to have a wide range of different technology providers to choose from for frequency control/regulation. Therefore, more effort is needed to make IBRs like wind and solar generation provide PFR and SFR on a daily basis.

Regarding the integration of weather-dependent DERs such as rooftop PV which do not contribute to frequency control or regulation [81–83], the question is what will happen from a frequency control/regulation point of view when, for example, almost all load is met by distributed PV. Here, reference [84] suggests that MW range DERs connected to MV/HV levels are required to better provide frequency control compared to their smaller and more embedded counterparts.

Therefore, TSOs need to decide how to utilize DERs. For example, an option is the centralized concept of Virtual Power Plant (VPP) [82], to provide all the necessary frequency control/regulation. Key concerns to keep in mind here include the absence of appropriate market access such as the design of the right incentives for DERs to provide frequency support or mandatory minimum frequency control/regulation requirements as part of the DER connection processes. Another technical issue to consider is what should key frequency characteristics of DERs look like in terms of response time, dead-band, and droop. According to lab tests [81], DERs can modify their output power very quickly (on a sub-second time scale), with droop and dead-band values between 3–5% and ±17-36 mHz (could provide frequency regulation), respectively. Nevertheless, real-world field trials are needed to build the confidence and trust of the TSOs.

Demand is becoming more volatile due to the electrification of heat and transport. For example, increased penetration of Electric Vehicles (EVs) and their associated charging infrastructure along with heat pumps will make it harder to maintain the electricity demand-supply balance (see Chap. 7), hence, raising the risk of black-outs [85, 86].

Connection of new large converter-based demand such as traditional and Artificial Intelligence (AI) data centers with cyclical fluctuation of their demand are also impacting frequency quality negatively as reported in the AIPS and US power grids [58] (see Chap. 7).

The lack of Fault Ride Through (FRT) capability from converter-based demand and generation such as data centers and DERs [87] is expected to lead to trips of IBRs/DERs when disturbances or the loss of LSI/LSO unit occur in the power system. These trips or Consequential Losss (CLs) will then worsen the impact on contingency size [88–90]. In this case, TSOs need to adapt their minimum inertia/reserve requirements to account for such additional active power losses (see Chap. 12).

Lack of observability, controllability, and forecasting of DERs. For instance, AEMO has recently reported significant challenges in managing high penetration of distributed solar PV such as roof-top solar that have pushed out other generation sources such as large synchronous units that provide frequency control [91]. The Irish TSOs expect similar challenges in the near future for the AIPS [92].

There is a need to implement remote management and curtailment of distributed PV as an emergency backstop to maintain system security. Moreover, increased visibility and forecasting of DERs is still in the early stages which means potential negative impacts on both short and long-term frequency quality. For instance, if TSOs do not have good DERs visibility, they could operate the system outside operational limits such as the System Non-Synchronous Penetration (SNSP) limit. Therefore, there is a clear need to develop and enhance/improve the three mentioned DER areas above.

Lack of accurate dynamic modeling of DERs and converter-based demand. The vast majority of TSOs still generally apply simplified modeling of DERs such as negative demand, thus, neglecting their dynamic behavior and impact on stability. Similarly, there are no accurate dynamic models for large converter-based demand such as traditional and AI data centers. For instance, AI workload dynamics can exhibit large load swings on millisecond time scales (see Fig. 7.1). Therefore, there is a need to capture the dynamic behavior of these nonlinear fluctuating loads through improved modeling and, thus, make studies accurate so that the TSOs can better anticipate potential issues.

As traditional loads such as industrial motors and pumps are becoming rare in power systems, the inertia being provided by them, known as "load relief", is consequently reduced. For instance, AEMO has recently reduced the load relief assumption considered when determining total reserve requirements (size of largest credible contingency minus load relief) from 1.5% in 2019 to 0.5% in 2023 [93].

There is evidence that indicates that, unlike synchronous generators that recover their active power on the order of ms, some wind farms can show much higher delayed active power recovery following a fault or voltage dip [90, 94]. This is a concern at high levels of RES and low-inertia conditions as the delayed active power recovery will cause the system frequency to fall significantly causing RoCoF problems as well. This is also known as the Voltage Dip Induced Frequency Dip (VDIFD) phenomenon. Two relevant solutions to this issue are: (i) introduce ancillary services to tackle VDIFD such as the ones in the AIPS namely Synchronous Inertial Response (SIR),

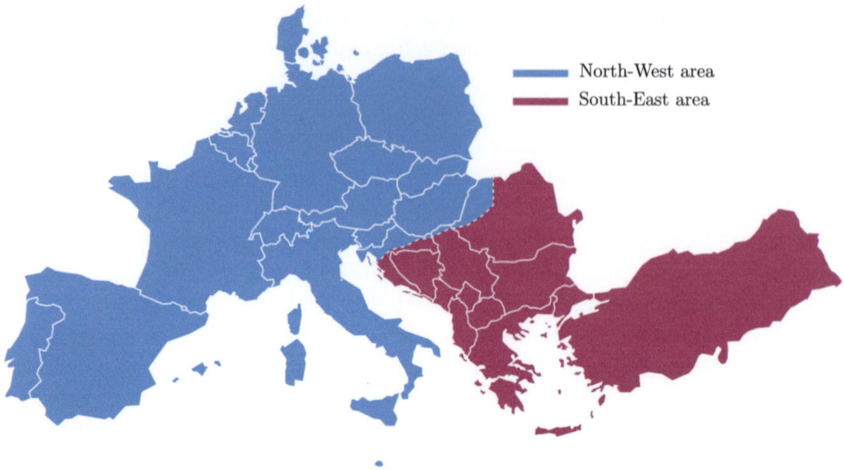

Fig. 3.15 Resulting synchronous areas after the CE system split [96]

FFR, fast dynamic reactive response to mitigate fault propagation, and the fast post-fault active power recovery; and (ii) review active vs reactive power priority of IBRs such as wind farms during the fault. For example, in IE, active power provision is prioritized by wind farms. The former is a market-based solution while the latter is a mandatory grid code requirement. In any case, whatever solution is implemented, it should not exclusively address frequency-related issues and neglect other security issues. For example, voltage issues are anticipated to become more prominent in the future.

System split events are low-probability high-impact events and difficult to model. For this reason, they are one of the most difficult ones to deal with from a frequency quality but also security/reliability point of view as depending on system conditions in the different areas, frequency limits can be easily breached and thus lead to potential blackouts.

The three most recent and significant worldwide system split events occurred in CE where Spain and Portugal were disconnected from the rest of the CE system on April 28, 2025, Chile on February 25, 2025 and in Europe on January 8, 2021, where the CE power system was split into two synchronous areas. The Chilean event was driven by protection systems malfunctioning that tripped a double circuit 500 kV transmission line connecting north and south of the country which then resulted in a system separation and nationwide blackout [95]. Another relevant event include the CE system split on January 8, 2021, where the CE power system was divided into two main synchronous areas namely OF in the South-East area and an UF in the North-West area, see Fig. 3.15 [96]. Motivated by potential system split incidents and their impact on frequency stability, ENTSO-E recently proposed new RoCoF standards to be implemented in the EU legislation [97]. It is worth mentioning that RoCoF is not a concern when the CE system is interconnected.

As discussed above, DFDs are the main source of frequency deviations in the CE and Nordic systems. While there has been significant effort from ENTSO-E to address the issue more work is needed to find the best solution(s) for a sustainable (from the frequency dynamic performance point of view) development and design of electricity markets, at an affordable cost for the end consumer.

The coupling of multinational electricity markets is also a source of concern for frequency quality. For example, price differences between two different control areas/bidding zones may lead to significant flow changes over ICs (see GB power system) that, in turn, lead to significant frequency deviations. Technical solutions include: (i) enforcing strict ramping limits on the IC; (ii) reducing IC capacity for market purposes; and (iii) developing a dispatch tool that could vary the ramp rate based on real-time system conditions. The first two solutions are not desirable from a market and environmental point of view as ICs create value for electricity systems by enabling electricity imports from markets with lower prices as an alternative to higher-priced indigenous electricity generation and reduce CO_2 emissions [98, 99]. Then, fast IC ramping may be desirable as that can help to avoid RES curtailments. Hence, other market-based solutions could be considered such as increased reserve and ramping volumes to keep up with more complex market dynamics.

Similar to increased cross-border trading, local flexibility and congestion markets run by the TSOs and Distribution System Operators (DSOs) [100] are another potential source of frequency deviations. This is an issue if there does not exist proper pre/post market gate-closure TSO-DSO coordination in place. For example, DSOs should inform TSOs about their total procured flexibility services and the respective activation times and duration. Notably, post-fault flexibility services which involve time scales of minutes, for example, 5 to 15 min for service activation, can negatively impact the overall balance of the system and, in turn, long-term frequency quality.

The aging of the conventional generating portfolio will lead to several units that are unable to keep up with faster frequency dynamics. Then, remaining conventional units, even if capable of a faster response, may not be able to provide the best frequency performance due to, for example, limited capacity or strict physical limits.

Related to the above are insufficient volumes and response time of manual balancing reserves and trips/unavailability of large generation plants, and poor frequency control and regulation provided by TSOs. For example, the ENTSO-E report finds different control areas not doing their job properly as the main causes of poor frequency quality in the CE power system in 2023.

Ongoing climate change is growing the occurrence and impact of extreme weather events on power systems, including both transmission and distribution systems [101, 102]. These events are generally classified as low-probability high-impact events due to their rare occurrences but potentially significant impact on, for example, frequency stability [103]. TSOs employ additional measures to manage these events such as: (i) keeping online additional generators to provide more inertia and reserves; (ii) increasing control room operators; (iii) canceling planned outages and bringing back transmission lines/components that were in outage; and (iv) increase TSO-DSO coordination. Despite these measures, it is never entirely guaranteed that there will not be adverse impacts on the safe and secure operation of power systems.

Poor coordination, cooperation, and information sharing between key stakeholders such as industry, RAs, and TSOs as well as lack of training and mentoring of relevant practitioners involved in power system frequency quality/control. It appears particularly important that RAs are educated more on technical issues/challenges so that they can make informed decisions.

Depending on the size of the power system, some of the above factors may impact more or less on frequency quality. Notably, higher inertia systems will experience lower RoCoFs. On the other hand, low-inertia systems with relatively large infeed/outfeed will be more susceptible to, for example, CLs.

Finally, depending on how frequency and RoCoF are defined and estimated (see Chap. 1), one may get different values which may or may not exceed relevant frequency quality limits. This may appear only a semantic issue, but inaccurate definitions or estimations may have a significant impact on how a TSO invests money to reinforce the system, which ultimately, has an impact on the electricity price for the end consumer.

3.6 Monitoring and Reporting

This section focuses on how TSOs monitor and report on different short- and long-term frequency quality metrics. With this aim, we examine relevant European (ENTSO-E, EirGrid/SONI, and NESO) and Australian (AEMO) cases.

3.6.1 Monitoring Short-Term Frequency Quality

To ensure that frequency quality indices stay within limits in real-time, TSOs carry out stability analyzes of the system. These analyzes provide critical decision-support information for control room engineers and radar-like guidance on how to operate the power system in a safe and secure manner. For example, EirGrid and SONI have developed the Look-ahead Security Assessment Tool (LSAT), which is a state-of-the-art stability analysis tool that ensures that frequency quality indices stay within limits in real-time. LSAT is utilized by other TSOs as well such as ERCOT in the US [104]. LSAT carries out rotor-angle, frequency, and voltage stability analyzes and is implemented in the control centers of EirGrid and SONI. In addition to stability analysis, the Irish TSOs utilize a real-time frequency oscillation monitoring tool that utilizes PMU data. This tool informs control room engineers regarding power fluctuations and is useful for postmortem analysis. Frequency, instantaneous RoCoF (see Eq. 1.67), and time error (see Eq. (2.41)) are also monitored in real-time in the Energy Management System (EMS).

Figure 3.16 shows the main features of the real-time LSAT. This tool was committed in 2010 and has been operating since then to assess the security of the AIPS in terms of rotor angle stability, frequency security, and voltage security [105]. Security

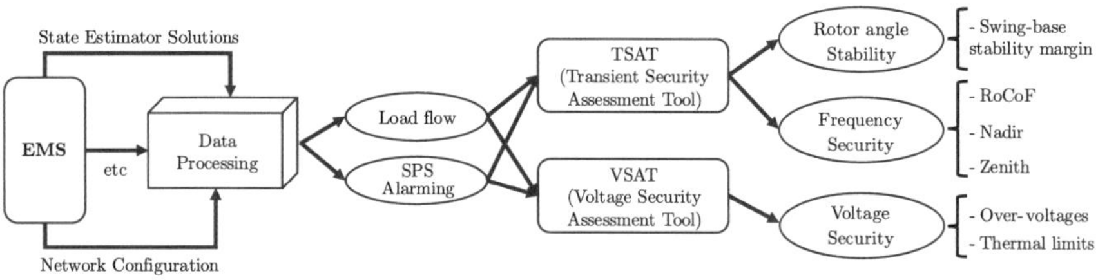

Fig. 3.16 Real-time LSAT developed by EirGrid and SONI

is a broad term that includes stability [106]. For instance, a system may be stable but not necessarily secure, but not the other way around. From December 2020, the look-ahead functionality was integrated into LSAT with the aim of assessing the security of the network 10 h ahead. Such forward-looking dynamic assessment capability was identified as a critical requirement to increase SNSP above 65% [105]. LSAT is composed of two major components, namely, Transient Security Assessment Tool (TSAT) and Voltage Security Assessment Tool (VSAT). TSAT conducts dynamic analyzes to assess the rotor angle stability, based on the swing-base stability margin criterion [107], and frequency security, based on nadir, zenith, and RoCoF criteria. On the other hand, VSAT checks voltage issues by conducting power flow analysis to assess the voltage security in quasi steady-state, based on over-voltage and thermal limit criteria. The assessment is carried out both for real-time and near future (look-ahead) time horizons.

Regarding real-time assessment, every 5 min, LSAT receives the network configuration and state estimator solutions, together with other data, from EMS. This data is processed and fed into TSAT and VSAT. These components, in parallel, run around 800 $N-1$ contingencies in less than 5 min and notify about insecure contingencies.

3.6.1.1 Monitoring Long-Term Frequency Quality Indices in Real-Time

To illustrate how EirGrid and SONI ensure that frequency stays within the ±100 mHz and ±200 mHz ranges in real-time, Fig. 3.17 shows the same frequency variations for the AIPS as shown in Fig. 3.8 but now adding two limits, that is, ±100 mHz and ±200 mHz ranges. As mentioned in Chap. 2, the Irish TSOs do not implement an AGC but rather rely on PFR from conventional units with ±15 mHz dead-band and sometimes from wind farms as well through APC and manual dispatch actions whenever frequency drifts away from ±100 mHz range.

Note also that while SOGL, the EU legislation, requires to keep frequency within the ±200 mHz range, in IE, the RA incentives EirGrid to maintain frequency within an even tighter range namely ±100 mHz range, for more than 98% of the time during a year. This means that the minutes outside the ±200 mHz range will likely never violate the target (15,000 min in Table 3.1) as the TSOs manage frequency within the ±100 mHz range. To fulfill the incentive, the TSOs monitor in real-time

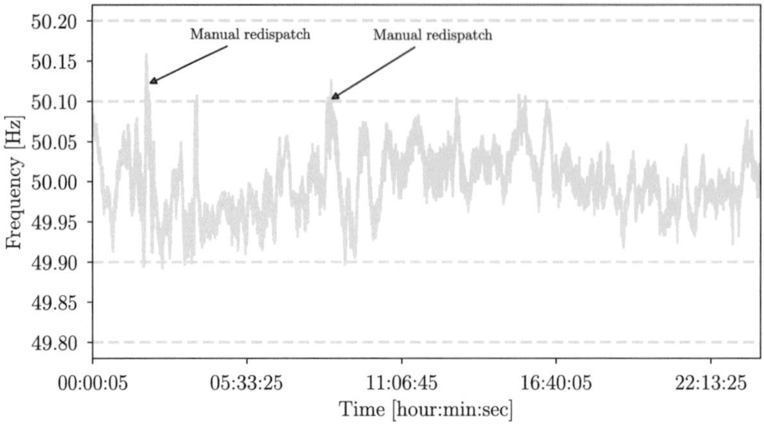

Fig. 3.17 Long-term frequency quality management in real-time

frequency variations. Specifically, the TSOs display on a monitor (video wall) in the control room frequency variations similar to Fig. 3.17 with a ±100 mHz range limit. Then, whenever the frequency moves away from this limit, an alarm is triggered which means the dispatchers/operators need to take action to redispatch conventional generation to bring the frequency back within this range.

These steps are clearly visible in Fig. 3.17, where it can be seen that frequency quickly recovers after going outside the ±100 mHz range. For this tighter frequency regulation, frequency never went outside the ±200 mHz range. Based on the first author's experience working in the EirGrid control room, and supported by frequency measurements (see Chap. 11), frequency tends to stay within the ±100 mHz range quite well throughout the day with just PFR from online conventional plants. Of course, there are situations when the frequency becomes quite volatile due, for example, to high wind/solar production, and the operators will have to take more manual adjustments. In these cases, operators prefer to turn on APC from wind power plants as well, which is an effective way to provide PFR [54].

3.6.2 Reporting of Frequency Quality

This section provides a brief overview of how TSOs monitor and report on the long-term performance of frequency quality parameters. From an Irish perspective, EirGrid and SONI monitor on a monthly basis the performance of some of the parameters such as RoCoF and minutes outside the ±100 mHz range. This performance is part of the joint Operational Policy Review Committee (OPRC) meetings every month [108]. OPRC governs the process of operational policy changes. If there is a degradation of the performance of these metrics EirGrid and SONI can take the necessary measures to address such issues. For example, if there are significant issues

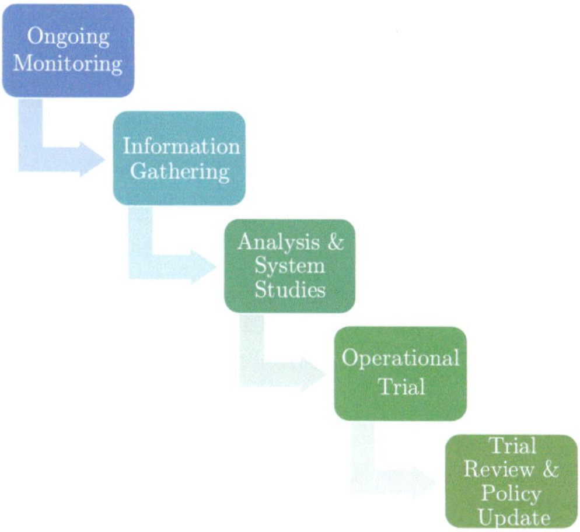

Fig. 3.18 EirGrid and SONI operational policy change process

with RoCoF, OPRC may decide to change the operational policy regarding the minimum inertia levels required at all times. Similarly, if there are issues with the ±100 mHz range criteria, it may decide to carry on more PFR reserves. Figure 3.18 depicts the operational policy change process in place in the AIPS.

This prudent process includes five main stages as follows.

- *Ongoing Monitoring*. In this phase, the TSOs monitor the system parameters such as RoCoF and minutes outside ±100 mHz range and analyze events and disturbances to assess system performance and generator compliance relative to operational policy parameters and metrics.
- *Information Gathering*. In this phase, the current status of the policy and parameters are assessed, and consultation is held with operations specialists on the drivers, requirements and need for changes.
- *Analysis and System Studies*. Here, subject matter experts and operations policy specialists study the system under an extensive and detailed range of conditions. In particular, the impact of the proposed policy change is studied. This is then followed by recommendations on the conditions of the operational trial. The OPRC approve or reject the proposed trial, based on the studies and in-depth discussions.
- *Operational Trial*. Subject to OPRC approval, an operational policy trial will commence under strict operational criteria and parameters, including hours of operation. Operations staff retain the authority to suspend the trial at any point should adverse impacts arise.
- *Trial Review and Policy Update*. If the operational trial period concludes without adverse impacts, subject matter experts will analyze the resulting parameters and criteria. This includes an in-depth examination of the trial period which is documented and reported to the OPRC. In particular, special focus is given to conditions and events that led to any trial-related issues. The OPRC will then review the findings and report to decide whether to:

– Stop the trial noting adverse impacts.
– Continue the trial to collect more data and information for a clearer decision.
– Approve the proposed policy change as the enduring operational policy.

The above process is internal which means the TSOs do not publicly report this information. What they do instead is publish quarterly the performance of their frequency quality metrics per month in the ENTSO-E transparency platform [109]. An example of the frequency quality parameters submitted/published by EirGrid and SONI is provided in Table 3.4 for the third quarter (Q3) of 2023. The Irish TSOs submit every month quite detailed statistical information on frequency deviations, including the percentiles that provide insights on extreme events. EirGrid and SONI maintain high-frequency quality standards. For instance, the fact that 1-percentile is around 49.9 Hz means that just 1% of frequency data are less or equal to 49.9 Hz and 99% of the frequency values are greater than 49.9 Hz.

Similar information is also reported by the other two synchronous areas namely the Nordic and CE [109]. Note that not all parameters listed in Table 3.1 are reported in these monthly reports. Additionally, EirGrid and SONI report on the overall frequency performance as part of the annual All-Island transmission system performance report [110]. For example, details include information on UFLS, lowest system frequency during the last decade, minutes outside the ±100 mHz range in the previous year, and relevant historical frequency excursions.

The monitoring and reporting process for frequency quality of EirGrid and SONI is comprehensive and useful to the industry in terms of transparency. Other TSOs provide similar monitoring and reporting of frequency quality namely AEMO. Specifically, AEMO publishes quarterly a frequency monitoring report that covers the following topics:

- How to improve frequency control performance;
- Achievement, or lack thereof, of operating standards such as number of minutes outside standard range of ±150 mHz;
- RoCoF;
- ACE;
- Reviewable incidents that led the frequency to go outside the standard range of ±150 mHz;
- Recent credible generation/load events such as generation or load reduced by more than 200 MW; and
- Detailed methodology on how to assess frequency events.

Although producing such a report means more work and resources for AEMO, it is a useful information for the Australian power industry, RA, and, in general, for all power system engineers/researchers worldwide because it provides detailed and continuous real-world evidence of frequency performance in (low-inertia) power systems.

GB is another interesting jurisdiction for frequency quality monitoring and management. Similar to EirGrid/SONI, NESO produces an annual frequency performance report in which it provides information on the weekly evolution of σ_f and

Table 3.4 EirGrid and SONI frequency quality data for Q3 2023

Data item	July	August	September
Mean (Hz)	49.99	50.00	50.00
σ_f (Hz)	0.0429	0.0431	0.0417
1-percentile (Hz)	49.904	49.904	49.907
5-percentile (Hz)	49.93	49.931	49.934
10-percentile (Hz)	49.945	49.945	49.947
90-percentile (Hz)	50.055	50.056	50.055
95-percentile (Hz)	50.07	50.071	50.069
99-percentile (Hz)	50.097	50.096	50.097
Time > 200 mHz (minutes)	0	0	0
Time < –200 mHz (minutes)	0	0.1333	1.1
Time > 1000 mHz (minutes)	0	0	0
Time < –1000 mHz (minutes)	0	0	0
No. of events: freq. deviation > 400 mHz and > 200 mHz 15 min later	0	0	0
No. of events: freq. deviation < –400 mHz and < –200 mHz 15 min later	0	0	0
No. of events freq. deviation > 500 mHz and > 500 mHz 1 min later	0	0	0
No. of events freq. deviation < –500 mHz and < –500 mHz 1 min later	0	0	0
No. of events: FRCE > 500 mHz and > 50 mHz after 15 min	0	0	0
No. of events: FRCE < –500 mHz and < –50 mHz after 15 min	0	0	0

relevant frequency excursions [111]. Additionally, NESO organizes weekly operational transparency forums where it updates the industry on the recent operational actions taken in the control center and answers any questions [112].

Similarly, NESO is one of the most advanced TSOs in what concerns a clear methodology on how to set the optimal balance between frequency quality risk and cost forecast for the following year. NESO does this through an annual Frequency Risk and Control Report (FRCR) which includes [113]:

- Information on relevant transient frequency deviations;
- The forecast impact of those deviations;
- The cost to secure the system for those deviations; and
- Confirmation on which risks NESO will or will not secure.

The FRCR report is extremely useful to the industry from a transparency point of view as it sets clear objectives on how NESO balances between risk and cost. Based on the FRCR assessment, NESO provides relevant recommendations such as whether to reduce the minimum inertia requirements or not.

3.7 Examples

3.7.1 Deterioration of Frequency Quality in CE

This example discusses poor frequency quality in one of the biggest power systems in the world namely the CE power system [114]. We argue that, if poor frequency quality occurs in such big systems with high inertia and substantial frequency regulation, then it is very likely that the situation for smaller power systems can be much worse.

The quality of the CE power system frequency has gradually decreased since more than 20 years now despite measures being implemented by different TSOs such as in Germany and Italy [47]. Indeed, researchers had predicted back in 2015 through an investigation of frequency control adequacy of the CE power system that long-term frequency quality would significantly decrease if the amount of active power reserves were not adjusted, that is, more intermittent generation causes the increasing need for active power reserves [115].

Figure 3.19 depicts the evolution of σ_f and total minutes outside ±50 mHz which is the standard frequency range in the CE system (see Table 3.1) in recent years. There is an almost perfect linear relationship between these two frequency quality indices and, most importantly, the fact that the second parameter exceeded, for the first time, its limit of 15,000 min in a year in 2023. This trend is expected to continue in the years to come if no effective solutions are found to deal with the increasing frequency deviations, especially the DFDs and long-lasting frequency deviations.

ENTSO-E reports that the main reason why the target for 2023 was not reached is due to long-lasting frequency deviations in the first half of 2023 [53], mostly in the direction of OF [61]. While most control blocks contributed to not fulfilling the

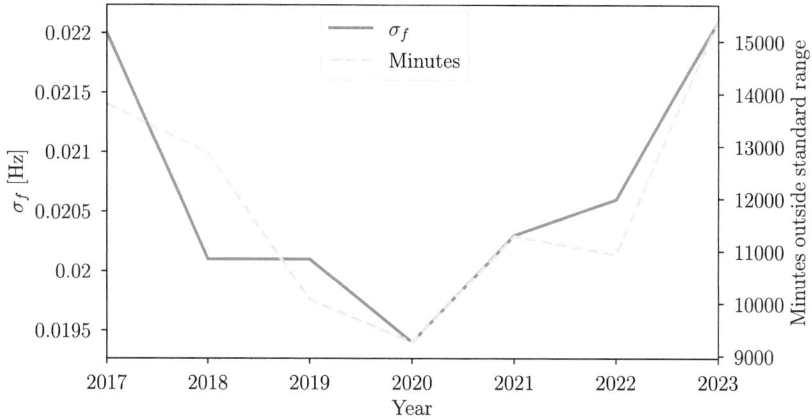

Fig. 3.19 Evolution of minutes outside standard frequency range (±50 mHz) and σ_f in recent years in the CE power system

frequency quality target for 2023, France and Bulgaria were the main control blocks that had the largest impact. The TSOs of France and Bulgaria state that the reasons include market and technical ones such as limited frequency regulation reserves and trips/unavailability of large generation plants. These long-lasting deviations can lead to a depletion of FCR reserves and, if combined with other events, such as a large outage of generation or a significant DFD, can contribute to the deterioration of grid security.

Another potential reason why frequency is spending more time outside the ±50 mHz range in the CE power system could be related to the penetration of RES generation. To support this statement, Fig. 3.20 shows the minutes outside ±50 mHz and total (onshore and offshore) wind generation in 2023. These data suggest that there exists a correlation for the vast majority of the year, that is, for months from February to September. But the other months (October, November, December, and January) seem not to show this correlation. This can be explained by the fact that, despite these months having a significant wind penetration, total demand tends to be higher during these months which, in turn, means more conventional generators committed online to meet the additional demand. Note that the introduction of RES has, so far, mainly been policy-driven with the legislation requiring priority dispatch of RES. This means that the vast majority of RES are still today not used to providing frequency regulation on an ongoing basis as this would generally mean pre-curtailment for them to be able to provide, for example, UF support. However, looking ahead, it is critical that RES such as, for example, wind generation provide frequency regulation, including both PFR and SFR, to help keep long-term frequency quality indices within limits.

Motivated by the decline of frequency quality, ENTSO-E established back in December 2024 a dedicated project group on mitigation actions against poor frequency quality with members from different TSOs where the first author is part of

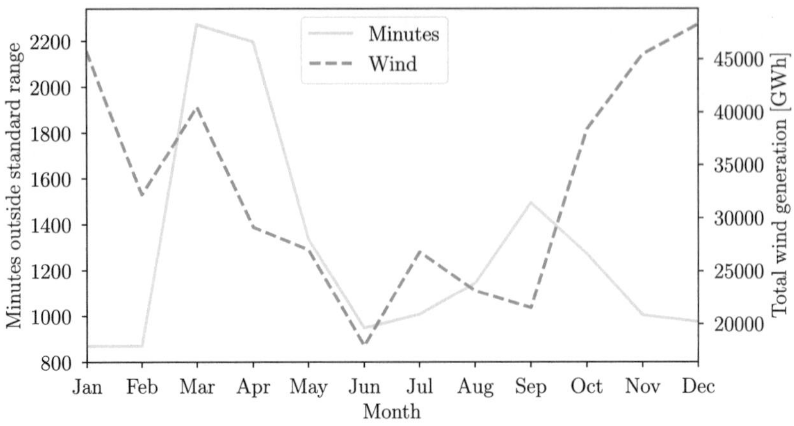

Fig. 3.20 Correlation between minutes outside standard frequency range (±50 mHz) and wind generation in the CE power system in 2023

representing EirGrid and SONI. The group is investigating possible operational and structural means and countermeasures that can help prevent the occurrence of poor frequency quality or, at least, allow TSOs to better manage it in real-time.

One might ask whether exceeding the 15,000 min limit is a big problem at all given that the CE has a very narrow standard frequency range namely ±50 mHz compared to the Nordic (±100 mHz) and GB/AIPS (±200 mHz). However, in practice, this is not an issue. For example, there is no issue for electrical machines to operate safely if the frequency is slightly above ±50 mHz. However, more reserves will be activated/used, in particular, PFC reserves (or FCR in European terminology), and thus leave the system short if a contingency were to happen. The latter can be addressed by increasing the size of reserves, with the drawback of increased costs, to cover the dimensioning incident plus some more to deal with normal operating conditions.

3.7.2 *Real-World IBR-Driven Oscillation*

Other types of frequency quality issues include oscillations. Oscillations and oscillatory stability have always been experienced in power systems and presented a challenge for TSOs. For instance, it has been well-known for quite a long time that in the Nordic power system frequency oscillates constantly around 50 Hz with a period of around 40 to 90 s, or equivalently, in the frequency range from 0.011 to 0.025 Hz [59]. Similar low-frequency oscillations in the range from 0.03 to 0.08 Hz have been observed on the AIPS for a number of years [116, 117].

However, the increasing use of IBRs such as wind, solar PV, and BESS has led to oscillations with a wider range of characteristics and root causes [76, 118]. These

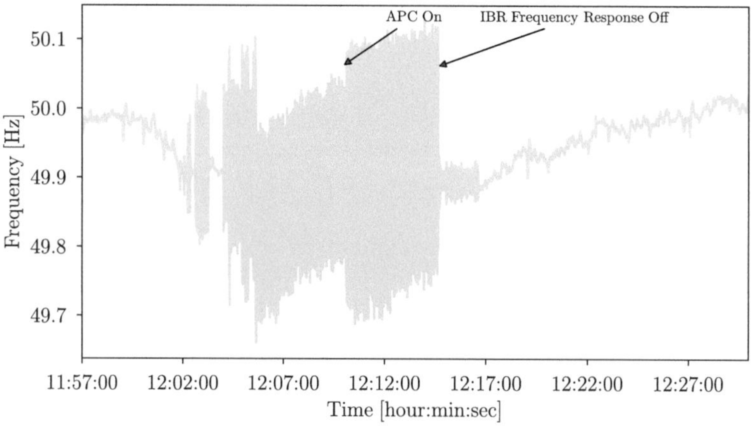

Fig. 3.21 AIPS frequency trace on October 5 and 6, 2024

could lead to unwanted equipment disconnections and damage, supply interruptions, and other violations of reliability criteria. Therefore, careful consideration should be given to how to model and study oscillations in order to better manage them.

We illustrate this with a recent sub-synchronous oscillation with an oscillation frequency of 5 Hz and peak oscillation magnitude of around 0.5 Hz that occurred on the AIPS between October 5 and 6, 2024. Figure 3.21 depicts the frequency trace for this event.

Sustained oscillations start when the frequency drops below 49.9 Hz at around 12:00:00. These led to a generating unit tripping and frequency decaying further. Oscillations then persisted again at around 12:04:00 which led to another unit being tripped and frequency dropping below 49.8 Hz. At this point, and based on defined operational policy measures [54], control room operators turned on APC (at around 12:10:00) on wind farms to assist in damping these oscillations. Surprisingly, the oscillation magnitude increased until at around 12:14:00 when the operators switched off the frequency response of a suspected IBR unit and later at around 12:16:00 opened the circuit breaker of the relevant substation/bay where the unit(s) where connected. Later, it was confirmed by the TSOs that aggressive FFR settings of two IBR units were the cause of these oscillations. What is interesting though is that while APC on wind farms have historically demonstrated the effectiveness to damp low-frequency oscillations [119], this was not the case with sub-synchronous oscillations. Therefore, it is fair to say that TSOs are still adapting and learning how to best deal with these types of IBR-driven oscillations.

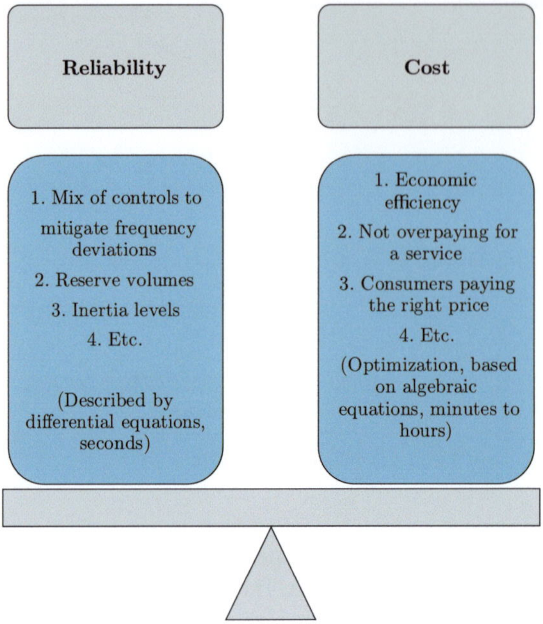

Fig. 3.22 Balance between reliability and cost

3.8 Concluding Remarks

As for any power system problem, it is not enough to propose technical solutions to address frequency quality problems. The second, equally crucial, aspect is to present and discuss these solutions to industry, in particular, to the national RAs. Such solutions need to be very clear, simple, and transparent to "convince" the relevant stakeholders and decision-making authorities (RAs) that the solutions are the right ones and at the least cost (see Fig. 3.22). Based on our experience, the engagements with industry and the RAs are often overlooked.

Then, it should not be forgotten that the practical implementation of technical solutions has a long lead time. It can easily take, for example, a couple of years or more for a central dispatch TSO to include a new constraint in the market scheduling systems such as the inertia floor constraint. It can take even more if one wants to implement a solution that involves both software and hardware implementations. And even more if this solution also impacts regulations and the market. For example, the implementation of the AGC required around a decade to be implemented in India (from concept/planning, and trials to running in all areas) [120].

Changes tend to be complex and time-consuming, especially if the changes have not been comprehensively studied and tested, the implementation of proposed solutions can be more than a challenge. For this reason, TSOs generally prefer "evolution" rather than "revolution." Therefore, power system engineers, especially those working for TSOs, need to keep in mind that coming up with a solution is the first step of a long and complex journey toward, hopefully, implementing it in practice.

An alternative to finding and implementing technical solutions is to consider market-based mechanisms and incentive structures that could address the same specific power system problems. For instance, back in 2020 NESO in GB successfully introduced a new service, which was approved by the national RA, called "Optional Downward Flexibility Management" as a temporary solution to manage the system over the summer period with low demand and high RES generation to help balance the system [121]. The service consisted of calling, in particular, small-scale RES such as wind and solar PV to turn down or turn off their generation of electricity.

In Brief

- Frequency quality was not an issue 30 years ago.
- Frequency quality has started to deteriorate mainly due to RES penetration and continuous market development and integration.
- Large interconnected power systems focus on long-term than short-term frequency quality. Small islanded power systems focus on both phenomena.
- There is no clear or widely-accepted definition of frequency quality.
- We propose to define frequency quality as a measure of power quality, security and reliability.
- The introduction of electricity markets and increased trading activities implies a schedule-based grid operation instead of classic real-time load following. This has, and continues to contribute, to a decrease of long-term frequency quality.
- Oscillations are an increasing concern for TSOs.
- Current industry approach to frequency quality problems is more reactive than proactive.

Open Questions

The main cause of the poor frequency quality observed by TSOs in recent years is the integration of IBR-based renewable generation. TSOs are currently in the process of learning how to manage the causes negatively impacting on frequency quality and there are many open questions. For example, relevant unresolved issues for now are how to deal with situations where TSOs have exceeded the amount of SFC reserves; what are the most adequate frequency quality indices to be used; and what preventive and corrective measures should be taken to prevent frequency oscillations.

An important operational aspect to be solved is how to balance between high-frequency quality standards and cost. While markets contribute to ensuring electricity is provided at an efficient cost to consumers, they are also contributing to a deterioration of frequency quality. Therefore, it is still not clear how to avoid market-driven imbalances and their effect on frequency quality at an affordable cost for the end consumer.

References

1. ENTSO-E, 'NAG – frequency quality report,' 2015. [Online]. Available: www.statnett.no
2. EPRI, 'EPRI power system dynamics tutorial,' Electric Power Research Institute, United States, Tech. Rep., 2009
3. Lamme BG (1918) The technical story of the frequencies. Proceedings of the American Institute of Electrical Engineers 37(1):1–21
4. Owen E (1997) The origins of 60-Hz as a power frequency. IEEE Ind Appl Mag 3(6):8–14
5. R. C. Dugan, S. Santoso, M. F. McGranaghan *et al.*, *Electrical Power Systems Quality*, ser. McGraw-Hill professional engineering. McGraw-Hill Education, 2002. [Online]. Available: books.google.ie/books?id=Obv2GVmNlt0C
6. M. H. J. Bollen, *Understanding power quality problems*. IEEE press New York, 2000, **3**
7. M. H. J. Bollen and I. Y. H. Gu, *Signal Processing of Power Quality Disturbances*. John Wiley & Sons, Ltd, 2006. [Online]. Available: onlinelibrary.wiley.com/doi/abs/10.1002/9780471931317.fmatter
8. S. L. Engelbrektsson, 'Evaluation of key performance indices for frequency quality: A method for evaluating frequency stability in the Nordic power system,' M.E. thesis, School of Electrical Engineering and Computer Science, Stockholm, Sweden, 2023
9. Oxford English Dictionary, 'Flexibility,' 2023. [Online]. Available: www.oed.com
10. European Commission, 'Establishing a guideline on electricity transmission system operation,' 2017. [Online]. Available: eur-lex.europa.eu
11. Mohammadi Y, Polajžer B, Leborgne RC et al (2024) Quantifying power system frequency quality and extracting typical patterns within short time scales below one hour. Sustainable Energy Grids and Networks 38:101359
12. Kundur P, Paserba J, Ajjarapu V et al (2004) Definition and classification of power system stability IEEE/CIGRE joint task force on stability terms and definitions. IEEE Trans Power Syst 19(3):1387–1401
13. NERC Rs Committee (2011) Balancing and frequency control: A technical document prepared by the NERC resources subcommittee. Tech. Rep, NERC Technical Report
14. C. Tan, Y. Li, Q. Lan *et al.*, 'Multi-area automatic generation control scheme considering frequency quality in Southwest China grid: Challenges and solutions,' *IEEE Access*, **8**, 199 813–199 828, 2020
15. Cao Q, Chen L, Yan F et al (2018) 'Research on AGC assessment criteria's adaptability', in. International Conference on Power System Technology (POWERCON) 2018:404–409
16. Jayawardene I, Venayagamoorthy GK, Zhong X (2022) Resilient and sustainable tie-line bias control for a power system in uncertain environments. IEEE Transactions on Emerging Topics in Computational Intelligence 6(1):205–219
17. ENTSO-E, 'Supporting document for the Network Code on Load-Frequency Control and Reserves,' 2013. [Online]. Available: https://eepublicdownloads.entsoe.eu/
18. Li ZW, Samuelsson O, Garcia-Valle R (2011) 'Frequency deviations and generation scheduling in the Nordic system', in. IEEE Trondheim PowerTech 2011:1–6
19. NERC Rs Committee (2021) Balancing and frequency control: Reference document. prepared by the NERC resources subcommittee. Tech. Rep, NERC Technical Report
20. FINGRID, 'Frequency quality analysis,' 2018. [Online]. Available: www.fingrid.fi
21. Kuivaniemi M, Ruokolainen P, Modig FN et al (2021) Requirement for minimum inertia in the Nordic power system. Tech. Rep, ENTSO-E
22. M. Hurtado, M. Jafarian, T. Kërçi *et al.*, 'Stability assessment of low-inertia power systems: A system operator perspective,' in *IEEE PES General Meeting*, 2024, 1–5
23. J. Matevosyan, 'Inertia monitoring,' 2022. [Online]. Available: www.esig.energy
24. B. Liu, 'Advanced ROCOF protection of distribution systems,' Ph.D. dissertation, University of Nottingham, 2012
25. N. Miller, D. Lew, S. Barnes *et al.*, 'Advisory on equipment limits associated with high RoCoF,' *GE Energy Consulting*, 2017

26. M. Xiang, Z. Yang, J. Yu *et al.*, 'Real-time dispatch with secondary frequency regulation: A pathway to consider intra-interval fluctuations,' *IEEE Systems Journal*, 1–12, 2021
27. Electricity Authority, 'Addressing more frequency variability in New Zealand's power system. Consultation paper,' 2024. [Online]. Available: www.ea.govt.nz
28. B. J. Kirby, 'Frequency control concerns in the North American electric power system,' Oak Ridge National Lab.(ORNL), Oak Ridge, TN (United States), Tech. Rep., 2003
29. M. Stewart, 'Frequency control frameworks review-Engineers Australia submission-24 april 2018,' 2018
30. T. Kërçi, M. Hurtado, M. Gjergji *et al.*, 'Frequency quality in low-inertia power systems,' in *IEEE PES General Meeting*, 2023, 1–5
31. A. Dixon, *Modern aspects of power system frequency stability and control*. Academic Press, 2019
32. D. Razmi, T. Lu, B. Papari *et al.*, 'An overview on power quality issues and control strategies for distribution networks with the presence of distributed generation resources,' *IEEE Access*, **11**, 10 308–10 325, 2023
33. ENTSO-E, 'Report on deterministic frequency deviations,' 2020. [Online]. Available: eepublicdownloads.entsoe.eu
34. ENTSO-E, 'Overview of frequency control in the Nordic power system,' 2024. [Online]. Available: https://eepublicdownloads.blob.core.windows.net/
35. Mosca C, Bompard E, Aluisio B et al (2019) 'HVDC for frequency stability under RES penetration: the Sardinia island case', in. AEIT HVDC International Conference (AEIT HVDC) 2019:1–6
36. EirGrid & SONI, 'Day-ahead system services auction (DASSA) volume forecasting methodology consultation paper v1.0,' 2024. [Online]. Available: consult.eirgrid.ie
37. Kosanić M, Ilić M, Baker D et al (2023) 'Data-based optimal estimation of frequency bias: The case of Southwest power pool', in. IEEE Belgrade PowerTech 2023:1–6
38. E. Ørum, M. Laasonen, and H. Kuisti, 'Frequency quality, phase 2,' *Project Report-Version*, **1**, 2017
39. M. H berg and G. Doorman, 'Classification of balancing markets based on different activation philosophies: Proactive and reactive designs,' in *2016 13th International Conference on the European Energy Market (EEM)*, 2016, 1–5
40. M. H berg and G. Doorman, 'Proactive planning and activation of manual reserves in sequentially cleared balancing markets,' in *2017 IEEE Electrical Power and Energy Conference (EPEC)*, 2017, 1–6
41. De Vos K, Stevens N, Devolder O et al (2019) Dynamic dimensioning approach for operating reserves: Proof of concept in Belgium. Energy Policy 124:272–285
42. P. Mancarella, 'Grid and market challenges with ultra-deep penetration of renewables and DER: Experiences from Australia,' 2022. [Online]. Available: www.researchgate.net
43. Chen Y (2023) Addressing uncertainties through improved reserve product design. IEEE Trans Power Syst 38(4):3911–3923
44. O. Yurdakul, E. Ela, and F. Billimoria, 'Flexible ramping product procurement in day-ahead markets,' *IEEE Transactions on Energy Markets, Policy and Regulation*, 1–18, 2024
45. Ju P, Li H, Pan X et al (2018) Stochastic dynamic analysis for power systems under uncertain variability. IEEE Trans Power Syst 33(4):3789–3799
46. Adeen M, Milano F (2021) On the impact of auto-correlation of stochastic processes on the transient behavior of power systems. IEEE Trans Power Syst 36(5):4832–4835
47. ENTSO-E, 'Continental Europe significant frequency deviations – January 2019 ,' 2019. [Online]. Available: eepublicdownloads.entsoe.eu
48. FINGRID, 'Frequency quality analysis,' 2023. [Online]. Available: www.fingrid.fi
49. M. de la Torre Rodr guez, M. Scherer, D. Whitley, and F. Reyer, 'Frequency containment reserves dimensioning and target performance in the European power system,' in *IEEE PES General Meeting*, 2014, 1–5
50. ENTSO-E, 'Deterministic frequency deviations – root causes and proposals for potential solutions,' 2011. [Online]. Available: eepublicdownloads.entsoe.eu

51. J. Kruse, B. Sch fer, and D. Witthaut, 'Exploring deterministic frequency deviations with explainable AI,' in *2021 IEEE International Conference on Communications, Control, and Computing Technologies for Smart Grids (SmartGridComm)*, 2021, 133–139
52. Rousseau J, Vorwerk J, Avramiotis I et al (2023) 'Employing imbalance forecasts to proactively counteract deterministic frequency deviations', in. IEEE Belgrade PowerTech 2023:1–6
53. ENTSO-E, 'Long lasting frequency deviations: Measures taken by Continental European TSOs to address long lasting frequency deviations,' 2021. [Online]. Available: eepublicdownloads.entsoe.eu
54. EirGrid & SONI, 'Active power control groups,' 2020. [Online]. Available: www.sem-o.com
55. Energinet, 'Outlook for ancillary services 2023-2040,' 2018. [Online]. Available: en.energinet.dk
56. H. Olsson, 'Future frequency control services in the Nordic power system,' 2017
57. T. Kërçi, M. Hurtado, S. Tweed *et al.*, 'Emerging challenges of integrating solar PV in the Ireland and Northern Ireland power systems,' in *IEEE PES General Meeting*, 2024, 1–5
58. T. Kërçi, C. Duggan, U. Farooq *et al.*, 'Impact of converter-based demand on frequency quality in the Ireland and Northern Ireland power systems,' in *CIGRE Paris*, 2024, 1–11
59. J. Seppänen, M. Lehtonen, M. Kuivaniemi *et al.*, 'Long term characteristics of ultra low frequency oscillations in the Nordic power system,' in *2022 IEEE PES Innovative Smart Grid Technologies Conference Europe (ISGT-Europe)*, 2022, 1–5
60. B. Pinheiro, M. Llerena, G. Faria *et al.*, 'Evaluation of the impact of IBR on the frequency dynamics in the brazilian power system,' 2025
61. ENTSO-E, '2023 annual load-frequency control report,' 2024. [Online]. Available: eepublicdownloads.entsoe.eu
62. E. Weihs, M. Persson, and P. Chen, 'Frequency quality in the Nordic system 2040,' in *2020 IEEE PES Innovative Smart Grid Technologies Europe (ISGT-Europe)*, 2020, 809–813
63. T. K r i and F. Milano, 'Asymmetry of frequency distribution in power systems: Sources, estimation, impact and control,' *IEEE Open Access Journal of Power and Energy*, **12**, 135–145, 2025
64. Khan A, Hosseinzadehtaher M, Shadmand MB et al (2020) On the stability of the power electronics-dominated grid: A new energy paradigm. IEEE Ind Electron Mag 14(4):65–78
65. Peng Q, Yang Y, Wang H et al (2018) 'On power electronized power systems: Challenges and solutions', in. IEEE Industry Applications Society Annual Meeting (IAS) 2018:1–9
66. J. E. de Haan, C. Escudero Concha, M. Gibescu *et al.*, 'Stabilising system frequency using HVDC between the Continental European, Nordic, and Great Britain systems,' *Sustainable Energy, Grids and Networks*, **5**, 125–134, 2016
67. F. Milano, F. D rfler, G. Hug *et al.*, 'Foundations and challenges of low-inertia systems (invited paper),' in *Power Systems Computation Conference (PSCC)*, 2018, 1–25
68. Liang X (2017) Emerging power quality challenges due to integration of renewable energy sources. IEEE Trans Ind Appl 53(2):855–866
69. H. T. Nguyen, G. Yang, A. H. Nielsen *et al.*, 'Challenges and research opportunities of frequency control in low inertia systems,' in *E3S Web of Conferences*, **115**. EDP Sciences, 2019, p. 02001
70. Fresia M, Minetti M, Rosini A et al (2024) A techno-economic assessment to define inertia needs of the Italian transmission network in the 2030 energy scenario. IEEE Trans Power Syst 39(3):5050–5061
71. J. Kruse, B. Schäfer, and D. Witthaut, 'Revealing drivers and risks for power grid frequency stability with explainable AI,' *Patterns*, **2** (*11*), 2021
72. Hong Q, Khan MAU, Henderson C et al (2021) Addressing frequency control challenges in future low-inertia power systems: A Great Britain perspective. Engineering 7(8):1057–1063
73. Johnson SC, Rhodes JD, Webber ME (2020) Understanding the impact of non-synchronous wind and solar generation on grid stability and identifying mitigation pathways. Appl Energy 262:114492
74. M. N. H. Shazon, Nahid-Al-Masood, and A. Jawad, 'Frequency control challenges and potential countermeasures in future low-inertia power systems: A review,' *Energy Reports*, **8**, 6191–6219, 2022

75. Tielens P, Van Hertem D (2016) The relevance of inertia in power systems. Renew Sustain Energy Rev 55:999–1009
76. ESIG, 'Diagnosis and mitigation of observed oscillations in IBR-dominant power systems. a practical guide,' ESIG, Tech. Rep., 2024
77. A. Boricic, 'Investigation of persistent sub-synchronous oscillations in the Netherlands,' ESIG 2025 Spring Technical Workshop, Tech. Rep., 2025. [Online]. Available: www.esig.energy
78. M. Garcia, J. Kyle Skolfield, F. Wilches-Bernal *et al.*, 'Co-optimization of ancillary services providing primary frequency control in theday-ahead market,' *IEEE Access*, 1–1, 2025
79. D. Al Kez, A. M. Foley, F. Ahmed *et al.*, 'Overview of frequency control techniques in power systems with high inverter-based resources: Challenges and mitigation measures,' *IET Smart Grid*, **6** (*5*), 447–469, 2023
80. R. Quint and D. Ramasubramanian, 'Impacts of droop and deadband on generator performance and frequency control,' in *IEEE PES General Meeting*, 2017, 1–5
81. A. Hoke, R. Mahmud, A. Nelson *et al.*, 'Fast grid frequency support from distributed energy resources,' National Renewable Energy Lab.(NREL), Golden, CO (United States), Tech. Rep., 2021
82. T. Kërçi, W. Zhong, A. Moghassemi *et al.*, '6 - frequency control and regulating reserves by VPPs,' in *Scheduling and Operation of Virtual Power Plants*, A. Zangeneh and M. Moeini-Aghtaie, Eds. Elsevier, 2022, 131–162
83. AEMO, 'AEMO NEM virtual power plant demonstrations,' 2021. [Online]. Available: aemo.com.au
84. B. Badrzadeh and Z. Emin, *Power System Dynamic Modelling and Analysis in Evolving Networks*. Springer Nature, 2024
85. Gilleran M, Bonnema E, Woods J et al (2021) Impact of electric vehicle charging on the power demand of retail buildings. Advances in Applied Energy 4:100062
86. Ehsan A, Preece R (2022) Quantifying the impacts of heat decarbonisation pathways on the future electricity and gas demand. Energy 254:124229
87. MacIver C, Bell K, Nedd M (2021) An analysis of the August 9th 2019 GB transmission system frequency incident. Electric Power Systems Research 199:107444
88. NERC, 'Odessa disturbance,' North American Electric Reliability Corporation, Tech. Rep., September, 2021
89. S. Singh and R. K. Jha, 'IBR performance issues: Potential threat to grid stability and security,' in *2024 IEEE Third International Conference on Power Electronics, Intelligent Control and Energy Systems (ICPEICES)*, 2024, 649–653
90. H. W. Qazi, P. Wall, M. Val Escudero *et al.*, 'Impacts of fault ride through behavior of wind farms on a low inertia system,' *IEEE Transactions on Power Systems*, **37** (*4*), 3190–3198, 2022
91. AEMO, 'Supporting secure operation with high levels of distributed resources,' 2024. [Online]. Available: aemo.com.au
92. A. Moshari, S. Aldahmor, M. Hurtado *et al.*, 'Tackling solar energy integration challenges on the Ireland and Northern Ireland power system,' in *23rd Wind & Solar Integration Workshop (WIW 2024)*, **2024**, 2024, 1023–1030
93. AEMO, 'Load relief,' 2025. [Online]. Available: aemo.com.au
94. Weise B (2015) Impact of K-factor and active current reduction during fault-ride-through of generating units connected via voltage-sourced converters on power system stability. IET Renew Power Gener 9(1):25–36
95. Wikipedia, '2025 Chile blackout,' 2025. [Online]. Available: en.wikipedia.org/w
96. ENTSO-E, 'Continental Europe synchronous area separation on 08 January 2021,' 2021. [Online]. Available: eepublicdownloads.entsoe.eu
97. ENTSO-E, 'ENTSO-E workshops on connection network code amendments on (RoCoF and Grid Forming) & stability management,' 2022. [Online]. Available: eepublicdownloads.entsoe.eu
98. G. Castagneto Gissey, B. Guo, D. Newbery *et al.*, 'The value of international electricity trading,' UCL Institute for Sustainable Resources, Tech. Rep., 2019

99. Pean E, Pirouti M, Qadrdan M (2016) Role of the GB-France electricity interconnectors in integration of variable renewable generation. Renewable Energy 99:307–314
100. Kërçi T (2023) 'On the definition and classification of power system flexibility', in. IEEE PES Innovative Smart Grid Technologies Europe (ISGT EUROPE) 2023:1–5
101. Panteli M, Mancarella P (2015) Influence of extreme weather and climate change on the resilience of power systems: Impacts and possible mitigation strategies. Electric Power Systems Research 127:259–270
102. Hawker G, Bell K, Bialek J et al (2024) Management of extreme weather impacts on electricity grids: an international review. Progress in Energy 6(3):032005
103. K. Das, F. Guo, E. Nu o *et al.*, 'Frequency stability of power system with large share of wind power under storm conditions,' *Journal of Modern Power Systems and Clean Energy*, **8** (*2*), 219–228, 2020
104. P. Du, *Renewable Energy Integration for Bulk Power Systems: ERCOT and the Texas Interconnection*. Springer Nature, 2023
105. M. Val Escudero, 'Look ahead security assessments: Operations radar to navigate high IBR waters in the journey to net zero,' 2023. [Online]. Available: www.esig.energy
106. Capitanescu F (2023) Are we prepared against blackouts during the energy transition?: Probabilistic risk-based decision making encompassing jointly security and resilience. IEEE Power Energ Mag 21(3):77–86
107. Kundur P (1994) Power System Stability and Control. McGraw-Hill, New York
108. EirGrid & SONI, 'Operational policy roadmap 2023-2030,' 2022. [Online]. Available: https://www.eirgrid.ie/
109. EirGrid & SONI, 'Measurements of frequency quality (legacy),' 2023. [Online]. Available: transparency.entsoe.eu
110. EirGrid & SONI, 'All-island transmission system performance report 2023,' 2024. [Online]. Available: https://www.eirgrid.ie/
111. NESO, 'National electricity transmission system performance report,' 2024. [Online]. Available: www.neso.energy
112. NESO, 'NESO operational transparency forum,' 2025. [Online]. Available: www.neso.energy
113. NESO, 'Frequency risk and control report,' 2024. [Online]. Available: www.neso.energy
114. ENTSO-E, '2017-2023 load-frequency control annual reports,' 2024. [Online]. Available: eepublicdownloads.entsoe.eu
115. Scherer M, Andersson G (2015) 'How future-proof is the Continental European frequency control structure?' in. IEEE Eindhoven PowerTech 2015:1–6
116. Modi N, Escudero MV, Aramaki K et al (2024) High inverter-based resource integration: The experience of five system operators. IEEE Power Energ Mag 22(2):78–88
117. Duggan C, Liu X, Brogan P et al (2022) Very low-frequency oscillation source localization on Ireland's power system. IEEE Open Journal of Industry Applications 3:192–201
118. You S, Kou G, Liu Y et al (2017) Impact of high PV penetration on the inter-area oscillations in the U.S. Eastern interconnection. IEEE Access 5:4361–4369
119. C. Duggan, X. A. Liu, R. Best *et al.*, 'Active power control from wind farms for damping very low-frequency oscillations,' *Frontiers in Energy Research*, **Volume 10 - 2022**, 2022
120. GIZ, 'Proposed roadmap for implementing automatic generation control (AGC),' 2019. [Online]. Available: energyforum.in
121. NESO, 'Managing reduced demand for electricity - what is our new ODFM service, and why do we need it?' 2020. [Online]. Available: www.neso.energy/news

Chapter 4
Frequency Distribution

Intuitively, the higher the symmetry in the dynamic response of a dynamical system, the higher the predictability of the system behavior and controllability. Thus, a precise evaluation of the asymmetries present in a system contributes toward increased power system stability and resilience. The devices and controllers that form a power system are for the most part symmetrical. However, it has been recently observed by some Transmission System Operators (TSOs) that power systems are becoming increasingly asymmetrical [1, 2]. The aim of this chapter is to study the causes of Frequency Probability Distribution (FPD) asymmetry in power systems, provide a qualitative theoretical appraisal of this asymmetry; and, finally, show how asymmetry can be compensated through modified control.

The chapter is organized as follows. Section 4.1 provides a rationale on why asymmetry reduces the "quality" of the frequency control and system operation. Section 4.2 provides a qualitative appraisal of the causes of the asymmetry in the frequency distribution of power systems and presents the proposed control strategy to compensate this asymmetry. Section 4.3 defines various metrics, to evaluate the asymmetry of the frequency distribution. Section 4.4 illustrates two real-world examples, namely the Irish and Australian transmission systems, where the FPD asymmetry appears to be a direct consequence of the high penetration of wind power generation. Section 4.5 illustrates metrics and compensation control through several numerical examples. Section 4.6 provides a comparison of the FPD of the Continental Europe (CE), All-Island Power System (AIPS), Great Britain (GB) and Nordic power systems.

4.1 Overview

The topic of FPD in power systems and the various sources and parameters that influence it has recently received attention in the literature, in particular, in light of the integration of uncertain and variable Renewable Energy Sources (RES) such as

T. Kërçi and F. Milano, *Frequency Quality in Power Systems*, Power Systems,
https://doi.org/10.1007/978-3-032-04721-2_4

wind and solar generation. The main focus, however, has been on the modeling and study of how to reproduce the frequency distribution observed in real grids such as the bimodal distribution. However, the effect of losses, control limits and RES providing Primary Frequency Control (PFC) has not been considered so far.

With regard to the latter, there is a concern in industry that enforcing RES providing PFC with narrow dead-band is contributing to new phenomena arising in power systems (so far unexplained) such as low-frequency oscillations and asymmetry in the frequency distribution [2]. The asymmetry of FPD has, for example, a direct impact on legacy devices and industrial processes that rely on the use of synchronous electric clocks for time keeping as frequency will tend to spend more time below or above the nominal. Specifically, Australian Energy Market Operator (AEMO) has recently acknowledged feedback from industry that: *the universal application of very narrow governor dead-bands may be contributing to unexplained oscillations from some plant and asymmetry in the National Electricity Market (NEM)'s frequency characteristic* [2]. In addition, the Australian Energy Council (AEC) and the power system industry in Australia are concerned that this asymmetry is decreasing power system resilience as reported in [1]: *Furthermore, since the introduction of mandatory Primary Frequency Regulation (PFR), power system frequency has been exhibiting behavior that suggests resilience has decreased. As confirmed in the report prepared by Provecta commissioned by the AEC as part of its submission to the Draft Determination, system-wide frequency is displaying:*

- A "wobble" in terms of a slow frequency cycling with a period 18–24 s.
- A "skew" in terms of an asymmetry in the distribution.

Thus, it is difficult to see how any contribution to improved power system resilience is realized particularly in view of the above comments.

These excerpts as well as recent academic studies such as [3–5] indicate that the changing dynamic behavior of power systems (which can manifest as the appearance of new phenomena such as asymmetry of the FPD) is yet to be fully understood.

4.2 Qualitative Appraisal of Asymmetry

In this section, first we show that the cause of the skewness of the distribution of the states of a power system is due to the nonlinearity of the equations of the system itself (Sect. 4.2.1). This section shows that only for linear systems a Gaussian noise source leads to normal distributions of the system variables. This conclusion applies to systems of any size. Then, we discuss two specific examples of nonlinearity that lead to the asymmetry of the FPD, namely network losses (Sect. 4.2.2) and wind frequency control obtained with variable pitch angle blades (Sect. 4.2.3). As the source of the asymmetry is the nonlinearity, considering a simple system for the theoretical appraisal helps understanding the problem without lack of generality. We also briefly discuss the effect of regulator hard limits (Sect. 4.2.4) as well as introduce the proposed control-based nonlinear compensation to reduce FPD asymmetry (Sect. 4.2.5).

4.2.1 Probability Distribution of System Variables

For the developments discussed in this section, we consider the power system model (A.1), which is repeated here for convenience:

$$\begin{aligned} \boldsymbol{x}' &= \boldsymbol{f}(\boldsymbol{x}, \boldsymbol{y}, \boldsymbol{u}, \boldsymbol{z}, \boldsymbol{\eta}), \\ \boldsymbol{0}_m &= \boldsymbol{g}(\boldsymbol{x}, \boldsymbol{y}, \boldsymbol{u}, \boldsymbol{z}, \boldsymbol{\eta}), \\ \dot{\boldsymbol{\eta}} &= \boldsymbol{a}(\boldsymbol{\eta}) + \mathcal{B}(\boldsymbol{\eta})\,\boldsymbol{\zeta}, \end{aligned} \tag{4.1}$$

where, for simplicity, we have assumed matrix $\boldsymbol{T} = \boldsymbol{I}_n$, that is, the identity matrix of order n. The model includes system dynamics through the differential equations $\boldsymbol{f}$ and state variables $\boldsymbol{x}$, lumped network equations through the algebraic equations $\boldsymbol{g}$ and variables $\boldsymbol{y}$, as well as noise, through the stochastic processes $\boldsymbol{\eta}$. More details on (4.1) are given in Appendix A.

It is possible to show that, for a linear system, if the stochastic processes $\boldsymbol{\eta}$ are normally distributed, then also $\boldsymbol{x}$ and $\boldsymbol{y}$ are normally distributed. Linearization of (4.1) leads to:

$$\begin{aligned} \Delta\boldsymbol{x}' &= \boldsymbol{f}_x \Delta\boldsymbol{x} + \boldsymbol{f}_y \Delta\boldsymbol{y} + \boldsymbol{f}_\eta \Delta\boldsymbol{\eta}, \\ \boldsymbol{0}_m &= \boldsymbol{g}_x \Delta\boldsymbol{x} + \boldsymbol{g}_y \Delta\boldsymbol{y} + \boldsymbol{g}_\eta \Delta\boldsymbol{\eta}, \\ \Delta\boldsymbol{\eta}' &= \boldsymbol{a}_\eta \Delta\boldsymbol{\eta} + \mathcal{B}(\boldsymbol{\eta}_o)\,\boldsymbol{\zeta}, \end{aligned} \tag{4.2}$$

where we have used the notation $\boldsymbol{h}_u = \nabla_u^T \boldsymbol{h}(\boldsymbol{u})$ to represent the Jacobian matrices of the differential and algebraic equations as well as of the drift of the stochastic processes. The Jacobian matrices are assumed to be calculated at an equilibrium point, say $(\boldsymbol{x}_o, \boldsymbol{y}_o, \boldsymbol{\eta}_o)$, of the deterministic version of system (4.1), that is, by imposing $\mathcal{B} = \boldsymbol{0}$. This system can be rewritten as:

$$\begin{aligned} \begin{bmatrix} \Delta\boldsymbol{x}' \\ \Delta\boldsymbol{\eta}' \end{bmatrix} &= \begin{bmatrix} (\boldsymbol{f}_x - \boldsymbol{f}_y \boldsymbol{g}_y^{-1} \boldsymbol{g}_x) & (\boldsymbol{f}_\eta - \boldsymbol{f}_y \boldsymbol{g}_y^{-1} \boldsymbol{g}_x) \\ \boldsymbol{0}_{p,n} & \boldsymbol{a}_\eta \end{bmatrix} \begin{bmatrix} \Delta\boldsymbol{x} \\ \Delta\boldsymbol{\eta} \end{bmatrix} + \begin{bmatrix} \boldsymbol{0}_{n,q} \\ \mathcal{B}(\boldsymbol{\eta}_o) \end{bmatrix} \boldsymbol{\zeta} \\ &= \boldsymbol{A}\,\Delta\boldsymbol{\xi} + \boldsymbol{B}\,\boldsymbol{\zeta}, \end{aligned} \tag{4.3}$$

where $\Delta\boldsymbol{\xi} = [\Delta\boldsymbol{x}^T, \Delta\boldsymbol{\eta}^T]^T$ and the algebraic variables $\Delta\boldsymbol{y}$ have been eliminated using the expression:

$$\Delta\boldsymbol{y} = -\begin{bmatrix} \boldsymbol{g}_y^{-1}\boldsymbol{g}_x & \boldsymbol{g}_y^{-1}\boldsymbol{g}_\eta \end{bmatrix} \Delta\boldsymbol{\xi}. \tag{4.4}$$

Then, the Fokker-Planck equation allows writing the following probability distribution of all state variables in stationary condition [6]:

$$\Pr(\Delta\boldsymbol{\xi}) = (\det \mid 2\pi \boldsymbol{C} \mid)^{-1/2} \cdot \exp\left(-\frac{1}{2} \Delta\boldsymbol{\xi}^T \boldsymbol{C}^{-1} \Delta\boldsymbol{\xi} \right), \tag{4.5}$$

where $\boldsymbol{C}$ is the covariance matrix of the state variables in (4.3) obtained as the solution of the following Lyapunov equation:

$$\boldsymbol{A}\boldsymbol{C} + \boldsymbol{C}\boldsymbol{A}^T = -\boldsymbol{B}\boldsymbol{B}^T . \tag{4.6}$$

The covariance matrix is symmetrical and positive and it can be interpreted as a linear operator that, if applied to a vector of random variables, maps the linear combination of these random variables. Thus, for linear(ized) systems, the distribution of the states $\Delta\boldsymbol{x}$ can be obtained directly from the distribution of $\Delta\boldsymbol{\eta}$, as follows:

$$\Delta\boldsymbol{x} = \boldsymbol{C}_{x\eta} \boldsymbol{C}_{\eta\eta}^{-1} \Delta\boldsymbol{\eta} , \tag{4.7}$$

where we have clustered matrix $\boldsymbol{C}$ as:

$$\boldsymbol{C} = \begin{bmatrix} \boldsymbol{C}_{xx} & \boldsymbol{C}_{x\eta} \\ \boldsymbol{C}_{x\eta}^T & \boldsymbol{C}_{\eta\eta} \end{bmatrix} , \tag{4.8}$$

where $\boldsymbol{C}_{xx}$ and $\boldsymbol{C}_{\eta\eta}$ are the covariance matrices of $\Delta\boldsymbol{x}$ and $\Delta\boldsymbol{\eta}$, respectively, and $\boldsymbol{C}_{x\eta}$ represents the covariance matrix between deterministic $\Delta\boldsymbol{x}$ and stochastic $\Delta\boldsymbol{\eta}$ state variables.

In conclusion, for a linear system, $\Delta\boldsymbol{x}$ are normally distributed if $\Delta\boldsymbol{\eta}$ are normally distributed. From (4.4), also $\Delta\boldsymbol{y}$ are normally distributed. However, the power system model (4.1) is not linear. Thus, it has to be expected that nonlinearity *deforms* the distribution of states and algebraic variables. If the nonlinearity is not symmetrical, it has also to be expected that the distribution of (at least some) variables is skewed. The following two sections demonstrate this statement through two relevant cases, which are also further illustrated in the case study.

4.2.2 *Asymmetry Due to Network Losses*

Let us consider a simple one-machine one-load power system as shown in Fig. 4.1. Let us also assume that the system is at an equilibrium point. The power generated by the machine compensates both the load consumption and the losses in the transmission system:

$$p_G = p_D + p_{\rm loss} , \tag{4.9}$$

where losses are proportional to the square of the current. Assuming negligible the variations of the voltage, we can write:

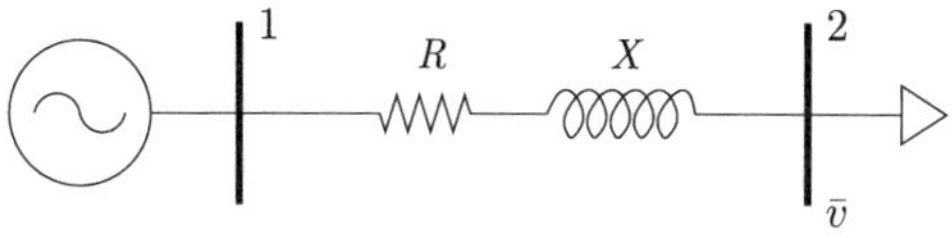

Fig. 4.1 One-machine one-load system

$$p_G \approx p_D + \frac{R}{v^2}\, p_D^2 \,, \tag{4.10}$$

where R is resistance of the line and the internal losses of the generator and v is the voltage, which, for simplicity of illustration, we assume constant. Moreover, we assume that initially the frequency of the system is equal to the synchronous reference, namely, $\omega = 1$ pu.

Starting from the initial balance in (4.10), consider a variation of the load consumption Δp_D:

$$\begin{aligned} p_G + \Delta p_G &= p_D + \Delta p_D + \frac{R}{v^2}\,(p_D + \Delta p_D)^2 \,, \\ &= p_D + \frac{R}{v^2}\, p_D^2 + 2\frac{R}{v^2}\, p_D \Delta p_D + \frac{R}{v^2}\, \Delta p_D^2 \,, \end{aligned} \tag{4.11}$$

where Δp_G is the active power variation of the generator due to the Turbine Governor (TG). Assuming that Δp_G is a function of the droop-control of the TG of the machine, in steady-state, one has:

$$\Delta p_G = -\frac{1}{\mathcal{R}}\, \Delta\omega \,, \tag{4.12}$$

where $\mathcal{R}$ is the droop coefficient of the TG. For simplicity, we have neglected the contribution of automatic Secondary Frequency Control (SFC) in (4.12). The case study section shows that Automatic Generation Control (AGC) helps reducing asymmetry but cannot remove it as it does not remove the nonlinearity of the system. Merging (4.10) and (4.12) in (4.11), we obtain:

$$\begin{aligned} \Delta\omega &= -\left(2\frac{R\mathcal{R}}{v^2}\, p_D\right) \Delta p_D - \left(\frac{R\mathcal{R}}{v^2}\right) \Delta p_D^2 \,, \\ &= -K_1\, \Delta p_D - K_2\, \Delta p_D^2 \,, \end{aligned} \tag{4.13}$$

which shows that the effect of losses represented with the quadratic term on the frequency deviation $\Delta\omega$ is always positive and thus has an asymmetrical effect on the steady-state value of the frequency. This effect, in turn, impacts the probability distribution of the frequency which is skewed toward the side with $\Delta\omega > 0$ pu, as the effect of losses mitigates negative variations of the load power consumption.

4.2.3 Asymmetry Due to Wind Frequency Control

The mechanical power p_w extracted from the wind is a nonlinear function of the wind speed ν, the rotor speed ω_m and the pitch angle θ_p. The mechanical power p_w of the wind turbine can be approximated as:

$$p_w = \frac{\rho}{2}\pi r^2 c_p(\lambda, \theta_p)\, \nu^3 \,, \tag{4.14}$$

where ν is the wind speed; r is the radius of the turbine disk; θ_p and $\lambda = \omega_t r/\nu$ are the blade pitch angle and tip speed ratio, respectively, where ω_t is the angular speed of the turbine; and ρ is the air density. The turbine efficiency function $c_p(\lambda, \theta_p)$ can be approximated as follows [7]:

$$c_p = 0.22\left(\frac{116}{\lambda_i} - 0.4\,\theta_p - 5\right)e^{-\frac{12.5}{\lambda_i}} \,, \tag{4.15}$$

with

$$\frac{1}{\lambda_i} = \frac{1}{\lambda + 0.08\,\theta_p} - \frac{0.035}{\theta_p^3 + 1} \,, \tag{4.16}$$

where numerical coefficients are determined empirically. There exist several other alternative empirical expressions for c_p [8]. In conventional wind turbines, where the main goal of the generator is to maximize the power extracted from the wind, $\theta_p = 0$ in normal operating conditions and $\theta_p \neq 0$ is utilized exclusively to limit the power of the wind turbine at high wind speeds.

A way to regulate the frequency through wind turbines, equivalent to a power reserve, is to introduce an offset of the pitch angle in normal operating conditions [9]. Then the frequency can be regulated through a droop control:

$$\theta_p' = \frac{1}{\mathcal{R}}\Delta\omega - \theta_p \,. \tag{4.17}$$

Considering again the load variation Δp_D and assuming for simplicity constant wind speed and thus constant tip speed ratio, and that the power injected into by the wind turbine coincides with the mechanical power p_w, that is, neglecting losses and considering an ideal converter control, in steady-state and considering small variations around the equilibrium, one has:

$$\Delta\theta_p = -\frac{1}{\mathcal{R}}\Delta\omega \,, \tag{4.18}$$

and hence:

$$\begin{aligned} \Delta p_G &= c_1\,\Delta\theta_p + c_2\,\Delta\theta_p^2 + O(\Delta\theta_p^3)\,, \\ &= -\frac{c_1}{\mathcal{R}}\,\Delta\omega + \frac{c_2}{\mathcal{R}^2}\,\Delta\omega^2 + O(\Delta\omega^3)\,, \end{aligned} \tag{4.19}$$

where the coefficient c_1 and c_2 are obtained from the Taylor series expansion of c_p centered at the initial operating point and $O(\Delta\omega^3)$ is the residual terms of order higher than 2. Neglecting high-order terms, (4.13) obtained in the previous section can be rewritten as follows:

$$c_1\,\Delta\omega - \frac{c_2}{\mathcal{R}}\,\Delta\omega^2 = -K_1\,\Delta p_D - K_2\,\Delta p_D^2\,, \tag{4.20}$$

which can be solved for $\Delta\omega$ and is, as expected, a nonlinear expression linking $\Delta\omega$ and Δp_D. Note that $c_1 \approx c_2$, and hence $c_1 \ll c_2/\mathcal{R}$ as the droop coefficient $\mathcal{R}$ is of the order of 10^{-2}. Then, (4.20) can be simplified as:

$$\Delta\omega^2 \approx \frac{\mathcal{R}\,K_1}{c_2}\,\Delta p_D + \frac{\mathcal{R}\,K_2}{c_2}\,\Delta p_D^2\,. \tag{4.21}$$

The examples presented in the case study show that the nonlinearity due to wind frequency regulation obtained using pitch angle control is "stronger" than that due to system losses and thus leads to higher asymmetry.

4.2.4 Asymmetry Due to Regulator Hard Limits

Another important source of asymmetry in any controller are the hard limits. If a controller operates close to one of its limits, in fact, every time that the limit is binding the effect the regulator becomes inactive, thus, leading to a substantially different behavior of the controlled system. This effect is particularly common in wind generation (with or without PFC) and has been observed specifically in the Irish system when narrow dead-bands are enforced in the PFC of wind turbines. This aspect is further discussed in Sects. 4.4.1 and in the case study.

4.2.5 Compensation of Asymmetry Through Control

In this section, we discuss a nonlinear dead-band function to reduce asymmetry which uses the frequency deviation $\Delta\omega$ as an input signal (see Fig. 4.2), as follows:

$$u(\Delta\omega) = (1 - \gamma\Delta\omega)^{\alpha}\,\Delta\omega^{\beta}\,, \tag{4.22}$$

where γ and α are adjustable parameters. Substituting the expression (4.22) into (4.12) one obtains:

$$\Delta p_G = -\frac{1}{\mathcal{R}}u(\Delta\omega)\,, \tag{4.23}$$

and, hence, (4.13) becomes:

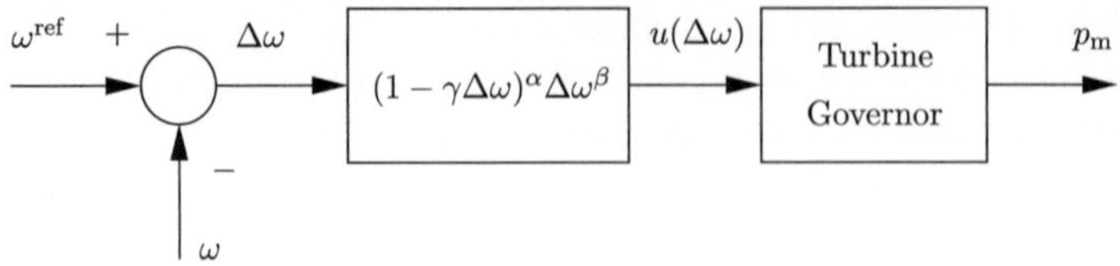

Fig. 4.2 Nonlinear compensation for TGs

$$(1-\gamma\Delta\omega)^{\alpha}\Delta\omega^{\beta} = -K_1\Delta p_D - K_2\Delta p_D^2 \,. \tag{4.24}$$

If one chooses $\alpha = \beta = 1$ and $\gamma = K_2$, then:

$$\Delta\omega = -K_1\Delta p_D \,, \tag{4.25}$$

which leads to a linear (symmetric) controller where nonlinearity due to losses is fully compensated.

Similarly, if in (4.18) one implements the proposed controllers, and sets $\alpha = \beta = 0.5$ and $\gamma = \mathcal{R}\,K_2/c_2$, (4.21) can be rewritten as:

$$\Delta\omega \approx -\frac{\mathcal{R}\,K_1}{c_2}\,\Delta p_D \,, \tag{4.26}$$

which, again, is a linear expression and makes the pitch-angle frequency control symmetric.

As the expressions (4.13) and (4.21) are obtained through approximated models where higher order terms are neglected, and as the coefficient K_2 depends on the operating point, it is not possible, in general, to obtain a full compensation of the losses. Based on several tests, for example, we have determined empirically that $\alpha = 0.5$ and $\beta = \gamma = 1$ work well in the simulated-based scenario considered in the case study for TGs of conventional power plants.

4.3 Metrics

To measure asymmetry, we propose to calculate the left and right-hand side standard deviations of the FPDs, namely "negative" (σ_{f-}) for when the frequency is below the nominal (f_n) (sample size N_-) and "positive" (σ_{f+}) when frequency is above f_n (sample size N_+). Then, we calculate the frequency standard deviation σ_f using a weighted-average method of the two standard deviations (σ_{f-}, σ_{f+}). Next, the asymmetry $\Delta\sigma_f$ is defined as the difference between σ_{f-} and σ_{f+}. These equations are shown below:

$$\sigma_{f-} = \sqrt{\frac{1}{N_-}\sum_{i=1}^{N_-}(f_i - f_n)^2}\,, \quad \{f_i : f_i < f_n\}\,, \tag{4.27}$$

$$\sigma_{f+} = \sqrt{\frac{1}{N_+}\sum_{i=1}^{N_+}(f_i - f_n)^2}\,, \quad \{f_i : f_i > f_n\}\,, \tag{4.28}$$

$$\sigma_f = \sqrt{\frac{1}{N}\sum_{i=1}^{N}(f_i - f_n)^2} \approx \sqrt{\frac{N_+\sigma_{f+}^2 + N_-\sigma_{f-}^2}{N_+ + N_-}}\,, \tag{4.29}$$

$$\Delta\sigma_f = |\sigma_{f-} - \sigma_{f+}|\,. \tag{4.30}$$

Note that the expressions in (4.29) are substantially equal as, in practice, very few measurements are exactly equal to f_n.

As there are other existing asymmetry-based metrics used in the literature such as the skewness, we calculate and compare it, where relevant, with $\Delta\sigma_f$. The skewness parameter, γ_f, is defined as follows [5]:

$$\gamma_f = \frac{1}{N}\sum_{i=1}^{N}\left(\frac{f_i - \mu_f}{\sigma_f}\right)^3\,, \tag{4.31}$$

where μ_f is the mean of system frequency. Note that if γ_f equals to zero the distribution is normal and, hence, symmetric. In contrast, a non-zero skewness implies a distribution that is not symmetric.

Another key metric of long-term/annual frequency quality used by TSOs is the so-called ± 100 mHz criteria which measures the minutes frequency spends outside the ± 100 mHz range [10]. We calculate these minutes in relevant scenarios discussed in Sect. 4.5.

4.4 Asymmetry in Real-World Power Systems

This section provides evidence of FPD asymmetry based on measurements obtained from EirGrid and AEMO.

4.4.1 Irish Grid

We first show the appearance of asymmetry in a real-world system with high shares of wind generation, namely the Irish power system [11]. With this aim, we select three relevant hours with the following details:

Table 4.1 Scenarios of the Irish power system

Scenario	Wind generation	APC	db_{wind} [mHz]	db_{conv} [mHz]	AGC Conv/wind	Wind ramps	Load noise	Wind noise	Losses	Saturation	db_{model}
1	Yes	Off	±200	±15	No	Yes	Yes	Yes	Normal	No	–
2	Yes	On	±15	±15	No	Yes	Yes	Yes	Normal	No	–
3	Yes	Off	±200	±15	No	Yes	Yes	Yes	Normal	No	–

- *Scenario 1*: dead-band of wind farms is ±200 mHz and thus Active Power Control (APC) is turned Off.
- *Scenario 2*: dead-band of wind farms is reduced to ±15 mHz and thus APC is turned On.
- *Scenario 3*: Represents conventional power systems with near-zero wind generation and APC Off (±200 mHz dead-band).

Further details on each scenario can be found in Table 4.1. Note that currently in the Irish power system APC is normally disabled. The APC is enabled in special circumstances, for example, during periods of high export or if there are severe frequency oscillations in the system [12].

In the remainder of this section, we consider three measurement data sets obtained from the TSOs historical information system (coming from Supervisory Control and Data Acquisition (SCADA) and stored in 1 s resolution) and calculate σ_f, $\Delta\sigma_f$ and minutes outside the ±100 mHz range. Note that AGC is not utilized in the Irish power system. Table 4.5 summarizes these scenarios. Further information on the operating conditions for each scenario can be found online in [13].

Scenarios 1 and 2. These two scenarios correspond to January 27, 2024, namely 6 consecutive hours, 3 when APC was Off and 3 when APC was turned On. The Irish power system experienced high wind generation around 3.5 GW. Note that the peak demand of the Irish system is approximately 7.5 GW and valley demand is less than 2.5 GW. In these 6 h, the System Non-Synchronous Penetration (SNSP) was around 72% on average and system conditions remained almost the same, at least, in terms of wind generation, number of conventional units online (7) and demand.

Figure 4.3 depicts the FPDs for both scenarios, namely with and without APC. APC Off leads to a normal distribution and symmetric behavior of the FPD. It can be inferred from Table 4.2, in fact, that the asymmetry for the 3 h when APC was Off (Scenario 1) is very small and similar to Scenario 4 in the Western Systems Coordinating Council (WSCC) test system (see Table 4.6 below). The minutes outside the ±100 mHz range are also small. It is worth pointing out that whenever the frequency drifts away from the ±100 mHz range, the operators take manual actions sich as conventional generation redispatch to bring back frequency within the range. On the other hand, Fig. 4.3 shows that when APC is enabled (Scenario 2) to reduce the frequency oscillations (all 6 APC groups [12]) it seriously increases the asymmetry of the FPD (this scenario is equivalent to Scenario 10 in Table 4.6 below). Specifically, it can be seen from the results in Table 4.2 that while the average standard deviation (σ_f) is more or less the same with Scenario 1, the asymmetry $\Delta\sigma_f$ has increased

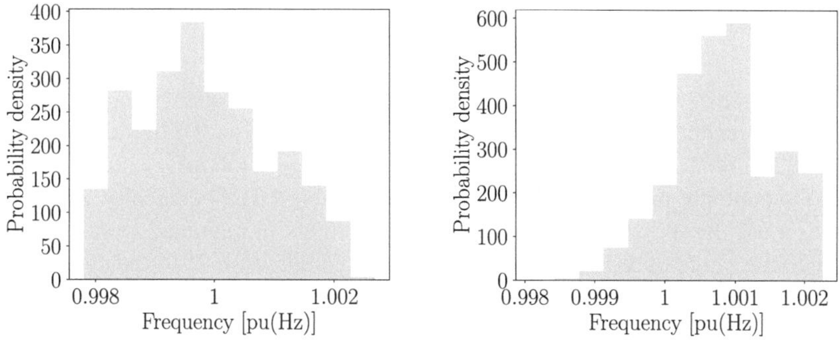

(a) without APC (b) with APC

Fig. 4.3 FPD of the Irish power system with and without APC

Table 4.2 Results for the Irish power system

Scenario	σ_f [Hz]	σ_{f-} [Hz]	σ_{f+} [Hz]	$\Delta\sigma_f$ [Hz]	Minutes outside ± 100 mHz	Minutes outside ± 100 mHz	Minutes outside -100 mHz	p_{losses} [pu(MW)]	q_{losses} [pu(MVAr)]
1	0.0558	0.0557	0.0560	0.0003	6.6	3.8	2.8	–	–
2	0.0547	0.0259	0.0575	0.0316	7	7	0	–	–
3	0.030	0.0359	0.0152	0.0207	0	0	0	–	–

dramatically. In fact, the asymmetry is very similar to that considered in Scenario 10 for the WSCC test system. The asymmetry can also be observed in the minutes outside the ± 100 mHz range. We also tested other real-world scenarios when APC was Off and On and obtained similar results. Note also that while wind turbines may operate based on asymmetric droop characteristic [14], that is not the case in the Irish power grid (symmetric droop is used instead).

Scenario 3. This scenario refers to April 20, 2024 and represents a conventional power system with near-zero wind generation as wind levels in the Irish power system during this particular period were around 50 MW. This scenario thus allows for an interesting comparison with respect to the Scenarios 1 and 2 with high (3.5 GW) wind shares.

Results suggest that σ_f, $\Delta\sigma_f$ and minutes outside the ± 100 mHz range are reduced compared to Scenario 2. However, compared to Scenario 1, the asymmetry ($\Delta\sigma_f$) is increased significantly. This can be explained by the fact that being a near-zero wind generation condition means that the conventional generators online and Interconnectors (ICs) should operate closer to their maximum limits [13].

4.4.2 Australian Grid

As anticipated in the introduction, the asymmetry in the Australian (mainland) power system has recently raised concerns. We select three years with different dead-band implementations namely 2010 ($\pm$30-50 mHz), 2019 ($> \pm$150 mHz) and 2023 ($\pm$15 mHz) to compare the asymmetry evolution (see Table 4.3). Specifically, it is worth pointing out that prior to 2015 the dead-band settings in mainland were generally set within a range of $\pm$30 mHz to $\pm$50 mHz [1]. Whereas in 2016 the decisions of the Australian Energy Regulator and Market Commission not to enforce generator participants a specific dead-band made the latter widening their dead-bands, effectively resulting in no PFC in the standard frequency range of $\pm$150 mHz. Finally, in 2020 the mandatory PFC rule with $\pm$15 mHz dead-band was enforced motivated by a deterioration of frequency quality.

Table 4.4 presents all the relevant results, while Fig. 4.4 depicts the FPD in 2023 and 2019 (frequency recordings are made publicly available by AEMO). While σ_f and minutes outside $\pm$150 mHz have dramatically decreased in 2023 compared to 2010 and 2019, that is not the case for the asymmetry. Specifically, the asymmetry in 2023 equals $\Delta\sigma_f = 0.00267$ Hz compared to just 0.00032 Hz and 0.0011 Hz in 2019 and 2010, respectively. As mentioned above, the main change after 2020 is the introduction of mandatory PFC provision with $\pm$15 mHz dead-band.

In each case, the asymmetry can be also observed in the minutes outside $\pm$150 mHz range. In particular, the degradation of frequency performance in 2019, which led to a significant increase in the minutes outside $\pm$150 mHz, is not surprising given the wider dead-band implementation by generators. These results from the Australian power system further support the theoretical insights and observations in

Table 4.3 Scenarios of the Australian power system in 2010, 2019 and 2023

Year	Wind generation	APC	db_{wind} [mHz]	db_{conv} [mHz]	AGC	Wind Ramps	Wind Noise	Losses	Saturation
2010	Yes	Off	$\pm 30 - 50$	$\pm 30 - 50$	Yes	Yes	Yes	–	–
2019	Yes	Off	$> \pm 150$	$> \pm 150$	Yes	Yes	Yes	–	–
2023	Yes	On	± 15	± 15	Yes	Yes	Yes	–	–

Table 4.4 Asymmetry results for the Australian power system in 2010, 2019 and 2023

Year	σ_f [Hz]	σ_{f-} [Hz]	σ_{f+} [Hz]	$\Delta\sigma_f$ [Hz]	Minutes outside $\pm$150 mHz	Minutes outside +150 mHz	Minutes outside −150 mHz	γ_f [Hz]
2010	0.0286	0.0294	0.0282	0.0011	137.13	4.4	132.73	–0.18
2019	0.0635	0.0647	0.0643	0.00032	4059	841	3218	–0.047
2023	0.0256	0.0271	0.0244	0.00267	3.4	0.133	3.266	–0.221

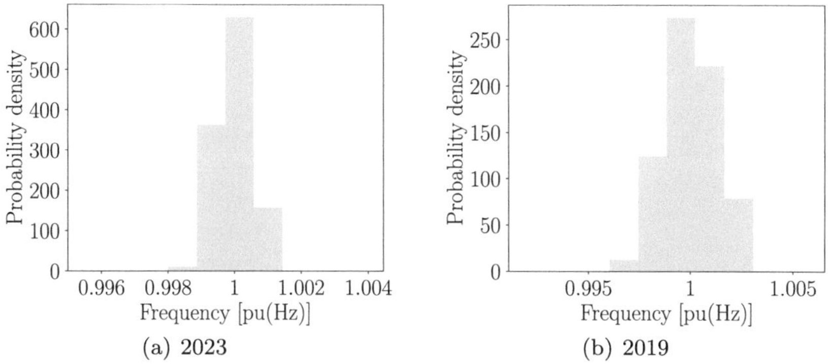

Fig. 4.4 FPD of the Australian power system in 2023 and 2019

the Irish grid (and also the conclusions of the case study below) that narrow frequency dead-bands from RES lead to substantially increase the asymmetry of the FPD.

Finally, we calculate the γ_f and compare it with $\Delta\sigma_f$. Table 4.4 suggests that while γ_f is at least two orders of magnitude higher than $\Delta\sigma_f$, it is consistent with $\Delta\sigma_f$ in terms of it being higher in 2023 (-0.221 Hz) compared to 2019 (-0.047 Hz) and 2010 (-0.18 Hz), respectively. It is also worth pointing out that all γ_f values take a negative sign indicating a leftward skew with a longer tail for negative values of the frequency distribution. This is also consistent with the results of the standard deviation-based metrics, that is, σ_{f-} being higher than σ_{f+} in all cases.

4.5 Examples

To analyze the statistical properties of the frequency we run long-term Time Domain Simulations (TDSs), namely 48 h, based on the stochastic dynamic model (4.1). All simulations are solved with the software tool Dome [15]. The results shown in this section are obtained using the WSCC 9-bus system (see Appendix B.1), adapted for the various considered scenarios for a comprehensive evaluation of the different sources of asymmetry. A short description of each scenario is provided below. Table 4.5 provides relevant information on setup and controllers and Table 4.6 shows simulation results for each scenario.

- *Scenario 4*: Conventional power systems without wind generation. Synchronous generators have a ± 15 mHz governor dead-band and are under AGC.
- *Scenario 5*: Same as Scenario 4 but with increase of network losses.
- *Scenario 6*: Same as Scenario 5 but with inclusion of the proposed nonlinear compensation in the PFC of synchronous generators.
- *Scenario 7*: Studies the effect of saturation on the FPD.

Table 4.5 Scenarios of the WSCC test system

Scenario	Wind generation	APC	db_{wind} [mHz]	db_{conv} [mHz]	AGC conv./ wind	Wind ramps	Wind distr.	Losses	Saturation	Compensation Eq. (4.22)
4	–	–	–	±15	Conv.	–	–	Medium	–	–
5	–	–	–	±15	Conv.	–	–	High	–	–
6	–	–	–	±15	Conv.	–	–	High	–	Yes
7	–	–	–	–	–	–	–	Medium	Yes	–
8	Yes	Off	±200	–	–	–	Normal	Medium	–	–
9	Yes	On	±15	–	–	–	Normal	Medium	–	–
10	Yes	On	±15	–	–	Yes	Normal	Medium	–	–
11	Yes	On	±15	–	Conv.	Yes	Normal	Medium	–	–
12	Yes	Off	±200	–	Conv. and wind	Yes	Normal	Medium	–	–
13	Yes	On	±15	–	Conv. and wind	Yes	Normal	Medium	–	–
14	Yes	On	±15	–	Conv. and wind	Yes	Normal	Medium	–	Yes
15	Yes	On	±15	–	Conv. and wind	Yes	Weibull	Medium	–	–

- *Scenario 8*: Power system with wind generation providing PFC with ±200 mHz dead-band (APC Off).
- *Scenario 9*: Same as Scenario 8 but with ±15 mHz dead-band (APC On).
- *Scenario 10*: Same as Scenario 9 and with wind ramps modeled based on realistic data from the Irish system.
- *Scenario 11*: Same as Scenario 10 but with AGC installed (only conventional power plants).
- *Scenario 12*: Wind farms provide AGC functionality with APC Off.
- *Scenario 13*: Wind farms provide AGC functionality with APC On.
- *Scenario 14*: Same as Scenario 13 and APC includes the proposed nonlinear compensation to reduce asymmetry.
- *Scenario 15*: Same as Scenario 13 with Weibull-distributed wind speed fluctuations.

For all scenarios, load consumption includes Gaussian noise modeled as mean-reverted Ornstein-Uhlenbeck processes and stochastic jumps as described in [16, 17], respectively.

Table 4.6 Results for the WSCC test system

Scenario	σ_f [Hz]	σ_{f-} [Hz]	σ_{f+} [Hz]	$\Delta\sigma_f$ [Hz]	Minutes outside ±100 mHz	Minutes outside +100 mHz	Minutes outside −100 mHz	p_{losses} [pu(MW)]	q_{losses} [pu(MVAr)]
4	0.0107	0.0108	0.0107	0.0001	0	0	0	0.0409	–0.9452
5	0.0314	0.0330	0.0294	0.0036	5.28	1.27	4.00	0.5097	–0.6151
6	0.0350	0.0349	0.0350	0.00007	10.8	6.44	4.35	0.5097	–0.6151
7	0.0236	0.0245	0.0227	0.0018	0.2233	0.0183	0.205	0.0409	–0.9452
8	0.0602	0.0606	0.0598	0.0008	280.49	136.38	144.10	–	–
9	0.0803	0.0841	0.0762	0.0079	611.41	287.69	323.71	–	–
10	0.1085	0.1254	0.0868	0.0386	1073.61	456.67	616.94	–	–
11	0.0794	0.0845	0.0745	0.01	575.79	267.32	308.47	–	–
12	0.0635	0.0629	0.0641	0.0012	314.91	152.70	162.20	–	–
13	0.0591	0.0630	0.0555	0.0074	–	–	–	–	–
14	0.0649	0.0660	0.0638	0.0021	–	–	–	–	–
15	0.1044	0.110	0.0994	0.0105	–	–	–	–	–

4.5.1 Power Systems with Conventional Generation

In this first section, we focus on the asymmetry that exists in conventional power systems. Two potential sources of asymmetry are considered, namely, losses and control limiters.

Scenario 4: This scenario represents conventional power systems with synchronous machines providing both PFC and AGC [18]. The main source of volatility in this scenario is noise in loads modeled as a stochastic process that incorporates both continuous and event-driven (jumps) dynamics. This is based on real-world behavior of loads in the Irish power system [17]. Figure 4.5a depicts the center of inertia FPD. As expected, the FPD shows a normal distribution and symmetric behavior. These results are confirmed in Table 4.6, which shows that the asymmetry calculated using (4.30) is small ($\Delta\sigma_f = 0.0001$ Hz) when compared to power systems with Inverter-Based Resource (IBR)-based generation (see below). The minutes outside the ± 100 mHz range are zero. Note that since the focus of this chapter is on long-term frequency distribution, it does not matter what frequency measurement is used, that is, the center of inertia or, for example, the frequency measured at a local bus. In fact, in the Irish grid example, we utilize the frequency measurements at a specific bus. However, the asymmetry is still present and consistent when using the center of inertia method in this section.

With regard to AGC, its main goal is to eliminate the frequency error, that is, difference between the reference frequency and the measured frequency at a pilot bus of the system, through an integral controller as shown in Fig. 2.19 [18]. As it shrinks the frequency distribution [19], the AGC is beneficial to reduce the effects of the asymmetries present in the system.

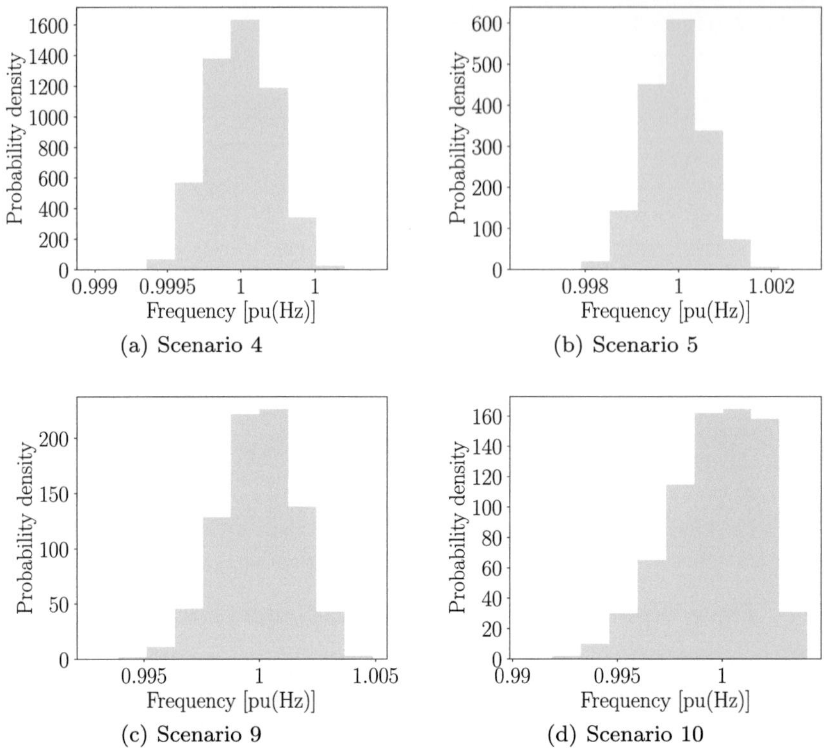

Fig. 4.5 FPD for relevant scenarios

Scenario 5: Here we focus on the impact that the increase of system losses have on the FPD. For illustration purposes, we increase the resistance of transmission lines by ten times (this can represent distribution systems). Figure 4.5b shows that losses have a significant impact on the FPD and its asymmetry. Moreover, Table 4.6 shows that, compared to Scenario 4, $\Delta\sigma_f$ has increased from $\Delta\sigma_f = 0.0001$ Hz to $\Delta\sigma_f = 0.0036$ Hz. Consequently, the standard deviation and the ± 100 mHz criteria have also increased.

Scenario 6: This scenario explores the effectiveness of the proposed nonlinear compensation model presented in Sect. 4.2.5 in removing/reducing the asymmetry. With this aim, we select the case with high losses (Scenario 5) and set the values of γ and α to 1 and 0.5, respectively (of course, other values might be used). Results from Table 4.6 indicate that the asymmetry is reduced significantly (from $\Delta\sigma_f = 0.0036$ Hz to $\Delta\sigma_f = 0.00007$ Hz). Therefore, this compensation can be considered as a viable option for TSOs operating grids with high asymmetry of the frequency distribution.

Scenario 7: In this scenario, we simulate the effect of saturation/limits (highly loaded systems). Specifically, we assume that the maximum powers of synchronous generators 1 and 2 are reduced by approximately 40% and 50%, respectively. As expected, the results of Table 4.6 indicate an increase in the asymmetry in the FPD compared to, for example, Scenario 4, but lower compared to Scenario 5 that considers high network losses.

4.5.2 Power Systems with IBR-Based Generation

This section discusses the effect of wind generation and its control under different scenarios. In particular, the aim is to see whether the introduction of wind and its frequency control/regulation capability in the power system affects the asymmetry of the FPD. With this aim, we consider again the WSCC 9-bus system (see Appendix B.1) and replace the synchronous generator 3 with a wind power plant modeled as a double-fed induction generator [20]. We assume that the wind power plant provides primary frequency regulation through droop-based PFC with tight dead-band namely ±15 mHz (same used by the governors of conventional generators), as is the case in the Irish power system [10]. This is also known as wind/solar farm APC [12]. To be able to provide up and down regulation, we assume wind is operating 20% below its Maximum Power Point Tracking (MPPT). The description of Scenarios 8–15 is provided in Table 4.5.

Scenario 8: We assume that wind farms provide frequency support only to respond to contingency events (dead-band of ±200 mHz). Note that since in practice frequency rarely goes outside the ±200 mHz range during a day (except in very small island power systems and Microgrids (MGs)), it can be said that wind farms provide regular wind generation. We also model noise from wind and loads as Gaussian noise ($\boldsymbol{b}$ is constant) but no jumps ($\boldsymbol{c} = \mathbf{0}$) in (4.1). Results in Table 4.6 indicate that the inclusion of wind farms increase σ_f, the asymmetry ($\Delta\sigma_f = 0.0008$ Hz) and minutes outside the ±100 mHz range compared to conventional power systems without wind (see Scenario 4).

Scenario 9: This scenario considers the effect of wind farms providing APC functionality through the ±15 mHz dead-band. This means they are continuously providing dynamic primary frequency regulation since frequency moves outside the ±15 mHz dead-band on a regular basis during a day. Noise is same as in Scenario 8. Figure 4.5c and Table 4.6 show relevant results. Surprisingly, the APC leads to increase σ_f and the asymmetry of the FPD (from $\Delta\sigma_f = 0.0008$ Hz to $\Delta\sigma_f = 0.0079$ Hz). These results are a consequence of the nonlinearity of the wind turbine, namely, the cubic relationship between wind speed and turbine torque.

Scenario 10: In addition to wind farms providing APC functionality, here we also consider wind ramps modeled based on realistic profile from the Irish power system. Ramps are obtained through stochastic jumps in (4.1) as described in [17]. Results are shown in Table 4.6. The standard deviation ($\sigma_f = 0.1085$ Hz), asymmetry ($\Delta\sigma_f = 0.0386$ Hz) and minutes outside the ±100 mHz range (1073.61) have dramatically increased compared to the previous scenario. To illustrate this, we plot in Fig. 4.5d the FPD. It is striking to see that the behavior of the FPD is quite asymmetric.

Scenario 11: This is the same scenario as Scenario 10 but now with the inclusion of AGC (only conventional generators). Looking at the results in Table 4.6, we can see now that AGC significantly reduces σ_f (from 0.1085 Hz to 0.0794 Hz), the asymmetry (from $\Delta\sigma_f = 0.0386$ Hz to $\Delta\sigma_f = 0.01$ Hz) and minutes outside the ± 100 mHz range (from 1073.61 to 575.79). Thus, the AGC appears as a viable solution to reduce the frequency asymmetry in power systems and help keep frequency quality within limits.

Scenario 12: The asymmetry of the FPD and dynamic behavior of the system can be improved if wind is also providing AGC functionality (but with APC Off). This is shown in Table 4.6 where we can see that σ_f, $\Delta\sigma_f$ and minutes outside the ± 100 mHz range are improved significantly with the inclusion of wind farms in AGC.

Scenario 13: This is the same as Scenario 12 but now we reduce the dead-band of wind farms to ± 15 mHz to see its impact on asymmetry. Interestingly, making wind farms adjust their MW output much more dynamically to control frequency under normal, pre-contingency, conditions leads to an increase of $\Delta\sigma_f$ namely from 0.0012 Hz (Scenario 12) to 0.0074 Hz (this Scenario). These results support the observations in the Irish and Australian power grids where higher asymmetry is observed when RES provide APC-type frequency regulation.

Scenario 14: In this scenario, we check the effectiveness of the nonlinear dead-band implementation in wind farms in reducing the asymmetry. With this aim, we set the values of γ and α to 2 and 0.2, respectively. Compared to the previous scenario, it can be seen that $\Delta\sigma_f$ is reduced by more than three times namely from 0.0074 Hz to 0.0021 Hz. Therefore, such a dead-band implementation might be considered as a potential solution by TSOs.

Scenario 15: Finally, we study the effect of different wind speed noise distributions on the frequency asymmetry, namely, Gaussian (symmetric) and Weibull distribution (asymmetric). Details of the model of the wind speed based on Weibull distribution are given in [21]. Table 4.6 shows that the Weibull distributed noise leads to a higher asymmetry ($\Delta\sigma_f = 0.0105$ Hz) compared to the Gaussian noise ($\Delta\sigma_f = 0.0074$ Hz), as expected. We note however that, short-term wind speed fluctuations are better represented as Gaussian noise around an average value [22]. Such an average, calculated across periods of, say, an hour, is distributed as a Weibull or other non-symmetrical distributions. The main source of asymmetry (nonlinearity) due to wind generation is thus not the wind speed but its turbine and its frequency control.

4.6 Comparison of FPD for Real-World Power Systems

We conclude the chapter with a comparison of FPDs of the AIPS, GB, CE and Nordic power systems. Figure 4.6 shows frequency data for these systems as measured on April 29, 2025.

The CE and Nordic power systems show a normal distribution while the AIPS and GB display non-Gaussian features, that is, skewed and bimodal distributions, respectively. The main reason why the power grid frequency of AIPS and GB sys-

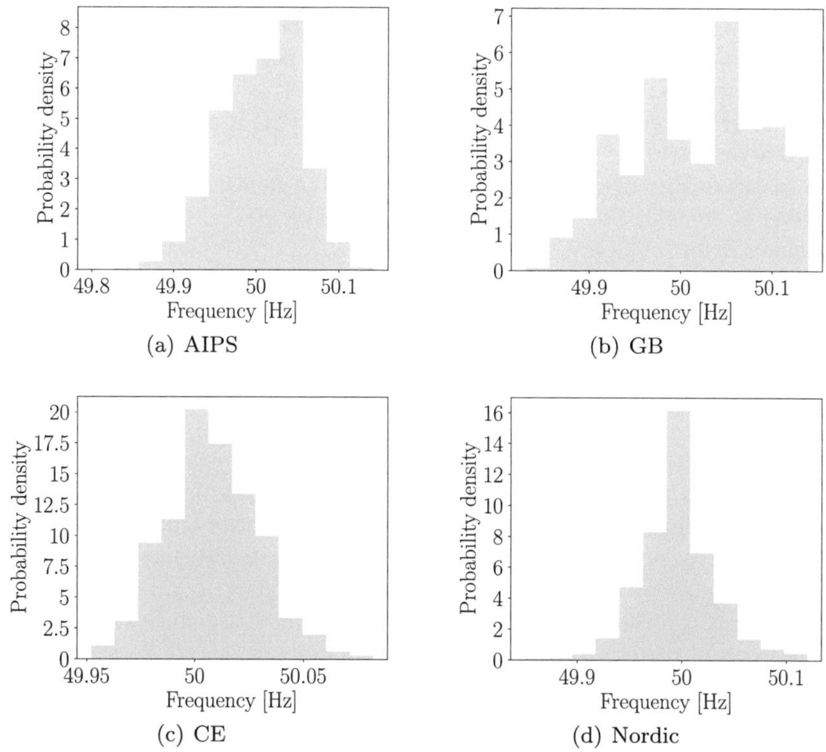

Fig. 4.6 FDP of four power systems on April 29, 2025

tems is not following a normal distribution is the lack of a perfect tracking SFC in these jurisdictions. On the other hand, the TSOs of CE and Nordic system utilize an AGC/Automatic Frequency Restoration Reserve (aFRR) with mainly conventional units participating under it. Some literature suggests that the different PFC dead-band settings may be contributing to these differences [23]. However, Table 2.8 shows that these four jurisdictions apply very similar dead-band settings. Moreover, to the best of our knowledge, except for the AIPS that occasionally reduces the PFC dead-band on RES from $\pm$200 mHz to $\pm$15 mHz the rest of jurisdictions do not utilize a narrow dead-band on RES.

In Brief

- A metric to quantify the level of asymmetry in power systems is the difference between left and right standard deviations of the FPD. This metric allows evaluating the "quality" of frequency control of different power systems.

- The main cause of asymmetry in frequency distribution is nonlinearity (losses, control limits, and wind generation providing PFC and APC).
- AGC is beneficial to reduce FPD asymmetry.
- Nonlinear compensation of the frequency error in PFC is effective to reduce the FPD asymmetry.
- In Irish and Australian transmission grids, RES is a source of asymmetry.
- Power systems that do not employ perfect tracking SFC display non-Gaussian features in the frequency distribution.

Open Questions

The link between nonlinearity and asymmetry in power systems is an emerging topic that needs to be further investigated. In practice, it is never a good idea to have asymmetries in a system as they generally imply lower efficiency and, thus, higher costs for consumers. However, how asymmetry, in particular asymmetry in frequency control of RES, impacts power system efficiency is still not fully clear. Once this aspect is clarified, TSOs can define what level of asymmetry is acceptable for the dynamic performance, operation, and electricity markets of the system. This understanding will likely lead to design proper controllers to reduce or completely remove the asymmetry.

References

1. CS Energy (2023) Submission: draft determination—review of the frequency operating standard. [Online]. www.aemc.gov.au
2. AEMO (2023) Primary frequency response requirements. [Online]. www.aemo.com.au
3. Milano F, Dörfler F, Hug G et al (2018) Foundations and challenges of low-inertia systems (invited paper). In: Power systems computation conference (PSCC), pp 1–25
4. Wen X, Anvari M, Gorjão LR et al (2023) Non-standard power grid frequency statistics in Asia, Australia, and Europe. arXiv:2308.16842
5. Schäfer B, Beck C, Aihara K et al (2018) Non-Gaussian power grid frequency fluctuations characterized by Lévy-stable laws and superstatistics. Nat Energy 3(2):119–126
6. Carmichael HJ (2013) Statistical methods in quantum optics 1: master equations and Fokker-Planck equations. Springer Science & Business Media
7. Heier S (1998) Grid integration of wind energy conversion systems. John Wiley & Sons, Chichester, England
8. Slootweg J, de Haan S, Polinder H et al (2003) General model for representing variable speed wind turbines in power system dynamics simulations. IEEE Trans Power Syst 18(1):144–151
9. Wang T, Ding L, Yin S et al (2015) A new control strategy of DFIG-based wind farms for power system frequency regulation. In: IEEE PES Asia-pacific power and energy engineering conference, pp 1–5
10. Kërçi T, Hurtado M, Gjergji M et al (2023) Frequency quality in low-inertia power systems. In: IEEE PES general meeting, pp 1–5

11. Hurtado M, Kërçi T, Tweed S et al (2023) Analysis of wind energy curtailment in the Ireland and Northern Ireland power systems. In: IEEE PES general meeting, pp 1–5
12. EirGrid & SONI (2020) Active power control groups. [Online]. www.sem-o.com
13. EirGrid & SONI (2022) Operational policy roadmap 2023–2030. [Online]. https://www.eirgrid.ie/
14. Ela E, Gevorgian V, Fleming P et al (2014) Active power controls from wind power: bridging the gaps. National Renewable Energy Lab. (NREL), Golden, CO (United States), Tech. Rep
15. Milano F (2013) A Python-based software tool for power system analysis. In: IEEE PES general meeting, pp 1–5
16. Milano F, Zárate-Miñano R (Nov2013) A systematic method to model power systems as stochastic differential algebraic equations. IEEE Trans Power Syst 28(4):4537–4544
17. Mele FM, Zárate-Miñano R, Milano F (2019) Modeling load stochastic jumps for power systems dynamic analysis. IEEE Trans Power Syst 34(6):5087–5090
18. Kërçi T, Murad MAA, Dassios I et al (2021) On the impact of discrete secondary controllers on power system dynamics. IEEE Trans Power Syst :1–1
19. Mele FM, Ortega Á, Zárate-Miñano R et al (2016) Impact of variability, uncertainty and frequency regulation on power system frequency distribution. In: Power systems computation conference (PSCC), pp 1–8
20. Milano F (2010) Power system modelling and scripting. Springer, London
21. Zárate-Miñano R, Anghel M, Milano F (2013) Continuous wind speed models based on stochastic differential equations. Appl Energy 104:42–49
22. Jónsdóttir GM, Milano F (2019) Data-based continuous wind speed models with arbitrary probability distribution and autocorrelation. Renew Energy 143:368–376
23. Vorobev P, Greenwood DM, Bell JH et al (2019) Deadbands, droop, and inertia impact on power system frequency distribution. IEEE Trans Power Syst 34(4):3098–3108

Part II
Control and Stability Analysis

Chapter 5
Frequency Strength

As has been discussed so far in the book, in electric power systems, the balance of generation and consumption is ensured through frequency control. This chapter aims at introducing the concept of "frequency control strength" as a novel approach to understanding how different real-world power systems compare to each other in terms of the effectiveness and performance of system-wide frequency control. We present a comprehensive comparison, based on measurement data, of the frequency control strength of four real-world, renewable-based, synchronous island power systems, namely Great Britain (GB), All-Island Power System (AIPS), Australia (AUS), and Tasmania (TAS). The strength is evaluated by means of different short- and long-term frequency quality metrics (see Table 3.1).

In Sect. 5.1 we introduce the concept of frequency control strength and its current common understanding. As there is no currently general definition of *frequency strength*, we provide one. Section 5.2 delves into the main characteristics and provide an overview of frequency control of the four selected power systems, identifying where relevant similarities and differences. Section 5.3 describes all the frequency strength metrics. Then, Sect. 5.4 applies the respective strength metrics to a real-world case study of the four power systems, including normal and abnormal operating conditions. Final remarks and possible mitigation measures are given in Sect. 5.5.

5.1 Overview

Frequency control of power systems is an emerging area of research due to the increasing penetration of variable Inverter-Based Resource (IBR)-based Renewable Energy Sources (RES) such as wind and solar Photovoltaic (PV) generation [1]. For example, as the inertia decreases with the displacement of conventional synchronous generators there is a concern that frequency excursions become faster, and therefore the likelihood of instability occurring earlier increases [2]. While there are signifi-

T. Kërçi and F. Milano, *Frequency Quality in Power Systems*, Power Systems,
https://doi.org/10.1007/978-3-032-04721-2_5

cant ongoing efforts from both academia and industry on how to best deal with the frequency control challenges (during both normal and abnormal system conditions), it is not clear how different real-world RES-dominated power systems compare in terms of the "strength" of frequency control. This chapter fills this gap by comparing the strength of frequency control of four real-world synchronous islands power systems namely the GB, AIPS, and AUS mainland,[1] and TAS. The frequency control strength of each system is evaluated by means of various metrics of frequency quality.

It is generally understood that small (islanded) synchronous power systems exhibit rapid and substantial frequency excursions compared to large power systems following typical generation loss [4, 5]. For instance, when comparing the different frequency control of GB, AIPS, and AUS power system, reference [6] states: *Ireland is chosen because its power system has, arguably, greater challenges than GB's, owing to its small size, limited interconnection and high penetration of wind. AUS was chosen because of its National Electricity Market's comparable size and renewable penetration to the GB grid.* The measurement data discussed in this chapter, however, demonstrate that size may not necessarily mean greater or lower frequency control challenges. Specifically, despite GB being bigger/much bigger than AIPS, AUS, and TAS systems (see next section), it appears, based on this study, to face greater frequency regulation challenges. We elaborate on these differences by means of the concept of *frequency control strength.*

There is currently no commonly accepted definition of "frequency strength" and how it may relate and overlap with system strength definition(s) [7]. For instance, reference [8] suggests that frequency strength exhibits its meaning in two dimensions, namely the inertia support capability, which defines the initial Rate of Change of Frequency (RoCoF)—Eq. (5.1)—after an active power disturbance, and the Primary Frequency Control (PFC) capability, which defines the amount of active power that the power system can absorb or release during the frequency deviation. However, this definition is incomplete as it assumes that frequency strength only deals with contingency events and does not include the ability of power systems to counteract frequency deviations during normal operating conditions.

Motivated by this confusion and gap in the literature we propose the following definition of frequency strength:

> **Frequency strength** is the ability of power systems to resist and control changes in frequency during normal and abnormal operating conditions.

In this context, references [9, 10] propose real-time frequency strength evaluation indices based on a unified transfer function structure to theoretically quantify the frequency strength in terms of nadir and the average RoCoF. In the same vein,

[1] AUS consists of Queensland, New South Wales, Victoria, and South AUS [3].

a frequency security index to evaluate power system frequency performance is proposed in [11]. These metrics, however, aim at quantifying frequency strength only during abnormal operating conditions. On the other hand, reference [12] focuses only on normal system conditions and proposes frequency regulation performance requirement constraints into a generation planning model to ensure frequency quality. Frequency quality is also studied in [13] with a focus on sub-15-min and sub-1-hr time scales and the highest data resolution used is 10 s.

References [14, 15] provide an overview of frequency control challenges in the GB and AUS power systems, respectively. Reference [16] discusses the efficacy of the proposed new frequency response services in GB using a month-long case study. Similarly, references [17, 18] study the volatility of GB frequency using 1 s frequency resolution data and demonstrate the existence of the relationship between increased frequency events and RES penetration, and rate of change of demand, respectively. Reference [19] utilizes real-world frequency data of different countries in Asia, AUS, and Europe and compares the statistical properties of the frequency, that is, asymmetry of Frequency Probability Distribution (FPD).

This chapter compares the strength of frequency control under normal and abnormal operating conditions of four power systems that are at the forefront of the integration of RES, namely, GB, AIPS, AUS and TAS.

5.2 System Characteristics and Frequency Control

This section provides a detailed comparison, from the frequency control point of view, of the main characteristics of the GB, AIPS, AUS, and TAS power systems. These systems are at the forefront internationally to integrate pioneering levels of IBR-based RES. Table 5.1 compiles all the relevant information. Note the different sizes of the systems in terms of peak demand and inertia floors. Specifically, the GB system is, from a peak demand and inertia perspective, between 5.2–6.3, 1.3–3.3, and 22–37 times bigger than AIPS, AUS and TAS systems, respectively.

On the other hand, the four power systems are very similar in terms of instantaneous RES penetration and electricity met by RES. Concerning the PFC provision, the GB system differs from AIPS, AUS, and TAS systems as it procures/pays units to reserve headroom (but PFC bids are mandatory) [6]. Motivated by a decline in the frequency control performance under normal conditions, in 2020 the Australian Energy Market Commission made a final (mandatory) rule to require all generators (scheduled and semi-scheduled) in the National Electricity Market (NEM) to provide PFC response (droop-based) with narrow dead-band ($\pm$15 mHz) and thus support the secure operation of the power system. This change led to a significant improvement in frequency quality in the NEM which will be discussed in detail in the case study section. Next, all four systems have similar requirements for dead-band ($\pm$15 mHz) and droop (3–5%). Regarding other relevant services, AIPS includes an SIR service while the rest do not. This is not the case for the FFR service where all four systems have recently introduced it with a Full Activation Time (FAT) between 0.15–2 s.

Table 5.1 Main characteristics of GB, AIPS, AUS and TAS power systems

Item	GB	AIPS	AUS	TAS
Peak demand [GW]	44	7.5	34	2
Inertia floor [GWs]	120	23	35.8	3.2
Instantaneous RES [%]	87.6	75	72.1	100
Electricity from RES [%]	51	42	39.4	93.4
PFC provision	Market/Mandatory	Mandatory	Mandatory	Mandatory
dead-band [mHz]	$\leq\pm15$	$\leq\pm15$	$\leq\pm15$	$\leq\pm15$
Droop [%]	3–5	3–5	≤5	≤5
SIR	No	Yes	No	No
FFR	Yes (1 s)	Yes (0.15–2 s)	Yes (0.5–1 s)	Yes (0.5–1 s)
AGC	No	No	Yes	Yes
RoCoF ride-through requirement [Hz/s]	0.125–1	1	1	3
LSI loss [MW]	1400	500	–	144
LSO loss [MW]	1400	500	–	144
Number of ICs (ac/HVDC)	10	3	6	1
IC (ac/HVDC) ramp rate [MW/min]	100	5	40	40
Dispatch model	Self	Central	Central	Central
Flexibility markets	Mature	Early stage	Early stage	Early stage

Note that FFR is a class of PFC that is more needed at low levels of synchronous inertia [20]. Battery Energy Storage Systems (BESS) are one of the most favorable candidates to provide the FFR service due to their fast responsive time and flexibility of operation [21]. With respect to AGC, both GB and AIPS do not currently implement it while AUS and TAS do.

With regard to RoCoF ride-through requirement, all systems have a limit ranging from 0.125 Hz/s in GB (the most sensitive RoCoF protection on the GB system) 3 Hz/s in TAS. This makes sense since TAS is a much smaller system and thus needs to deal with higher RoCoFs. In terms of LSI and LSO loss, National Energy System Operator (NESO) operates and designs the system assuming a 1400 MW loss (the capacity of an IC) to ensure the resulting RoCoF would not exceed 0.5 Hz/s while AIPS a 500 MW and TAS a 144 MW loss (AUS does not include this criterion). There is also a significant difference in the number of ac and dc ICs. Specifically, this number ranges from 10 for GB to just 1 for TAS. Note that while more ICs between countries mean better market integration, it could also have the negative effect on frequency control if several ICs ramp at the same time. In this context, GB has a much higher IC ramp rate (100 MW/min) compared to AIPS (5 MW/min), and AUS/TAS (40 MW/min). For comparison, the AIPS system expects to have a combined IC ramp rate of 40 MW/min by 2030 which has to be compared to the current limit of 15 MW/min. These aspects are critical to frequency variations and are discussed below.

Table 5.2 summarizes all the relevant frequency control products of the selected jurisdictions. More specifically, Table 5.2 suggests that reserve products in AIPS namely Primary Operating Reserve (POR), Secondary Operating Reserve (SOR), Tertiary Operating Reserve 1 (TOR1), Tertiary Operating Reserve 2 (TOR2), and Replacement Reserve (RR) are designed and procured to deal with Under-frequency (UF) events (i.e., upward reserves), while in GB and AUS/TAS reserves are procured to deal with both under and Over-frequency (OF) events (PFC-based upward/downward reserves). Note that Dynamic Containment (DC)/Dynamic Regulation (DR)/Dynamic Moderation (DM) services (all with a $\pm$15 mHz dead-band and predominately being provided by BESS) are faster than the two legacy response products mandatory frequency response and static firm frequency response services.

One may argue, and we certainly agree, that there is no need for so many frequency reserve products. For instance, as shown in Fig. 2.3, European Network of TSOs for Electricity (ENTSO-E) has defined and recommended the use of four standard reserve products namely Frequency Containment Reserve (FCR), Automatic Frequency Restoration Reserve (aFRR), Manual Frequency Restoration Reserve (mFRR), and RR. In addition to these products, Transmission System Operators (TSOs) operating low-inertia grids might need to introduce an FFR product.

5.3 Frequency Control Strength Metrics

5.3.1 Nadir/Zenith and Minutes-Based Metrics

Minutes-based metrics such as "minutes outside the normal operating band" are, more often than not, used by ENTSO-E as a measure of long-term/annual frequency quality. Table 5.3 presents various frequency quality parameters for the four power systems. TSOs continuously monitor and periodically report on these indices [3]. In particular, it is worth pointing out that the standard frequency range in GB and AIPS is $\pm$200 mHz while in AUS and TAS is $\pm$150 mHz. In addition, AIPS and GB have a target to maintain frequency within an even tighter range namely $\pm$100 mHz for $\geq$98% of the time, while AUS and TAS have a target to keep frequency within $\pm$150 mHz for $\geq$99% of the time. We apply these metrics to operational data to evaluate and quantify the strength of frequency control under both normal and abnormal system conditions.

Similarly, TSOs operate and design power systems based on predefined maximum instantaneous frequency deviations namely nadir and zenith, and if these limits

Table 5.2 Comparison of frequency services of GB, AIPS and AUS/TAS systems

Jurisdiction	Service	Direction (Upward/Downward)	FAT [s]	Purpose
AIPS	FFR	Upward	0.15–2	Post-fault contingency
AIPS	POR	Upward	5	Post-fault contingency
AIPS	SOR	Upward	15	Post-fault contingency
AIPS	TOR1	Upward	90	Post-fault contingency
AIPS	TOR2	Upward	300	Post-fault contingency
AIPS	RR	Upward	1200	Post-fault contingency
GB	DC	Upward/Downward	1	Post-fault contingency
GB	DM	Upward/Downward	1	Pre-fault continuous
GB	DR	Upward/Downward	10	Pre-fault continuous
GB	Mandatory frequency response	Upward/Downward	< DC/DM/DR	Post-fault continuous
GB	Static firm frequency response	Upward	10–30	Post-fault contingency
GB	Commercial frequency response	Varies	Varies	Post-fault contingency
GB	Slow reserve	Upward/Downward	900	Post-fault contingency
GB	Quick reserve	Upward/Downward	60	Pre-fault continuous
GB	Fast reserve	Upward/Downward	120	Pre-fault continuous
GB	Balancing reserve	Upward/Downward	600	Pre-fault continuous
AUS/TAS	FFR	Upward/Downward	0.1–1	Post-fault contingency
AUS/TAS	Fast reserve	Upward/Downward	6	Post-fault contingency
AUS/TAS	Slow reserve	Upward/Downward	60	Post-fault contingency
AUS/TAS	Delayed reserve	Upward/Downward	300	Post-fault contingency
AUS/TAS	AGC	Upward/Downward	300	Pre-fault continuous

Table 5.3 Relevant frequency quality parameters of the CE, GB, AIPS, AUS and TAS power systems

Parameter	CE	GB	AIPS	AUS (mainland)	TAS
Standard frequency range [mHz]	±50	±200	±200	±150	±150
Frequency key performance indicator [mHz]	–	±100 (≥98%)	±100 (≥98%)	±150 (≥99%)	±150 (≥99%)
Maximum instantaneous frequency deviation [mHz]	800	800	1000	1000	2000
Maximum steady-state frequency deviation [mHz]	200	500	500	500	500/1000
Time to restore frequency [min]	15	15	15	–	–
Frequency restoration range [mHz]	not used	±200	±200	–	–
Maximum number of minutes	15,000	15,000	15,000	–	–
outside the standard frequency range					
Minutes outside normal operating frequency band	–	–	–	≤5 (5)	≤5 (10)
during normal (abnormal) system conditions					

are exceeded then defense measures may be in place (for example, load/generation shedding) to avoid system blackouts.

5.3.2 Δf and RoCoF-Based Metrics

Frequency stability is generally evaluated based on three key metrics namely RoCoF, nadir and zenith [22]. RoCoF measures how fast frequency changes following imbalances between generation and demand. RoCoF is important during frequency transients as it is used by protections such as, for example, loss-of-mains protection settings by generators.

In Sect. 1.3, we have discussed the analytical expression of RoCoF, as time derivative of the instantaneous frequency as well as the various practical estimation methods utilized by system operators and defined in network codes. Of all these expressions, in the following, we utilize (1.67) as it is the formula mostly utilized by TSOs and, for convenience, we repeat it here:

$$\mathrm{RoCoF}|_{t=0_+} = \frac{\Delta p_{\mathrm{imbalance}}}{p_{\mathrm{load}}} \frac{\omega_o}{4\pi H_{\mathrm{agg}}}, \tag{5.1}$$

where $\Delta p_{\mathrm{imbalance}}$ is the size of the infeed/outfeed outage event, p_{load} is the current system load consumption; H_{agg} is the system aggregated inertia constant; ω_o is the synchronous reference frequency. $\mathrm{RoCoF}|_{t=0_+}$ is used by the four TSOs that we examine in this chapter to determine the minimum inertia floors to maintain RoCoF within limits [15, 23, 24].

This way to calculate the RoCoF is motivated by the swing equation of synchronous machines but gives only a snapshot and only at the initial instant after a major event. Moreover, $\mathrm{RoCoF}|_{t=0_+}$ is a conservative calculation as it neglects the FFR being provided in the inertial time frame. Thus, it appears useful to have additional information over a mobile window in the seconds after the event. With this aim, we define first the frequency window of interest as:

$$\Delta f_i(t) = \left| f(t) - f(t-i) \right|, \tag{5.2}$$

where $f(t)$ is a suitable estimation of the instantaneous frequency at time t, assuming that this estimation can be carried out without measurement biases and with all the limitations and caveats that have been discussed in Chap. 1; and i is a given time period. As it stands, Δf_i is a useful metric for normal operating conditions and we will use it in the remainder of this chapter to compare long-term frequency deviations of real-world power systems.

We are now ready to define RoCoF-based metrics that are complementary to (5.1), as follows:

$$\text{RoCoF}_i(t) = \frac{\Delta f_i(t)}{i}, \tag{5.3}$$

$$\overline{\text{RoCoF}_i}(t) = \frac{1}{N}\sum_{h=0}^{N-1} \text{RoCoF}_i(t - h\Delta t), \tag{5.4}$$

$$\text{RoCoF}_{\text{max},i}(t) = \max_{h=0,\dots,N-1} \left\{\text{RoCoF}_i(t - h\Delta t)\right\}. \tag{5.5}$$

Equation (5.3) is RoCoF calculated for a given period i, whereas Eqs. (5.4) and (5.5) are the average and maximum RoCoFs, respectively, calculated over the time period i with $\Delta t = i/N$. While these RoCoF-based metrics can be applied during both normal and abnormal system conditions, these three metrics best characterize abnormal operating conditions.

The metrics above are "absolute." To be able to compare the strength of frequency control of different power systems it is also useful to consider "relative" RoCoF-based calculations that can take into account the size of the power systems. With this aim, we propose two relative metrics, as follows:

$$\text{RoCoF}_{\text{r1}} = \text{RoCoF}_{\text{max},i} \frac{p_{\text{total}}}{\Delta p_{\text{imbalance}}}, \tag{5.6}$$

$$\text{RoCoF}_{\text{r2}} = \text{RoCoF}_{\text{max},i} \frac{p_{\text{conv}}}{\Delta p_{\text{imbalance}}}, \tag{5.7}$$

where p_{total} is the total active power in the system; and p_{conv} is the amount of conventional generation in the system. These metrics can be interpreted as follows. Two systems should have same RoCoF_{r1} (RoCoF_{r2}) if their inertia and control are equally proportional to p_{total} (p_{conv}). For example, the product $\text{RoCoF}_{\text{max},i} \cdot p_{\text{total}}$ should be the same for two systems controlled in the same way. On the other hand, a weak system will show a higher RoCoF_{r1} than a strong one.

5.3.3 *Frequency Standard Deviation-Based Metrics*

Frequency standard deviation-based metrics are widely and for a long-time used in power systems. For instance, the TSOs in the US utilize the standard deviation of the frequency, say, σ_f (calculated based on 1-minute frequency deviation averages over a year), as a long-term frequency Control Performance Standard (CPS)1 [25]. In the case of Electric Reliability Council of Texas (ERCOT), CPS1 compliance is assumed if σ_f (year) $\leq$30 mHz [25].

In this context, we recently proposed a new metric based on σ_f (see Eqs. (4.27)–(4.30)) to calculate and measure the asymmetry of the frequency distribution ($\Delta\sigma_f$) in power systems [26].

In an ideal scenario $\Delta\sigma_f = 0$. However, this is impossible in practice due to losses and nonlinearity. It makes sense thus to use the asymmetry index as a measure of strength of frequency control as a perfect frequency control would lead to $\Delta\sigma_f = 0$.

5.4 Real-World System Data

In this section, we apply the metrics described in the previous section to real-world data of the four power systems for both normal and abnormal (contingency) operating conditions. These have been made publicly available by the relevant TSOs. In particular, and if not otherwise stated, we use 1 s frequency resolution time series data for GB and AIPS and 4 s for AUS and TAS power systems. For this reason, and where possible, we calculate the relevant frequency control strength metrics at 4 s resolution for GB and AIPS to allow a direct comparison with AUS and TAS systems.

5.4.1 Normal System Conditions

This section focuses on normal operating conditions and applies various minutes-based, Δf_i-based and σ_f-based metrics to quantify and evaluate frequency strength of the GB, AIPS, AUS and TAS power systems.

5.4.1.1 Minutes-Based Comparison

The first comparison that we look at are the minutes outside the relevant standard frequency ranges. For this, we select year 2023 and present all the results in Table 5.3. For illustration and comparison purposes, we present these minutes for the CE power system as well which has a peak demand of around 440 GW and thus approximately ten times bigger than GB. While the standard frequency range for GB and AIPS is ±200 mHz, we calculate the minutes outside ±100 mHz as well as this is an even tighter frequency band that the TSOs of GB and AIPS systems aim at maintaining and reporting on a continuous basis. On the other hand, while the standard frequency range for AUS and TAS is ±150 mHz, we calculate the minutes outside ±100 mHz as well to allow a direct comparison with GB and AIPS power systems. Regarding the CE statistics, we show the minutes outside ±50 mHz.

Table 5.4 suggests that despite CE being the biggest power system, it is the one that has exceeded, for the first time, its annual frequency quality target, namely, frequency outside the ±50 mHz range for more than 15,000 min. Table 5.4 also suggests that, by far, frequency in GB is spending much more minutes outside the ±200 mHz (710.6) and ±100 mHz (80,131.36) ranges compared to the AIPS, AUS and TAS power systems. It should be noted, however, that GB is still within frequency quality limits in terms of minutes outside standard frequency range (±200 mHz) in Table 5.3 (15,000 min). As a matter of fact, in 2014 frequency was inside the ±100 mHz range for 94% of the time (or approximately 534 h outside) compared with 90% in 2021 (or approximately 832 h outside) [27]. It seems, though, that frequency spent even more hours outside limits during 2023 (approximately 1335) compared to 2014 (534)

Table 5.4 Frequency quality in 2023 for the power systems of CE, GB, AIPS, AUS and TAS

Minutes outside	CE	GB	AIPS	AUS	TAS
±50 mHz	15,389	–	–	–	–
±200 (±100) mHz	–	710.6 (80,131.36)	(3.56) 6,796	–	–
±150 (±100) mHz	–	–	–	3.4 (134.26)	2,074.6 (7,157.66)

and 2021 (832). This is interesting considering that GB recently launched two new Primary Frequency Regulation (PFR) products namely DR and DM (see Fig. 2.7) to tackle the challenge of frequency regulation. It appears, though, that frequency regulation is, as expected by NESO, still a major challenge to be addressed [27]. In fact, NESO anticipates *"this exposure to increase in the future as the system is getting more volatile (more renewable connected, low inertia, large uncertainty)*, and suggests *the need for faster response and reserve products"* [27].

Looking at the AIPS, AUS and TAS statistics in Table 5.4, one can see that AUS outperforms the rest of power systems (only 3.4 min outside ±150 mHz and 134.26 min outside ±100 mHz in 2023). This could be explained by the much bigger size of the AUS system compared to the AIPS and TAS systems and the fact that the AUS system utilizes an AGC compared to the AIPS system. What is interesting, though, is that despite being slightly a smaller system than GB, the AUS system shows a significant better frequency performance during normal operating conditions. One of the main differences with GB relate to the fact that the PFC provision (±15 mHz dead-band and droop/proportional response) is fully mandatory in the AUS system whereas that is not the case in GB (see Table 5.1). This can be seen as the frequency regulation task is distributed among many units in AUS while in GB it is concentrated onto a few units. In this context, it has been shown in the literature that the former approach leads to a better frequency performance [28]. Another key reason why AUS shows a better frequency regulation performance is that AUS utilizes and AGC whereas GB does not.

The AUS system, in fact, was showing a decline in frequency control performance under normal conditions (similar to GB) before 2020 where frequency regulation was predominantly managed through AGC and relatively wide dead-band (for example, ±150 mHz). However, managing frequency regulation only through AGC seemed a hard task for Australian Energy Market Operator (AEMO). For this reason, in 2020 the Australian Energy Market Commission introduced mandatory PFC rule for all generators, including from IBRs such as wind power but excluding Distributed Energy Resources (DERs) such as roof-top PV [29]. The mandatory PFC provision with narrow dead-band (±15 mHz) and proportional droop response led to a significant improvement of frequency performance in AUS. This is illustrated in Fig. 5.1 where frequency traces of the GB and AUS systems are compared for the same hour and day (May 21, 2019 and 2024). Frequency variations in AUS have

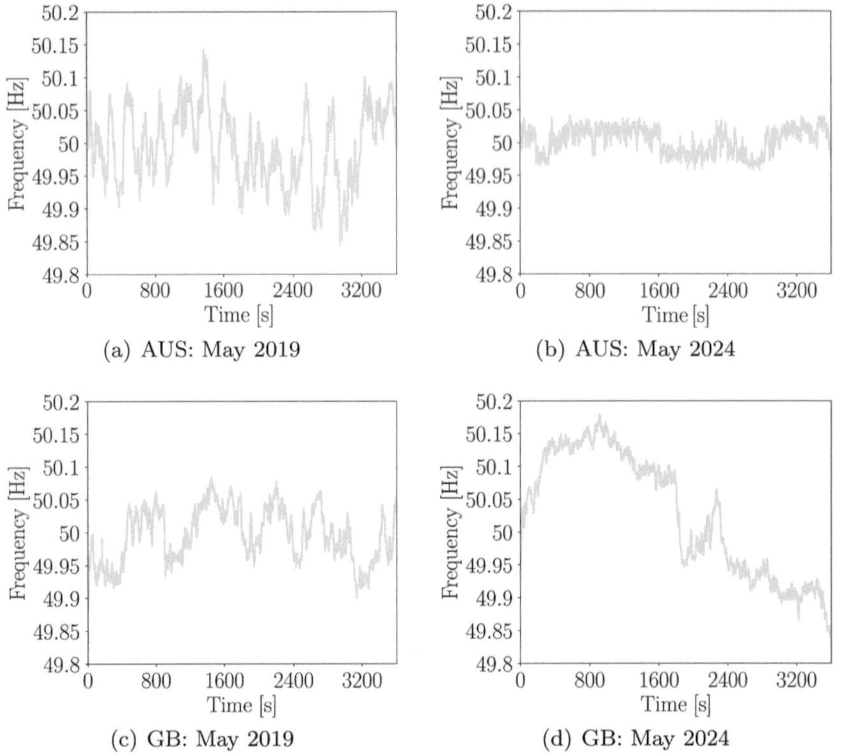

Fig. 5.1 Comparison of frequency traces of GB and AUS power systems for 1 h on May 21, 2019 and 2024, respectively

dramatically improved compared to 2019 but that is not the case in GB. This suggests that a potential solution for GB is to impose PFC with narrow dead-band, even though this might be difficult to implement in practice by all generators as narrow dead-band increases wear and tear. For instance, large dead-bands may be preferred for nuclear units to avoid movement in active power output from scheduled values caused by changes in frequency [30].

If the mandatory PFC rule with narrow dead-band is not possible, then implementing an AGC (well-proven frequency regulation capability) should be another viable solution for GB to consider.[2] We believe this is important considering the fact that the IC ramp rates in GB are significantly higher than in AIPS (100 MW/min vs. 5 MW/min). In other words, several ICs ramping at the same time exacerbates the control of system frequency [32] and, thus, an AGC may be best to deal with it. In addition, despite PFC being a fast-acting and continuous method of control, it is not perfect tracking and thus will not bring the frequency error to zero in steady-state

[2] For instance, the Nordic TSOs identified aFRR/AGC as one of the main measures to stop the weakening trend of the frequency quality and introduced it back in 2013 [31].

Table 5.5 DR & DM requirements increase

	DR-Low [MW]	DR-High [MW]	DM-Low [MW]	DM-High [MW]
Before February 2, 2025	330	330	170	200
Since February 2, 2025	480	480	300	300

in contrast to AGC (includes an integrator term) [33]. Another solution to better manage the real-time power imbalance could be fast generation dispatch but it has been shown in the literature that AGC outperforms it in terms of frequency regulation performance [34]. The GB might also consider increasing the volumes of DR and DM products. As a matter of fact, in February 2025, NESO increased these volumes to better manage significant MW movements observed in recent weeks, see Table 5.5 [35]. However, these new volume requirements will lead to additional costs.

5.4.1.2 Δf-Based Comparison

To get a better insight into the dynamics of frequency variations under normal conditions of the four power systems, we calculate Δf_i over different time periods. With this aim, we select March 2024 as the reference month. Specifically, Figs. 5.2, 5.3, 5.4 and 5.5 show Δf_i calculated for $i = 4$ s, $i = 60$ s, $i = 300$ s and $i = 900$ s time periods. As expected, considering its small size, TAS system shows the highest Δf_i across all the considered time periods compared to GB, AIPS and AUS systems.

GB follows TAS with the second highest Δf_i experienced across 4 s, 60 s, 300 s and 900 s, respectively. While this is a counter intuitive result from the system size point of view, it is somehow expected following the frequency quality deterioration of GB discussed in the previous section. Figure 5.2 shows another insightful result, that is, it appears that Δf_i gets higher for higher time periods, for example, Δf_{300} higher than Δf_{60}. These results are consistent with the observation from NESO that states that during 2021 around 65% of the total time outside ± 100 mHz is due to events lasting 60 s or more, and only around 15% of the total time outside ± 100 mHz is due to events lasting 5 min or more [27]. This observation is not as obvious for AIPS and AUS systems. This means that events that lead to frequency variations in AIPS and AUS last less time and are compensated more quickly than in GB.

While one might argue that Δf_i in the AUS system are lower compared to GB due to AGC in AUS, that is not the case when comparing GB with AIPS system. Specifically, since both AIPS and GB systems have a manual AGC, then the main difference between the two systems is the PFC rule, that is, mandatory in AIPS and non-mandatory in GB. Another potential reason (difference) why frequency variations last more in GB could be related to market-driven variations that might be significantly higher in GB due to, for example, higher IC ramp rates and more

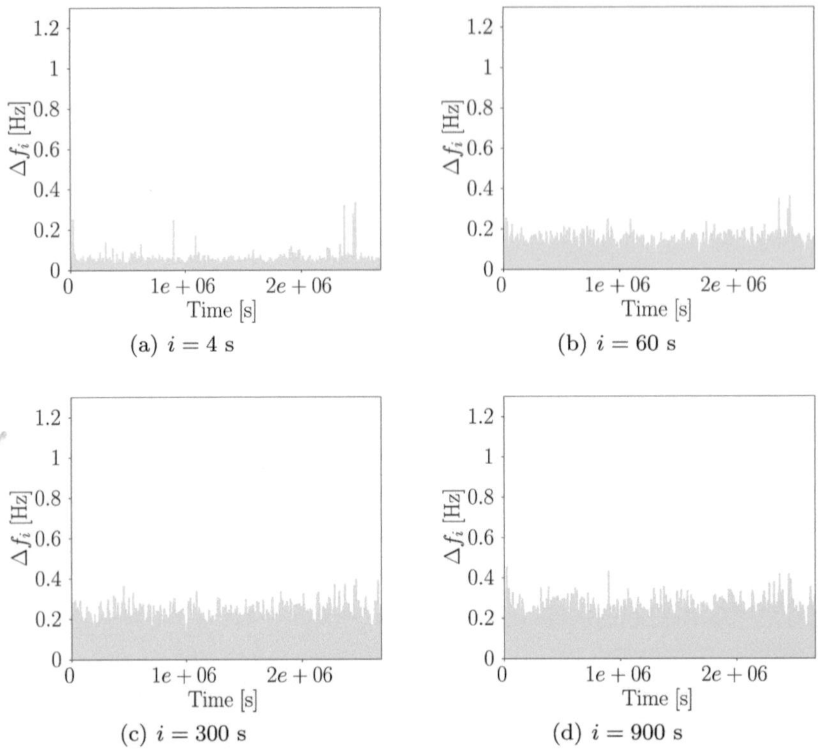

Fig. 5.2 Δf_i variations in the GB power system

demand volatility due to GW level flexibility services being procured by NESO and Distribution System Operators (DSOs) in GB [36]. For example, reference [18] suggest that one of the causes of frequency events in GB is a high rate of change of demand. In fact, if one looks at the flexibility service volumes contracted in GB in Table 5.6, in particular, post-fault products ("Dynamic" and "Restore") that have response times in time scale of minutes, it might be concluded that demand response has increased significantly in recent years [37].

5.4.1.3 Frequency Standard Deviation-Based Comparison

We conclude the frequency control strength comparison for normal system conditions by comparing the various σ-based metrics presented in Sect. 5.3.3. Results are presented in Table 5.7. Similar to the frequency quality and Δf_i-based results, GB has the highest σ_f (0.076 Hz) compared to AIPS (0.042 Hz), AUS (0.025 Hz), and TAS (0.042 Hz) systems. These different σ_f in March 2024 match quite well the σ_f in 2023 for the four power systems. Specifically, the σ_f in 2023 for GB, AIPS, AUS and TAS are 0.069, 0.042, 0.025, and 0.040 Hz, respectively. These results sup-

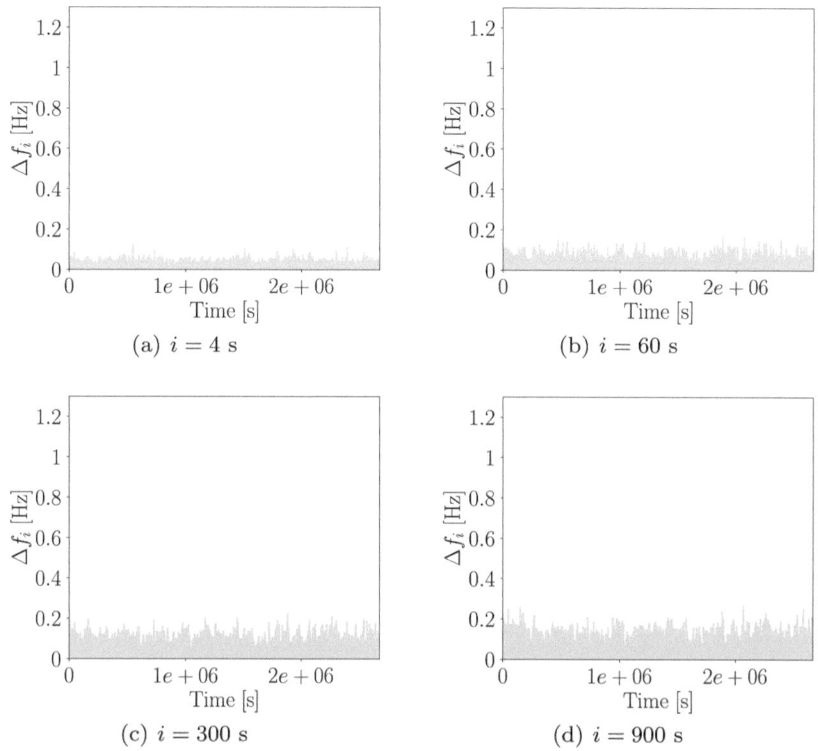

Fig. 5.3 Δf_i variations in the AIPS

port the idea that one month frequency analysis/data (March 2024) is representative enough of the overall frequency performance of the selected power systems. In fact, σ_f has recently constantly increased in the GB power system as shown in Fig. 5.6. As mentioned above, because of this increased frequency volatility, NESO increased the volumes of procured PFR reserves namely DM and DR reserve volumes.

Regarding the asymmetry of the FPD, $\Delta\sigma_f$, it is probably non-intuitive to see that AIPS shows lower asymmetry compared to AUS, GB, and TAS systems. This is despite TAS and AUS systems having an AGC installed.

Figure 5.7 shows the FPDs of the frequency of the four systems for March 2024. Arguably, the source of asymmetry might come from IBRs providing mandatory PFC with ±15 mHz dead-band in AUS and TAS similar to what is reported in AIPS in [26]. As mentioned in Chap. 4, AEMO recognizes that the observed asymmetry in the NEM's frequency characteristic could be due to the application of narrow dead-bands in some power plants [38].

Note that in GB 50 Hz is not the most common frequency. For this reason, NESO procures frequency response assuming a pre-fault frequency different 50 Hz [27]. This approach is not common among other TSOs that instead assume nominal frequency. In any case, NESO maintains frequency within operational limits mainly

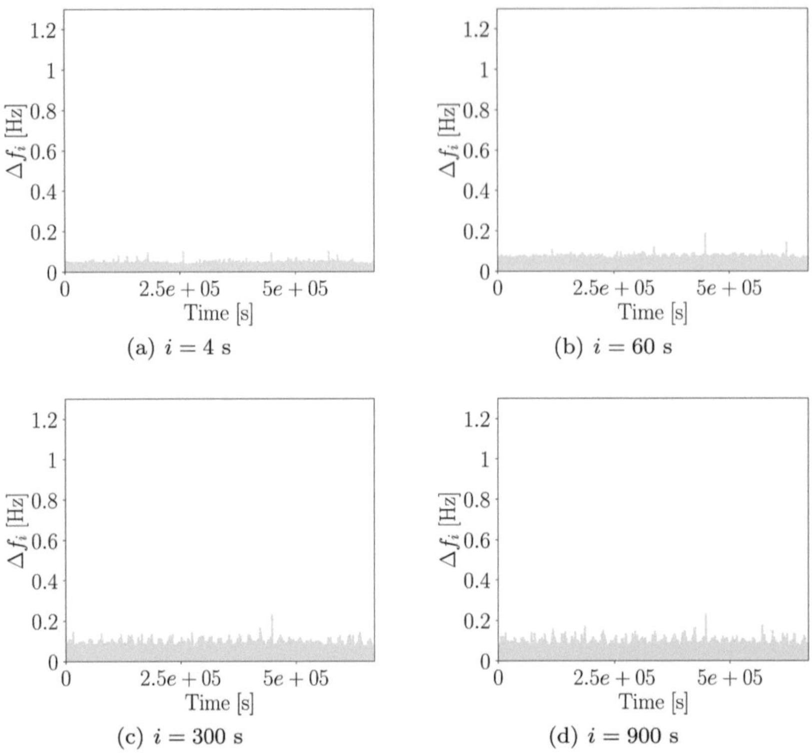

Fig. 5.4 Δf_i variations in the AUS power system

because of the large size, that is, aggregated inertia, of its power system. For example, it has been shown in the literature that the size of the grid serves as a controlling factor to make grid dynamics more robust [39, 40]. In other words, "size" can be seen as part of the frequency control and, in this case, might "hide" the strength, or lack thereof, of control. This aspect is further discussed in the next section.

5.4.2 Abnormal System Conditions

We now discuss the frequency control strength for abnormal operating conditions. Tables 5.8 and 5.9 compare: (i) information on recent relevant trips of the four systems; (ii) operating conditions, namely, inertia, total demand at the time and power imbalance due to the outage; and (iii) various frequency strength metrics as well as the five RoCoF-based metrics described above.

GB and AIPS trips on May 14, 2024

On May 14, 2024, at around 1 am, the two HVDC ICs that connect AIPS with GB tripped one 20 s after the other with a total import (AIPS)/export (GB) capacity lost

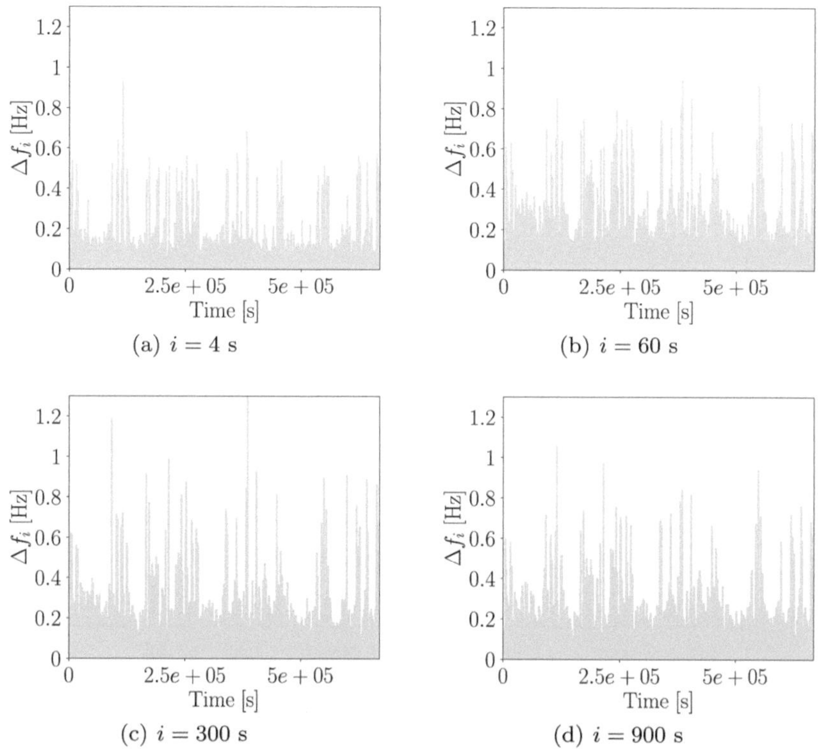

Fig. 5.5 Δf_i variations in the TAS power system

Table 5.6 Evolution of service products contracted across all GB utilities

Flexibility service	2018 [MW, %]	2019 [MW, %]	2020 [MW, %]	2021 [MW, %]	2021/2022 [MW, %]
Sustain	0, 0	0, 0	2, 0	13.3, 1	28.1, 2
Secure	23.8, 20	10.3, 4	105.1, 9	262.6, 16	375.2, 20
Dynamic	33.8, 29	120.8, 47	555.9, 48	729.7, 45	925.7, 50
Restore	58.5, 50	125.1, 49	502.5, 43	603, 37	538, 29
Total	116.1, 100	256.2, 100	1165.5, 100	1608.5, 100	1867, 100

of 912 MW (530 and 382 MW). This allows an excellent frequency control strength comparison, as it represents the same contingency for the two systems. Tables 5.8 and 5.9 show system conditions and all relevant results for the AIPS 2024 and GB 2024 cases, while Fig. 5.9 shows the relevant frequency transients following the ICs trips. In particular, note that the inertia level in GB was just above the limit (121 GWs) while in AIPS was quite above the limit (32.3 GWs) as wind levels were particularly low (534 MW) which means many conventional units online (2153 MW) to meet

Table 5.7 Summary of standard deviation-based results in March 2024

Power System	σ_f [Hz]	σ_{f-} [Hz]	σ_{f+} [Hz]	$\Delta\sigma_f$ [Hz]
GB	0.076	0.078	0.074	0.0043
AIPS	0.042	0.0423	0.0433	0.0010
AUS	0.025	0.0267	0.024	0.0026
TAS	0.042	0.0458	0.0386	0.0072

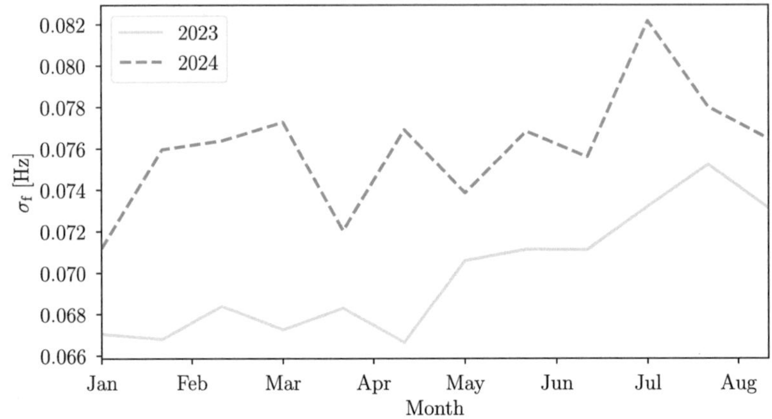

Fig. 5.6 Evolution of σ_f in the GB power system in 2023 and 2024

demand (3699 MW). Also, note that only the MW trip of the first IC ($\Delta p = 530$ MW) is shown in the table as that is important for the different RoCoFs calculations. The second IC, in fact, trips 20 s later and less active power is lost.

Tables 5.8 and 5.9 show that frequency zenith/nadir for the GB/AIPS systems reached 50.154 and 49.593 Hz, respectively. The difference in the maximum frequency deviations is expected as the total active power trip represented around 25% of total demand for AIPS system while only around 4% for GB. Also, nadir happens quickly in AIPS (2 s for the first trip and 26 s for the second trip) than zenith in GB (4 s and 27 s, respectively) due to much lower inertia. Frequency took around 969 s to recover within ±100 mHz for AIPS system while for GB only 393 s. Results also show that the maximum calculated RoCoFs ($\text{RoCoF}_{\text{max},i}$) based on 1 s (4 s) resolutions are 0.066 Hz/s (0.034 Hz/s) and 0.16 Hz/s (0.06) Hz/s for GB and AIPS systems, respectively. These absolute values are expected considering the different inertia levels for both systems namely 121 GWs and 32.3 GWs for GB and AIPS, respectively. What is interesting, though, is the fact that the $\text{RoCoF}_{\text{max},i}$ for GB (0.066 Hz/s) almost matches the initial $\text{RoCoF}|_{t=0_+} = 0.063$ Hz/s. This suggest that the inertial level of GB appears to be a good indication of the $\text{RoCoF}_{\text{max},i}$ experienced even when using 1 s resolution data.

For the AIPS system, on the other hand, $\text{RoCoF}_{\text{max},i} = 0.16$ Hz/s (while the one calculated from 20 ms resolution data and over a rolling 500 ms period is higher

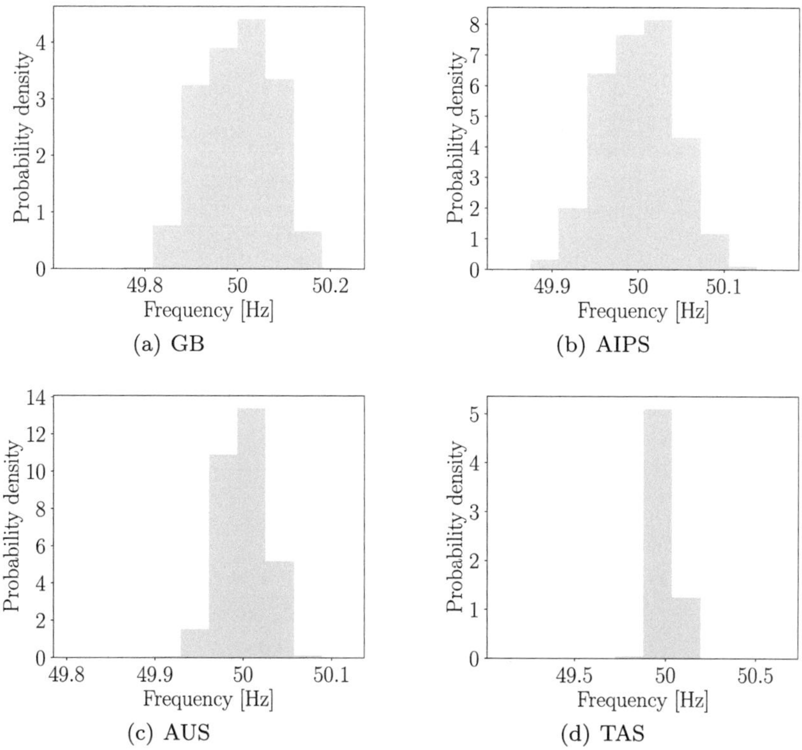

Fig. 5.7 Comparison of FPD of GB, AIPS, AUS and TAS power systems in March 2024

Table 5.8 Frequency strength for contingency events in GB, AIPS, AUS and TAS

Power system year		GB 2024	AIPS 2024	GB 2019	AIPS 2022	AUS 2024	TAS 2024
Inertia	[GWs]	121	32.3	201	34.7	90	6
p_{conv}	[MW]	9,649	2,153	15,980	3,077	–	–
p_{total}	[MW]	21,773	3,699	28,029	4,847	23,122	1,111
$\Delta p_{\text{imbalance}}$	[MW]	530	530	1,000	530	660	114
Nadir/zenith	[Hz]	50.154	49.593	49.62	49.674	49.8	49.3
Time to nadir/zenith	[s]	4/27	2/26	6	3	8	8
Time to recover within ±100 mHz	[s]	393	969	255	780	64	48

Table 5.9 RoCoF-based metrics for contingency events in GB, AIPS, AUS and TAS

Power system year		GB 2024			AIPS 2024			GB 2019		
i	[s]	1	4	0.02	1	4	0.02	1	4	0.02
$\text{RoCoF}\vert_{t=0_+}$	[Hz/s]		0.063			0.41			0.124	
$\overline{\text{RoCoF}_i}$	[Hz/s]	0.0021	0.0012	–	0.0012	0.0010	–	0.0026	0.0022	–
$\text{RoCoF}_{\text{max},i}$	[Hz/s]	0.066	0.034	–	0.16	0.06	0.27	0.118	0.085	–
RoCoF_{r1}	[Hz/s]	2.71	1.39	–	1.11	0.42	1.88	3.30	2.38	–
RoCoF_{r2}	[Hz/s]	1.20	0.62	–	0.65	0.24	1.09	1.88	1.36	–
Power system year		AIPS 2022			AUS 2024			TAS 2024		
i	[s]	1	4	0.02	1	4	0.02	1	4	0.02
$\text{RoCoF}\vert_{t=0_+}$	[Hz/s]		0.38			0.18			0.47	
$\overline{\text{RoCoF}_i}$	[Hz/s]	0.0018	0.0012	–	–	0.0024	–	–	0.0072	–
$\text{RoCoF}_{\text{max},i}$	[Hz/s]	0.242	0.086	–	–	0.023	0.14	–	0.23	0.46
RoCoF_{r1}	[Hz/s]	2.21	0.78	–	–	0.80	4.90	–	2.24	4.48
RoCoF_{r2}	[Hz/s]	1.40	0.50	–	–	–	–	–	–	–

Table 5.10 FFR volumes in GB, AIPS and AUS systems

Direction	Units	GB (DC)	AIPS (FFR)	AUS/TAS (FFR)
Upward	[MW]	~1,300	~1,800	250
Downward	[MW]	~1,300	<100	125

namely 0.27 Hz/s, as expected). The initial RoCoF calculated with (5.1) equals $\text{RoCoF}|_{t=0_+} = 0.41$ Hz/s (still higher than 0.27 Hz/s from 20 ms resolution data). The fact that $\text{RoCoF}|_{t=0_+} > \text{RoCoF}_{\text{max},i}$ means that there is more frequency support in the inertial time frame in the AIPS than in GB. This can be explained by significant volumes of FFR being provided in less than a second in the AIPS system and thus helping with addressing RoCoF as well (instead GB has a full FFR delivery requirement of 1 s). For example, Table 5.10 and Fig. 5.8 show that the AIPS system has around 1,800 MW of upward FFR available (tariff-based procurement) to deal with UF events. This volume is significantly higher compared to its size and the rest of the power systems and thus comes with a significant cost. In this context, NESO has developed a clear and transparent methodology to determine the right balance between the two competing objectives of reliability and cost, focusing on the risks, impacts and controls for managing the frequency [23].

For instance, the 2025 Frequency Risk and Control Report (FRCR) from NESO recommends reducing the inertia floor from 120 to 102 GWs due to significant cost savings (see Table 5.11 [41]). To improve system risk, NESO recommends procuring 200 MW additional DC-Low (or upward FFR) as the most cost-effective solution.

In fact, because of high reserve costs, the AIPS system is introducing more competitive arrangements to procure reserves (auction-based) in the future including FFR (see Chap. 12). Another factor could be that there might be slightly more inertia available in the AIPS system than the 32.3 GWs figure (assumed coming only from

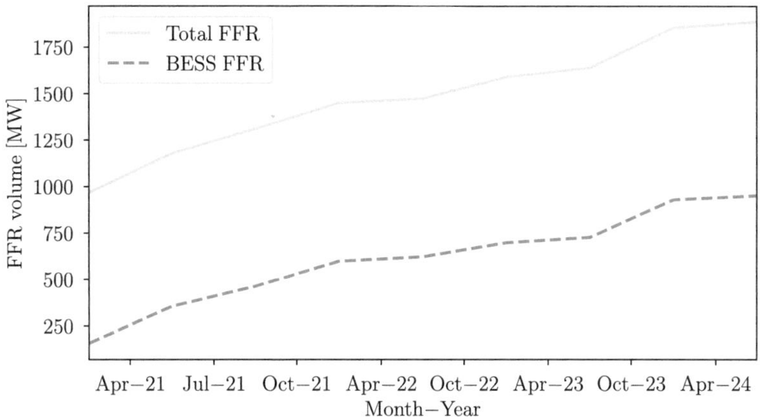

Fig. 5.8 Evolution of FFR contracted volumes in the AIPS

Table 5.11 Assessment of minimum inertia requirements in GB

Inertia floor [GWs]	140	120	110	102
Cost [Łm]	524	266	198	170

conventional generators and neglecting inertia from demand, for example), while for GB the 121 GWs figure might be representing better the actual inertia in the system.

Based on the discussion above and the considered metrics, it appears that the GB power system is stronger than the AIPS system in absolute terms. However, if we calculate the relative frequency strength of the two power systems using Eqs. (5.6) and (5.7) the situation changes. Specifically, RoCoF_{r1} values for GB and AIPS systems are 2.71 Hz/s (1.39 Hz/s) and 1.11 Hz/s (0.42 Hz/s), respectively, and thus lower for AIPS. Note that the values before (within) brackets are calculated using $\text{RoCoF}_{\max,i}$ from 1 s (4 s) data resolution. Similarly, the RoCoF_{r2} values are lower for AIPS than GB namely 0.65 Hz/s (0.24 Hz/s) and 1.20 Hz/s (0.62 Hz/s).

GB trip on July 1, 2019 and AIPS trip on August 9, 2022

To further validate the above results and conclusions, we select two more relevant contingencies for GB (GB_{2019}) and AIPS (AIPS_{2022}) namely IC trip (1,000 MW import) on July 1, 2019 at 08:27 in GB and IC trip (530 MW import) on August 9, 2022 in AIPS [42]. The respective system operating conditions and results are given in Table 5.8. It can be seen that the results are very similar to the previous comparison. Specifically, the nadir happens quicker in AIPS than in GB and similarly the frequency restoration in AIPS takes much more time. The GB system performs better than the AIPS system in terms of frequency recovery following contingencies.

Next, notice again that $\text{RoCoF}_{\max,i}$ and $\text{RoCoF}|_{t=0_+}$ for GB are very similar (0.118 and 0.124 Hz/s, respectively) while that is not the case for AIPS (0.242 and 0.38 Hz/s, respectively). As mentioned above, two contributing factors here are the bigger support from FFR in the AIPS system (response time down to 150 ms) and potentially more inertia available than the calculated inertia from conventional units in AIPS (34.7 GWs inertia figure might be higher in reality) compared to GB (201 GWs seems to be close to actual inertia). It is also fascinating to see that the difference between $\text{RoCoF}_{\max,i}$ and $\text{RoCoF}|_{t=0_+}$ has increased for AIPS over the last years (0.138 Hz/s in 2019 and 0.25 Hz/s in 2024). Basically, this means that the frequency support in the inertial time frame has increased steadily (see Fig. 5.8). This, in turn, has led to significant frequency stability improvement in recent years in the AIPS system (see Fig. 12.7).

With regard to the relative RoCoF calculations, again, AIPS shows lower values than GB. Specifically, RoCoF_{r1} values for GB and AIPS systems are 3.30 Hz/s (2.38 Hz/s) and 2.21 Hz/s (0.78 Hz/s), respectively. RoCoF_{r2} values are also lower for AIPS (1.40 Hz/s (0.50 Hz/s)) than GB (1.88 Hz/s (1.36 Hz/s)). This lead to confirm that the AIPS is stronger than the GB system relative to its size. Considering both jurisdictions are moving towards lower levels of inertia (Table 5.11) and competitive procurement of reserves (Chap. 12), it would be interesting to perform a similar comparison in a few year's time to see whether the above conclusion will still hold.

AUS trip on March 21, 2024 and TAS trip on March 6, 2024

We conclude the comparison for abnormal operating conditions by selecting two relevant trips for AUS and TAS systems namely March 21, 2024 (trip of Bayswater Power Station Unit 1 at 660 MW at 18:00) and March 6, 2024 (trip of Comalco at 114 MW at 8:07), respectively [3]. It is important to note that the frequency measurements for these two events are provided in 4 s resolution and the inertia values are guessed based on time series graphs provided in [3]. While these minor limitations in data might lead to small discrepancies, we note that it does not affect the main conclusions drawn.

The system operating conditions and results are provided in Table 5.8, while Fig. 5.9 illustrates the relevant frequency transients. Due to the size of the TAS system (around 6 GWs compared to 90 GWs of AUS) and the size of the contingency at the time of incident (114 MW representing approximately 10% of demand), frequency nadir reached the lowest value (49.3 Hz). Regarding the time to nadir, both the AUS and TAS systems show a similar value of 8 s. Note, however, that since only 4 s data is available these times may be slightly lower and thus similar to AUS and AIPS systems. On the other hand, in contrast to GB and AIPS, the AUS and TAS power systems show a much better frequency recovery, namely, tens of seconds vs hundreds of seconds for GB and AIPS. This could be explained by the fact that both AUS and TAS systems employ an AGC while that is not the case for GB and AIPS systems (see Table 5.1).

Regarding different RoCoFs calculations, one can see that the average $\overline{\text{RoCoF}_i}$ ($i = 4$) shows a higher value for AUS (0.0024 Hz/s) and TAS (0.0072 Hz/s) systems as compared to the GB (0.0012 and 0.0022 Hz/s) and AIPS (0.0010 Hz/s and 0.0012) systems. This makes sense considering that both the AUS and TAS systems recover their frequency quite quickly to ± 100 mHz (see Fig. 5.9). With regard to the maximum $\text{RoCoF}_{\max,i}$, since there is no 1 s resolution data, we utilize the value provided

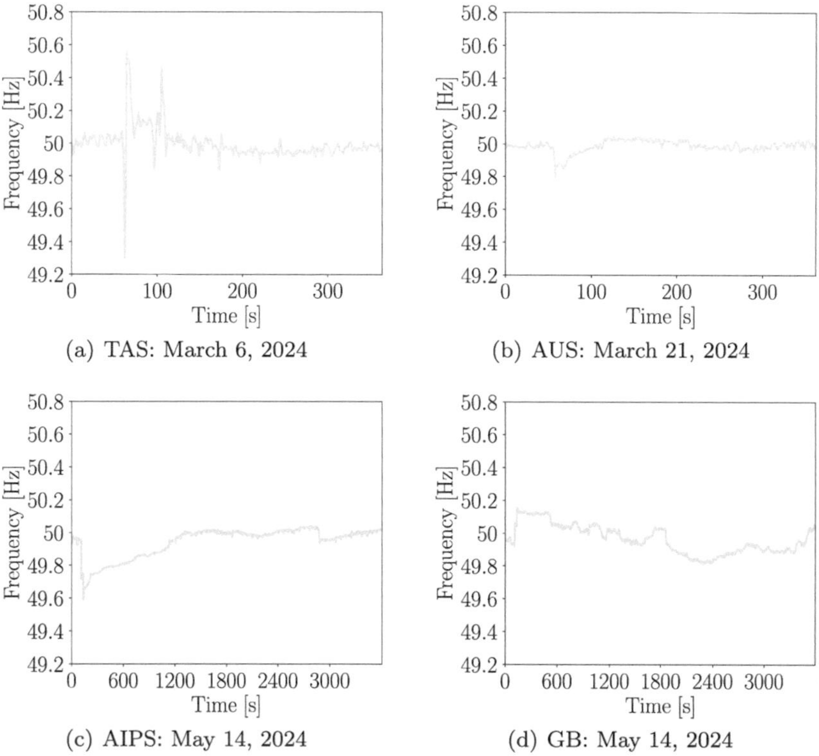

Fig. 5.9 Comparison of frequency traces of the GB, AIPS, AUS, and TAS power systems for relevant contingencies in 2024

in [3] by AEMO (0.14 and 0.46 Hz/s for AUS and TAS, respectively). AEMO calculates these from high-resolution data (20 ms) and 500 ms rolling window and filtering short-term transients [3]. But if we are to calculate $\text{RoCoF}_{\max,i}$ from 4 s resolution data then these values are 0.023 and 0.23 Hz/s for AUS and TAS, respectively. This means that the TAS system experiences the worst $\text{RoCoF}_{\max,i}$ (0.23 Hz/s) compared to GB (0.034 and 0.085 Hz/s), AIPS (0.06 and 0.086 Hz/s), and AUS (0.023 Hz/s) systems. Similar to the nadir explanation above, this is to be expected considering the size of the TAS system and contingency.

Looking at the $\text{RoCoF}|_{t=0_+}$ calculations, we can see that they are very similar to those calculated using 20 ms data and 500 ms rolling window namely 0.14 and 0.18 Hz/s for AUS, and 0.46 and 0.47 Hz/s for TAS system. Similar to the GB cases (GB_{2024} and GB_{2019}), the similarity in the values of $\text{RoCoF}|_{t=0_+}$ and $\text{RoCoF}_{\max,i}$ indicate that the inertial levels of the AUS and TAS systems are a good indication of the actual experienced RoCoF and that there is little FFR being provided in the inertial time frame. Note that even if the inertia values of the AUS and TAS systems might be slightly different than the current values (90 GWs for AUS and 6 GWs for

TAS), the value of the $\text{RoCoF}|_{t=0_+}$ will not change significantly and thus will not affect the main conclusions.

The RoCoF_{r1} results mean that the AUS (0.80 Hz/s) shows higher relative frequency control strength compared to GB (1.39 and 2.38 Hz/s) and TAS (2.24 Hz/s) systems and lower strength compared to the AIPS (0.42 and 0.78 Hz/s) system. Therefore, it can be concluded that based on the relative RoCoFs comparisons namely RoCoF_{r1} and RoCoF_{r2}, the AIPS system shows a higher relative frequency control strength than GB, AUS and TAS systems, in fact, relative RoCoF_{r1} and RoCoF_{r2} are lower for AIPS. Note that if one considers other frequency strength metrics such time to recover within ± 100 mHz, then the above conclusion may completely change. However, we think that since RoCoF-based results, in particular, RoCoF_{r1} and RoCoF_{r2}, are more critical/important than the recovery period in terms of frequency stability, it makes sense to reach the conclusion based on those results.

5.5 Remarks

The common understanding is that the bigger the capacity of a power system, the bigger its robustness with respect to events and contingencies. Data discussed in this chapter show that this is not always the case in the context of frequency control. Specifically, the key findings of this chapter are summarized below.

Normal operating conditions
Our analysis indicates that despite being the second-biggest power system, AUS performs better in frequency regulation during normal system conditions. This is a counterintuitive result from a system size point of view [39] and, in particular, because NESO has recently introduced two new dynamic frequency regulation products (but DC also helps slightly in normal conditions by having ± 15 mHz dead-band, see Fig. 2.7). A possible mitigation for GB is to consider a similar path to AUS regarding mandatory PFC provision with a narrow dead-band. If this is the case, it is suggested to focus on a potential increase in the asymmetry of frequency distribution like in AUS. Alternative solutions include implementing an AGC and revising frequency regulation products. Despite the AUS system showing, overall, the best frequency regulation performance, it appears that frequency asymmetry ($\Delta\sigma_f$) is higher than in AIPS. It is suggested to study more in detail $\Delta\sigma_f$ and its sources in the AUS system.

Abnormal operating conditions
The frequency control strength comparison is more complex during contingency events due to the different stages of frequency control such as FFR/PFC and AGC. The AIPS power system appears the "strongest" to arrest frequency and RoCoF relative to its size. This is mainly due to the significant procurement of FFR volumes and thus comes to a higher cost compared to the rest of the power systems. The AUS and TAS show better frequency recovery compared to the AIPS and GB power systems. A possible solution to this problem could be that GB and AIPS consider implementation of automatic Secondary Frequency Control (SFC) such as AGC similar to AUS and TAS.

In Brief

- We provide a simple yet practical defintion of "frequency strength."
- A comprehensive review of the main characteristics of four real-world power systems and an overview of their frequency control shows that frequency strength depends on several factors.
- For this reason, several metrics are needed to evaluate and enable a systematic comparison of frequency control strength on actual historical frequency measurements/events.
- The Australian power system shows higher frequency regulation strength, while the AIPS power system shows the "strongest" frequency performance to arresting frequency and RoCoF relative to its size.
- New grid codes and market rules are needed to help improve frequency performance.

Open Questions

There is no general accepted definition for "frequency strength" and how it may relate and overlap with system strength definition(s). It is also still not clear whether it is better from a frequency performance point of view to have a single PFC reserve product to deal with contingencies and normal conditions (by having narrow deadband), or it is better to split it into two, that is, one for disturbances and one for normal operation such as in the Nordic (Frequency Containment Reserve for Disturbances (FCR-D) and Frequency Containment Reserve for Normal Operation (FCR-N)) and GB (DC and DR/DM) power systems. A proper tradeoff between mandatory grid code requirements and market-based solutions to address frequency control strength issues remains to be defined.

References

1. Bevrani H, Golpra H, Messina AR et al (2021) Power system frequency control: an updated review of current solutions and new challenges. Electr Power Syst Res 194:107114
2. Hatziargyriou N, Milanovic J, Rahmann C et al (2021) Definition and classification of power system stability—revisited and extended. IEEE Trans Power Syst 36(4):3271–3281
3. AEMO, Frequency monitoring—Quarter 1 2024. www.aemo.com.au
4. Martnez-Barbeito M, Gomila D, Colet P (2023) Dynamical model for power grid frequency fluctuations: application to islands with high penetration of wind generation. IEEE Trans Sustain Energy 14(3):1436–1445
5. Al Kez D, Foley AM, Ahmed F et al (2023) Overview of frequency control techniques in power systems with high inverter-based resources: challenges and mitigation measures. IET Smart Grid 6(5):447–469
6. Nedd M, Browell J (2020) Operating a zero carbon GB power system in 2025: Frequency and fault current [annexes-review of system and network issues, frequency stability, power electronic devices and fault current, and market needs]. University of Strathclyde

7. Badrzadeh B, Emin Z (2024) Power system dynamic modelling and analysis in evolving networks. Springer Nature
8. Xu Z, Zhang N, Zhang Z et al (2023) The definition of power grid strength and its calculation methods for power systems with high proportion nonsynchronous-machine sources. Energies 16(4)
9. Gao H, Xin H, Huang L et al (2022) Common-mode frequency in converter-integrated power systems: definition, analysis, and quantitative evaluation. IEEE Trans Power Syst 37(6):4846–4860
10. Hu G, Gao H, Xin H et al (2023) Online frequency strength quantification for power systems with more renewable energy sources. In: IEEE PES general meeting, pp 1–5
11. Delkhosh H, Seifi H (2021) Power system frequency security index considering all aspects of frequency profile. IEEE Trans Power Syst 36(2):1656–1659
12. Cai X, Zhang N, Du E et al (2024) Low inertia power system planning considering frequency quality under high penetration of renewable energy. IEEE Trans Power Syst 39(2):4537–4548
13. Mohammadi Y, Polajžer B, Leborgne RC et al (2024) Quantifying power system frequency quality and extracting typical patterns within short time scales below one hour. Sustain Energy Grids Netw 38:101359
14. Hong Q, Khan MAU, Henderson C et al (2021) Addressing frequency control challenges in future low-inertia power systems: a great Britain perspective. Engineering 7(8):1057–1063
15. Alhelou HH, Bahrani B, Ma J et al (2024) Australia's power system frequency: current situation, industrial challenges, efforts, and future research directions. IEEE Trans Power Syst 39(3):5204–5218
16. Homan S, Brown S (2021) The future of frequency response in Great Britain. Energy Rep 7:56–62; 5th annual CDT conference in energy storage and its applications
17. Homan S, Brown S (2020) An analysis of frequency events in Great Britain. Energy Rep 6:63–69; 4th annual CDT conference in energy storage and its applications
18. Homan S, Mac Dowell N, Brown S (2021) Grid frequency volatility in future low inertia scenarios: challenges and mitigation options. Appl Energy 290:116723
19. Wen X, Anvari M, Gorjão LR et al (2023) Non-standard power grid frequency statistics in Asia, Australia, and Europe. arXiv:2308.16842
20. Milano F, Drfler F, Hug G et al (2018) Foundations and challenges of low-inertia systems (invited paper). In: Power systems computation conference (PSCC), pp 1–25
21. Akram U, Nadarajah M, Shah R et al (2020) A review on rapid responsive energy storage technologies for frequency regulation in modern power systems. Renew Sustain Energy Rev 120:109626
22. Hurtado M, Jafarian M, Kërçi T et al (2024) Stability assessment of low-inertia power systems: a system operator perspective. In: IEEE PES general meeting, pp 1–5
23. NESO (2024) Frequency risk and control report. Security and quality of supply standards. www.nationalgrideso.com
24. Hurtado M, Kërçi T, Tweed S et al (2023) Analysis of wind energy curtailment in the Ireland and Northern Ireland power systems. In: IEEE PES general meeting, pp 1–5
25. Chávez H, Baldick R, Matevosyan J (2015) The joint adequacy of AGC and primary frequency response in single balancing authority systems. IEEE Trans Sustain Energy 6(3):959–966
26. Kri T, Milano F (2025) Asymmetry of frequency distribution in power systems: Sources, estimation, impact and control. IEEE Open Access J Power Energy 12:135–145
27. NESO (2022) Reserve Reform Show & Listen #3 23 June 2022. www.nationalgrideso.com
28. Kërçi T, Murad MAA, Dassios I et al (2021) Contribution of energy storage systems to long-term power system dynamic performance. In: IEEE madrid powerTech, pp 1–5
29. AEMO (2021) Enduring primary frequency response requirements for the NEM. www.aemc.gov.au
30. Quint R, Ramasubramanian D (2017) Impacts of droop and deadband on generator performance and frequency control. In: IEEE PES general meeting, pp 1–5
31. ENTSO-E (2024) Nordic balancing philosophy. www.eepublicdownloads.entsoe.eu
32. NESO (2021) National trends and insights. www.nationalgrideso.com

33. Kërçi T, Murad MAA, Dassios I et al (2021) On the impact of discrete secondary controllers on power system dynamics. IEEE Trans Power Syst 1
34. Gao N, Gao DW, Fang X (2023) Manage real-time power imbalance with renewable energy: fast generation dispatch or adaptive frequency regulation? IEEE Trans Power Syst 38(6):5278–5289
35. NESO (2025) NESO operational transparency forum. www.neso.energy
36. Kërçi T (2023) On the definition and classification of power system flexibility. In: IEEE PES innovative smart grid technologies Europe (ISGT EUROPE), pp 1–5
37. EPRI (2022) Procuring distribution services from distributed energy resources: Progress reports from United Kingdom and France: White paper. www.epri.com
38. AEMO (2023) Primary frequency response requirements. www.aemo.com.au
39. Schäfer B, Beck C, Aihara K et al (2018) Non-Gaussian power grid frequency fluctuations characterized by Lévy-stable laws and superstatistics. Nat Energy 3(2):119–126
40. Tchuisseu EBT, Dongmo E-D, Prochazka P et al (2023) Secondary frequency control stabilising voltage dynamics. Eur J Appl Math 34(3):467–483
41. NESO (2025) Frequency risk and control report (frcr). https://www.neso.energy/
42. NESO (2019) Technical report on the events of 9 August 2019. www.nationalgrideso.com

Chapter 6
Frequency Stability

Retaining frequency stability is one of the most important and, at the same time, challenging problems Transmission System Operators (TSOs) operating low-inertia power systems have to face on a daily basis. This was not necessarily a major problem for the majority of power systems in the past as they were being operated quite far from the stability limits. In this chapter, we provide real-world evidence on how TSOs are dealing with the frequency stability challenge.

Section 6.1 recalls the definitions of power system stability for power systems and Sect. 6.3 presents an analysis of conventional stability phenomena, namely, rotor angle, voltage and frequency stability, based on a real-world large-scale low-inertia system, namely the All-Island Power System (AIPS) of Ireland (IE) and Northern Ireland (NI). This system currently accommodates up to 75% of System Non-Synchronous Penetration (SNSP), which is a world record and is planning to increase to 80% by end of 2025. The chapter discusses one-month results obtained with the state-of-the-art stability tool called Look-ahead Security Assessment Tool (LSAT). This tool carries out rotor-angle, frequency and voltage stability analyzes and is implemented in the control centers of the TSOs. Next, Sect. 6.4 of the chapter discusses frequency stability in 100% inverter-based generation power systems. This is based on a futuristic model of the AIPS model where all synchronous machines are substituted with Inverter-Based Resources (IBRs). Finally, Sect. 6.5 discusses the impact of different power system topologies on frequency stability.

6.1 Overview

Large-scale low-inertia power systems are characterized by high penetration of IBRs such as wind and solar PV. The power system community have recently updated the classification of power system stability to account for the changing dynamic behavior mainly due to the introduction of such technologies [1]. However, the dynamics of

T. Kërçi and F. Milano, *Frequency Quality in Power Systems*, Power Systems,
https://doi.org/10.1007/978-3-032-04721-2_6

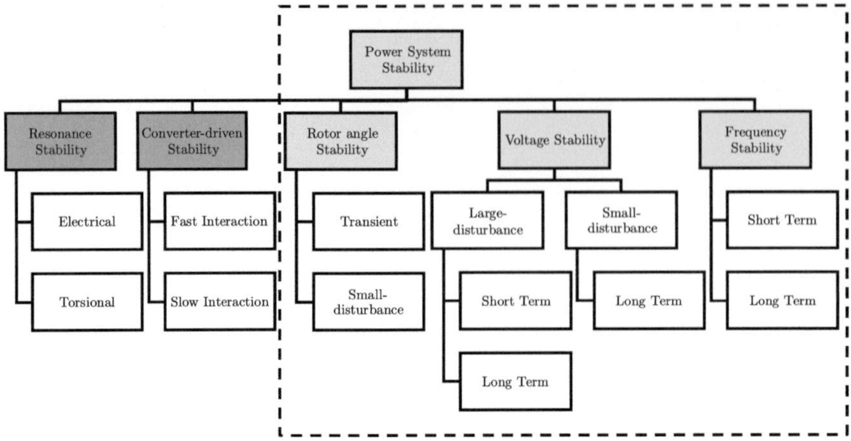

Fig. 6.1 Classification of power system stability types [1]. The dashed rectangle indicates conventional stability types that characterize rotating machines and ac transmission systems

such systems are still to be fully studied and understood [2]. In particular, in the literature, there is no final conclusion on what is the main stability constraint that limits the penetration of non-synchronous generation in low-inertia power systems. This chapter attempts to answer this question by running dynamic stability studies on a real-world large-scale low-inertia power system, namely the AIPS of IE and NI which allows up to 75% of SNSP [3].

The topic of low-inertia power systems and, in particular, technical (dynamic) constraints associated with it has been growing fast in recent years [4–7]. For example, the study in [8] considers the small-signal stability of low-inertia systems and shows that depending on the power system network and the mix of generators considered, one can obtain completely different penetration levels of IBR units, ranging from 60 to 93%. Reference [9] study the large-signal (transient) stability problem in low-inertia systems in terms of the coupling between the electromagnetic dynamics of the IBRs and the electro-mechanical dynamics of the synchronous generators. The hypothetical scenario of 100% IBRs using an AIPS model is studied, for example, in [10].

This chapter addresses the above limitations by assessing transient, frequency and voltage stability in low-inertia systems using a real-world, fully-fledged dynamic power system model, namely the AIPS (see Fig. 6.1). The dynamic performance of the grid is evaluated by means of the state-of-the-art tool LSAT. With regard to the new stability phenomena in Fig. 6.1, the AIPS has so far experienced a few cases of very-low frequency oscillations and sub-synchronous torsional interactions. However, these phenomena are identified post-events as currently the TSOs do not have the real-time capability to evaluate them pre-event. For this reason, these phenomena are not considered in this chapter.

6.2 Definition of Frequency Stability

In [11], frequency stability is defined as follows:

Frequency stability refers to the ability of a power system to maintain steady frequency following a severe system upset resulting in a significant imbalance between generation and load. It depends on the ability to maintain/restore equilibrium between system generation and load, with minimum unintentional loss of load. Instability that may result occurs in the form of sustained frequency swings leading to tripping of generating units and loads.

In turn, frequency stability is linked exclusively to power balance and is, thus, a byproduct of the behavior of synchronous machines discussed in Chap. 1 and that, in its simplest form can be expressed through Eq. (1.1), which we repeat here for convenience:

$$2H_{\rm agg}\,\omega'(t) = p_{\rm gen}(t) - p_{\rm dem}(t) - p_{\rm loss}(t)\,. \tag{6.1}$$

The reference in the definition to sustained frequency swings is also a byproduct of the dynamic coupling of synchronous machines, which can lead to inter-area oscillations. The classification introduced in [11] also distinguished between short-term (order of seconds) and long-term (order of tend of seconds to minutes) frequency stability. The former is a consequence of systems with small inertia and the latter is caused by the over-speed control of Turbine Governors (TGs).

It is also interesting to note the reference to unintentional loss of load, which makes this definition very similar to that of power system resilience. Indeed, one can argue whether the whole concept of "resilience" is substantially the same as frequency stability in conventional systems.

The definition above has remained unaffected in [1], which is the revisited version of [11]. Reference [1] also includes a discussion on the impact of converter-based generation (CIG) and indicates that, while these devices do not inherently link power unbalances to frequency variations, CIG can be regulated in order to improve the overall system dynamic performance.

Reference [1] mentions new kinds of stability directly linked to or caused by CIG, namely, converter-driven stability, which can be fast or slow; and weak system stability (due to the dynamic coupling of Phase-Locked Loops (PLLs) with long distribution lines). These are considered to be distinct phenomena with respect to frequency stability because they are not linked to a power unbalance.

While these classifications are quite practical, there are two observations that may require in the future to revisit the definition of frequency stability.

The first remark is that CIG might provide very fast control, including very fast frequency control following an unbalance. In a scenario where CIG can fully compensate the power unbalance in a very short time (say, less than a second), the frequency would not vary as the synchronous machines would not have the time to change their rotor speed.

The second remark is related to the fact that CIG is often connected to distribution systems where the link between frequency and reactive power is not negligible. In a future scenario where the vast majority of the generation is connected at the distribution system level, thus, it might be possible to cause frequency variations through reactive power unbalances, and, on the other way round, regulate the frequency through reactive power support. For example, in [12], it is shown how a "dual" grid-forming converter can change the conventional (6.1) to the following:

$$\tilde{M}\,\varrho'(t) = p_{\rm gen}(t) - p_{\rm dem}(t) - p_{\rm loss}(t)\,, \tag{6.2}$$

where ϱ is the instantaneous bandwidth or, equivalently, the real term that appears in the definition of the complex frequency discussed in Chap. 1—see Eq. (1.15). ϱ depends on the variations of bus voltage magnitude rather than on the variation of the bus voltage angle. This quantity is thus a sort of "dual" frequency with respect to ω. The formulation of the dual grid-forming converter is described in Sect. 6.4.2.

The scenarios described above are futuristic for now. However, the main point is that the definition of frequency stability will have to be revisited both in terms of its time scale and of its link to the active power balance. The need for a system power balance, on the other hand, will remain unchanged. It is thus to be expected that we will talk about "power balance stability" or maybe "complex frequency balance" rather than "frequency balance." We illustrate both scenarios in the examples discussed in Sect. 6.4 at the end of this chapter.

6.2.1 *Impact of Reduced Inertia*

Due to its direct/close link with the power system balance—see Eq. (6.1) reduced inertial levels have, as it has to be expected, a significant impact on frequency stability. In fact, the lower the value of the system aggregated inertia $H_{\rm agg}$, the higher the Rate of Change of Frequency (RoCoF) for the same power imbalance. This issue can make protections inadequate, in particular, RoCoF protections, for very low values of the inertia, and thus impact the overall stability of the system. Moreover, note that synchronous generators have very tight over-speed and under-speed protections that disconnect the generator in case of severe variations of the system frequency. The disconnection of synchronous machines further reduces the inertia and can lead to the cascade phenomenon.

Less immediate is the impact of low inertia on other kind of stability. This statement is supported by a recent Australian Energy Market Operator (AEMO) study looking at the future role of inertia [13]: *But given the focus around inertia requirements has largely been on frequency control to date, AEMO is concerned that this ultra-low inertia levels could result in adverse system stability outcomes beyond frequency, particularly if the remaining synchronous inertia on the power system is geographically scattered in a way likely to give rise to regional shortfalls in stability margins.* While the total aggregated inertia can be significantly low, the individual

Table 6.1 Impact of reduced inertia on power system stability

Stability area	Type of impact	Level of impact
Frequency stability	Direct	High
Voltage stability	Indirect	Low
Rotor angle stability	Indirect	Low
Converter-driven stability	Indirect	?
Resonance stability	Indirect	?

inertia of each conventional power plant remains the same, so transient and small-signal stability studies for synchronous machine remain the same. Indirectly, however, low inertia, means that a high number of inverter-based generation is connected to the grid. This may lead to undesirable couplings among the controllers of the converters and between converters and synchronous machines [14]. In particular, poorly damped oscillations that results form the coupling of Automatic Voltage Regulators (AVRs) and machine electromechanical models can be triggered also by the coupling of the voltage controllers of converters and the machine electromechanical modes. Of course, converters can also give rise to new types of oscillations and resonances that do not involve synchronous machines.

For extremely high levels of inverter-based generations (above 75%), transient instability is more like to occur as reported in some studies, for example, [1], as well as by EirGrid [15]. On the other hand, inertia is not expected to directly influence long term voltage stability due to its faster time scale. This is also shown in Sect. 6.3.2.

The discussion above is summarized in Table 6.1. The question marks indicates that it is still not fully understood whether a power system with high or extremely high inverter-based generation will experience significant oscillatory/resonance phenomena. In particular, it is not clear whether these phenomena can have catastrophic effects such as a system-wide blackout.

6.3 Stability Assessment of the AIPS

Irish TSOs manage dynamic stability in the operational time frames using different dynamic operational constraints [3], and by performing online dynamic assessment through LSAT. With regard to the operational constraints, the TSOs have in place four operational constraints/limits namely [16]: (i) SNSP limit; (ii) Minimum Number of Conventional Units Online (MUON); (iii) a RoCoF limit; and (iv) a minimum inertia floor. The TSOs are also planning to introduce a new dynamic stability metric called "System Strength" to maintain stable operation of IBRs during high SNSP scenarios, among others [16]. The current and expected evolution of these constraints are shown in Table 6.2 (see Chap. 11 for more details). In particular, RoCoF ± 1 Hz/s became an enduring operational constraint in May 2023. Because of this, we provide below a

Table 6.2 Evolution of operational policy constraints in the AIPS [17]

Year	SNSP (%)	RoCoF	Inertia	MUON	System strength
2025	75	±1 Hz/s	23 GWs	7	Under development
2030	~95	±1 Hz/s	TBD	~3	Enduring policy
2035	~100	±1 Hz/s	TBD	~0	Enduring policy

brief description of the rationale and practical issues behind the RoCoF limit change (from ±0.5 to ±1 Hz/s). Note that an overview of the real-time component of LSAT which is used to evaluate the dynamic stability in the AIPS is provided in Chap. 3.

6.3.1 *RoCoF Setting Change*

More than a decade ago an ambitious target of 40% of electricity to originate from renewable generation by 2020 was set, most of this from wind turbines. The 2010 published Facilitation of Renewables (FoR) study identified that having 40% of IBRs would have a significant impact on RoCoF experienced in the network. For this reason, it was essential to increase the maximum RoCoF from ±0.5 to ±1 Hz/s. The FoR study also identified that during times of high wind generation and following the loss of the single largest credible contingency, RoCoF values greater than ±0.5 Hz/s but no greater than ±1 Hz/s could be experienced.

Following the conclusions of the FoR, generator market participants and distribution system operators worked toward changing the RoCoF setting on all assets to ±1 Hz/s. This change involved real field testing of generators to showcase the withstand capability. After running RoCoF compliance tests for all large conventional generators, a RoCoF trial started in June 2020. It was envisaged the trial project would broadly comprise detailed technical studies and simulations to identify potential RoCoF-related vulnerabilities in the system and a trial of operating the power system with an increased RoCoF limit. The expectation was following successful completion of the trial, and subject to appropriate mitigation strategies being established, the RoCoF operational policy range would be widened from the threshold of ±0.5 Hz/s to ±1 Hz/s on an enduring basis. In May 2023, the RoCoF trial was concluded and approved (following a successful trial and detailed studies), and is now allowing operating the AIPS with a RoCoF limit of ±1 Hz/s. This enduring limit is enabling higher levels of renewable energy in the AIPS.

In summary, the whole process of changing the RoCoF took more than a decade. The roadmap of RoCoF settings change in the AIPS is summarized in Fig. 6.2.

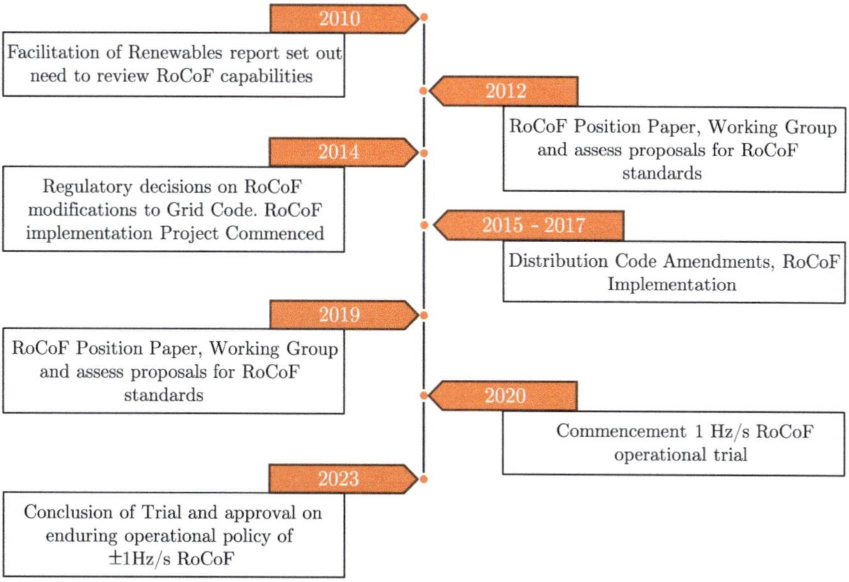

Fig. 6.2 Roadmap of the changes of RoCoF settings in the AIPS

6.3.2 AIPS Real-World Example

To assess the frequency, voltage and transient stability of the AIPS, we use the real-time LSAT analysis and results for one relevant month, in this case, June 2023. This month is particularly relevant as the network experienced low levels of inertia. However, the study shows that the trend extracted from the results of this month is consistent to that in previous months.

We evaluate each stability problem based on certain predefined criteria. These are presented in Table 6.3. In particular, for security reasons and to account for potential uncertainties such as in modeling and data accuracy, the Irish TSOs have build up different margins for different stability metrics. For example, a RoCoF of ± 0.9 Hz is used instead of the operational limit of ± 1 Hz/s. RoCoF is calculated using a rolling period of 500 ms and filtering is applied to eliminate high frequency transients. Similarly, the zenith limit set in LSAT is 50.8 Hz instead of the 51 Hz operational limit. All the results presented are with respect to these margins. On the other hand, rotor-angle stability is evaluated by means of the negative margin metric which captures the maximum rotor angle difference between any pair of generators across the network [18]. Note that if any of the limits are reached during the simulation, the security limit is flagged to the operators but the system is not unstable per se. Control room operators will then take the necessary mitigating actions in a timely manner to resolve frequency insecurities.

Table 6.3 Rotor-angle, voltage and frequency security limits in LSAT

Binding constraint	Security criteria
Rotor-angle	Negative margin [18]
Voltage	Outside Grid Code ranges [19]
RoCoF	±0.9 Hz/s
Zenith	50.8 Hz
Nadir	49.0 Hz

Table 6.4 Summary of total rotor-angle, frequency and voltage cases with binding constraints in June 2023

Binding constraint	Total cases	% of all cases	Comparative %
Rotor-angle	67	0.78%	16.03%
Voltage	160	1.86%	38.28%
RoCoF	116	1.35%	27.75%
Zenith	49	0.57%	11.72%
Nadir	26	0.30%	6.22%

6.3.2.1 Overview of Transient, Frequency and Voltage Insecurities

The total number of LSAT cases run for June 2023 is 8,594. As mentioned above, each case evaluates around 800 N-1 contingencies. Table 6.4 presents all the relevant statistics for the three stability problems namely rotor-angle, voltage and frequency. Note that out of 8,594 cases only 418 cases (or approximately 4.86%) are reported as insecure. In all cases, the system was secure in the base case where no contingency was applied to the network. This number of insecurities is to be expected as the Irish TSOs push the operational boundaries of the AIPS. In fact, the number of insecure cases is minimal considering the original stability margins set by the TSOs, for example, ±0.9 Hz/s instead of ±1 Hz/s. Note also that power systems cannot operate 100% securely as it is cost-prohibitive if at all physically feasible [20]. While comparing the % of the three different stability problems, it can be seen that frequency is the main problem with 2.22% of total insecure cases. In particular, it appears that RoCoF is the main problem with 1.35% of total cases. However, in practice, frequency stability has significantly improved in the AIPS in terms of nadir and zenith in recent years [21].

Among those security issues, our investigation reveals that all rotor-angle stability issues were local, e.g., due to two units oscillating against the whole system, and hence, they were not related to the system being low in inertia. Similarly, voltage insecurities were mainly concerned with not having enough resources to absorb extra reactive power generation in a certain part of the system and then again, not related to the inertia of the system.

As frequency stability appears to be the dominant stability problem for this particular month (more than 45% of the total insecure cases), we analyze in more detail each frequency stability metric and in which system conditions they appear more in the next section. In terms of the best technologies to prevent high RoCoF events, reference [22] suggests the following best suitable technologies ranked from the most to the least effective.

- Synchronous condensers;
- Reduction in the minimum active power generation;
- Rotating stabilizers;
- Pumped hydro;
- Flexible thermal power plant;
- "Parking"[1];
- Battery technology;
- Compressed Air Energy Storage (CAES);
- Flywheels;
- Wind turbines;
- High Voltage Direct Current (HVDC) Interconnector (IC);
- Demand Side Management (DSM);
- AC ICs.

The ranking is based on a high level assessment that included different criteria and metrics such as effectiveness, technology maturity, lead time, ability to provide additional system services, and geographical flexibility between the technologies. In particular, the first six technologies are based on synchronous rotating mass (inertia) and without converters.

6.3.2.2 Correlation Between Frequency Insecurity and Operating Conditions

In this section, we are interested to identify the system-wide contributing factors (specific variables of the system) to insecurities in order to have a better understating of the trends that lead to those insecurities.

Inertia

Figure 6.3 shows the correlation between total inertia in the AIPS and RoCoF-, nadir, RoCoF+, zenith and all cases, respectively. Light shade means secure cases while dark shade represents insecure cases. There is a strong correlation between RoCoF- and nadir and high all-Island inertia. While this may be counter-intuitive, the system conditions in specific areas of the AIPS following a contingency can be such that support this correlation. For example, it is well-known that in case of the loss of the North-South (N-S) Tie-line between IE and NI (system separation) and depending

[1] "Parking" of conventional generators means operating generation plant at low active power output levels but with reduced or no capability to provide system services at the lower output levels.

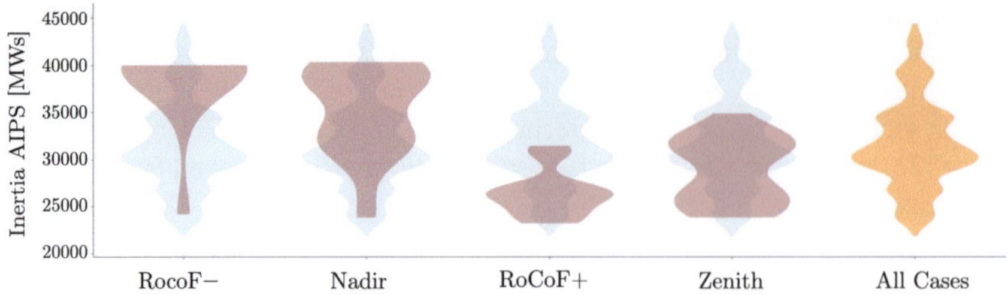

Fig. 6.3 Correlation between inertia and frequency insecurities

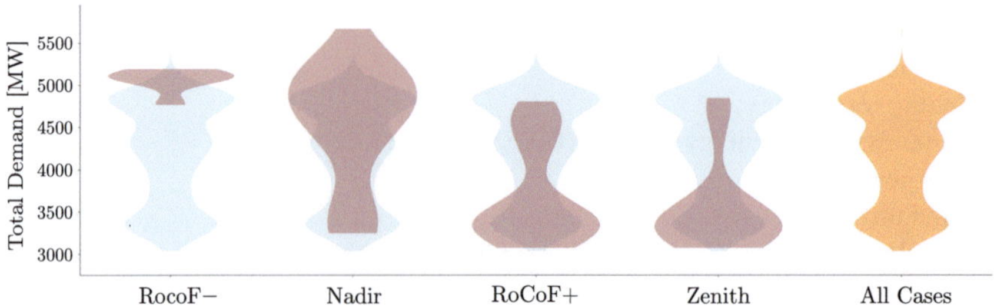

Fig. 6.4 Correlation between total demand and frequency insecurities

on the infeeds/outfeeds in IE and NI, frequency stability issues can appear [16]. In fact, this is one of the main reasons why the TSOs have a requirement for a minimum number of large conventional units online in both jurisdictions [3]. To further support this, the TSOs have recently identified through detailed dynamic studies that there is a need to procure 4 GWs of inertia in NI (incentivizing one zone within NI) and 6 GWs of inertia in IE (incentivizing two zones within IE) [23]. On the other hand, as expected, there is strong correlation between low inertia and RoCoF+ and zenith. Overall, there is a medium correlation between inertia and all cases.

Demand

Figure 6.4 shows the results of the correlation between demand and all frequency insecurities. Similar to inertia, there is a strong correlation between high demand and RoCoF– and nadir. On the other hand, low demand scenarios appear to lead to RoCoF+ and zenith issues. For example, during night hours with a lot of wind generation and low demand can lead to such potential insecurities. Overall, demand has a strong impact in all cases.

Wind

The correlation results between wind and all frequency insecurities are shown in Fig. 6.5. Wind appears to have the strongest correlation with the frequency insecurities compared to inertia and demand. In the vast majority of cases low wind leads to RoCoF– and nadir insecurities. Similar to inertia, these insecurities are driven by specific network conditions following N–1 contingencies. On the other hand, relatively medium and high wind scenarios lead to potential zenith and RoCoF+ insecurities.

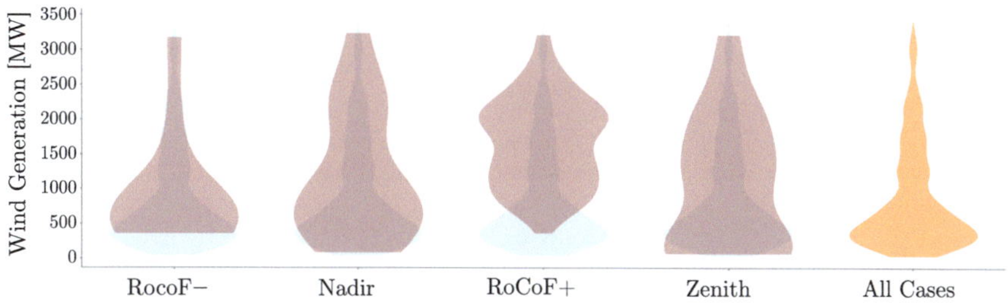

Fig. 6.5 Correlation between wind generation and frequency insecurities

6.3.2.3 Frequency Stability

To have a better understanding of the relationship between different system conditions and frequency insecurities, we plot inertia, total demand, and wind generation against each other and all LSAT cases with scatter plots.

Frequency nadir and zenith
Figure 6.6 shows scatter plots for frequency nadir and zenith, respectively. The total LSAT cases including the insecure cases are plotted as a function of: (i) demand vs wind generation; (ii) inertia vs wind generation; and (iii) demand versus inertia. Frequency nadir happens generally during periods of high demand, low wind and relatively high inertia. While these results seem somewhat counter-intuitive, they can be explained by the fact that in high demand scenarios and low wind, all the generators are at high outputs and the ICs are also importing, thus, leading to potential extreme nadirs. The frequency zenith constraint is usually binding during periods of low demand and relatively high wind generation and low inertia. As opposed to nadir results, these are expected as, for example, during nights with high wind generation, low demand and low inertia, tripping of ICs exporting leads to Over-frequency (OF).

RoCoF+ and RoCoF−
Figure 6.7 show the results for RoCoF including both positive and negative values. RoCoF+ insecurities happen predominately during periods of high wind and low inertia and demand scenarios. Similar to zenith results, this is expected as a light power system with high non-synchronous penetration (wind) is prone to frequency related problems in case of large-scale contingencies. Similar to nadir results, RoCoF− insecurities occur more often during periods of low wind, and high demand and inertia scenarios.

6.4 Frequency Stability for 100% IBR-Based Generation

An extreme case that has not been yet observed in large real-world power systems is a system with 100% inverter-based generation. As a closure to Sect. 6.2, we consider

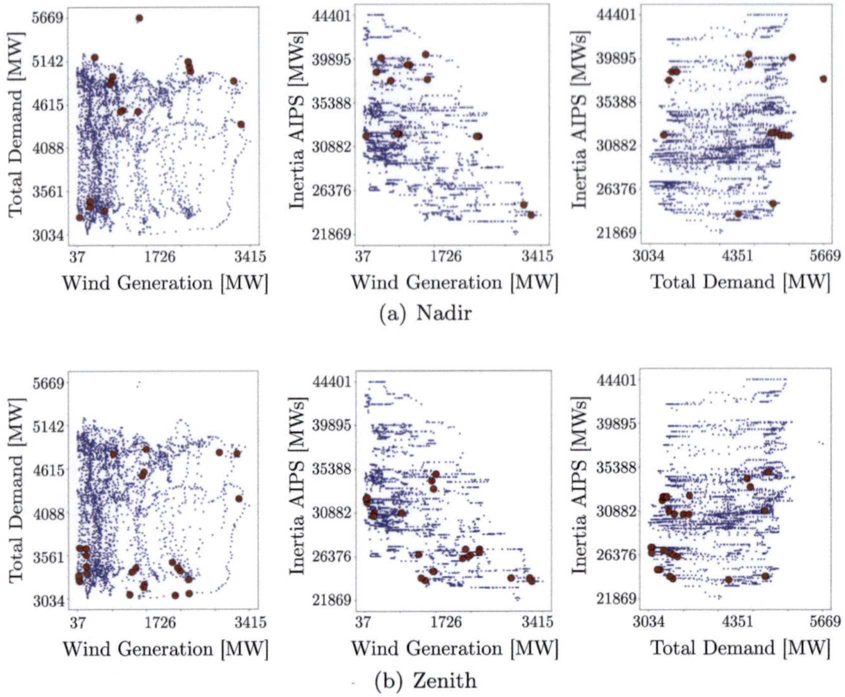

Fig. 6.6 Relationship between system conditions (inertia, demand, and wind) and nadir and zenith insecurities

this operating condition and illustrate its performance based on a modified AIPS model where all synchronous machines are substituted with IBRs (see Appendix C).

6.4.1 Frequency-Based Power Balance Control

The current common implementation of IBRs assumes that the deviation of the frequency with respect to the synchronous reference is the best metric for the determination of power imbalances. This assumption is based on the experience with synchronous machines and while it is still not clear whether frequency is still relevant for a power system with 100% inverter-based generation, it is certainly a reasonable assumption.

We increase the generation of wind power plants by gradually replacing the synchronous power plants with IBRs. We assume that wind power plants are connected to the grid through full-size direct-drive converters and energy storage systems. Finally, we consider Grid-Following Converter (GFL)-droop and Grid-Forming Converter (GFM)-Virtual Synchronous Generator (VSG) control strategies for the converters.

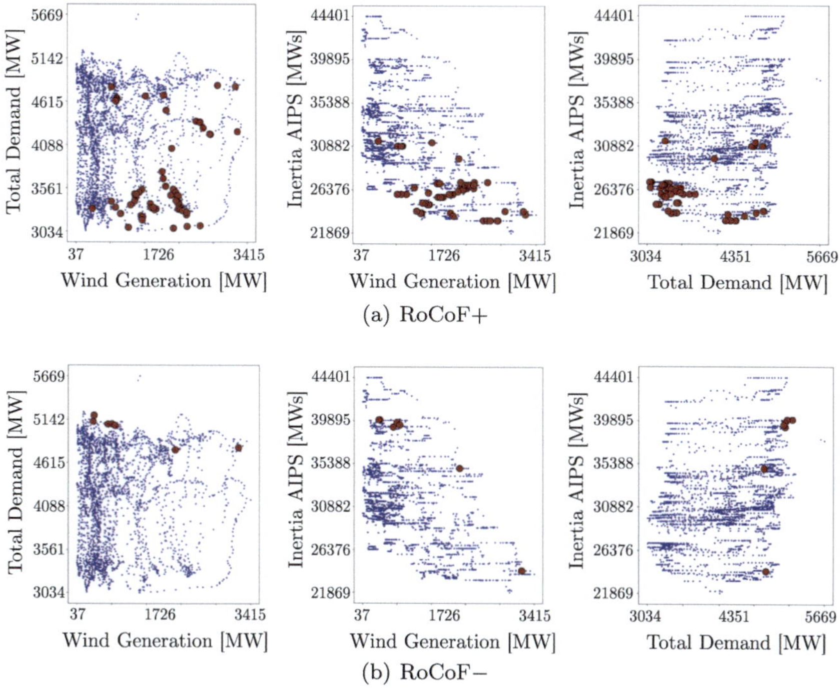

Fig. 6.7 Relationship between system conditions (inertia, demand, and wind) and RoCoF+ and RoCoF− insecurities

Figures 6.8 and 6.9 show the frequency nadir and RoCoF, respectively, measured at 500 ms after the contingency parametrized with respect to the percentage of IBR penetration in the system. Results shows that VSG-controlled converters improves the overall system frequency transients by increasing the frequency nadir and reducing the RoCoF. Droop-controlled GFLs, on the other hand, have the opposite effect and make the system unstable above 80% of penetration. This is expected as substituting synchronous machines with GFLs reduces the inertia of the system. The droop-based GFLs scenario becomes unstable for IBR generation higher than 80%, whereas the VSG-based scenario is not only stable but shows a better performance than the original system that includes conventional generation. These results confirm the importance of the inertia to retain the stability of the system and suggest that this inertia can be obtain through proper control of energy storage systems. Physical inertia of the machines does not appear to be fundamental for the safe operation of the system.

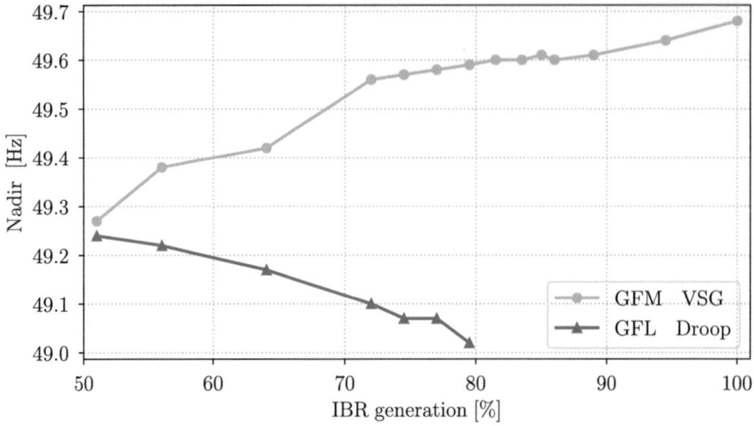

Fig. 6.8 Frequency nadir for the AIPS with increasing penetration of IBR generation

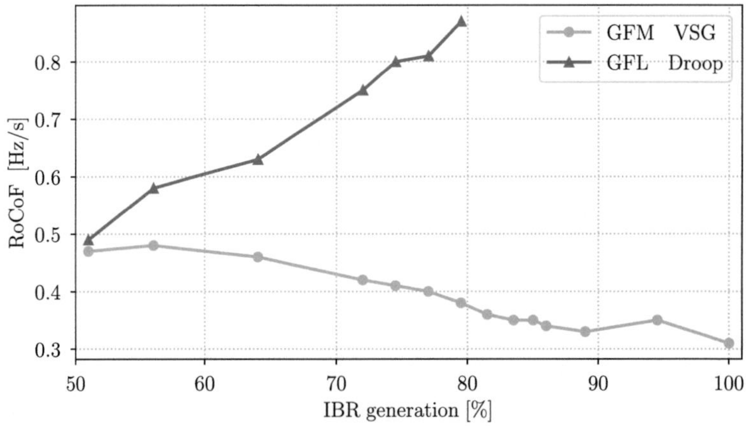

Fig. 6.9 RoCoF of the AIPS with increasing penetration of IBR generation

6.4.2 Voltage-Based Power Balance Control

In this second example, we show that inertia is not necessarily associated with frequency in a system with 100% inverter-based generation. To illustrate this point, consider the power injections of the lossy electromechanical model of the synchronous machine shown in Fig. 6.10:

$$p = \frac{[ev\cos(\delta-\theta) - v^2]R_a + [ev\sin(\delta-\theta)]X_{d1}}{R_a^2 + X_{d1}^2}, \tag{6.3}$$

$$q = \frac{[ev\cos(\delta-\theta) - v^2]X_{d1} - [ev\sin(\delta-\theta)]R_a}{R_a^2 + X_{d1}^2}. \tag{6.4}$$

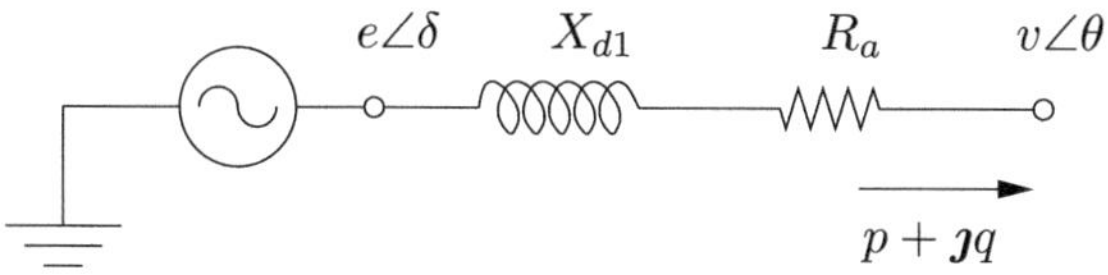

Fig. 6.10 Classical transient model of the synchronous machine

In synchronous machines, the armature resistance r_a is small with respect to x'_d and is often neglected, thus leading to the well-known equations:

$$p = \frac{ev \sin(\delta - \theta)}{X_{d1}}, \tag{6.5}$$

$$q = \frac{ev \cos(\delta - \theta) - v^2}{X_{d1}}. \tag{6.6}$$

On the other hand, if we focus on the *dual* parts of (6.3) and (6.4), that is, the terms that depend on the armature resistance and suppose that the reactance X_{d1} is zero or negligible, we obtain:

$$\tilde{p} = \frac{ev \cos(\delta - \theta) - v^2}{R_a}, \tag{6.7}$$

$$\tilde{q} = -\frac{ev \sin(\delta - \theta)}{R_a}. \tag{6.8}$$

Of course, this model does not make sense for a physical generator but we assume that the relations above are obtained with a converter, the control of which allow a much bigger degree of freedom than conventional electrical machine. For convenience, let us also assume that the virtual parameter that represents the armature resistance is negative, say $K = -1/R_a$, thus leading to:

$$\tilde{p} = K v^2 - K ev \cos(\delta - \theta), \tag{6.9}$$

$$\tilde{q} = K ev \sin(\delta - \theta). \tag{6.10}$$

Equations (6.9) and (6.10) describe a device where the active power strongly depends on the magnitude of the internal emf e, while the reactive power strongly depends on the phase angle δ.

The next step is to define a swing equation *dual* with respect to the swing equation of conventional synchronous machines. With this aim, consider the complex quantity:

$$\bar{e} = e \exp(\jmath\, \delta), \tag{6.11}$$

where e and δ are assumed to be time varying. Then, defining $\varsigma = \ln(e)$, $e \neq 0$, (6.11) becomes:

$$\bar{e} = \exp(\varsigma + \jmath\,\delta)\,, \tag{6.12}$$

and its time derivative is:

$$\dot{\bar{e}} = (\varsigma' + \jmath\,\delta')\exp(\varsigma + \jmath\,\delta) = (\varrho + \jmath\,\omega)\,\bar{e}\,, \tag{6.13}$$

where ω is the usual angular frequency of the voltage and ϱ is:

$$\varrho = \varsigma' = e'/e\,. \tag{6.14}$$

The quantity $\varrho + \jmath\,\omega$ is the *complex frequency* of $\bar{e}$ (see also Eq. (1.15) in Chap. 1). Finally, the dual swing equation in terms of the instantaneous bandwidth is:

$$\begin{aligned} \varsigma' &= \varrho\,, \\ \tilde{M}\,\varrho' &= p^{\text{ref}} - \tilde{p}(\varsigma, v, \delta, \theta) - \tilde{D}\,\varrho\,, \end{aligned} \tag{6.15}$$

or, equivalently

$$\begin{aligned} e' &= \varrho\,e\,, \\ \tilde{M}\,\varrho' &= p^{\text{ref}} - \tilde{p}(e, v, \delta, \theta) - \tilde{D}\,\varrho\,, \end{aligned} \tag{6.16}$$

where $\tilde{M}$ and $\tilde{D}$ are parameters that resemble a virtual inertia and a virtual damping, respectively; p^{ref} is the active power as defined by the control of the converter; and $\tilde{p}$ is given in (6.9). Note that, in steady state, $\varrho = 0$ as bus voltage magnitudes are required to be constant.

To complete the model of the dual-GFM, we need to define primary active and reactive power controllers. The primary active power control for the is required to track ϱ, for example:

$$\tilde{T}_m\,(p^{\text{ref}})' = \frac{1}{\tilde{\mathcal{R}}}(\varrho^{\text{ref}} - \varrho) + p_o^{\text{ref}} - p^{\text{ref}}\,, \tag{6.17}$$

where $\varrho^{\text{ref}} = \varrho_o = 0$ and p_o^{ref} is the converter power set point. As ϱ must be zero in steady state, to avoid a drift of e, the active power primary control of the dual-GFM control can be perfect tracking ($\tilde{\mathcal{R}} \to 0$). This is a significant difference with respect to the conventional Primary Frequency Control (PFC) of synchronous machines which cannot be perfect tracking in general (see discussion in Sect. 2.3.3.4 of Chap. 2).

The reactive power can be regulated directly through the virtual angle δ:

$$\begin{aligned} T_q\,\delta' &= K_q(q^{\text{ref}} - \tilde{q}) - \delta\,, \\ \tilde{T}_r\,(q^{\text{ref}})' &= \tilde{K}_r(\omega^{\text{ref}} - \omega) - q^{\text{ref}}\,, \end{aligned} \tag{6.18}$$

where the expression of $\tilde{q}$ is given in (6.10) and the bus voltage phase angle and frequency are assumed to be obtained using a standard PLL. This control resembles

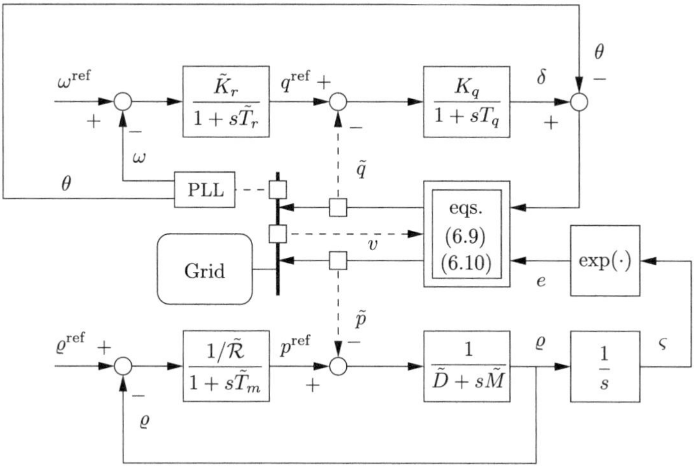

Fig. 6.11 Control scheme of the dual-GFM converter

the dual version of a typical primary voltage droop regulation. In fact, defining $\delta_r = K_q\, q^{\mathrm{ref}}$ and $\hat{K}_r = \tilde{K}_r / K_q$, one can rewrite (6.18) as:

$$
\begin{aligned}
T_q\, \delta' &= \delta_r - K_q\, \tilde{q} - \delta\,, \\
\tilde{T}_r\, \delta_r' &= \hat{K}_r(\omega^{\mathrm{ref}} - \omega) - \delta_r\,.
\end{aligned}
\tag{6.19}
$$

The resulting dual-GFM control scheme is shown in Fig. 6.11.

The virtual angular speed does not appear in the formulation of the dual-GFM control except for the time derivative of the internal signal δ in (6.18). This is consistent with synchronous machines, where ϱ or, which is substantially the same, e', does not appear anywhere except for the time derivative of the internal emf e in the primary voltage control. It is still necessary, of course, to fix the frequency in the system. As shown in Fig. 6.11, the dual-GFM converter imposes the frequency at the bus through the regulation of the reactive power.

We illustrate the dynamic performance of the dual-GFM using again the model of the AIPS. In this case, all synchronous machines are substituted with the dual-GFM described above. The following parameters for all dual-GFM are used: $K = 1$, $\tilde{M} = 15$ s, $\tilde{D} = 0.5$, $\tilde{T}_m = 2$ s, $\tilde{\mathcal{R}} = 0.05$, $K_q = 10$, $T_q = 5$ s, $\tilde{K}_r = 40$, $\tilde{T}_r = 1$ s. K, $\tilde{M}$, $\tilde{D}$ and $\tilde{\mathcal{R}}$ are in pu w.r.t. the capacity of the converter. The modified system, thus, not only does not include any conventional synchronous machine, but does not have any conventional GFL or GFM.

Figure 6.12 shows the dynamic performance of the modified AIPS following the outage of a large load. The dual-GFMs cope well with the power imbalance and recover a normal operating condition. These results lead to two emblematic conclusions: (i) the angular frequency is not the unique quantity that allows defining the power imbalance of a large interconnected ac system; and (ii) inertial response can

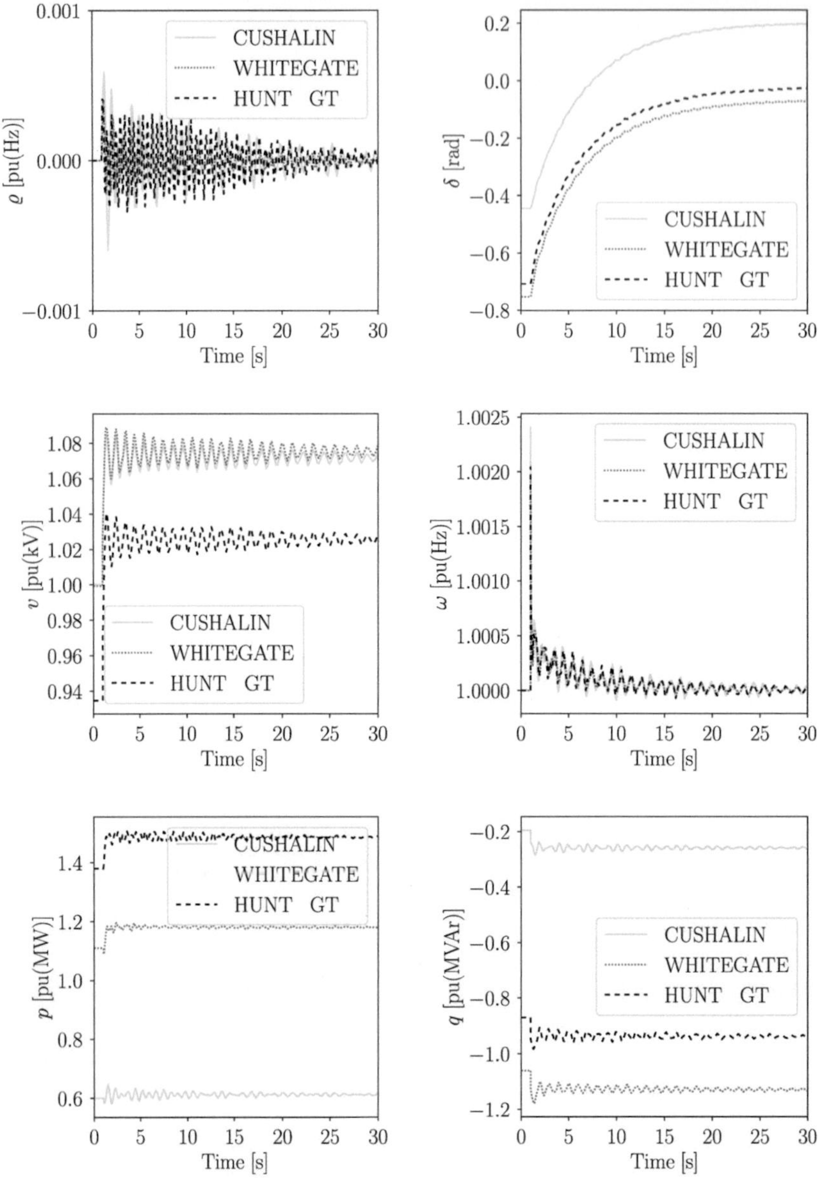

Fig. 6.12 Dynamic performance of relevant variables of three dual-GFMs following the outage of a large load

be obtained through voltage regulation alone. A byproduct of the first point is that the dual-GFM approach can potentially also be utilized in dc systems. Of course, all these conclusions are purely theoretical for now and more study and, crucially, the constructions of prototypes are needed to prove that the dual-GFM can actually work in practice. But the main message remains valid, that is, the swing equation of synchronous machines (virtual or real) is not the only possible mechanism way to keep an electric power system balanced.

6.5 Impact of Network Topology

Assessment of frequency stability has so far mainly focused on the quantity of connected synchronous kinetic energy in the system, or inertia and has neglected the impact of other factors such as network topology and its relevant parameters. Reference [24] demonstrates how network topology, for example, fully radial topology, negatively influences both the frequency nadir and the RoCoF. Similarly, in [25], it is argued that network topology parameters such as transmission line reactance have the same effect as inertia, decreasing the RoCoF magnitude.

In this section, we present a statistical study on the impact of topology of the transmission system on the transient and frequency stability of power systems. Four topologies are considered: random graph; small-world graph; nested small-world graph; and lattice graph as shown in Fig. 6.13. Apart from obvious visual differences, these graph differ from an important metric, namely, the average distance between any two nodes. The distance is calculated as the minimum number of branches that one has to go through to connect two given nodes. The random graph has the smallest average distance, and the lattice the largest. Small-world and nested small-world are close to the random, but better resemble the typical topology of power systems, where geographical constraints prevents purely random connections. The main difference between small-world and nested small world topologies is that the former represents better a single power system or region whereas the latter represents better large interconnected systems [26].

These graphs are utilized to generate hundreds of synthetic networks with given topology and adequately generated static and dynamic data. The interested reader can find more details on the construction and feature of each graphs and the heuristic rules to generate the dynamic data in [27]. Then, each network is simulated based on a Monte Carlo method and considering random contingencies and different levels of inertia. For each topology and scenario, 500 synthetic networks are generated. A simulation is assumed to be stable if 20 s after the contingency, the frequency of the center of inertia of the system is within a given range, namely [0.984, 1.016] pu(Hz). Figure 6.14 show the results for each topology for various levels of inertia. We note that the level of inertia, is not crucial for the system as the load outage is never causing a loss of synchronism. Results, however, indicates that the higher the average distance between nodes, the lower the number of stable simulations.

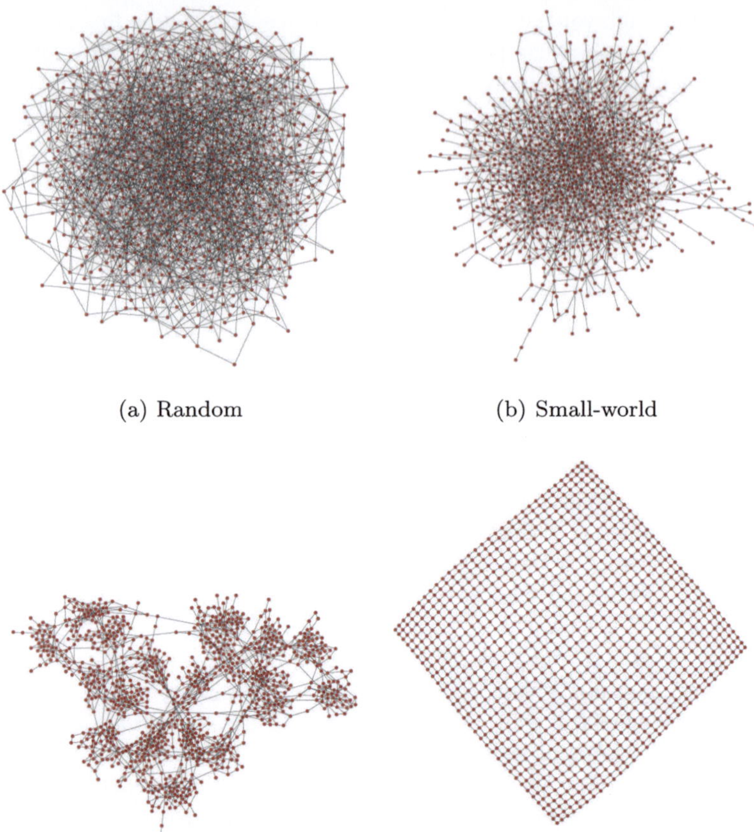

(a) Random (b) Small-world

(c) Nested small-world (d) Lattice

Fig. 6.13 Examples of network topologies

In Brief

- Frequency stability is linked to power balance.
- The main binding constraint in the AIPS is currently related to frequency stability and, in particular, to the limits on RoCoF.
- As fewer conventional generators are expected to be online in the future, it is critical that IBRs provide Fast Frequency Response (FFR) and PFC to support frequency stability.
- It is possible, at least in theory, to maintain the balance of large power systems through GFMs with a VSG control strategy.
- It is possible, at least in theory, to maintain the balance of large power systems through GFMs converters that implement a dual swing equation based on the voltage magnitude rather than on the frequency.

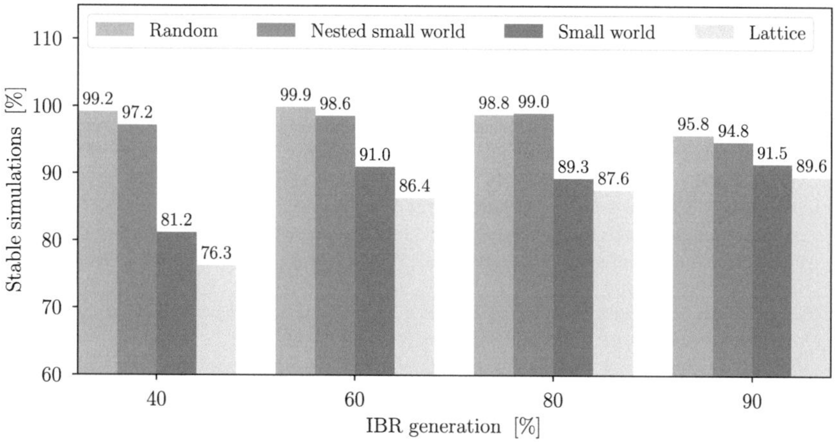

Fig. 6.14 Topology-based frequency stability analysis: percentages of stable simulations following a loss of generation [27]

- The average distance between two buses of a power system impacts on frequency stability. The higher the average distance, the poorer is the dynamic performance of the frequency control of the system.

Open Questions

The link between frequency and power balance has been the foundation of the operation and control of power systems since the very early stage of electrification. With the move toward 100% IBR-based generation, this link might disappear and with it, the whole concept of frequency stability. Understanding whether frequency is a fundamental quantity for the control and operation of future power systems, and in which time scale, is currently one of the most challenging open questions. Answering this question will also help solve other arising matters, for example, what is the correct ratio of GFL and GFM to ensure stability and what is the actual impact of converter-driven and resonance stability on real-world power systems.

References

1. Hatziargyriou N, Milanovic J, Rahmann C et al (2021) Definition and classification of power system stability—revisited and extended. IEEE Trans Power Syst 36(4):3271–3281
2. Milano F, Drfler F, Hug G et al (2018) Foundations and challenges of low-inertia systems (invited paper). In: Power systems computation conference (PSCC), pp 1–25

3. Hurtado M, Kërçi T, Tweed S et al (2023) Analysis of wind energy curtailment in the Ireland and Northern Ireland power systems. In: IEEE PES general meeting, pp 1–5
4. Ratnam KS, Palanisamy K, Yang G (2020) Future low-inertia power systems: requirements, issues, and solutions—a review. Renew Sustain Energy Rev 124:109773
5. Hong Q, Khan MAU, Henderson C et al (2021) Addressing frequency control challenges in future low-inertia power systems: a Great Britain perspective. Engineering 7(8):1057–1063
6. Ahmed F, Al Kez D, McLoone S et al (2023) Dynamic grid stability in low carbon power systems with minimum inertia. Renew Energy 210:486–506
7. Shakerighadi B, Johansson N, Eriksson R et al (2023) An overview of stability challenges for power-electronic-dominated power systems: the grid-forming approach. IET Gener Transm Distrib 17(2):284–306
8. Markovic U, Stanojev O, Aristidou P et al (2021) Understanding small-signal stability of low-inertia systems. IEEE Trans Power Syst 36(5):3997–4017
9. He C, He X, Geng H et al (2022) Transient stability of low-inertia power systems with inverter-based generation. IEEE Trans Energy Convers 37(4):2903–2912
10. Chen J, Liu M, Milano F et al (2020) 100% converter-interfaced generation using virtual synchronous generator control: A case study based on the Irish system. Electr Power Syst Res 187:106475
11. Kundur P, Paserba J, Ajjarapu V et al (2004) Definition and classification of power system stability IEEE/CIGRE joint task force on stability terms and definitions. IEEE Trans Power Syst 19(3):1387–1401
12. Milano F (2025) Dual grid-forming converter. IEEE Trans Power Syst 1–4
13. AEMO (2024) The role and need for inertia in a NEM-like system. www.aemo.com.au
14. Tielens P, Van Hertem D (2016) The relevance of inertia in power systems. Renew Sustain Energy Rev 55:999–1009
15. EirGrid & SONI (2024) All-island transmission system performance report 2023. https://www.eirgrid.ie/
16. EirGrid & SONI (2022) Operational policy roadmap 2023–2030. https://www.eirgrid.ie/
17. Eirgrid & SONI (2025) Operational policy roadmap 2025–2035. www.cms.eirgrid.ie
18. EirGrid & SONI (2021) Potential solutions for mitigating technical challenges arising from high RES-E penetration on the Island of Ireland. www.soni.ltd.uk
19. EirGrid (2024) EirGrid Grid Code. Version 14.2. https://www.eirgrid.ie/
20. Capitanescu F (2023) Are we prepared against blackouts during the energy transition?: Probabilistic risk-based decision making encompassing jointly security and resilience. IEEE Power Energ Mag 21(3):77–86
21. Kërçi T, Hurtado M, Gjergji M et al (2023) Frequency quality in low-inertia power systems. In: IEEE PES general meeting, pp 1–5
22. EirGrid & SONI (2015) RoCoF alternative solutions technology assessment. www.cms.eirgrid.ie
23. EirGrid & SONI (2022) Consultation on low carbon inertia service (LCIS) competitive procurement. https://www.eirgrid.ie/
24. Fradley J, Preece R, Barnes M (2020) The influence of network factors on frequency stability. IEEE Trans Power Syst 35(4):2826–2834
25. Farmer WJ, Rix AJ (2021) Understanding the impact of network topology on frequency stability considering continuous spatial-temporal disturbances from wind generation. Electr Power Energy Syst 129:106776
26. Wang Z, Scaglione A, Thomas RJ (2010) Generating statistically correct random topologies for testing smart grid communication and control networks. IEEE Trans Smart Grid 1(1):28–39
27. Ebrahimzadeh F, Adeen M, Milano F (2019) On the impact of topology on power system transient and frequency stability. In: IEEE international conference on environment and electrical engineering (EEEIC), pp 1–5

Chapter 7
Converter-Based Demand

In recent years, an increasing number of new kinds converter-interfaced loads has become prominent. Compared to traditional loads such as motor and resistive loads, converter-interfaced loads do not provide inherent automatic grid-stabilization mechanisms, that is, the power consumption does of converter-interfaced loads does not decrease if the frequency or voltage drops [1]. Moreover, newest electronic loads may disconnect from the grid following normally cleared disturbances. This is the case, for example, of data centers. Certain types of new loads, in particular EVs, also tend to introduce a lot of demand variability given their large numbers and the associated (especially uncontrolled) charging infrastructure.

The increased proportion of converter-based demands has recently raised concerns for TSOs in terms of stability challenges. For this reason, industry has started to discuss solutions. For example, ESIG has recently established a task force named the Large Loads Task Force (LLTF) looking to address various issues such as large loads modeling, performance requirements and data collection and forecasting [2]. The first author of this book is a member of LLTF (represents EirGrid/SONI) and recently attended the 2025 ESIG Spring technical workshop where he experienced exiting and, at times, heated discussions on large loads challenges and solutions. NERC has also recently established a LLTF to better understand the reliability impacts of these emerging loads [3].

In this chapter, we consider two technologies, namely, data centers and EVs, which presents different and complementary challenges for the grid. Data centers are substantially large constant power consumptions with limited FRT capability whose impact is critical in terms of short-term frequency deviations. EVs, on the other hand, are small stochastic, often highly correlated, processes, the impact of which is visible in the voltage stability and long-term frequency deviations. In different ways, both loads have an impact on the frequency quality and are thus relevant for the matter of this book.

Section 7.1 provides a brief overview on the main impacts of data centers. Section 7.1.1 presents a short background on data centers integration on the AIPS and

T. Kërçi and F. Milano, *Frequency Quality in Power Systems*, Power Systems,
https://doi.org/10.1007/978-3-032-04721-2_7

the associated challenges. Section 7.1.2 presents an overview of data center characteristics including their load supply and Uninterruptible Power Supply (UPS) schemes. Section 7.1.3 then provides a brief description of how the EirGrid and SONI model the data center load in their decision-making support tools such as dynamic stability assessment tools. Section 7.1.4 discusses the main results through examples based on operational data. Section 7.1.5 discusses different options considered by EirGrid and SONI to address the data center performance issues. Section 7.1.6 briefly focuses on the importance of the development of FRT standards for data centers. Section 7.2.1 provides a background on the integration of EVs and their associated challenges. Section 7.2.2 describes the models of the power system, the Plug-in Electric Vehicles (PEVs) fleets, and the control strategies analyzed. Finally, Sect. 7.2.3 presents the Additive-Increase-Multiplicative-Decrease (AIMD) algorithm and the proposed control schemes and Sect. 7.2.4 introduces our system set-up, and the main results obtained.

7.1 Impact of Data Centers

Data centers are one of the large critical demand facilities that store digital data. Some of main modern large loads include [4]:

- Hyper-scale data centers
- Could data centers
- AI/ML data centers
- High-performance computing data centers
- Cryptocurrency and Blockchain data
- Hydrogen/electro-fuel production facilities
- EVs charging stations

Due to their size up to GW levels and power characteristics, these loads can have a significant impact on power system dynamics:

- Frequency quality
- Angular stability
- Voltage stability
- Sub-synchronous resonance
- Harmonic stability

In particular, there is a concern that AI data centers may significantly impact on frequency quality due to their highly nonlinear fluctuating load (for example, 50 MW up and down in 1 s) that could lead to load-induced oscillations, as illustrated in Fig. 7.1 below. These sharp power surges and dips, and significant peak-idle power ratios can also induce voltage flicker concerns and introduce risk of transient instability. For example, they can contain frequency content at machine swing frequencies or interarea swing frequencies [5]. These *never-seen-before dynamics* make AI data

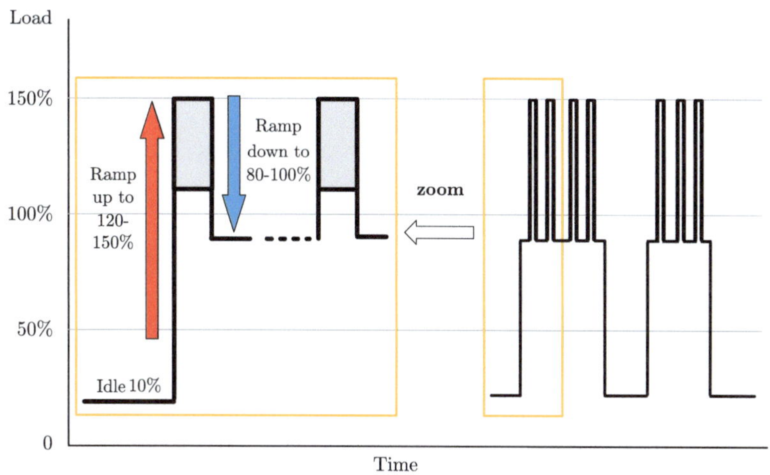

Fig. 7.1 Periodicity, variability and ramping of AI training loads (courtesy of Elevate Energy Consulting [9])

centers very unique loads that could pose threats to the power grid reliability and resilience [6]. In general, emerging large loads could also be price-responsive meaning they can ramp up and down in response to price in a non-coordinated way from the grid needs perceptive and, thus, cause frequency fluctuation concerns [7].

A potential and promising solution to these problems is, for example, the installation of non-GFM E-Static Synchronous Compensator (STATCOM) that could reduce these variations [8]. GFM BESS applications could also be utilized to address the same by buffering transients. Conventional generators are not a solution to these problems due to their slow response. Additionally, TSOs require that limits are set on load amplitude of variations and ramp-up and ramp-down rate. Figure 7.2 illustrates the degree of risk as a percentage of interconnection queues for various loads. At the time of writing, data centers are by far the highest risk to the Western Interconnection in US.

Thanks to our direct experience, in this section, we describe the impact of data centers on frequency quality in the AIPS. The demand response behavior of data centers during normal system conditions and contingency events is, in fact, causing real issues for the secure operation of the AIPS. Data is presented below correlating the cyclical power fluctuations in data centers' demand with frequency transients. Moreover, the response of data centers to system faults is exacerbating disturbances on the AIPS through "disconnection" and automatic reconnection of their demand. Such issues are analyzed and illustrated using data recorded during actual system events. Modeling such behaviors is a complex task and the current literature is scarce in that respect.

Most network codes have extensive requirements for generation performance, but less clear performance standards for demand users. Given our experience within the AIPS, and knowledge of similar issues being experienced by some other TSOs,

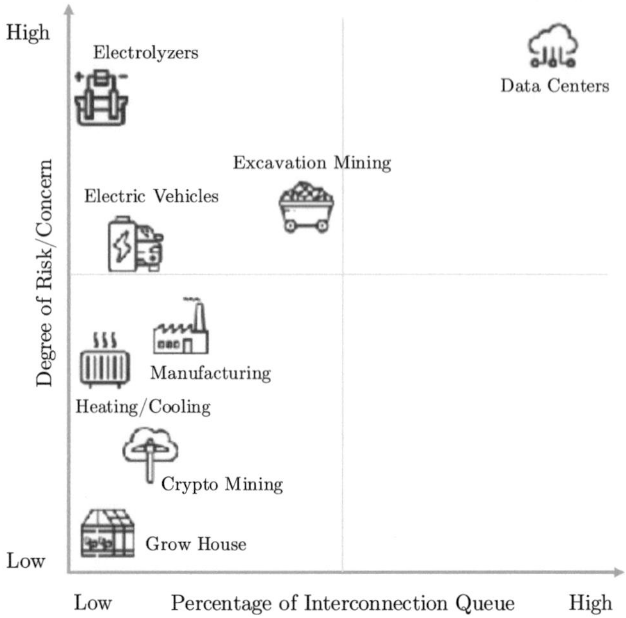

Fig. 7.2 Degree of concern relative to queue size (*Source* Elevate Energy Consulting [9])

we are of the view that the development of appropriate data center performance standards, such as those proposed by ENTSO-E, is required to address the issues outlined in this chapter.

It should be noted that while the focus here is on the challenges related to the performance of data center during system faults and impact on system frequency, data centers can be seen also as an opportunity to provide various energy and ancillary services such as FFR and emergency services.

7.1.1 Data Centers in Ireland

In previous chapters, we have mentioned how the generation landscape is changing dramatically, from synchronous to IBR-based in many countries around the world to accommodate renewable generation. A similar move towards power-electronics has started to happen with the demand side as well. It is expected to have around 1,000,000 EVs and 700,000 heat pumps installed in the AIPS by 2030 [10]. However, in IE, data centers are by far the most relevant type of converter-interfaced loads. At the time of writing, such demand is just above 800 MW over a peak demand of 7 GW, and is expected to reach above 1.6 GW by 2030 making it to account for around 30% of peak demand [11]. The boom in data center demand is not a peculiarity of the Irish system. It is associated with increased utilization of digital services, growth

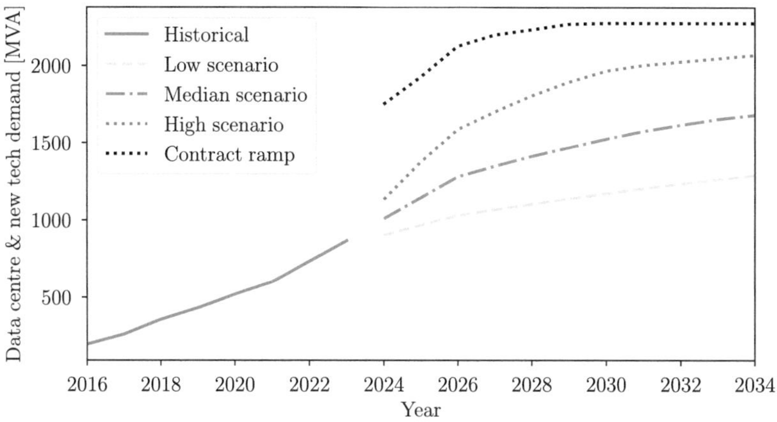

Fig. 7.3 IE demand expected from data centers and new technology loads [13]

of big data analytics, and popularity of cloud-based services [12]. Thus it should not constitute a surprise that North America leads the world in the total number of installed data centers with a total installed capacity of around 4,458 MW, with another 1,144 MW under construction [12]. ERCOT, which currently manages a peak demand of around 85 GW, is expecting to connect around 100 GW of large loads by 2030 [7].

With regard to IE, favorable climate conditions combined with high renewable electricity production levels and targets, have led to increasing power consumption from data centers within a relatively small geographical region (a significant proportion of this load is in the Dublin region). Figure 7.3 shows the evolution of data center demand installation and its expected evolution. The significant capacity coupled with decreasing inertia and frequency regulation from synchronous machines presents a risk to the frequency stability of the system. Due to voltage waveform quality having a significant impact on data center reliability and availability, their IT load has UPS protection systems in place to immediately disconnect from the grid when a disturbance is detected. This can cause large frequency and RoCoF excursions on the system adding additional stresses to generation and possibly creating contingencies of tripping wind generation if system frequency reaches 50.45 Hz or RoCoF exceeds ±1 Hz/s.

Another interesting phenomenon that data center load may introduce, and that has been observed in the AIPS, are minor consistent and sustained frequency transients, where frequency can jump in the range from 0.01 to 0.05 Hz. These frequency transients occur in sustained periods up to some minutes. It is important to mention that, at the time of writing, TSOs do not have explicit requirements for data center to deal with their FRT behavior and minor sustained frequency transients.

With respect to frequency quality, two types of data center demand behaviors are relevant, as follows.

- Regular, cyclical fluctuations in data center demand that are causing ongoing frequency quality issues such as the increase of σ_f [14].
- The demand response of data centers during and after transmission system faults that have a significant impact on the dynamic performance of the power system.

These two types of data centers create challenges for a power system accommodating data center connections while transmission systems are transitioning from a conventional synchronous machine generation-based system to a weather-dependent low inertia inverter-based system. These trends are expected to lead to data center demand causing significant transient frequency stability issues and additional constraints. In particular, the following are factors that impact on transient stability.

- Data center demand projected to increase to above 1,600 MW by 2030.
- Reducing the minimum number of conventional synchronous generating units to facilitate increased wind and solar generation causes a decrease in the inertia floor and frequency regulation.
- Data centers primarily being located in a relatively small geographic area making them collectively more susceptible to voltage disturbances on the grid.

7.1.2 Data Center Characteristics

High availability, and quality of electrical supply is of fundamental importance for data center owners. Voltage quality is important to ensure that issues such as harmonics, flicker, voltage dips, voltage unbalance does not damage or compromise critical IT load. To protect critical IT load, data centers utilize UPS systems to enhance power quality to their IT load during normal operating conditions. UPS systems are primarily designed to provide backup power from a few seconds to a few minutes until back up onsite power such as a gas or diesel generator comes online and provides power. In the event of voltage quality exceeding operational standards of the IT equipment, the UPS uses batteries to offer instantaneous power when the UPS detects the conformity of voltage has been compromised.

Figure 7.4 shows a typical internal load supply structure of a data center. Power is taken directly from the grid and supplies power directly to non-critical load, with critical IT load fed through a UPS. If the UPS system detects an abnormal system event such as a fault that exceeds operational characteristics of the IT load equipment, the UPS switches from using the grid for power to battery and if conditions do not return to normal operating conditions within a few seconds or minutes, it transfers to onsite generation until grid conditions return to normal operating limits.

An UPS is a short-term energy storage unit primarily used to maintain availability and redundancy to a data center operation in the event of grid frequency or voltage quality breaching UPS system tolerances. If an event breaches UPS tolerances for a given period, typically a few seconds, the power supply of the IT equipment in the data center may then switch from the short-term UPS energy storage capability

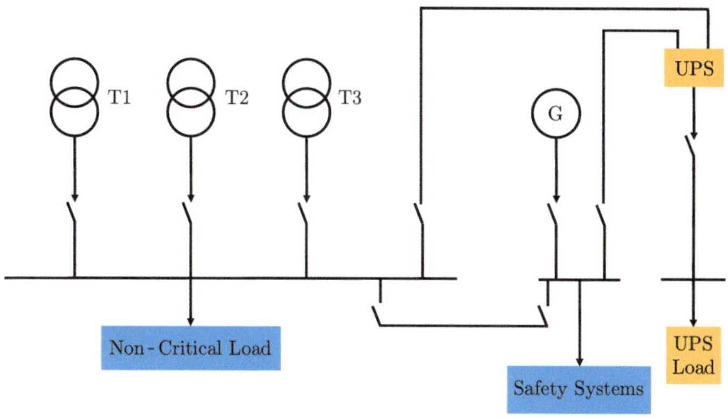

Fig. 7.4 Load supply structure of a typical data center

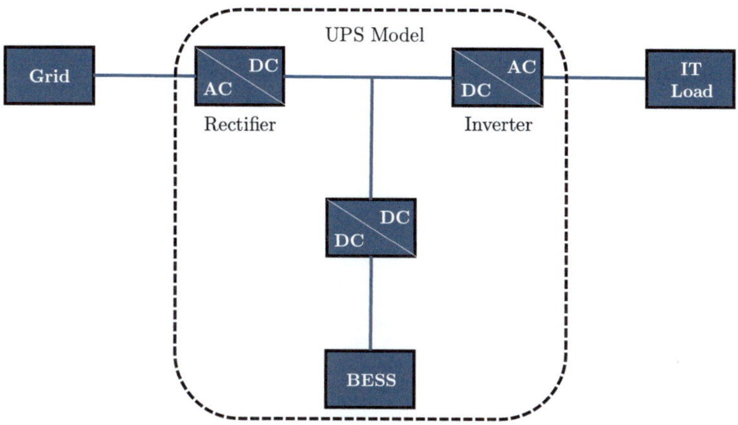

Fig. 7.5 Typical UPS schematic

to onsite generation. The UPS normally acts as a device to enhance power quality such as background harmonics on the system in ambient conditions that can extend the lifetime of IT equipment. Figure 7.5 below shows a typical UPS schematic that includes a rectifier and an inverter to feed the IT load. The rectifier is used to change the ac input power to dc and charge the BESS. The inverter switches the dc voltage from the rectifier or the BESS to an ac output for the critical load. The batteries in a UPS system provide emergency power in the event the data center protection settings switch their load from the grid connection point to their internal BESS and gas generation.

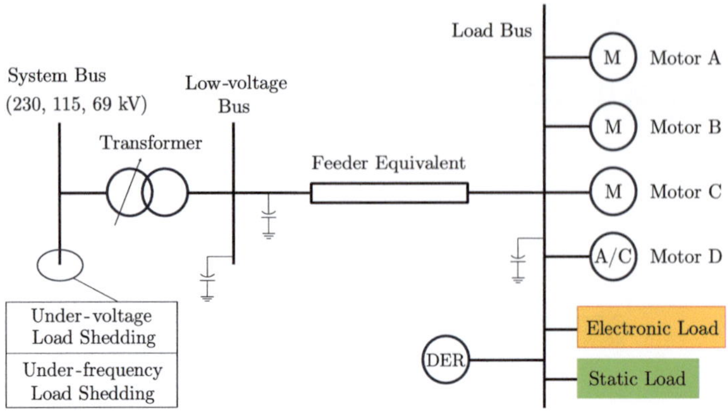

Fig. 7.6 Schematic diagram of CMPLDWG model

7.1.3 Data Center Modeling and Data Collection

7.1.3.1 Composite Load Model

Load modeling has always been important to power system stability but, at the same time, challenging due to the large number of diverse load components, time-varying compositions, and the lack of detailed load information and measurements. Load models consist of both static and dynamic versions. Static load models include include static constant impedance-current-power (ZIP) model and exponential model. Dynamic models aim at representing the active and reactive power as functions of both voltage and time such as, for example, induction motor and exponential recovery load model.

For dynamic studies involving large loads, most system operators around the world, especially in the US, utilize the well-known WSCC composite load model known as CMPLDWG (see Fig. 7.6). This model consists of several static and dynamic elements, including DERs [15, 16], and is implemented in most commercial software tools.

While it is widely utilized in industry, the CMPLDWG model often fails to accurately represent the behavior of data centers during transmission faults and does not allow reproducing actual system dynamics. The literature suggests several dynamic models to represent the behavior of power electronic loads such as the active distribution network model [17]. However, to the best of our knowledge, there is currently no TSO using these models in practice.

The load consumption of data centers can be modeled as the combination of a ZIP, an electronic load and an induction motor. We assume that the reader is familiar with the model of the induction motor, a description of which can also be found in [16]. We report below only the ZIP model and, most relevantly, the recovery logic of the electronic load.

Table 7.1 Calculation of coefficient $c_{\rm E}$ of the electronic load

Value of $c_{\rm E}$	Condition	Mode
0	$v(t) < V_{d2}$	1
$\dfrac{v(t) - V_{d2}}{V_{d1} - V_{d2}}$	$V_{d2} < v(t) < V_{d1}$, $v(t) \le V_{\min}(t)$	2
$\dfrac{V_{\min}(t) - V_{d2} + \alpha(v(t) - V_{\min}(t))}{V_{d1} - V_{d2}}$	$V_{d2} \le v(t) < V_{d1}$, $v(t) > V_{\min}(t)$	3
1	$v(t) \ge V_{d1}$, $V_{\min}(t) \ge V_{d1}$	4
$\dfrac{V_{\min}(t) - V_{d2} + \alpha(V_{d1} - V_{\min}(t))}{V_{d1} - V_{d2}}$	$v(t) \ge V_{d1}$, $V_{\min}(t) < V_{d1}$	5

The model of the static ZIP load is:

$$
\begin{aligned}
p_{\rm ZIP}(t) &= P_o \left[\alpha_p \left(\frac{v(t)}{V_n} \right)^2 + \beta_p \frac{v(t)}{V_n} + \gamma_p \right], \\
q_{\rm ZIP}(t) &= Q_o \left[\alpha_q \left(\frac{v(t)}{V_n} \right)^2 + \beta_q \frac{v(t)}{V_n} + \gamma_q \right],
\end{aligned}
\tag{7.1}
$$

where

$$
\begin{aligned}
1 &= \alpha_p + \beta_p + \gamma_p \, , \\
1 &= \alpha_q + \beta_q + \gamma_q \, ,
\end{aligned}
$$

where v is the bus voltage magnitude; $p_{\rm ZIP}$ and $q_{\rm ZIP}$ are the active power and reactive powers at the bus of interest, respectively; P_o and Q_o represent the rated active and reactive power, respectively; and V_n is the load nominal voltage.

The model of the electronic load is:

$$
\begin{aligned}
p_{\rm E}(t) &= c_{\rm E}(t)\, P^o_{\rm E} \, , \\
q_{\rm E}(t) &= c_{\rm E}(t)\, Q^o_{\rm E}
\end{aligned}
\tag{7.2}
$$

where $p_{\rm E}$ and $q_{\rm E}$ are the active and reactive powers of the electronic load, respectively; and $P^o_{\rm E}$ and $Q^o_{\rm E}$ are the rated active and reactive powers, respectively. Finally, $c_{\rm E}$ is a voltage-dependent coefficient that models the recovery of the load following faults and shown in Table 7.1.

In particular, V_{d1} and V_{d2} are thresholds and α represents the fraction of the electronic load that recovers from low voltage trip. If $\alpha > 0$, the electronic load is reconnected linearly as the voltage recovers. $V_{\min}$ is calculates as:

$$
V_{\min}(t) = \max\{V_{d2}, \min\{v(t), V_{\min}(t - \Delta t)\}\} \, , \tag{7.3}
$$

where Δt is the time interval at which $V_{\min}$ is refreshed.

In terms of the different modes, they depend on the terminal voltage v with the following rules:

- If $v(t) > V_{d1}$, the active and reactive powers of the electronic load are constant active and reactive powers.
- If $v(t) < V_{d2}$, the active and reactive powers of the electronic load are constant active and reactive powers.
- If $v(t)$ is between two threshold values V_{d1} and V_{d2}, where $V_{d1} > V_{d2}$, the active and reactive powers of the electronic load are linearly reduced to zero.

7.1.3.2 Data Center Modeling in the AIPS

The Irish TSOs model the response of data center protection in dynamic stability assessment tools by means of simple "dynamic" models based on protection settings data as provided by the data center owners. A typical model is a constant PQ load with protection-driven response. The protections represent various relays and UPS initiated voltage and frequency protection schemes. These include over- and under-voltage relays as well as over- and under-frequency relays. Whenever a limit is violated a pre-defined percentage of the data center load trips. Then, the data center models are incorporated in transient stability analysis tool running in the TSOs' control centers. This tool, called LSAT (see Sects. 3.6.1 and 6.3.2), provides warning of any potential insecurities resulting from the response of data centers to faults on the grid. The automatic active power demand recovery is currently not modeled in LSAT as LSAT simulates 15 s while demand recovery takes about 30 s.

An important aspect of data center modeling is also the need or precise information on protection settings. This is key from a dynamic model point of view as different data centers and, more in general, large loads, have different control structures and different protection settings. For example, there exist centralized and distributed architectures of the UPS. Then, similar to generator dynamic model validation and quality tests, large load models need to be periodically validated, to ensure they represent the correct behavior. This validation often requires the use of high-resolution data which means of, for example, PMUs or DRs, which have to be installed at or close to these demand facilities.

7.1.4 Real-World Examples

We are now ready to illustrate, through some real-world examples, the impact of data center demand behavior during normal and abnormal (faults) system conditions, respectively. Unless stated otherwise, in the remainder of this section, we use operational data obtained from TSO SCADA for the data center load data (see, for example, Fig. 7.11). Moreover, frequency transients shown below use PMUs data with 20 ms output resolution.

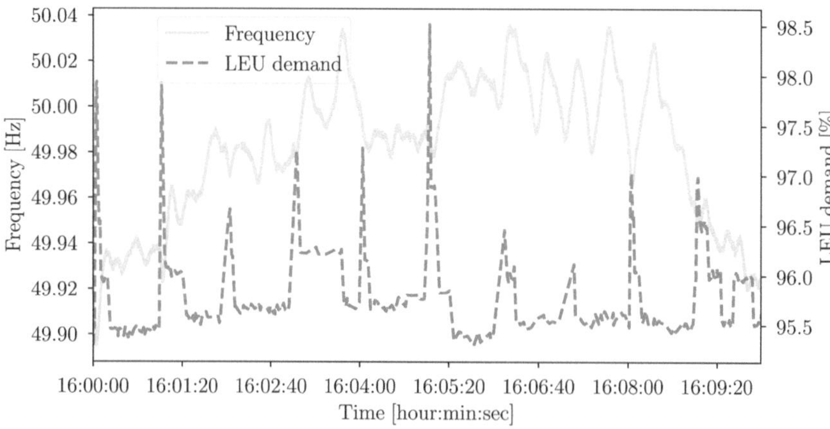

Fig. 7.7 Periodic frequency transients and active power surge from a data center demand site in the AIPS. Measurements were taken in 2025

7.1.4.1 Impact of Regular, Cyclical Fluctuations in Data Center Demand on System Frequency

Figure 7.7 shows the measurements of power variations of a data center connected to the Irish system. The power variations are clearly reflected in the system frequency. The transients occur in the range of 10 mHz to 50 mHz due to total MW surges from multiple data-center demand sites. These oscillations lead to a drop of 5-15% of nominal steady-state system frequency, that is, 5-15% of 0.2 Hz, where 0.2 Hz corresponds to the frequency range 49.9-50.1 Hz.

Moreover, power variations occur at regular intervals, thus making data centers a sort of periodic, although not fully deterministic, load. These cyclical fluctuations indicate potential digital processes run by data center customers in their sites. It is also important to highlight that the magnitude of load perturbations are not consistent. The differences in magnitude explain the varying size of the frequency variations that occur between 10 and 50 mHz.

There is a good correlation between minor frequency transients and load surges. The size of the perturbations is dependent on the number and size of data centers connected to the system at a given time. This is having an impact on frequency quality, potentially exciting system oscillations on the system and is interacting with governors and APC from wind and solar. This is a concern as the increase in data center demand in the near future and the fact that the system will get lighter in terms of inertia as the utilization of conventional synchronous generation decreases.

Table 7.2 Transmission faults in the AIPS occurred in 2022

Fault	Date	Hour	Type	Fault clearing Time [ms]	Min. Retained Voltage [%]	RoCoF [Hz/s]	Inertia [MWs]
1	Jan. 7	01:50	Line to line	61	40	−0.156	23,361
2	Dec. 13	16:57	Single line to ground	80, 90	28	0.119	47,249

7.1.4.2 Impact of Data Centers Response During and After Transmission System Faults

The behavior of data centers during transmission faults also concerns for system operators. Data centers have UPS that can transfer the load on to a local battery in case of voltage sags at the terminal. This behavior is driven by sensitive voltage protection settings. Here we consider only severe faults and not large generation loss events as the former leads to significant voltage dips in the system. All data center demand reduction in this section are due to low voltage following transmission faults. For illustration purposes, we select two of the most recent significant faults in the AIPS that have led data center UPS systems to activate within the Dublin region. The details of these faults are listed in Table 7.2. These faults allow assessing frequency deviation and RoCoF as well as voltage propagation at data center substation and their load response.

Transmission Fault 1: Killpaddoge-Killonan 220 kV

On January 7, 2022 at 01:50 h, the Killpaddoge-Killonan 220 kV line tripped for a phase-phase fault causing a nearby generator to trip as it fails to ride through the fault from 171 MW combined with 80 MW of data center demand disconnection. Table 7.2 shows a summary of the fault. At the time of the incident, the AIPS had an inertia value of 23,361 MWs which represents a low-inertia scenario.

Figure 7.8 shows the system frequency and RoCoF. The tripping of the generator led to a frequency drop. In this case, the demand reduction from data centers helped the system frequency recover. This may not always be the case and adds to risk and uncertainty in actual and modeled system performance during these events. System frequency is calculated using the derivative of the angles of the three phases then mean averaged. RoCoF is then calculated based on the averaged system frequency using a 500 ms window (Eq. 1.66) with a cut-off frequency of 3 Hz low pass filter due to the fault condition. While how to precisely calculate the RoCoF in low voltage conditions is currently an open question, a low pass filter to remove the high-frequency noise from frequency will suffice for this example.

For a better understanding of the severity of the fault in terms of voltage dip propagation, Fig. 7.9 shows a voltage contour map of the AIPS for Killpaddoge-

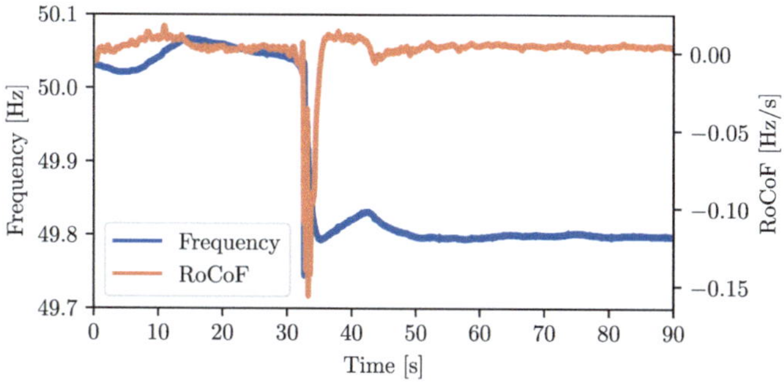

Fig. 7.8 Frequency and RoCoF for Killpaddoge-Killonan 220 kV event

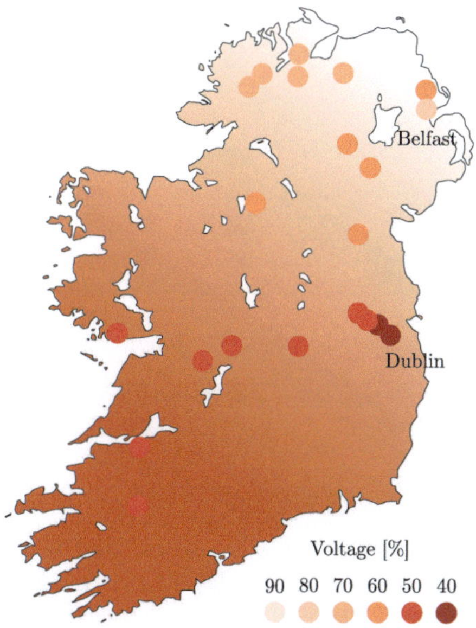

Fig. 7.9 Voltage contour map for Killpaddoge-Killonan 220 kV fault

Killonan 220 kV fault. The dots represent PMUs and DRs connected in the AIPS. The fault led to a significant voltage dip propagation throughout the AIPS. Despite its location being far from Dublin, where there is the biggest concentration of data centers in IE, the fault led to the reduction of 80 MW of data center demand.

Large voltage dip propagation phenomena are a concern considering the reduction in utilization of conventional synchronous generators that have been providing dynamic voltage support during faults. It appears crucial to replace this capability with other resources such as, for example, STATCOMs and synchronous condensers.

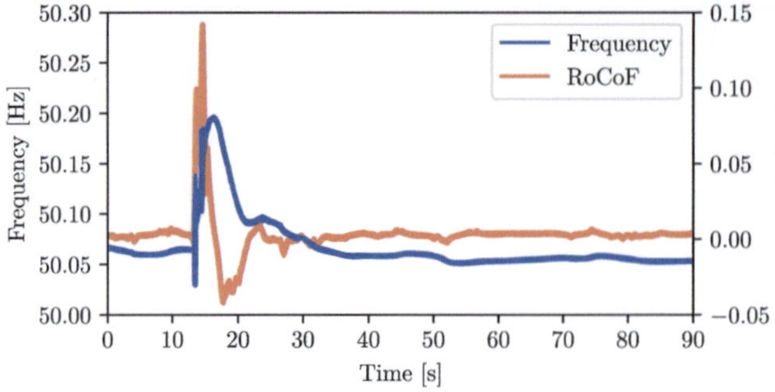

Fig. 7.10 Frequency and RoCoF for Kellystown-Woodland 220 kV event

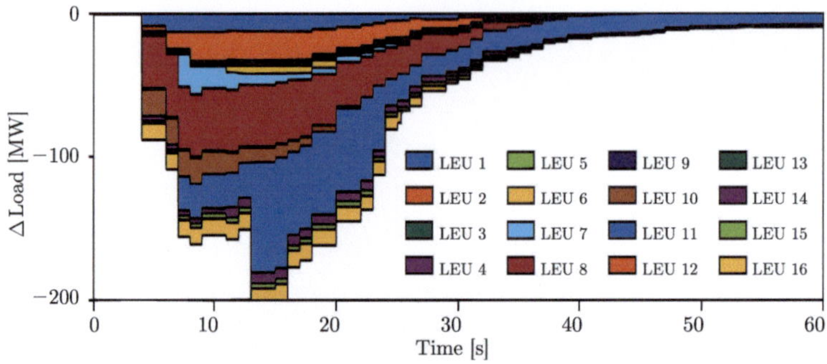

Fig. 7.11 Example of data center response (demand reduction) to a transmission fault

Transmission Fault 2: Kellystown-Woodland 220 kV

Another recent transmission fault that led to undesirable data center behavior is the following. Specifically, at 16:57 h on Tuesday 13 December 2022, the Kellystown-Woodland 220 kV line tripped, reclosed and tripped for a single phase to ground fault. This fault caused one of the largest data center demand reduction to date with 204 MW of data center demand being disconnected from the grid (approximately 50% of total data center demand at the time of incident) causing a RoCoF of 0.12 Hz/s and a zenith of 50.22 Hz, as shown in Figs. 7.10 and 7.11, respectively. It is worth highlighting that from a data center perspective, the load is riding though the fault by switching to UPS and then coming back when the grid is "healthy."

Considering the data center demand levels expected in the coming years, this behavior can potentially lead to significant system incidents. If nothing is done the FRT issue will continue to grow as data center load ramps up. This issue also has impacted, for example, the ability of the Irish TSOs to evolve operational policy to facilitate higher levels of renewable generation.

Figure 7.11 shows that, due to sensitive UPS voltage protection settings, for example, 10% from nominal, certain data centers can be seen to off-load at the point of connection within milliseconds of a transmission or distribution fault. Further issues

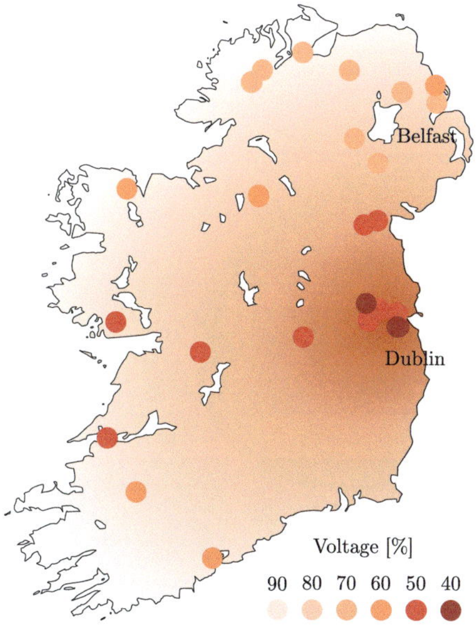

Fig. 7.12 Voltage contour map for Kellystown-Woodland 220 kV fault

are also evident when data centers reconnect, which occurs shortly after the fault has been cleared. This behavior has been seen to have a significant impact on system frequency and poses a challenge to the TSOs in matching or increasing generation during such events. Regarding the voltage dip propagation, Fig. 7.12 below shows that the propagation is lower compared to the fault above. However, because of the proximity with Dublin it lead to a high demand drop from data centers.

The lack of FRT capability at data centers has the potential to exacerbate the impact of transmission system faults, leading to cascade trips and a significant system incident. For instance, in January 2025 a 220 kV fault near Dublin led to a total of around 320 MW data center demand reduction (highest demand reduction to date). Unless resolved, this behavior will materially impact on the ability to operate the power system securely, and evolve operational policy to facilitate higher levels of renewable generation.

It is noteworthy that the AIPS is more susceptible to CLs due to its relatively small size when compared to other synchronous areas in Europe, see Table 7.3.[1] Having to deal with this scale of imbalance in the AIPS, namely, up to 30% of its minimum demand, would be unprecedented from an international power system perspective. It is important, thus, that relatively small systems such as AIPS have the necessary

[1] CE: Assuming 3,000 MW largest imbalance, 200,000 MW minimum demand and 1,000 GWs minimum inertia. GB: Assuming 1,800 MW largest imbalance, 25,000 MW minimum demand and 120 GWs minimum inertia. AIPS: Assuming 1,000 MW largest imbalance, 3,000 MW minimum demand and 23 GWs minimum inertia.

Table 7.3 Comparison of power system sizes with respect to potential largest imbalance and minimum demand, and minimum inertia

Synchronous Area	$\frac{\text{Largest imbalance}}{\text{Minimum demand}}$ [%]	$\frac{\text{Largest imbalance}}{\text{Minimum inertia}}$ [MW/GWs]
CE	≈ 1.5	≈ 3
GB	≈ 7	≈ 15
AIPS	> 30	≈ 43

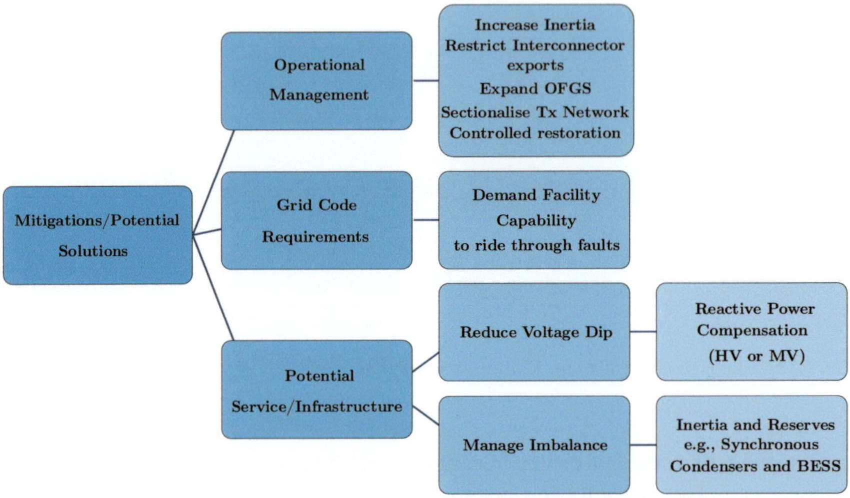

Fig. 7.13 Mitigation and potential solutions to the data center issues in the AIPS

solutions and mitigation actions in place as soon as possible (see next section) to ensure system security.

7.1.5 *Mitigation and Potential Solutions*

So far, we have discussed emerging issues caused by data centers to the transmission system. In this section, we present options considered by EirGrid and SONI to address the data center performance issues in the AIPS. Figure 7.13 shows the various solutions and mitigation strategies that have been considered by these system operators.

Operational Management

The operational actions TSOs could take to mitigate the impact of CL of data center demand as follows:

- *Increase inertia.* Keeping more synchronous inertia on the system could mitigate potential large (positive) RoCoFs due to the instantaneous loss of demand following faults (voltage dips). While this action is considered effective it is not desirable as it means TSO need to keep more conventional units online and thus limit the penetration of RES namely wind and solar power.
- *Restrict ICs export.* Since ICs usually set the LSO during high exports reducing these basically means a lower potential RI (sum of LSO and CL of demand, see Chap. 12) that TSOs need to deal with. However, while this action makes sense from a system security perspective, it is not desirable from a cross-border market point of view and could result in economic inefficiency.
- *Expand OFGS.* This third option considers using, for example, wind farms OFGS to deal with potential OF events due to load lost. OFGS and load shedding schemes are considered as the last defense measures and thus may not be ideal to be implemented to tack such issues. For instance, relying on such schemes means that potential malfunctions of OFGS protection could lead to an even worse situation to manage. Additionally, from a system dynamic point of view disconnecting a large number of resources during an event may originate cascading issues that could lead to total power system collapse/blackout.
- *Sectionalize the transmission network.* This solution consist in splitting up the transmission network or, in other words, making the transmission network more radial than meshed and thus increase the impedance between the fault and data center demand. This is a possible but not practical solution as it makes the system weak from an interconnection point of view.
- *Controlled restoration of demand.* Currently, data center demand is automatically restored to the grid following clearance of the fault without visibility from the TSO. A simple scenario that could be thought here is the case where due to the demand reduction there could potential generation trips as part of OFGS schemes to manage the OF event. But if the demand is automatically reconnected then it could lead to an UF event as there is less generation now online due to OFGS trips. Therefore, it is imperative that TSOs have control over the reconnection of the demand. Such a controlled restoration could be required as part of the performance requirements in grid codes.

Grid Code Requirements

This consist in introducing performance requirements such as FRT in grid codes that all transmission-connected demand users need to meet. This is the preferred option from a security point of view but presents challenges for existing demand units as their equipments may have not been designed to withstand or ride through faults with, for example, voltage close to or even zero.

Potential Service/Infrastructure

Dealing with FRT issues that lead to an active power imbalance has generally two "simple" solutions, that is, or try to reduce the voltage dip during the fault or have enough inertia and reserves to compensate the loss of the demand. The first option could include installing dynamic reactive power compensation devices such as E-STATCOM at the MV/HV level. However, since there is little that can be done to sustain the voltage during a fault, this solution may not be effective. Current UPS systems, in fact, are designed with very sensitive protection settings in the range of 80-90% retained voltage. It is thus almost impossible to keep voltage above 80-90% of nominal during a fault. In this scenario, the demand should recover quickly following the clearance of the fault, say, between 500 ms to 1 s. If this fast active power recovery is not possible for the existing demand facilities then it is crucial that TSOs are equipped with the necessary inertia and reserve services to deal with the imbalance caused by the loss of the demand. But whilst having these services in place mean ensuring system security the key question that needs to be addressed then is who pays for these additional services. Ultimately, the latter is more of a RAs decision that could take into account the balance between increased ancillary/system services cost and reliability risk.

7.1.6 Fault Ride Through Standards

Power system industry is experiencing a massive influx of data center demand which is expected to grow significantly in the near future. For example, ENTSO-E expects France, Netherlands, IE, Spain, and Italy to have 3,500 MW, 4,500 MW, 2000 MW, 3,000-4,000 MW, and 4000 MW, respectively, of installed data center capacity by 2030. Considering the importance of system security, the expected increase of the installed capacity of data centers, and their response to grid disturbances, ENTSO-E developed proposals for technical requirements that would apply to data centers but were not ultimately approved by the EU Agency for the Cooperation of Energy Regulators (ACER). The proposed requirements were based on those proposed for Power to Gas Demand (PtG) units in the Demand Code Connection (DCC) [18].

In terms of requirements, ENTSO-E proposed introducing a new data center demand category (110 kV and above) and apply FRT, RoCoF and limited frequency response requirements. Figure 7.14 shows the FRT capability as proposed by ENTSO-E (same FRT profile of a power-generating module taken from the Requirements for Generators Network Code) and Table 7.4 presents the parameters for FRT capability in Fig. 7.14.

Unfortunately, ENTSO-E proposals were rejected by ACER with the argument that data center issues appear to be local rather than pan-European issues. Standardization and uniform technical minimum performance requirements such as FRT are crucial to prevent unexpected operational challenges and ensure consistency in different jurisdictions and, thus, improve bulk system reliability. For example, the IEEE 2800 standard establishes a uniform set of requirements for IBRs interconnection [19].

ENTSO-E recently suggested to all TSOs to introduce as soon as possible at the national level key requirements for PtG units due to the significant expected size of this type of demand by 2028 (30 GW), and because the new EU Network Code on

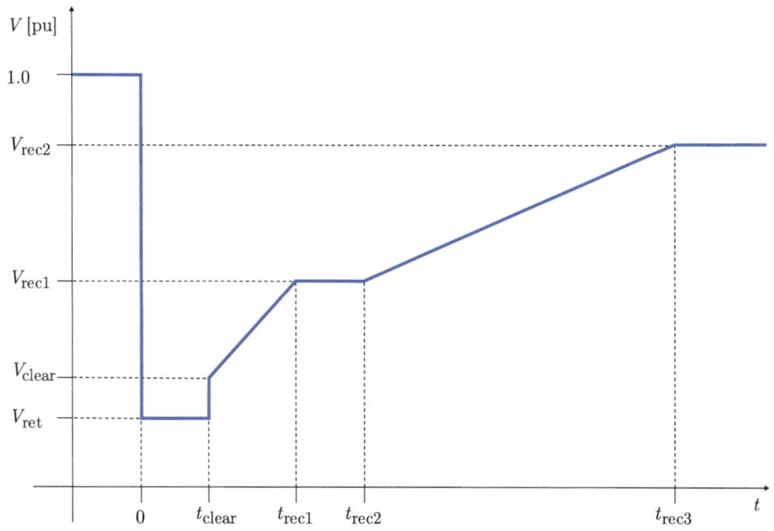

Fig. 7.14 Proposed ENTSO-E fault-ride-through profile for data centers

Table 7.4 Proposed ENTSO-E parameters for fault-ride-through capability data centers

Voltage parameters [pu]	Time parameters [s]
V_{ret}: 0	t_{clear}: 0.14–0.15 (or 0.14–0.25 if system protection and secure operation so require)
V_{clear}: V_{ret}	t_{rec1}: t_{clear}
V_{rec1}: V_{clear}	t_{rec2}: t_{rec1}
V_{rec2}: 0.85	t_{rec3}: 1.5–3

Demand Connection would only apply from three years after its publication in the Official Journal of the European Union [20]. The suggested requirements for all PtG demand facilities with a connection point of 110 kV include: (i) RoCoF robustness; (ii) FRT capability; and (iii) Active power recovery after faults.

After ACER rejection of FRT proposals, EirGrid and SONI started to develop similar standards to those proposed by ENTSO-E which are close to finalization and to be submitted to Grid Code review panel for discussion and then for approval with the RAs. The proposed voltage against time profile standard for transmission-connected demand facilities (including data centers) is shown in Fig. 7.15.

EirGrid and SONI propose other requirements for demand facilities such as, among others: (i) Following clearance of the fault, demand facilities should return to at least 95% of their pre-fault active power demand within 500 ms; (ii) RoCoF limit of 1 Hz/s; and (iii) Minimum time periods for demand facilities to remain operational without disconnecting (Table 3.2). In particular, the fast active power recovery time is

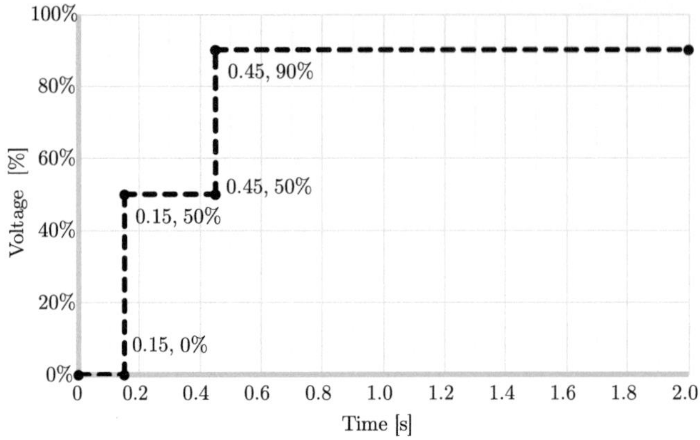

Fig. 7.15 Required voltage against time profile for data centers

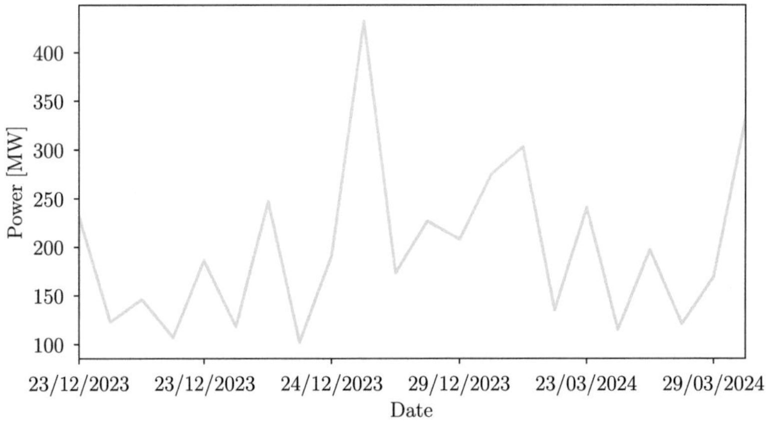

Fig. 7.16 Recent load loss events reported by ERCOT

a critical parameter that can greatly influence RoCoF. For instance, 1 s active power recovery time would lead to higher RoCoFs than 500 ms time.

Other TSOs in the world are introducing Large Energy Users (LEUs) FRT requirements. ERCOT is in the process of studying and implementing "large load voltage ride-through requirements" [21]. In this context "large" refers to loads of 75 MW or above. ERCOT state that "The voltage-ride through requirements in this NOGRR are necessary to ensure Large Loads do not present a reliability risk to the system by tripping when voltage excursions within a specified range occur". In fact, ERCOT faced the largest MW of load dropped in December 2022 when 1,560 MW of load dropped, which resulted in frequency rise to 60.235 Hz [21]. Since then, ERCOT has had quite a number of other events that have resulted in large load MW reductions, see Fig. 7.16 [22].

Considering the massive penetration of these loads, it is recommended that TSOs employ a robust compliance process for FRT. However, FRT issues are really a demonstration that planning "has not worked well". If planning is done correctly in terms of studying FRT capability requirements of users connecting to the power system, then FRT issues would not appear in operations/real-time. Therefore, it is critical that a proactive approach is implemented at the planning stage. Finally, it is important that planning and operations teams within a TSO talk to each other regularly to understand key assumptions and studies being performed and align them.

7.2 Impact of Electric Vehicles

In this section, we focus on the impact of the charge of a large number of Plug-in Electric Vehicles (PEVs) on the dynamic response of power systems. We have seen in the previous section that data centers impact frequency deviations mostly in the short term. The impact of uncontrolled PEV charging on the frequency dynamic performance of the grid is, on the other hand, in the long term. For this reason, it is important to control properly PEV. With this aim, we show how a decentralized stochastic control strategy can be an effective solution to manage the charge of a high number of PEVs with little communication effort. A modified version of the New England network is utilized to illustrate the impact of PEVs in a variety of scenarios and setups of the stochastic control.

7.2.1 Background

The projections of the future electricity demand indicate that PEVs will play a relevant role in power systems. It has been estimated that there will be over 1 billion electric vehicles by 2050, making electricity the primary and most critical energy carrier [23, 24]. PEVs are characterized by high flexibility, which can be a valuable resource for the system. However, an uncontrolled charging of large fleets of vehicles at peak times may pose challenges to the transmission grid [25]. The literature on the impact of PEVs on system dynamics can be roughly divided into two main groups: (i) studies that exploit the flexibility of PEVs to improve system dynamics; and (ii) studies that discuss how large fleets of PEVs affect the system and possibly lead it to collapse.

7.2.1.1 Harvesting Flexibility from PEVs

Smart charging strategies can be adopted by means of *ad hoc* electricity tariffs [26], and Vehicle-To-Grid (V2G)Vehicle-To-Grid power flows [27]. Some proposals focus on the ability of PEVs to provide control that can improve the performance of the grid.

For example, [28] lists a variety of positive aspects of the PEVs, such as their ability to provide ancillary services and facilitate the integration of RES. For short-term time scales, aggregated PEVs can be utilized for PFC [29]. The adoption of advanced centralized PEVs charging strategies can potentially also reduce the negative impact of the electric mobility penetration such as the congestion of distribution grids [30]. For longer time scales, PEVs can be utilized for SFC [31–33] as well as day-ahead generation scheduling to reduce grid operation costs and the impact of the variability of renewable generation [34].

7.2.1.2 Impact of PEVs on the Dynamic Performance of the Grid

The impact of PEVs on system dynamics is intrinsically tied to their charging strategy [35]. Without a suitable control, in fact, the charging of large fleets of PEVs may compromise the operation and security of a power system [28]. And some studies suggest that the evening ramp due to the PEVs domestic charge may cause a collapse of the system [36]. On the other hand, a study on the transition to grid-charged cars in Ontario concludes that about 6% of total vehicle fleet in Ontario (approximately 500,000 PEVs) can be charged without any additional power system investments, if the off-peak periods are smartly exploited [37].

These studies highlight the fact that the architecture of the control strategy of PEVs is critical. In particular, there are two main control approaches: centralized and decentralized. A centralized method is discussed for example in [38], where an event driven MPC approach is proposed for the management of PEVs charging in distribution grids. In [30], instead, a combination of local and centralized control methods is used to optimally coordinate the PEVs fleet charging. However, a centralized management of the charging process of a large number of PEVs may quickly become computationally intractable and impractical, therefore most charging algorithms are evolving toward a decentralized structure [39]. For this reason, decentralized approaches appear more promising.

In the remainder of the chapter, we present an example of fully decentralized control for the charge of PEVs. With this aim, we first describe the dynamic model of the grid and then a specific stochastic control technique, called AIMD. This technique was initially introduced to address internet congestion problems and optimally share bandwidth between connected computers. The AIMD works particularly well for resource sharing problems, where a group of agents with limited or no communication abilities at all wish to share a given resource in an optimal fashion [40]. In the case of PEV charging, the AIMD coordinates the power rates of the PEVs, so that the available power can be utilized by all vehicles in a fair and safe fashion. Finally, as we do not have measurements data as for the data centers, we present a case study that illustrates the main features of decentralized stochastic controls.

7.2.2 Modeling

For the purposes of illustrating the dynamic behavior of a stochastic control for the charging of PEVs, we consider a dynamic model of the power system based on stochastic DAEs, as described in Appendix A. Then, we define an adequate model for the PEV fleets. Many statistical analyzes have shown that in the near future in many countries a significant number of PEVs is expected to charge in a domestic scenario [41, 42], as soon as the drivers come back home after work [30, 43]. In this case, the number of connected PEVs may quickly increase in a short time, for example, right before dinner time, and may give rise to a new evening peak load. This peak of PEVs domestic charging can be modeled as many active power load ramps, connected to some buses of the network, where other conventional loads are present. The mathematical model of the loads at the buses where also PEVs are connected is:

$$\begin{aligned} p &= p_0 + R_{\text{PEV}}(t - t_0) + \eta_p, \\ \dot{\eta}_p &= -a\,\eta_p + b\,\zeta_p, \end{aligned} \tag{7.4}$$

where p is the total load; p_0 is the aggregated voltage-dependent pre-existing load at the bus; R_{PEV} is the ramp rate of the electric vehicles; t_0 is the time at which the evening ramp begins; η_p is a Gaussian mean-reverted stochastic process that reproduces load random fluctuations; a is the drift; b is the diffusion; and ζ_p is white noise [44].

We are now ready to introduce the model of the stochastic control based on the AIMD technique. This control solution have many significant advantages, most notably it is a fully decentralized approach, which implies that the solution is scalable and independent from the number of agents. In our case study, this implies that it is not required to count the number of PEVs connected for charging, and it is not required to know how much energy they need. Accordingly, neither PEVs, nor charging stations, require to communicate any information, contrary to the vast majority of other existing distributed charging algorithms. In fact, AIMD algorithms have been already proposed for PEV charging in the literature, for example, for the problem of charging a number of electric vehicles via limited capacity infrastructure [45]; to design a smart energy exchange algorithm based on bidirectional energy management [46]; and to fairly share energy and minimize queues [47].

The key point of any decentralized control strategy is the definition of an appropriate strategy for each individual PEV charger to modulate the charge rate so that the PEV gets charged as quickly as possible, without at the same time giving rise to power system stability issues. The AIMD is designed precisely to solve this task provided that appropriate (or simple to estimate) quantities are utilized to trigger a control action and decrease the charge rates of PEVs when getting close to the safety limits of the network.

As it can be expected, there are various alternative set-ups. Some examples are listed below.

- ω_i: the local frequency measured at the bus i. This quantity is indicative of the system power imbalance, and it is easier to estimate than $\omega_{\rm CoI}$ for a decentralized approach, as it can be locally measured.
- v_i: the voltage magnitude at bus i. This quantity is strictly linked to the transmission capacity of the lines.
- p_i: the PEVs power load at bus i.

It is in general not possible *a priori* which signal works the best for every scenario, but according to our studies [48], the control strategies that use p_i appears to be the most effective. In particular, the following facts are relevant:

- The time scale of the PEV ramping rate allows assuming that the frequency of the system is effectively the same in every point of the grid. As a matter of fact, all ω_i are always very close to $\omega_{\rm CoI}$, even when very steep ramp profiles of vehicles are simulated.
- Frequency-based control strategies (either using $\omega_{\rm CoI}$ or ω_i's, as they are equivalent as explained in the previous point) are ineffective as instability issues may occur without noticing any evident variation in the frequency signals (or they are noticed when it is too late to implement recovery strategies).
- Voltage-based control strategies are ineffective for the opposite reason, since significant voltage variations may be observed even in fully stable operations of power grids. This is due to the presence of many existing devices devoted at regulating voltages around the nominal values, including Under-Load Tap Changers (ULTCs) and static var compensators.
- It is reasonable to assume that grid operators can use historical data to estimate the maximum power allowed at each bus ($p_i^{\max}$) (following for instance the techniques proposed in [49, 50]), and that power-based control strategies can be used to maintain p_i below the value of $p_i^{\max}$.

7.2.3 *Decentralized Control of PEVs*

The AIMD algorithm consists of two alternating phases:

1. Additive increase (AI) phase: the resource consumption of each agent increases linearly in time with a certain step $\alpha > 0$;
2. Multiplicative decrease (MD) stage: each agent decreases its consumption in a multiplicative fashion, with a certain slope $\beta > 0$.

The MD step is performed when the so-called *Capacity Events (CEs)* occur, which means that the limit of the shared resource has been reached, for example, all the available power is fully shared by the connected PEVs.

Let $t_1 < t_2 < \cdots < t_k < \ldots$ denote the time instants when such CEs occur; if t_{k+1}^+ is the instant after the h-th agent performs an MD step at time t_k, its share of the resource $x_h(t_{k+1}^+)$ is:

$$x_h(t_{k+1}^+) = \beta \lim_{t \to t_{k+1}} x_h(t)\,, \tag{7.5}$$

and then h-th agent h will return to the AI phase, until the next CE occurs.

Therefore, the equation describing the behavior of the h-th agent is the following:

$$x_h(t) = \beta x_h(t_k) + \alpha(t - t_k), \quad t \in (t_k, t_{k+1}), \tag{7.6}$$

where t_k and t_{k+1} represent two consecutive time instants at which a CE occurs.

For the specific application of charging of PEVs, each bus of the network represents an independent system with maximum allowed power $p_s^{\max}$ that is shared among PEVs. Then, denoting with $p_{h,i}$ the power consumption of h-th agent at the i-th bus, a capacity event occurs at the s'th bus when the sum of the charge rates $p_{h,i}$ of all connected PEVs exceeds the maximum available power $p_i^{\max}$.

The version of the AIMD algorithm described so far is known as *synchronized* AIMD, as *all* PEVs reduce the charging rate in a multiplicative fashion when a CE is notified. In practice, only a subset of PEVs need to decrease their charge rates to fall between the safe thresholds. This observation makes possible to prioritize some PEVs over others [51]. This can be easily achieved by implementing a so-called *unsynchronized* version of the AIMD, where only a subset of PEVs react to a CE event by reducing the charge rates, in a probabilistic way.

Priorities of single buses can be formally stated by assuming that each PEV h has its own utility function $\psi_{h,i}(p_{h,i})$, where for convenience we shall consider convex, strictly differentiable utility functions $\psi_{h,i}(\cdot)$ that have a global minimum for the maximum allowed charge rate, which is in this case is 3.3 kW. Roughly speaking, this implies that the minimum value of the utility function of PEVs is achieved when PEVs are charged at the maximum power. With this aim, one can define an optimization problem for each bus. For example, in [48], the following optimization is proposed:

$$\begin{aligned} &\underset{p_{h,i} \in [0,3.3]\,\text{kW}}{\text{minimize}} \sum_{h=1}^{n(i)} \psi_{h,i}\big(p_{h,i}\big) \\ &\text{subject to} \sum_{h=1}^{n(i)} p_{h,i} \le p_i^{\max}\,, \end{aligned} \tag{7.7}$$

where $n(i)$ is the number of PEVs connected to the i-th bus. Roughly speaking, Problem (7.7) works as follows: when there are no stability issues, then the minimization of the sum of the utility functions is achieved when PEVs are charged at the maximum charging rate; when the aggregated vehicles load reaches the power limit, then PEVs are charged at a lower charging rate and safer system configurations are restored.

As proved in [52], the optimization problem (7.7) can be solved in a fully decentralized way if we let each vehicle asynchronously perform the decrease step with probability:

$$\varpi_{h,i}(\hat{p}_{h,i}(t)) = \gamma \frac{\psi'_{h,i}(\hat{p}_{h,i}(t))}{\hat{p}_{h,i}(t)}, \tag{7.8}$$

where $\psi'(\cdot)$ is the derivative of function $\psi(\cdot)$ with respect to time, and $\hat{p}_{h,i}(t)$ is the average charge rate of PEV h at bus i at previous time steps:

$$\hat{p}_{h,i}(t) = \frac{1}{t} \int_0^t p_{h,i}(\tau)\, d\tau, \tag{7.9}$$

and γ is a constant factor that maps $\varpi_{h,i}$ into a probability, that is, $0 \leq \varpi_{h,i} \leq 1$. Equation (7.8) means that each PEV h reacts to a CE independently from the others, with the customized probability $\varpi_{h,i}(\hat{p}_{h,i}(t))$, based on its historical power consumption. For the convenience of exposition, a pseudo-code to implement the AIMD algorithm at each bus is now given in Algorithm 1.

Algorithm 1 Unsynchronized AIMD algorithm at the s'th bus

Require: $k = 1$; γ is broadcast;
  while $k < k_{simulation}$ **do**
    if $\sum_h p_{h,i}(k) \geq p_i^{\max}(k)$ **then**

$$p_{h,i}(k+1) = \begin{cases} \beta\, p_{h,i}(k), & \text{with probability } \varpi_{h,i}(\hat{p}_{h,i}(k)) \\ p_{h,i}(k) + \alpha, & \text{with probability } 1 - \varpi_{h,i}(\hat{p}_{h,i}(k)) \end{cases};$$

    else
      $p_{h,i}(k+1) = p_{h,i}(k) + \alpha$;
    end if
    $k = k + 1$;
  end while

The AIMD algorithm has a number of significant advantages that are ideal for the charging problem: for instance, it can be implemented in a fully decentralized way without requiring PEVs to communicate anything to either other PEVs or charging stations; the solution is fully scalable independently from the number of agents; unlikely many other optimization approaches it is not required to solve a new optimization problem as new PEVs connect for charging, but a new optimized solution is obtained; all the communication is restricted to a CE event, which consists in one bit of communication.

Finally it is worth observing that, in a practical implementation of this algorithm in a power grid, it would be too complicated, and likely not necessary, to change the charging rates of PEVs too often. For this reason we assume that the occurrence of CEs only occurs every $T_p = 30$ s [53].

7.2.4 *Example*

We illustrate the model and stochastic control strategy described above though a case study based on the well-known New England power system (see Appendix B.2). The original network is modified by replacing the two conventional generators at buses 1 and 5 with wind farms with same capacity. The wind generators are assumed to be the doubly-fed induction machines. The remaining conventional generators are equipped with the base case TGs and AGC for PFC and SFC, AVRs and Power System Stabilizers (PSSs) as detailed in the Appendix.

All 16 loads in the New England system are connected to the grid through an ULTC with discrete control of the voltage. Moreover, aggregated models of the PEV charging stations are assumed to be located at buses 3, 4, 7, 8 and 12; for simplicity, all stations have the same maximum charging rate, which in this case is assumed to be 3.3 kW, that is, the typical nominal rate of domestic chargers [54]. Static VAR Compensators (SVCs) have also been included at these buses to prevent collapses due to shortage of reactive power in the grid.

The load consumption of the system has been increased by 10 pu with respect to the original data, and pre-existing loads (including industrial and residential loads) at each bus of the system are considered constant, which allows one to separate the effects of the PEVs from those of the conventional load.

It is important to note that, while the results presented in the remainder of this section refer to the specific power system described above, conclusions are general and may be extended to any other grid.

7.2.4.1 Uncontrolled Charging

As a first example and a motivation for the need of stochastic control, we discuss the case of uncontrolled charging of PEVs during the evening ramp (domestic charging). For this purpose, we assume that PEVs are equally distributed among the buses 3, 4, 7, 8 and 12, and that the arrival rate of PEVs for charging is about 120 veh/s. After 1 h of simulation, about 432,000 PEVs would be connected for charging. Considering that the total number of cars in New England can be estimated equal to $5 \cdot 10^6$ vehicles,[2] this corresponds to assuming that about 8-9% of the total number of cars connect for charging during the evening peak within an hour. Figure 7.17a, c, and e show that the system is not able to cope with such a large volume of PEVs and eventually collapses.

After one hour, frequency goes to 0.997 pu(Hz), which, for a 50 Hz system, corresponds to 49.85 Hz. This deviation is outside the 100 mHz criteria in AIPS and outside standard frequency range (±50 mHz) defined by CE. This deviation occurs despite having included in the system an AGC. Hence, even if the system eventually collapses for the load exceeding the maximum transfer capability, the impact of

[2] https://www.fhwa.dot.gov/policyinformation/statistics/2016/mv1.cfm#foot3.

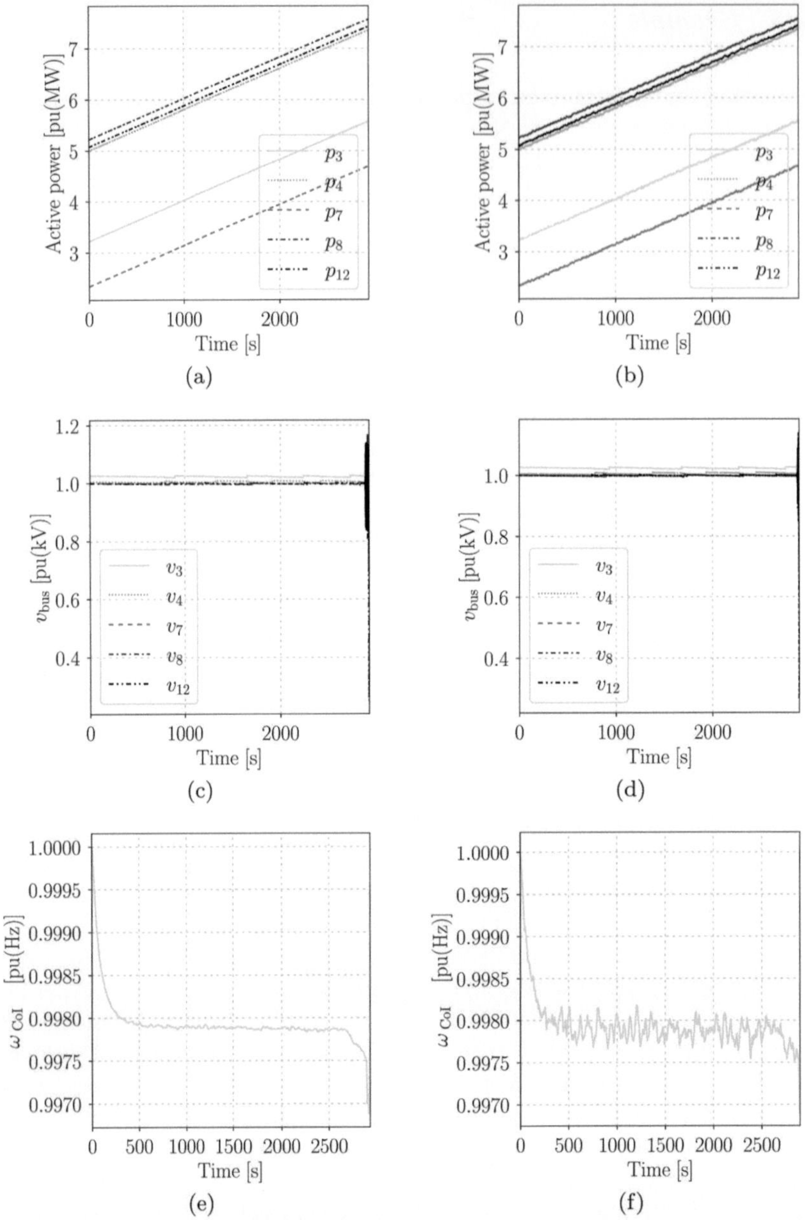

Fig. 7.17 Effect of uncontrolled charging of PEVs: **a** and **b** PEVs power consumption; **c** and **d** bus voltages; **e** and **f** ω_{CoI}. The right column shows results for a system with noise ten times higher than the left column

PEVs also impact negatively the frequency quality of the grid, namely, the minutes outside standard frequency range such as ±50 mHz or ±100 mHz.

A critical feature of the collapse illustrated above is that it may be hard to predict the upcoming instability, as the voltages and the frequency of the center of inertia ω_{CoI} remain always very close to their nominal value 1 pu in all scenarios, even very close to the collapse (see Fig. 7.17c, e. In particular, Fig. 7.17b shows that the behavior of the voltage values is mainly affected by the action of the ULTC devices. Note that the system behavior is independent of the Gaussian noise parametrization, which characterizes the load ramp equation—see (7.4). As a matter of fact, the results obtained with, say, ten times the standard deviation are very similar (see Fig. 7.17b, d and f). On the other hand, a careful handling of the reactive power is critical in the problem of charging PEVs in transmission networks. The SVCs included in the system prevent limit-induced bifurcations due to reactive power shortage. The instability shown in this scenario is due to to congestion in the transmission lines (saddle-node bifurcation) [55].

7.2.4.2 Synchronized AIMD

We now consider the power-based AIMD controller described in Sect. 7.2.3, where safety thresholds for the system are defined as a function of the maximum static power limits of load buses.

The synchronized version of AIMD may be regarded as the maximally fair solution to solve the problem, as all PEVs have the same priority and should be charged on average with the same power rate. In this section, we present the system behavior in the same settings of Sect. 7.2.4.1, with the action of our synchronized controller; the parameters of the AIMD are $\alpha = 0.34$ W, $\beta = 0.99$ and the threshold that triggers CEs is $p_i^{\max} = 180$ MW. The values of the AIMD parameters α and β are assumed to be the same in all simulations. In practice, the grid operator can tune these parameters as a tradeoff between optimal utilization of the available power and communication requirements (to prevent the system from frequently broadcasting CEs). Finally, the static limit $p_i^{\max}$ of each bus i is set equal to 80% of the actual bus power limit, assuming that grid operators will be conservative in estimating the maximum available power, for safety reasons.

Figure 7.18 summarizes the obtained results; the proposed control correctly manages to accomplish the charging task, as the control action prevents the power limit from being exceeded. Also, note that during the charging process, the frequency signal ω_{CoI} lies within a safe range of [0.996, 1.004] pu (see Fig. 7.18b), thus making this control strategy realistically feasible and safe for the system stability.

The determination of the maximum loading condition, namely $p_i^{\max}$, of a grid is an analysis that is commonly carried out by system operators, independently from the presence of PEVs in the grid. This analysis is aimed at determining the available loading condition or, equivalently, the voltage stability margin of the grid. System operators have a good knowledge of their grid and define $p_i^{\max}$ based on an analysis of the "available transfer capability", this takes into account an N-1 contingency

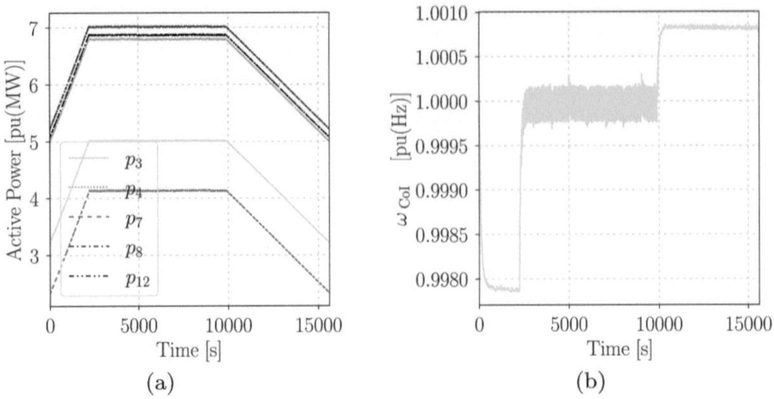

Fig. 7.18 Synchronized AIMD results; **a** active powers of loads at buses 3, 4, 7, 8 and 12; **b** frequency ω_{CoI} response

analysis and the "transmission reliability margin" as defined by NERC [56]. Of course, $p_i^{\max}$ varies for different topologies of the grid, and this requires updating AIMD parameters periodically. For simplicity, however, in all the examples, we assume that $p_i^{\max}$ does not change during the PEV ramp-up. It should also be noted that system operators often decrease the estimated value of $p_i^{\max}$ for security reasons. In the context of PEV charging, the lower the maximum loading condition limit $p_i^{\max}$, the longer the time required to charge all PEVs than strictly needed.

To support the statement given in Sect. 7.2.4.1 that a frequency-based control to charge PEVs is not adequate, we repeat the same simulation as above but this time the AIMD algorithm utilizes bus frequencies. In other words, the capacity event and consequent reduction of power consumption by the connected PEVs is triggered when the frequency falls below a safe threshold. AIMD parameters are $\alpha = 0.34$ W, $\beta = 0.99$ and $\omega^{\min} = 0.999$ pu(Hz) on a base of 60 Hz.

Figure 7.19 summarizes the obtained results. In particular, the charging process is very slow because a very conservative threshold for the frequency was chosen, and thus only very small changes of the load are allowed. Despite this conservative choice of the frequency threshold, instability occurs just the same, and when the system realizes of this event and starts reducing the charge rates of the connected vehicles, it is too late to prevent a network collapse from occurring. This results was to be expected as rate of change of power due to the PEV chargers is to slow to determine a measurable frequency variation.

7.2.4.3 Unsynchronized AIMD

The unsynchronized version of AIMD is more interesting than its synchronized counterpart, as it is possible to prioritize PEVs as desired. This allows, for example, implementing policies that favor certain PEVs (for instance because their owners

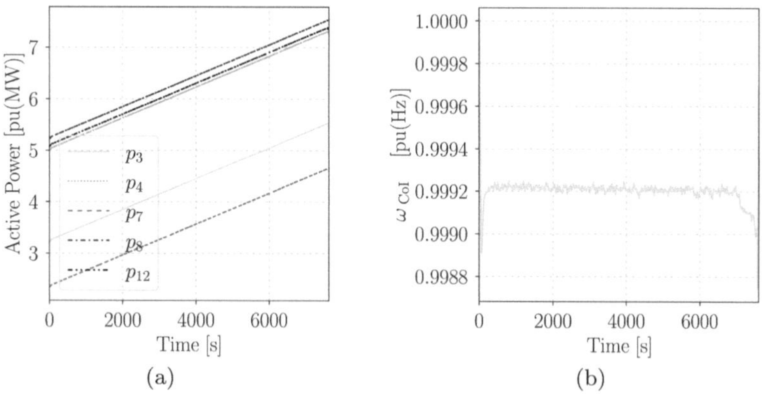

Fig. 7.19 Results of the synchronized frequency-based AIMD: **a** active powers of loads at buses 3, 4, 7, 8 and 12; **b** frequency ω_{CoI} response

are willing to pay more to be charged earlier, or also allow one to take into account energy requirements of the PEVs, or of the power grid). Let us assume again that PEVs are be equally distributed and connected to the same five different buses, that is, 3, 4, 7, 8 and 12. Then, we compare the following two scenarios.

S1: all PEVs at any i-th bus have the same objective function:

$$\psi_s(p_{h,i}) = -\frac{p_{h,i}^3}{3\, p_{\max}}, \tag{7.10}$$

where $p_{h,i}$ is the power load of the h-th PEV plugged at bus i, and $p_{\max}$ is the maximum charge rate of 3.3 kW.

S2: all PEVs that are connected at buses 3, 4 and 7 have the following utility function ψ_{i_1}, while PEVs connected at buses 8 and 12 have utility function ψ_{i_2}:

$$\psi_{i_1}(p_{h,i}) = -\frac{p_{h,i}^3}{3\, p_{\max}}, \tag{7.11}$$

$$\psi_{i_2}(p_{h,i}) = -\frac{p_{h,i}^4}{4\, p_{\max}^2}. \tag{7.12}$$

In both scenarios, up to $88 \cdot 10^3$ PEVs connect for charging at each bus. The AIMD parameters are $\alpha = 0.34$ W, $\beta = 0.99$, $p_i^{\max} = 180$ MW, and $\gamma = -1$.

Figure 7.20 summarizes the results for Scenario S1 (top row) and Scenario S2 (bottom row. The effect of the different objective functions can be noticed by comparing Fig. 7.20a, where the algorithm fairly manages to complete charging at all buses at about the same time, and Fig. 7.20c, where the vehicles associated with a

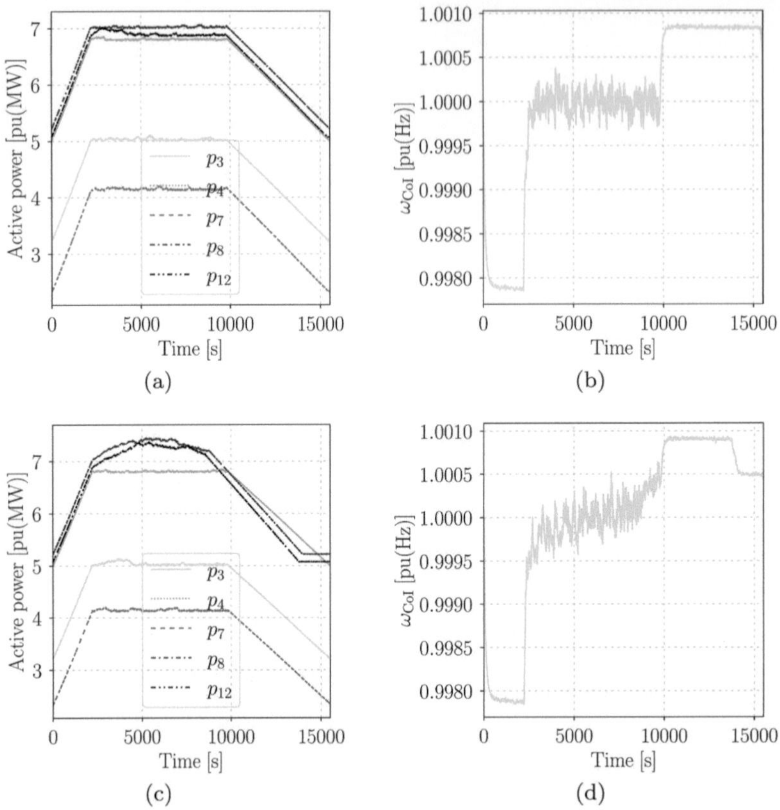

Fig. 7.20 Unsynchronized AIMD results: **a** active powers of loads at buses 3, 4, 7, 8 and 12, decentralized controller with no priorities; **b** ω_{CoI}, decentralized controller with no priorities; **c** active powers of loads at buses 3, 4, 7, 8 and 12 (where PEVs are), decentralized controller with priorities; **d** ω_{CoI}, decentralized controller with priorities

steeper utility function (prioritized vehicles) finish earlier than the vehicles of other buses. Hence, results confirm qualitatively our expectations. Moreover, Fig. 7.20b, d also show that the frequency always lies in a safe range [0.998, 1.001] pu. This indicates that by designing steeper, or less steeper utility functions, it is possible to prioritize some vehicles, if desired.

In Brief

- Data centers, including AI data centers, are becoming the dominant large loads and are expected to significantly increase in the next years.

- Traditional large loads do not usually disconnect following faults. However, power-electronic loads are more sensitive to faults and tend to switch critical load to back-up supply systems.
- Lack of FRT of large demands is becoming a critical operational issue for TSOs and threatens system security and RES targets. There is an urgent need to develop performance requirements for such loads.
- Significant ongoing efforts from industry to address performance requirements, in particular, the FRT problem of converter-based demand.
- Communication and transparency between TSOs and data center owners/customers is crucial.
- Uncontrolled charging of PEVs may have a significant impact on long-term frequency quality. Stochastic control appears as an effective strategy to reduce this impact.
- Automatic frequency-based, or voltage-based, control strategies for PEVs fail to preserve the stability of the network.
- Power-based control strategies can be used to automatically and seamlessly adjust charging rates of PEVs to optimize a desired cost function of interest.

Open Questions

An accurate dynamic model of data centers is missing. It is unclear whether existing load models such as the composite model and active distribution network model are adequate to represent the dynamic behavior of these loads. Additionally, it is not clear whether RMS or EMT-type models are needed.

There are also a number of open questions related to the regulation and dynamic performance of converter-based demand. The first and probably most fundamental question is whether converter-based demand should undergo same requirements as generators. How to deal with performance issues with existing converter-based demand and what are cost-effective short and medium-term actions the TSOs and demand customers can take are other relevant open question at the time of writing the book. As usual, when introducing new devices and controllers, real-world trials to demonstrate the effectiveness of EVs to provide frequency control and regulation are crucial.

References

1. Larkins A (2021) Power converters: a growing challenge to grid stability? [viewpoint]. IEEE Electrificat Mag 9(2):85–88
2. ESIG (2024) Large loads task force (LLTF). www.esig.energy
3. NERC (2024) Large loads task force (LLTF). www.nerc.com/comm

4. Chen K (2025) EPE perspective on modeling needs, ESIG 2025 Spring Technical Workshop, Technical report. www.esig.energy
5. Isaacs A (2025) Data center loads: analysis and mitigation, ESIG 2025 Spring Technical Workshop, Technical report. www.esig.energy
6. Li Y, Mughees M, Chen Y et al (2024) The unseen AI disruptions for power grids: LLM-induced transients. arXiv:2409.11416
7. Springer A (2025) Large loads in ERCOT—observations and challenges, ESIG 2025 Spring Technical Workshop, Technical report. www.esig.energy
8. Kynev S (2025) Application of E-STATCOM for AI data centers, ESIG 2025 Spring Technical Workshop, Technical report. www.esig.energy
9. Quint R, Zhao JJ, Thomas K (2025) An assessment of large load interconnection risks in the Western interconnection. Technical report, Elevate Energy Consulting
10. EirGrid and SONI (2023) Shaping our electricity future roadmap. https://www.eirgrid.ie/
11. EirGrid and SONI, All-island generation capacity statement 2021–2030 (2021). cms.eirgrid.ie
12. Mitra P (2023) Transmission planning and large data centers [view point]. IEEE Electrificat. Mag. 11(3):90–92
13. EirGrid and SONI (2025) All-island resource adequacy assessment 2025–2034. cms.eirgrid.ie
14. Seppänen J, Lehtonen M, Kuivaniemi M et al (2022) Long term characteristics of ultra low frequency oscillations in the Nordic power system. In: 2022 IEEE PES innovative smart grid technologies conference Europe (ISGT-Europe), pp 1–5
15. WECC (2015) WECC dynamic composite load model (CMPLDW) specifications. home.engineering.iastate.edu
16. Ma Z, Wang Z, Wang Y et al (2020) Mathematical representation of WECC composite load model. J Mod Power Syst Clean Energy 8(5):1015–1023
17. Arif A, Wang Z, Wang J et al (2018) Load modeling-a review. IEEE Trans Smart Grid 9(6):5986–5999
18. ENTSO-E (2023) Public consultation on the amendments to the electricity grid connection network codes. www.acer.europa.eu
19. IEEE standard for interconnection and interoperability of inverter-based resources (IBRs) interconnecting with associated transmission electric power systems. IEEE, pp 1–180 (2022)
20. ENTSO-E (2024) ENTSO-E position on urgent connection requirements for power-to-gas demand facilities. eepublicdownloads.blob.core.windows.net
21. ERCOT (2023) Registration, interconnection, and operation of customers with large loads; Information required of customers with loads 25 MW or greater. www.ercot.com
22. Cheng Y (2025) Large load impact on stability limits, ESIG 2025 Spring Technical Workshop, Technical report. www.esig.energy
23. International Renewable Energy Agency (IRENA), Innovation outlook: Smart charging for electric vehicles (2019). www.irena.org
24. International Renewable Energy Agency (IRENA), Global energy transformation: A roadmap to 2050 (2019 edition) (2019). www.irena.org
25. International Energy Agency (IEA) (2019) Global EV outlook 2019: scaling up the transition to electric mobility. www.iea.org
26. Biroon RA, Abdollahi Z, Hadidi R (2019) Effect of tariff on optimal electric vehicle connection to the grid in residential sector. IEEE Transp Electrificat Conf Expo (ITEC) 2019:1–6
27. Chandra Mouli GR, Kefayati M, Baldick R et al (2019) Integrated PV charging of EV fleet based on energy prices, V2G, and offer of reserves. IEEE Trans Smart Grid 10(2):1313–1325 (2019)
28. Richardson DB (2013) Electric vehicles and the electric grid: a review of modeling approaches, impacts, and renewable energy integration. Renew Sustain Energy Rev 19:247–254
29. Izadkhast S, Garcia-Gonzalez P, Fras P et al (2016) An aggregate model of plug-in electric vehicles including distribution network characteristics for primary frequency control. In: IEEE PES general meeting, pp 1–1
30. Lopes JAP, Soares FJ, Almeida PMR (2011) Integration of electric vehicles in the electric power system. Proc IEEE 99(1):168–183

31. Shi Y, Tuan HD, Savkin AV et al (2019) Model predictive control for smart grids with multiple electric-vehicle charging stations. IEEE Trans Smart Grid 10(2):2127–2136
32. Thran J, Marecek J, Shorten RN et al (2025) Reserve provision from electric vehicles: aggregate boundaries and stochastic model predictive control. IEEE Trans Power Syst 1–12
33. Wang J, Li F, Fang X et al (2024) Electric vehicles charging time constrained deliverable provision of secondary frequency regulation. IEEE Trans Smart Grid 15(4):3892–3903
34. Khodayar ME, Wu L, Li Z (2013) Electric vehicle mobility in transmission-constrained hourly power generation scheduling. IEEE Trans Smart Grid 4(2):779–788
35. Green RC, Wang L, Alam M et al (2011) Evaluating the impact of plug-in hybrid electric vehicles on composite power system reliability. North Am Power Sympos 2011:1–7
36. Moschella M, Murad MAA, Crisostomi E et al (2020) On the impact of PEV charging on transmission system: static and dynamic limits. IEEE PES Gen Meet 1–5
37. Hajimiragha A, Cañizares CA, Fowler MW et al (2010) Optimal transition to plug-in hybrid electric vehicles in Ontario, Canada, considering the electricity-grid limitations. IEEE Trans Industr Electron 57(2):690–701
38. Di Giorgio A, Liberati F, Canale S (2014) Electric vehicles charging control in a smart grid: a model predictive control approach. Control Eng Pract 22:147–162
39. Mukherjee JC, Gupta A (2015) A review of charge scheduling of electric vehicles in smart grid. IEEE Syst J 9(4):1541–1553
40. Corless M, King C, Shorten R et al (2016) AIMD dynamics and distributed resource allocation. SIAM
41. Haugneland P, Bu C, Hauge E (2016) The Norwegian EV success continues. EVS29 symposium. Quebec, Canada, Montreal, p 9
42. Hardman S, Jenn A, Tal G et al (2018) A review of consumer preferences of and interactions with electric vehicle charging infrastructure. Transp Res Part D: Transp Environ 62:508–523
43. Quir s-Tort s J, Ochoa LF, Lees B (2015) A statistical analysis of EV charging behavior in the UK. In: 2015 IEEE PES innovative smart grid technologies Latin America (ISGT LATAM), pp 445–449
44. Milano F, Zárate-Miñano R (2013) A systematic method to model power systems as stochastic differential algebraic equations. IEEE Trans Power Syst 28(4):4537–4544 Nov
45. Stüdli S, Crisostomi E, Middleton R et al (2012) A flexible distributed framework for realising electric and plug-in hybrid vehicle charging policies. Int J Control 85(8):1130–1145
46. Stüdli S, Crisostomi E, Middleton R et al (2014) Optimal real-time distributed V2G and G2V management of electric vehicles. Int J Control 87(6):1153–1162
47. Nisar Shah S, Incremona GP, Bolzern P et al (2019) Optimization based AIMD saturated algorithms for public charging of electric vehicles. European J Control 47:74–83
48. Moschella M, Murad MAA, Crisostomi E et al (2021) Decentralized charging of plug-in electric vehicles and impact on transmission system dynamics. IEEE Trans Smart Grid 12(2):1772–1781
49. Smon I, Verbič G, Gubina F (2007) Local voltage-stability index using tellegen's theorem. In: IEEE PES general meeting, pp 1–1
50. Parniani M, Vanouni M (2010) A fast local index for online estimation of closeness to loadability limit. IEEE Trans Power Syst 25(1):584–585
51. Crisostomi E, Shorten R, Stüdli S, Wirth F (2017) Electric and plug-in hybrid vehicle networks: optimization and control. CRC Press
52. Wirth FR, Stüdli S, Yu JY et al (2019) Nonhomogeneous place-dependent Markov chains, unsynchronised AIMD, and optimisation. J ACM (JACM) 66(4):1–37
53. Ferraro P, Crisostomi E, Shorten R et al (2018) Stochastic frequency control of grid-connected microgrids. IEEE Trans Power Syst 33(5):5704–5713
54. European Environment Agency (EEA) (2016) Electric vehicles in Europe: EEA report 20/2016. www.eea.europa.eu
55. Van Cutsem T, Glavic M, Rosehart W et al (2020) Test systems for voltage stability studies. IEEE Trans Power Syst 35(5):4078–4087
56. NERC (1999) Transmission capability margins and their use in ATC determination: white paper. www.nerc.com

Chapter 8
Flexible Resources

This chapter studies the role of flexible resources in supporting the system robustness against various disturbances and normal fluctuations of the frequency. In principle, any controllable resource is *flexible*. However, it is usually considered that flexible resources are distributed small devices which provide frequency control as an auxiliary service, for example, converter-interfaced energy storage systems, switchable loads and MGs.

The first part of the chapter discusses the ability of flexible loads to provide PFC by switching on and off on a probability function. This is similar to the stochastic control discussed in Chap. 7 however, in this case the loads are not connected to the grid through converters and do not constitute an issue for the system as, for example, EVs chargers. Section 8.1 provides a short background on the meaning of flexibility in power systems. Section 8.2 discusses the potential of BESS and flexible loads to provide PFC. The stochastic control strategy discussed in this section considers also the support of BESS and shows the benefit of a hybrid resources for the frequency support.

The second part of the chapter discusses a unified framework between the Optimal Energy Management (OEM) problem of grid-connected MGs and the frequency control of power systems through TDSs. MGs often include converter-interfaced resources, but these are, in general, DERs such as small solar or wind power plants as well as storage devices. MGs are not expected to provide control to the grid but can are valuable flexible resources. Section 8.3 illustrates how a properly adjusted OEMs and control strategies of MGs can improve the quality of system frequency.

8.1 Power System Flexibility

There is currently no well-established definition of *flexibility*. It can be argued that flexibility in power systems is the ability to manage demand and supply, solve network constraints, and avoid congestion. TSOs perform the first task of supply-demand

T. Kërçi and F. Milano, *Frequency Quality in Power Systems*, Power Systems,
https://doi.org/10.1007/978-3-032-04721-2_8

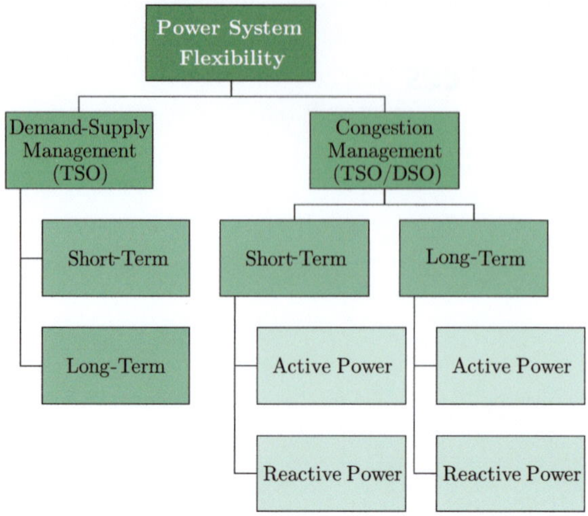

Fig. 8.1 Classification of power system flexibility

management, whereas TSOs and DSOs both perform congestion management. These tasks can be broadly classified into short-term and long-term. For supply-demand management, short-term spans from day-ahead to real-time and long-term from years ahead to day-ahead [1]. Congestion management, on the other hand, involves shorter time scales, with short-term representing real-time time scales and long-term meaning from week-ahead to day-ahead [2].

Congestion management involves both active power and reactive power products. These products are generally classified under the umbrella of "power system flexibility" as illustrated in Fig. 8.1 [3]. The management of supply-demand through reactive power services or local balancing by DSOs for reasons other than congestion are less intuitive and not yet well-developed and widespread in power systems and, as such, are left out of this classification.

8.2 Flexible Loads

DSM is considered an essential component for the successful evolution of power systems. As a matter of fact, DSM is already being used by TSOs to manage critical situations of the grid. For example, the TSOs of France and Italy used 1.7 GW of interruptible load as a response to the European network split on 8 January 2021 [4]. On a smaller scale, in AIPS, there are DSM units that perform ancillary services on behalf of the TSOs [5].

Converter-interfaced energy storage systems are another emerging key component of modern power systems. While there exist a large variety of energy storage technologies [6], in practice, system operators use BESS as a synonym of the whole class of energy storage systems. With a little abuse of notation, we will also use BESS in the remainder of this chapter to indicate converter-interfaced energy storage systems. In recent years, BESS have gained high interest in recent years due to their ability to add increased flexibility to power systems by facilitating the penetration of non-dispatchable power production mainly in the form of RES.

A primary reason for control based on flexible demand is due to the fact that the potential resource is quite large with the infrastructure already in place and not much additional investment is required to allocate loads to these controllers. Using flexible loads have also several other advantages [7]: (i) large groups of small loads have lower variability than a small group of large power plants; (ii) faster response than conventional power plants; and (iii) even distribution throughout the grid.

Among practical implementations of DSM, cooling and refrigeration units are particularly relevant as they are easily controllable due to their ability to store energy through gradients in temperature and the available resource is huge. For example, in California there is an estimated 4 GW of non-coincident load, that is, loads that are not used at the same time [8]. It is estimated that 50% of electricity consumption in the US comes from refrigeration units [9].

All this regulation capacity comes with some challenges. If not properly coordinated, flexible resources can become an issue for the regulation of power systems. Their discrete nature, in fact, gives rise to types of instability behaviors that do not arise for continuous controllers [10]. A very common type of instability due to discrete switching is *flapping*. This is a sustained oscillation that arises when uncoordinated resources switch on and off following a signal and these resources are big enough to modify the signal. For example, say that 20% of refrigerators in US are utilized to provide frequency control and that they all switch on/off following the deviation of the system frequency above/below a given threshold. Then 10% of load would suddenly switch on/off, thus leading not only to compensate the frequency variation but to cause a variation with opposite sign. Then, after while the refrigerators would all switch on again, causing another opposite frequency variation, and so on. This has already happened in real world (see Fig. 8.2). It is clear that proper regulation is needed. On the other hand, it is obviously impractical, if not impossible, to implement a centralized control that coordinates tens of millions of devices.

The stochastic control discussed in Chap. 7 is an option to avoid flapping but is not the definitive solution to all problems. Recent works show that, to avoid potential reappearance of flapping, stochastic frequency control of flexible resources in a power system cannot be truly decentralized [12]. There are also studies that show that the full potential of stochastic control has yet to be exploited. For example, a theoretical study shows that flexible discrete resources can be utilized to provide inertial response and could, in principle at least, substitute all conventional generation [13]. It is also important to note that one of the main challenges which load control programs face is not technical, but rather end user acceptance. Thorough discussions on this challenge are given in [7, 14].

A complementary approach to stochastic control is to couple DSM with BESS [15, 16]. The latter in fact can provide continuous regulation and thus prevent the insurgence of flapping. Another potential application is the utilization of EVs, which are the discrete counterpart of larger BESS. In this context, [17] studies the charging

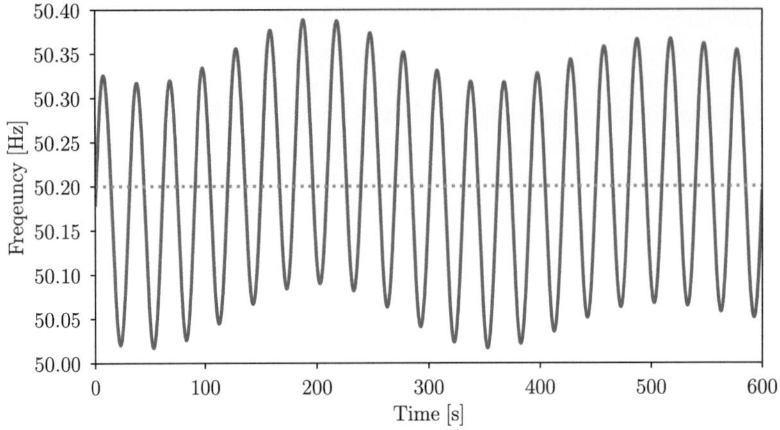

Fig. 8.2 Flapping caused by the so-called German 50.2 Hz problem. In 2007, the German regulator issued the directive EN50438:2007, which stipulated that all micro-generators must turn off, should the frequency exceed 50.2 Hz. In a specific case, when the frequency of the system increased above the 50.2 Hz threshold, the disconnection of large amount of PV generators caused frequency to drop down below the limit thus causing PVs reconnect. This led frequency to exceed 50.2 Hz again, and so on, causing the flapping [11]

cycles of an electric car using DSM with the aim of achieving financial savings and reduce peak load demand. In the same vein, the work in [18] discusses the idea of using EVs as DERs to support the power grid during severe system loading and outages.

In the following, we focus exclusively on PFC and describe a stochastic decentralized control similar to the one discussed in Chap. 7. In particular, we consider discrete flexible loads combined with a BESS. Loads are responsible for compensating larger disturbances and BESS are aimed at compensating smaller ones. Such an approach aligns well with operational practices where, for example, in the AIPS, BESS provide Fast Frequency Response (FFR) using a frequency dead-band between 15-200 mHz whereas DSUs utilize a wider dead-band in the range of 300–700 mHz (see Chap. 12).

8.2.1 Discrete Decentralized Control of Loads

We first describe the stochastic control of the flexible loads. This logic works switching on and off loads based on frequency measurements to provide frequency control to the system. The controller is decentralized and each load switches based on a local frequency measurement independently from the activity of all other loads within the system. The controller works by having N loads distributed within the system with the initial amount of loads which are connected to the system being N_0. For

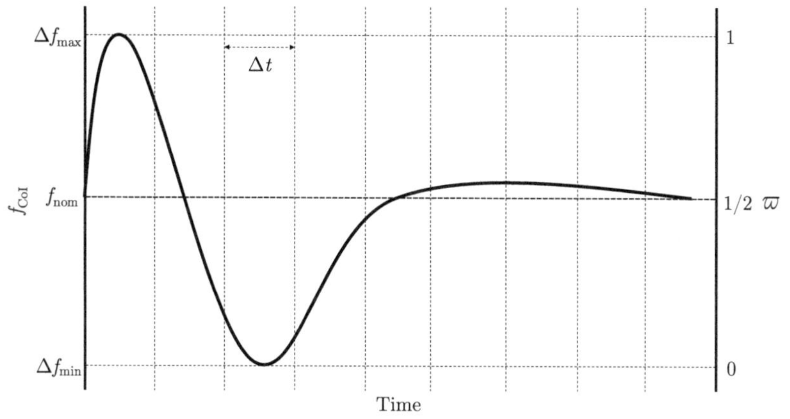

Fig. 8.3 Probability function of the switching logic of flexible loads

simplicity, in the following, N_0 is always chosen to be $N/2$. In this way, no matter if the frequency goes above or below the nominal frequency the controller will be able to switch loads on (above nominal) and shed load (below nominal). Then, a probability function is utilized at every time step Δt to decide if a load switches on or off. The value of this probability is proportional to the frequency deviation Δf. Mathematically, this can be written as follows. Let us define the quantity χ as:

$$\chi(t) = \frac{\Delta f(t) + \Delta f_{\max}}{2\Delta f_{\max}} \tag{8.1}$$

where $\Delta f_{\max}$ is the maximum allowable frequency change such that beyond this point full load reserve will be used. Then, the probability ϖ is given by (see Fig. 8.3):

$$\varpi(t) = \begin{cases} 0 & \text{if } \chi(t) \leq 0, \\ 1 & \text{if } \chi(t) \geq 1, \\ \chi(t) & \text{otherwise.} \end{cases} \tag{8.2}$$

Once the value of ϖ is determined, each load independently generates a random number, u, with a uniform distribution $\mathcal{U}_{[0,1]}$. If $u \leq \varpi$, the load will switch on, and switch off otherwise. In the following, we will assume that $\Delta f_{\max} = 0.2$ Hz, as in the German case (see Fig. 8.2), and $\Delta f_{\min} = -\Delta f_{\max}$. Then, outside the range (49.8, 50.2) Hz, the full amount of reserve is connected for the upper bound and disconnected for the lower bound.

We consider an example based on the WSCC (see Appendix B.1) with inclusion of flexible loads with stochastic frequency control. For all simulations, we assume that the model of the grid is that defined in Appendix A. Figure 8.4 compares the impact on the frequency of the Center of Inertia (CoI) of two different switching time

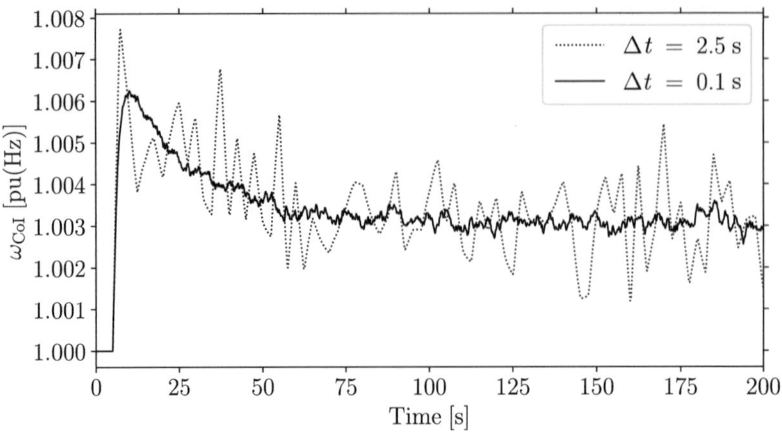

Fig. 8.4 Effect of time step of the flexible load control on frequency

steps, namely $\Delta t = 2.5$ s and $\Delta t = 0.1$ s. Both cases have 50×1 MW loads with N_0 being 25×1 MW. Finally, a 25 MW load outage is considered as a contingency.

The smaller the time step, in this case $\Delta t = 0.1$ s, makes the switching of the loads to happen more often and, as shown in Fig. 8.4, this is beneficial for the overall frequency response of the system. On the other hand, low granularity in terms of either power steps or time steps, distorts more the frequency. The level of granularity appears to be particularly relevant for systems with low inertia where the effect of large "jumps" is magnified [19, 20].

8.2.2 *Clusterized Control*

The example discussed in the previous section indicates the load-switching controller works the better the smaller the time steps with which the loads are switched. However, in practice, these loads represent sites which have to maintain a certain level of service continuity and quality for their customers/business. This means that the time step cannot be too small. This issue is discussed in this section.

With load frequency control schemes end user acceptance is an important factor. If a consumer is constantly being called upon to supply a load they may withdraw from the ancillary service program, possibly when the capacity is most needed. This phenomenon is known as response fatigue. This dual competing objective which exists is one of the main challenges which faces load control programs. The two objectives being, providing reliable support for the grid while at the same time the supplier providing an adequate level of service.

To overcome this issue, we now consider that loads are clusterized and that the number of clusters is small with respect to the total number of flexible loads. The amount of switching events is equal the amount of clusters available meaning that

Table 8.1 Parameters of the clusterized controllers

Parameter	Original controller	Clusterized controller
N	200	200
N_0	100	–
N_{size}	–	10
$N_{cluster}$	–	20
Δt [s]	0.05	$50/20 = 2.5$
Δf_{max} [Hz (pu)]	0.2 (0.0033)	0.2 (0.0033)

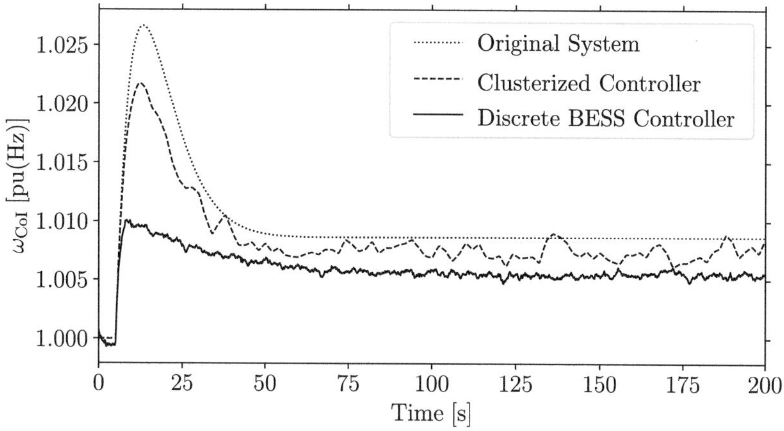

Fig. 8.5 Comparison on controller methods

each cluster can only switch once and therefore each load can only switch once. Over the time frame of a fault the amount of switching events is defined as $t_{fault}/N_{cluster}$, where $N_{cluster}$ is the amount of available clusters. The time taken for t_{fault} is 50 s as beyond this period the Secondary Frequency Control (SFC) will begin to act. A comparison is carried out between the original controller which works like a discrete BESS as the time step becomes sufficiently small and the clusterized controller. A 50 MW load outage is simulated with parameter controllers given in Table 8.1.

The comparison of the transient response of the frequency for the original and clusterized control is shown in Fig. 8.5. The original fault is shown for visualization purposes. As expected, the original controller has a substantially more favorable response with a very low frequency deviation considering the magnitude of the disturbance. The clusterized controller performs quite well for the initial disturbance over the first 50 s. The steady-state fluctuations introduced by the switching events is not favorable from a frequency and consumer satisfaction point of view. Eradicating this issue is addressed in the next section.

The clusterized controller assumes that the size of the clusters are 20×1 MW loads meaning that every time step 20 possible loads can switch. For larger faults such as this one the controller works well, but if this same cluster size is applied to a much smaller fault within the system it may not work as favorably.

8.2.3 Clusterized Control Coupled with a BESS

As was pointed out in the previous section, the method of using flexible loads which switch independently can be useful for larger faults within the system, but as the imbalance's in the system were smaller the controller did not work as well as its parameters must stay fixed over a given fault period. In this section, an BESS is used such that within a certain frequency range only the BESS will act and outside this range the controller will act alongside the BESS. This is done by using a frequency dead-band [21, 22]. As the capacity of the BESS becomes larger the speed at which it can discharge energy decreases but is less likely to run out of energy while compensating. Lower capacity BESS can act faster but the likelihood of hitting energy limits whilst discharging is increased. Combining the two methods of control attempts to exploit the respective weakness of each form of control.

The model of the BESS is shown in Fig. 8.6. The input signal ω is regulated through active power while voltage at the connection point is regulated by reactive power. The two lag blocks T_{BESS} and $T_{\text{Q,BESS}}$ describe the behavior of the dynamics of the BESS. A Proportional Integral (PI) controller is used to regulate active power while a lead-lag controller is used to regulate reactive power.

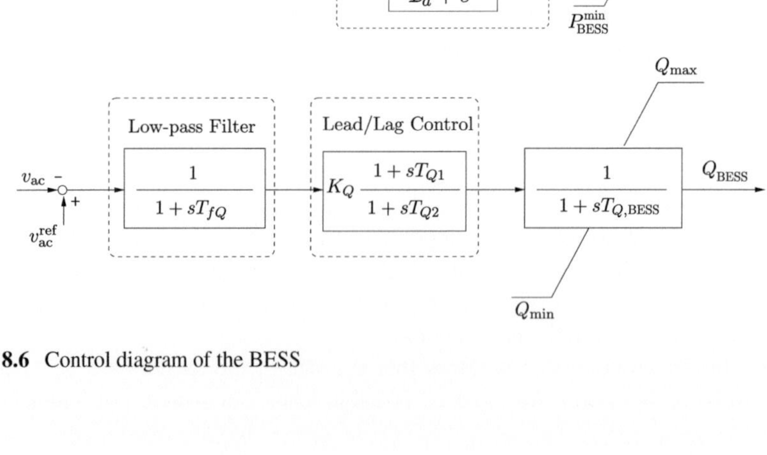

Fig. 8.6 Control diagram of the BESS

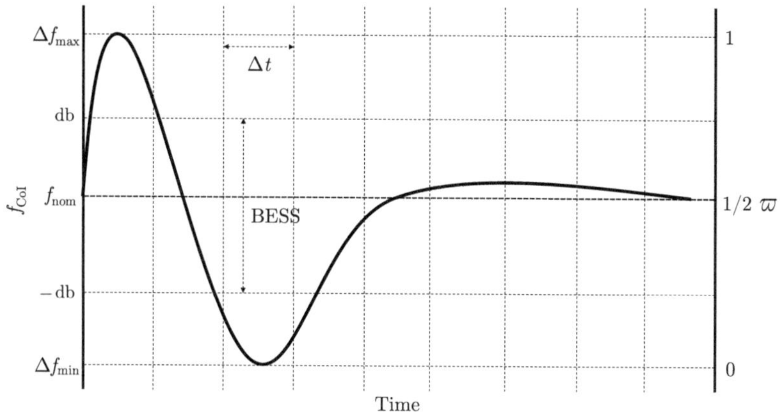

Fig. 8.7 Probability function with dead-band on frequency error. Note: Determination of the probability for the switching of the loads with inclusion of dead-band on the frequency error

Now we assume that the probability is updated if the frequency is outside the dead-band frequency range (see Fig. 8.7 and (8.3)).

$$\varpi(t) = \begin{cases} 0 & \text{if}\, \chi(t) \leq 0, \\ 1 & \text{if}\chi(t) \geq 1, \\ \varpi(t - \Delta t) & \text{if} - \text{db} \leq \chi(t) \leq \text{db}, \\ \chi(t) & \text{otherwise}. \end{cases} \tag{8.3}$$

where db is the frequency dead-band.

To test the coupled BESS and controller four different load outage sizes are simulated to show the different regions of operation, namely, 10 MW, 20 MW, 40 MW and 60 MW. First, we show the case of just the PFC (see Fig. 8.8). Next, Fig. 8.9 shows the case where the controller is included. Last, Fig. 8.10 shows the case when the Automatic Generation Control (AGC) is included. A substantial reduction has been made to the initial disturbances for all cases when the controller is included (see Fig. 8.9). Each case is described below:

- 10 MW: The frequency does not cross ±db therefore only the BESS acts.
- 20 MW: One switching event is only required to get frequency within ±db where BESS takes over
- 40 MW: A few switching events possibly 5 or 6 until BESS takes over.
- 60 MW: Many switching events with the steady state settling point being outside ±db but still the initial disturbance has been reduced significantly.

Figure 8.10 shows that the AGC is able to compensate the steady-state frequency error, which cannot be removed by the clusterized control with BESS.

Simulation results show that the load control can effectively improve frequency quality. However, other system variables such as voltage must not be neglected. For

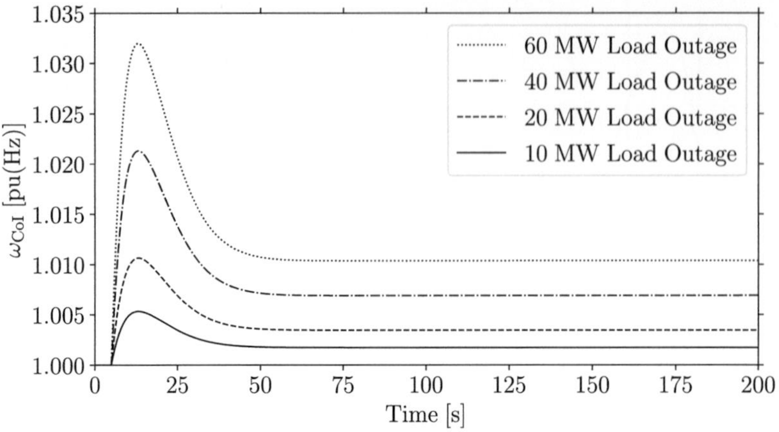

Fig. 8.8 Load outages with no load control and no BESS

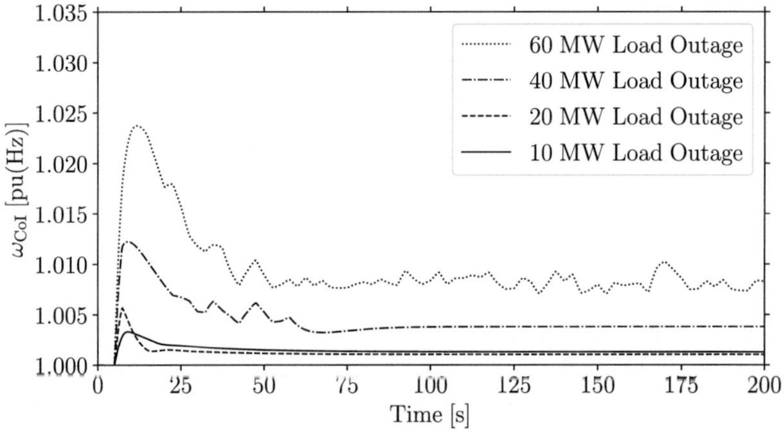

Fig. 8.9 Load outages with clusterised load control and BESS

the 20 MW load outage scenario, the voltage is shown for three controller setups in Fig. 8.11. These setups are: (i) clusterized controller with the BESS; (ii) BESS only; and (iii) clusterized controller without the BESS. As expected, the clusterized controller combined with an BESS has the best impact on the dynamic response of the system with the lowest initial disturbance and steady state settling point closest to the nominal value of 1 pu. The BESS alone shows a slightly higher steady state settling point whereas the clusterized controller alone shows a much lower steady-state value due to the increased switching events. Note also that the controller reduces the absorbed energy of the BESS.

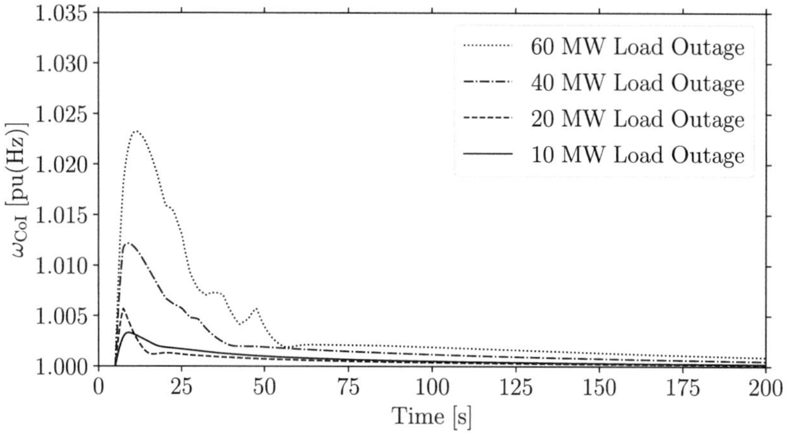

Fig. 8.10 Load outages with clusterized load control, BESS and AGC

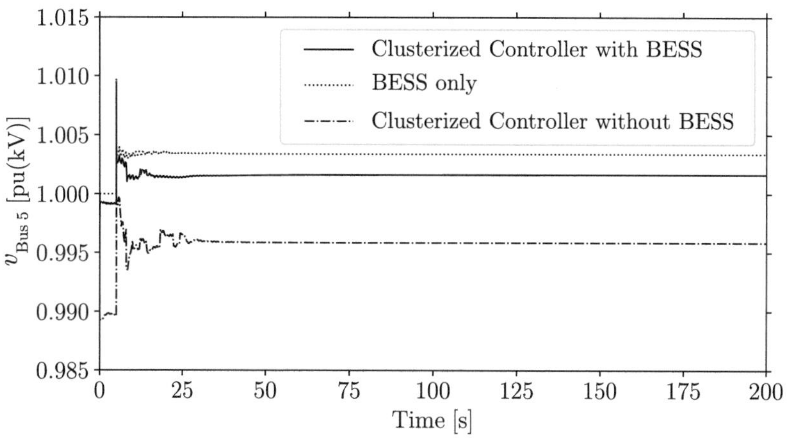

Fig. 8.11 Voltage at bus 5

8.3 Microgrid Control

We consider now the OEM problem of grid-connected MGs and show how the objective function of the OEM and the control strategies of the MGs affect both short- and long-term frequency quality [23]. With this aim, an aggregated MG dynamic model is introduced with the assumption that MG internal time constants are small against the ones of the high voltage transmission system.

Similarly to [24], the OEM model minimizes the MG main objective while ensuring energy reserves and operational limits. However, the MG energy reserves and the power at the Point of Common Coupling (PCC) are calculated using decentralized formulations solving two-stage stochastic optimization mathematical models with two different objectives:

- Minimizing local operational cost (*greedy*),
- Maximizing energy reserves (*cooperative*).

Uncertainties within each MG come from local loads and RES, and errors related to the operation are modeled using a traditional rolling horizon approach. The assessment of is based on the IEEE 39-bus test system, modified to include multiple MGs, a multi-objective OEM, and two control strategies to implement the OEM planning dispatch.

The OEM of a MG consists in finding the most advantageous operational set-points for dispatchable units to achieve selected objectives guaranteeing operational constraints. From the MG perspective, these objectives generally include the minimization of the operating costs, power losses, imported energy, gas emissions, among others [25].

At the same time, as the interaction between TSOs and MGs at the distribution level increases, it is expected that their cooperation assists to improve congestion problems at the PCC, relieving transmission lines overloading, providing voltage and frequency control support, among others [26]. Moreover, the insertion of energy sources with very limited to no rotational inertia, such as solar PV or wind turbines, add an additional layer of technical challenges to maintain the stability of the system [27] and frequency stability is one of the main concern [20].

Hence, considering the potential impacts of MGs on the transmission system and analyzing the effect that different OEM configurations might have on power system dynamics is of great interest for future power systems. This chapter focuses specifically on discussing such an impact through proper modeling and simulation of MGs, their OEM and the power grid.

The interaction between TSOs and distribution system operators to increase penetration of intermittent RES has drawn considerable attention in the last few years, from economic aspects like market architectures and business models [28], through operational coordination models and solution techniques [29], to game-theoretical mechanisms for flexibility aggregators [30]. However, the insertion of RES usually come with a reduction of the rotational inertia, which may lead to unintentional load/generation shedding or even to a system blackout [31].

An option to help increase RES penetration is the inclusion of BESS technologies acting in the PFC [32]. Numerous research works have been done in this direction, as in [33], which proposes a method to define the sizing and control of BESS to contribute in the PFC. A distributed control of BESS to improve frequency regulation is proposed in [34], and [35] provides a comprehensive review of fast-responding energy storage technologies was presented, including gaps, limitations, and recommendations for future research in frequency regulation.

Virtual Power Plants (VPPs) are another solution to address the RES penetration problem. This can be seen in [36], where VPPs are considered as participants of day-ahead energy and spinning reserve markets, or in [37], where the impact of

VPP in short-term transient response has been studied, showing that a coordinated control between BESS and DERs enhances the frequency response when compared to uncoordinated instances. We discuss this point in Sect. 9.5.

MGs have also be considered as an alternative for the provision of ancillary services to TSOs allowing the integration of RES. For example, energy reserves from multiple MGs have been proposed for frequency regulation implementing a coordinated centralized control [38]. Similarly, the effects of greedy MGs on the power system frequency regulation is studied in [39].

Finally it is important to consider also the impact of the OEM when studying the interaction of MGs with the grid. For instance, the optimal control of power flows between MGs and the main grid has been studied in [40], showing the advantages and benefits of applying coordinated control within a system of MGs or in [41], where the OEM of grid-connected unbalanced MGs has been studied. Another alternative is to consider operating flexibility regions as proposed in [42] for distribution networks, or calculating energy reserves as in [24].

8.3.1 Microgrid Optimal Energy Management

The stochastic optimization model introduced in this section considers multiple periods, BESS, and energy reserves [41]. Let $s \in \Omega_\text{S}$, as one scenario in a finite set of scenarios, where the vector of uncertain parameters, conformed by the load, solar irradiance, and wind speed, is represented by η_S. Similarly, the set of dispatchable energy sources, including the connection at the PCC can be represented as Ω_D, and the set of time periods as Ω_T. Thus, the vector of first-stage variables (*here-and-now*) $\boldsymbol{\xi}_\text{I} = \{p_{G,t},\ p_{D,t},\ \forall t \in \Omega_\text{T},\ D \in \Omega_\text{D}\}$ represents the exchange of active power between the microgrid and the main grid and the set point of dispatchable units at each time period; while the vector of second-stage variables (*wait-and-see*), $\xi_\text{II}(\eta_\text{S}) = \{\xi_{s,t},\ \forall t \in \Omega_\text{T},\ s \in \Omega_\text{S}\}$, embodies voltages, currents, state of charge (State of Charge (SOC)) from BESS, and deviations from set-points at each time period in each scenario. Consider $\text{a}_{d,t}$ as the generation costs of each dispatchable energy source (d) inside the microgrid and b_t as the energy cost of importing power from the main grid. Hence, the total energy cost of the microgrid is expressed as

$$\phi(\boldsymbol{\xi}_\text{I}) = \sum_{t\in\Omega_\text{T}} \sum_{D\in\Omega_\text{D}} \text{a}_{D,t}\, p_{D,t} + \sum_{t\in\Omega_\text{T}} \text{b}_t\, p_{G,t}\,, \tag{8.4}$$

The energy reserves can be expressed as a dead-band with a maximum and minimum SOC. The dead-band is calculated as

$$\mu_t^{\mathrm{SOC}} = \sum_{s\in\Omega_\mathrm{S}} \varpi_s \mathrm{SOC}_{t,s}, \quad \forall t \in \Omega_\mathrm{T} \tag{8.5a}$$

$$\left(\sigma_t^{\mathrm{SOC}}\right)^2 = \sum_{s\in\Omega_\mathrm{S}} \varpi_s \left(\mathrm{SOC}_{t,s}\right)^2 - \left(\mu_t^{\mathrm{SOC}}\right)^2, \quad \forall t \in \Omega_\mathrm{T} \tag{8.5b}$$

$$\mathrm{SOC}_t^{\max} = \mu_t^{\mathrm{SOC}} + \kappa\,\sigma_{t,}^{\mathrm{SOC}}, \quad \forall t \in \Omega_\mathrm{T}, \tag{8.5c}$$

$$\mathrm{SOC}_t^{\min} = \mu_t^{\mathrm{SOC}} - \kappa\,\sigma_t^{\mathrm{SOC}}, \quad \forall t \in \Omega_\mathrm{T} \tag{8.5d}$$

where ϖ_s stands for the probability or weighting factor of scenario s, μ_t^{SOC} represents the average of the SOC, σ_t^{SOC} the standard deviation, and κ a parameter defined by the microgrid operator to control the energy reserves.

Furthermore, the average cost of deviating from the set-point $\boldsymbol{\xi}_\mathrm{I}^*$ is penalized by $c_{p,t}$ and $c_{\mathrm{SOC},t}$ is considered as the payback from the TSO for the expected use of the energy reserves. These costs can be expressed as

$$C(\boldsymbol{\xi}) = \sum_{t\in\Omega_\mathrm{T}} \sum_{s\in\Omega_\mathrm{S}} \varpi_s \sum_{D\in\Omega_\mathrm{D}} c_{p,t}|p_{D,t,s}| - c_{\mathrm{SOC},t}(\mathrm{SOC}_t^{\max} - \mathrm{SOC}_t^{\min}), \tag{8.6}$$

where $\boldsymbol{\xi} = [\boldsymbol{\xi}_\mathrm{I}, \boldsymbol{\xi}_\mathrm{II}(\boldsymbol{\eta}_\mathrm{S})]$. Finally, the deterministic equivalent can be written as

$$\begin{aligned} \min.\quad & \mathrm{OF}(\boldsymbol{\xi}) = \gamma\,\phi(\boldsymbol{\xi}_\mathrm{I}) + (1-\gamma)\,C(\boldsymbol{\xi}) \\ \text{s.t.}\quad & \boldsymbol{k}(\boldsymbol{\xi}_\mathrm{I}) \le 0\,, \\ & \boldsymbol{h}(\boldsymbol{\xi}) \le 0\,, \end{aligned} \tag{8.7}$$

where $\boldsymbol{k}(\boldsymbol{\xi}_\mathrm{I})$ represents the constraints dependent on the first-stage variables only, e.g., unit operational limits, and $\boldsymbol{h}(\boldsymbol{\xi})$ the constraints dependent on both stage variables, e.g., power flow constraints and system operational constraints. Furthermore, it can be seen that the deterministic equivalent is a bi-objective mathematical model so the microgrid operator can define its priorities, namely, fully greedy ($\gamma = 1$) or fully cooperative ($\gamma = 0$).

The set of scenarios can be determined with any scenario generation technique, Markov chains, Monte Carlo methods, among others. For example, we assume that the $2m + 1$ point estimate method is used to generate the data trajectories as in [43].

8.3.2 *Microgrid Dynamic Control*

Figure 8.12 shows a MG connected to the transmission grid, where the exchange of active power between the microgrid and transmission system is p_G. We discuss below the dynamic model of the MG with two different mode of operation, namely, cooperative and greedy control approach.

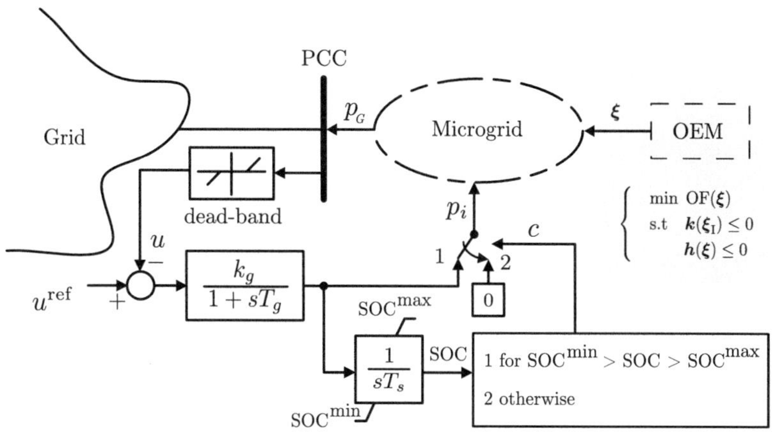

Fig. 8.12 A microgrid connected to the transmission system considering its OEM and dynamic control model

Cooperative Microgrid Control

In cooperative mode, the microgrid uses the energy reserves to participate in the frequency control. Therefore the measured signal is the frequency at the PCC, where the reference is defined as $u^{\mathrm{ref}} = \omega^{\mathrm{ref}} = 1$ pu ($u = \omega_{\mathrm{pcc}}$) and the frequency dead-band set to db $= 0.001$ pu. The controller output p_i in Fig. 8.12 is given by

$$T_g\, p_i' = k_g(u^{\mathrm{ref}} - u) - p_i\,, \tag{8.8}$$

where k_g and T_g are the controller gain and time constant, respectively.

During frequency control, the available energy capacity of the MG is limited by the SOC. Therefore, another controller tracks the SOC based on the real time active power (p_i) utilized by the controller. The SOC is obtained by an anti-windup integrator limited by the maximum and minimum SOC. Mathematically, one has:

$$\begin{aligned}
&\text{if SOC} \geq \text{SOC}^{\max} \text{ and SOC}' \geq 0: \\
&\qquad \text{SOC} = \text{SOC}^{\max} \text{ and SOC}' = 0\,, \\
&\text{if SOC} \leq \text{SOC}^{\min} \text{ and SOC}' \leq 0: \\
&\qquad \text{SOC} = \text{SOC}^{\min} \text{ and SOC}' = 0\,, \\
&\text{otherwise}: \\
&\qquad T_s\, \text{SOC}' = p_i\,,
\end{aligned} \tag{8.9}$$

where $T_s = \frac{1}{\alpha \mathrm{EC}}$ is the charging/discharging time constant; and EC is the BESS energy capacity and α the charging/discharging rate. Note that SOC$^{\min}$ and SOC$^{\max}$ are obtained from the OEM.

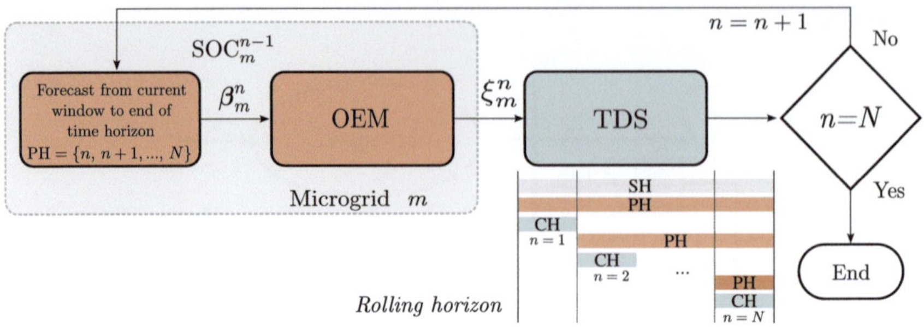

Fig. 8.13 Flowchart of the testing methodology for MGs considering a rolling horizon framework

Greedy Microgrid Control
The exchange of active power (p_G) between the MG and the transmission grid are held fixed from the value obtained in the OEM. In such a control scheme, the microgrid prioritizes its own operation and does not contribute to the frequency control.

8.3.3 Rolling Horizon and Simulation Approach

A rolling horizon framework is considered in this section to cope with changes in the system. These include generation set-points, tap positions, topological changes, and updated forecast data. Three main components are considered: a scheduling horizon (SH), a prediction horizon (PH), and a control horizon (CH). The SH represents the total time span to be analyzed, discretized in equal time windows defined as $\mathrm{SH} = \{1, 2, ..., N\}$. In the PH, random variables are assumed to be parameterized by known distribution functions and the OEM described in Sect. 8.3.1 is executed. The length of the PH is dynamic, and changes depending on the time window n, as $\mathrm{PH} = \{n, n+1, ..., N\}$. Finally, in the CH, the decisions found in the PH are applied, and the dynamic response is assessed using TDS during the time window n.

The interaction between the OEM and the dynamic control of a MG m is performed hierarchically at each time window n. First, the OEM of a microgrid m receives a set of input parameters ($\beta_m^n = \{p_m^f, \lambda_m, \mathrm{SOC}_m^{n-1}\}$), containing the forecast power from local RES and loads for the current prediction horizon, the forecast price for energy reserves, and the most recent SOC from the TDS (assumed as known for $n = 1$). The OEM is solved to obtain the input parameters for the dynamic control, including the power set-point, the SOC of the BESS, the energy reserves, and the control approach. This information is embodied by $\xi_m^n = \{p_G, \mathrm{SOC}, \mathrm{SOC}_n^{\min}, \mathrm{SOC}_n^{\max}, c\}$ and communicated to the dynamic control. Once all dynamic controls are updated, a TDS simulation is performed for the duration of the time window n, considering the transmission system behavior, similar to [44]. This process is repeated until the last control horizon has been considered, as depicted in Fig. 8.13.

8.3.4 Example

This case study is based on a modified version of the well-known IEEE New England 39-bus 10-machine system (see Appendix B.2). The system counts with 10 synchronous machines, 19 loads, 34 transmission lines and 12 transformers and the frequency of the system is 60 Hz. All synchronous generators are equipped with automatic voltage regulators, power system stabilizers and TGs. The TGs are coordinated through an automatic generation control. Fifteen MGs with different capacities are connected at different load buses, namely, bus 4, 7, 8, 12, 15, 16, 18, 20, 21 and 23-28 (see Fig. 8.14). The total accumulated capacity of the MGs is 45 MW and the total nominal load is around 6 GW; hence, the total contribution of the MGs in the system is around 1%. The parameters of the cooperative frequency controller of all MGs are considered same and are $k_g = 50$, $T_g = 0.1$. While the constant for the SOC controller of all MGs are based on individual MG capacities.

8.3.4.1 Deterministic Analysis

The deterministic case assumes perfect knowledge of the exogenous parameters in the system during the SH such as RES and loads. In this test case, two scenarios are analyzed: *conventional* and *low-inertia*. The difference between these scenarios is the number of conventional generators, being the low-inertia scenario the one with reduced total rotational inertia. Three different MG configurations are studied in each scenario, namely, no MGs, greedy MGs, and cooperative MGs. No stochastic variations of the loads and generation are considered in this section (see Sect. 8.3.4.2 for a stochastic analysis).

Conventional Scenario

This scenario considers the original 39-bus system with the connected MGs. The test system is simulated for an SH of 15 min with a contingency happening at $t = 1$ s consisting of a 25% instantaneous load increase at bus 3 from its nominal value. The CH was defined as 5 min; hence, the OEM is executed three times in total. The frequency response of the CoI is shown in Fig. 8.15, where it can be seen that the frequency excursion right after the contingency drops around 59.75 Hz if no MGs are considered. Note that Fig. 8.15 shows the primary and secondary frequency response.

An almost identical result was obtained for all MGs operating under the greedy control; this was expected, since the greedy control uses the MG internal energy reserves to maintain the power at the PCC close to the set-point given by the OEM and does not participate in the frequency control. On the other hand, when the MGs operate with the cooperative configuration, it can be seen that the maximum frequency excursion is around 59.8 Hz which represents an improvement of around 5 mHz over the other two configurations. Notice that even after a small contingency (1% of total load), the cooperative participation of the MGs would help to maintain the frequency within acceptable boundaries (±200 mHz).

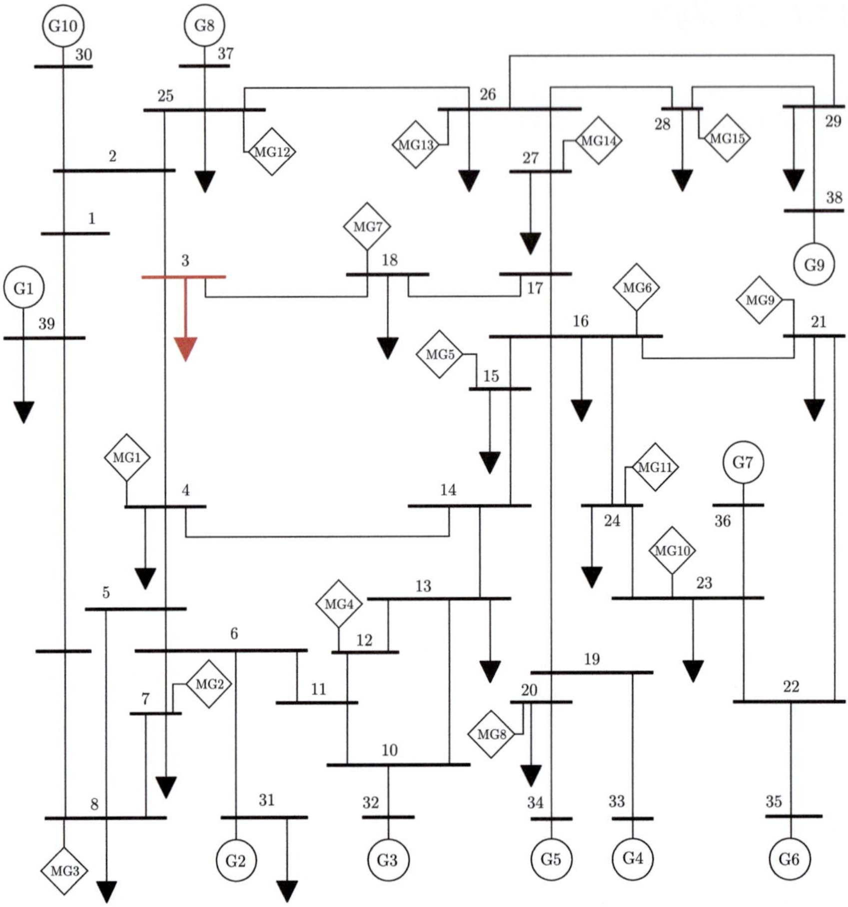

Fig. 8.14 IEEE 39-bus system with 15 MGs

Low-Inertia Scenario

This scenario considers replacing approximately 25% of the system conventional generation capacity (three synchronous machines) with wind farms in buses 30–32. This share of wind energy is more than expected in future electricity grids, considering the current trends and goals. The SH, CH, and contingency have been defined as in the previous test case. The results in Fig. 8.16 show that the reduction of rotational inertia in the system increases the frequency excursion after the contingency compared to the conventional scenario discussed in the previous section. In this case, the frequency drops lower than 59.7 Hz when no MGs are considered and with the greedy control configuration. Whereas for the conventional scenario, the minimum frequency reached with the cooperative control is around 59.75 Hz, which corresponds to a 5 mHz improvement. This similarity arises due to the local MGs OEM, that is, the optimal set-points and energy reserves.

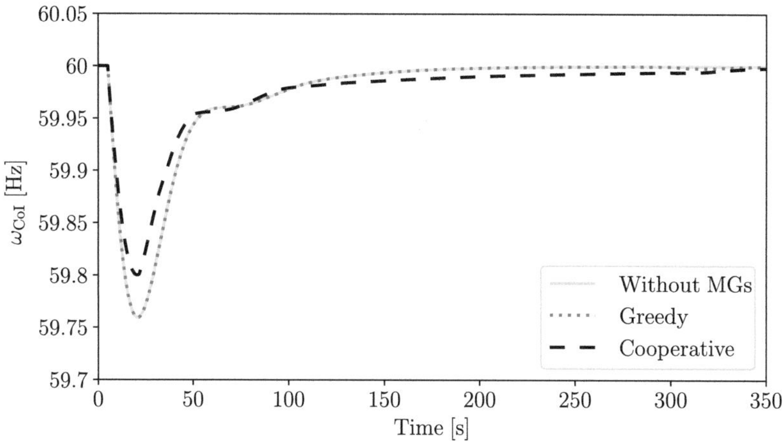

Fig. 8.15 Frequency response of the conventional scenario

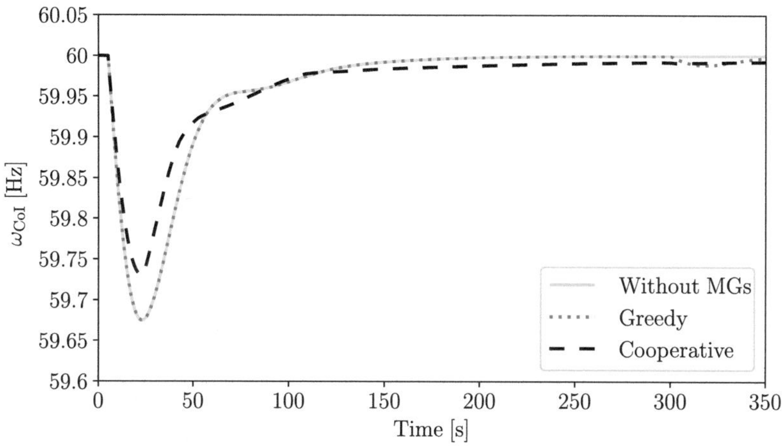

Fig. 8.16 Frequency response in low-inertia scenario

8.3.4.2 Stochastic Analysis

In this test case, a SH of 30 min was considered while keeping the CH to 5 min. The low-inertia scenario is considered. Uncertainty was included both in the PH and the CH steps of the rolling horizon framework. Hence, the objective of this test case is to assess the effect of the different control strategies considering the stochastic variations. Four scenario are taken into account: without MGs, greedy control, cooperative control, and a mix between cooperative and greedy.

Without Microgrids

This scenario does not consider any MGs in the system. The frequency of the CoI is shown in Fig. 8.17 where the gray lines represent each realization, the continuous

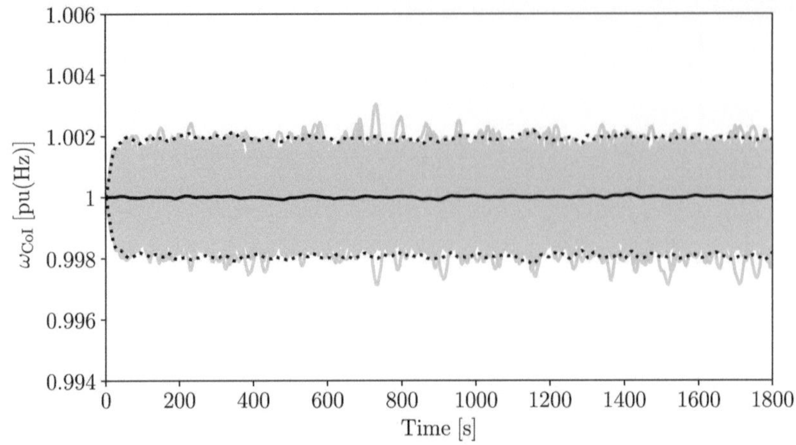

Fig. 8.17 Frequency of the CoI without MGs. The gray lines represent each realization, the black thick line represents the average of the process, while the dotted line represents the average ±3 times the standard deviation

black thick line represents the average of the process (mean value), while the dotted line represents the average ±3 times the standard deviation. This scenario is utilized as a base case to compare the results obtained with other scenarios.

Greedy Microgrids
In this scenario, all MGs were considered to be operating in the greedy mode. The frequency of the CoI is shown in Fig. 8.18. Notice that significant changes in the MGs set-points are relevant to the system frequency, as can be seen in Fig. 8.19, particularly in the time windows ($300 \leq t \leq 600$ s) and ($1200 \leq t \leq 1500$ s), where the set-points of three different trajectories are displayed for the MG connected at bus 4. The changes in the MGs set-point, in fact, make the dynamic behavior of the greedy approach (Fig. 8.18) slightly worse than the case without MGs (Fig. 8.17).

Cooperative Microgrids
This scenario assumes all MGs to be operating in the cooperative configuration. Figure 8.20 shows that the deviation of the frequency of the CoI during the SH is considerably lower in the cooperative case than in the greedy one. Figure 8.21 shows three trajectories of p_G for the MG connected at bus 4. It is relevant ot note that the power of the MG deviates from the set-point imposed by the OEM depending on the system requirements to provide frequency support. In the same vein, Fig. 8.22 shows how SOC varies depending on system conditions and the frequency support provided by the MGs.

Mixed Cooperative and Greedy Microgrids
In this last scenario, the first seven MGs were considered to be operating in a greedy configuration and the remaining eight MGs in cooperative mode. As expected, Fig. 8.23 and 8.24 show that the deviations of frequency the CoI and of the MG power generation improve compared to the greedy scenario but are bigger compared

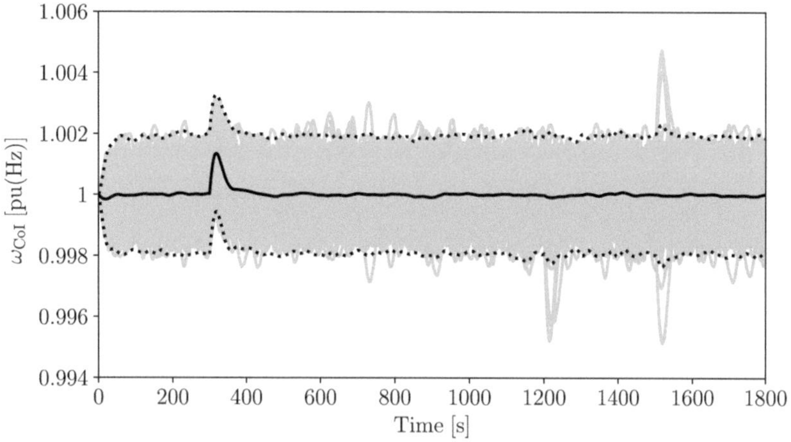

Fig. 8.18 Frequency of the CoI for the greedy approach. The gray lines represent each realization, the black thick line represents the average of the process, while the dotted line represents the average ±3 times the standard deviation

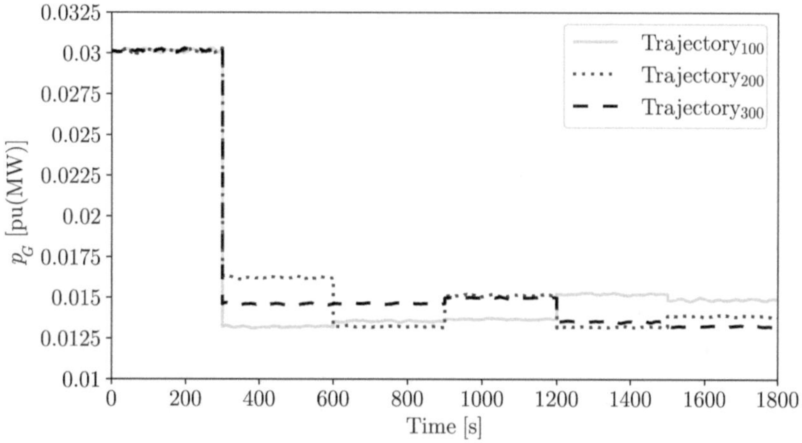

Fig. 8.19 Active power of MG connected at bus 4 for greedy approach

to the cooperative scenario. This result is also evident from Table 8.2 that compares the results of the four scenarios considered in this section.

In Brief

- The utilization of controllers based on discrete resources has a huge potential but it introduces new kinds of instabilities such as flapping.

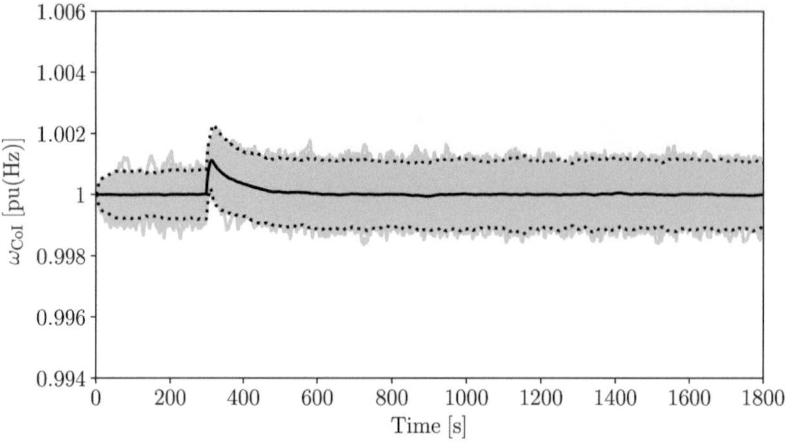

Fig. 8.20 Frequency of the CoI for the cooperative approach. The gray lines represent each realization, the black thick line represents the average of the process, while the dotted line represents the average ±3 times the standard deviation

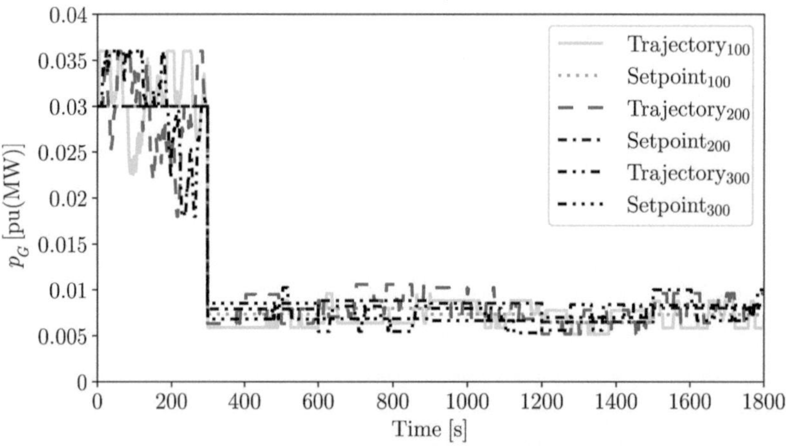

Fig. 8.21 Active power of MG connected at bus 4 for cooperative approach

- Stochastic control can be an effective solution to exploit small, granular flexible resources without the need of a central coordination.
- The discretization in size and time impact substantially on the dynamic performance of stochastic control.
- Energy storage such as BESS improves significantly the dynamic performance of stochastic control of discrete flexible loads.
- MGs can be a valuable resource for the frequency support of the grid but need to be properly coordinated.

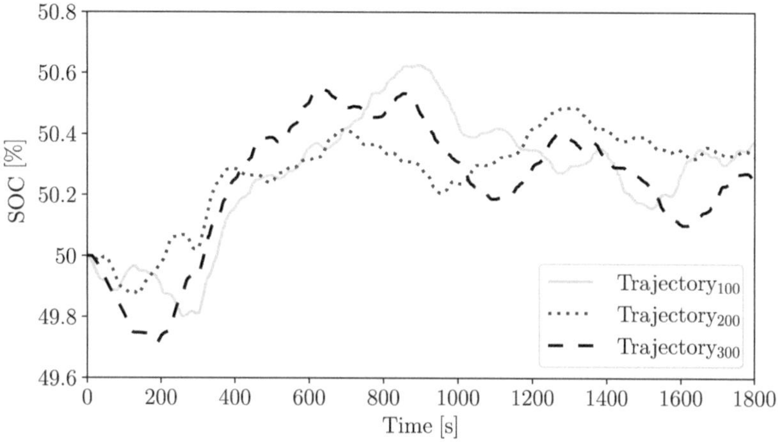

Fig. 8.22 State of charge of MG connected at bus 4 for cooperative approach

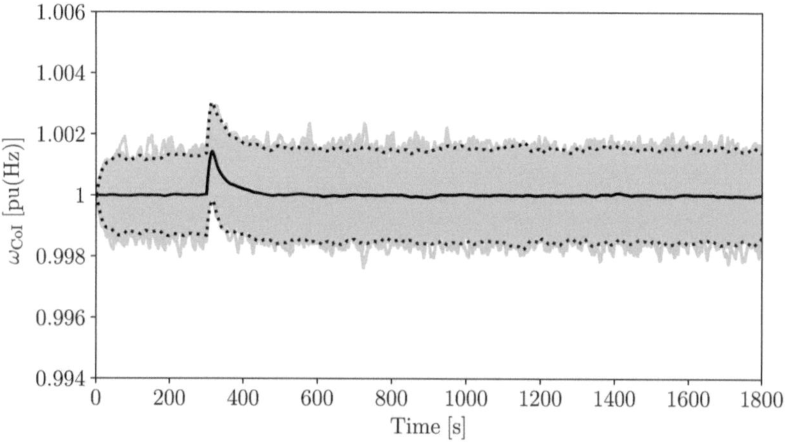

Fig. 8.23 Frequency of the CoI for the mixed case (cooperative/greedy). The gray lines represent each realization, the black thick line represents the average of the process, while the dotted line represents the average ±3 times the standard deviation

- Cooperative dispatch planning and stochastic optimization for OEM of MGs can help improve the frequency quality of the grid.

Open Questions

It is still not clear what are the "right" reserve volumes of flexible loads and BESS to achieve high transient and long-term frequency performance. It appears that

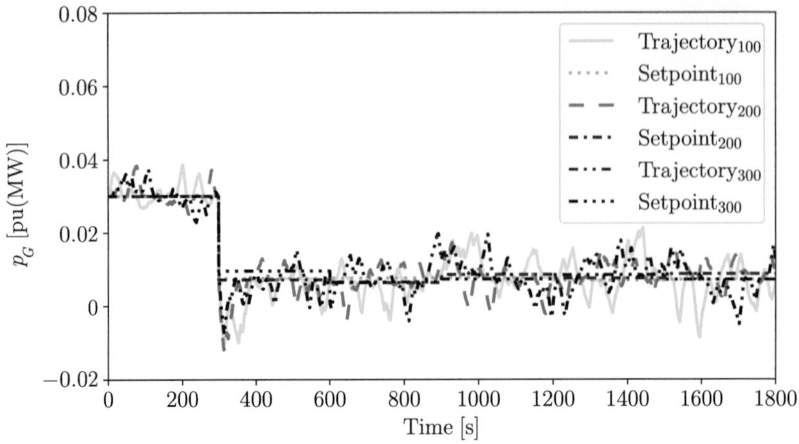

Fig. 8.24 Active power of MG connected at bus 4 for the mixed case

Table 8.2 Maximum standard deviation

Configuration	Std. Dev. Max [pu(Hz)]
No MG	0.0007275
Greedy	0.0007758
Cooperative	0.0004293
Mixed	0.0005866

substantial more research is needed to determine the best strategies to proper coordinate and utilize the huge control potential offered by flexible resources. Similarly, although a large number of studies and demonstrations have been carried out on the control and operation of grid-connected MG, their penetration level is still negligible to see any real-world evidence of their impact on large transmission grids such as on system frequency.

References

1. PNNL (2020) Grid and flexibility services: an overview of California ISO. www.ieahydro.org
2. ENA (2023) Flexibility services. www.energynetworks.org
3. Kërçi T (2023) On the definition and classification of power system flexibility. IEEE PES Innov Smart Grid Technol Europe (ISGT EUROPE) 2023:1–5
4. ENTSO-E (2021) Continental Europe synchronous area separation on 08 Jan 2021. eepublicdownloads.entsoe.eu
5. McMahon J, Kërçi T, Milano F (2021) Combining flexible loads with energy storage systems to provide frequency control. IEEE PES Innov Smart Grid Technol—Asia (ISGT Asia) 2021:1–5
6. Milano F, Ortega Á (2019) Converter-Interfaced energy storage systems: context. Cambridge University Press, Modelling and Dynamic Analysis

7. Callaway DS, Hiskens IA (2011) Achieving controllability of electric loads. Proc IEEE 99(1):184–199
8. Keep TM, Sifuentes FE, Auslander DM et al (2011) Using load switches to control aggregated electricity demand for load following and regulation. In: IEEE PES general meeting, pp 1–7
9. Zehir MA, Bagriyanik M (2012) Demand side management by controlling refrigerators and its effects on consumers. Energy Convers Manage 64:238–244
10. Chatzivasileiadis S, Aristidou P, Dassios I et al (2023) Micro-flexibility: challenges for power system modeling and control. Electr Power Syst Res 216:109002
11. Hermanns H, Wiechmann (2013) Demand-Response management for dependable power grids. Springer New York, NY, pp 1–22
12. Vaca Á, Molina Á, Milano F (2025) Challenges of decentralized stochastic control of discrete flexible resources. In: IEEE PES general meeting, pp1–5
13. Vaca Á, Milano F (2025) Synthetic discrete inertia. IEEE Trans Power Syst 1–4
14. Kim J-H, Shcherbakova A (2011) Common failures of demand response. Energy 36(2):873–880
15. Bruninx K, Dvorkin Y, Delarue E et al (2016) Coupling pumped hydro energy storage with unit commitment. IEEE Trans Sustain Energy 7(2):786–796
16. Zhang X, Hug G, Kolter JZ et al (2018) Demand response of ancillary service from industrial loads coordinated with energy storage. IEEE Trans Power Syst 33(1):951–961
17. Finn P, Fitzpatrick C, Connolly D (2012) Demand side management of electric car charging: benefits for consumer and grid. Energy 42(1):358–363
18. Pang C, Kezunovic M, Ehsani M (2012) Demand side management by using electric vehicles as distributed energy resources. IEEE Int Electr Veh Conf 2012:1–7
19. Ulbig A, Borsche TS, Andersson G (2014) Impact of low rotational inertia on power system stability and operation. IFAC Proc Vol 47(3), 7290–7297. 19th IFAC World Congress
20. Milano F, Drfler F, Hug G et al (2018) Foundations and challenges of low-inertia systems (invited paper). In: Power systems computation conference (PSCC), pp 1–25
21. Bhui P, Senroy N, Singh AK et al (2018) Estimation of inherent governor dead-band and regulation using unscented Kalman filter. IEEE Trans Power Syst 33(4):3546–3558
22. Liu M, Bizzarri F, Brambilla AM et al (2019) On the impact of the dead-band of power system stabilizers and frequency regulation on power system stability. IEEE Trans Power Syst 34(5):3977–3979
23. Giraldo JS, Murad MAA, Kërçi T et al (2022) Impact of decentralized microgrids optimal energy management on power system dynamics. Electr Power Syst Res 212:108337
24. Chen Z, Li Z, Guo C et al (2021) Fully distributed robust reserve scheduling for coupled transmission and distribution systems. IEEE Trans Power Syst 36(1):169–182
25. Katiraei F, Iravani R, Hatziargyriou N et al (2008) Microgrids management. IEEE Power Energy Magaz 6(3):54–65
26. Krechel T, Sanchez F, Gonzalez-Longatt F et al (2019) Transmission system-friendly microgrids: an option to provide ancillary services. In: Distributed energy resources in microgrids. Elsevier, pp 291–321
27. Mehigan L, Al Kez D, Collins S et al (2020) Renewables in the European power system and the impact on system rotational inertia. Energy 203:117776
28. Barbero M, Corchero C, Casals LC et al (2020) Critical evaluation of European balancing markets to enable the participation of Demand Aggregators. Appl Energy 264:114707
29. Givisiez AG, Petrou K, Ochoa LF (2020) A review on TSO-DSO coordination models and solution techniques. Electric Power Syst Res 189:106659
30. Tsaousoglou G, Giraldo JS, Pinson P et al (2021) Mechanism design for fair and efficient DSO flexibility markets. IEEE Trans Smart Grid 12(3):2249–2260
31. Tielens P, Van Hertem D (2016) The relevance of inertia in power systems. Renew Sustain Energy Rev 55:999–1009
32. Divya KC, Østergaard J (2009) Battery energy storage technology for power systems—an overview. Electr Power Syst Res 79(4):511–520

33. Oudalov A, Chartouni D, Ohler C (2007) Optimizing a battery energy storage system for primary frequency control. IEEE Trans Power Syst 22(3):1259–1266
34. Zhao T, Parisio A, Milanović JV (2020) Distributed control of battery energy storage systems for improved frequency regulation. IEEE Trans Power Syst 35(5):3729–3738
35. Akram U, Nadarajah M, Shah R et al (2020) A review on rapid responsive energy storage technologies for frequency regulation in modern power systems. Renew Sustain Energy Rev 120:109626
36. Mashhour E, Moghaddas-Tafreshi SM (2010) Bidding strategy of virtual power plant for participating in energy and spinning reserve markets—Part I: problem formulation. IEEE Trans Power Syst 26(2):949–956
37. Zhong W, Murad MAA, Liu M et al (2020) Impact of virtual power plants on power system short-term transient response. Electr Power Syst Res 189:106609
38. Yuen C, Oudalov A, Timbus A (2010) The provision of frequency control reserves from multiple microgrids. IEEE Trans Industr Electron 58(1):173–183
39. Ferraro P, Crisostomi E, Raugi M et al (2017) Analysis of the impact of microgrid penetration on power system dynamics. IEEE Trans Power Syst 32(5):4101–4109
40. Ouammi A, Dagdougui H, Sacile R (2014) Optimal control of power flows and energy local storages in a network of microgrids modeled as a system of systems. IEEE Trans Control Syst Technol 23(1):128–138
41. Giraldo JS, Castrillon JA, Castro CA et al (2019) Optimal energy management of unbalanced three-phase grid-connected microgrids. IEEE Milan PowerTech 2019:1–6
42. Contreras DA, Rudion K (2021) Computing the feasible operating region of active distribution networks: comparison and validation of random sampling and optimal power flow based methods. IET Gen Trans Distrib 15(10):1600–1612
43. Giraldo JS, Lpez JC, Castrillon JA et al (2019) Probabilistic opf model for unbalanced three-phase electrical distribution systems considering robust constraints. IEEE Trans Power Sys 34(5):3443–3454
44. Kërçi T, Giraldo JS, Milano F (2020) Analysis of the impact of sub-hourly unit commitment on power system dynamics. Electr Power Energy Syst 119:105819

Part III
Operation and Planning

Chapter 9
Automatic Generation Control

In this chapter, we study the impact of various AGC implementations on frequency stability and long-term performance of system frequency. We provide an analogy between the discrete AGC and a dynamic real-time electric market model (market-based AGC). Such an analogy then allows studying the impact of Market-based Automatic Generation Control (MAGC)-induced imbalances on the dynamic behavior of power systems. Finally, the chapter studies the effect of "memory" of market participants on short-term frequency quality. The memory effect is represented by employing proper mathematical tools from the theory of fractional calculus.

The chapter is organized as follows. Section 9.1 provides a general overview of AGC. Sections 9.2 and 9.3 discuss the impact of discrete AGC and MAGC on power system dynamics through various illustrative examples. Sections 9.4 and 9.5 describe the impact of wind generation and VPPs, respectively, on the long-term dynamic performance of the system frequency. Section 9.6 provides a case study on the effectiveness of BESS in mitigating MAGC-induced active power imbalances. Finally, Sect. 9.7 studies the impact of "memory effect" of market participants on transient frequency response of power systems by utilizing an MAGC model and employing fractional calculus.

9.1 Overview

As discussed in Chap. 2, the AGC is a discrete, centralized, automatic SFC implemented in the control centers of TSOs. Its main function is to adjust the generation of the dispatchable generating units, to keep frequency deviations and tie-line interchanges of each TSO control area within specified values [1]. Consequently, the effectiveness of this control depends on the characteristics of the controlled generating units.

T. Kërçi and F. Milano, *Frequency Quality in Power Systems*, Power Systems,
https://doi.org/10.1007/978-3-032-04721-2_9

AGC dynamics take place in the time scale of minutes and sends power signals to participating generating units at fixed intervals of, say, 5 s. This time interval has proved to be adequate in managing real-time power mismatches and, to the best of our knowledge, has not caused any power system instability. However, for the reasons that we explain in the following, it is relevant to investigate the effect of increasing this time interval.

The first part of this chapter discusses this scenario and shows that there is a range of values, from few tens of seconds to few minutes, of the execution cycles of conventional AGC that leads to a limit cycle. This result serves as a motivation to study the same phenomenon when considering real-time electricity markets with short trading intervals. In fact, these markets are currently moving to 5 minutes or even faster trading intervals in order to facilitate the integration of high shares of RES into power systems. AEMO, for example, manages a real-time electricity market with 5 minutes dispatch intervals and 5 minutes ISP [2]. These time scales are comparable with those of the AGC (see Table 2.9). Motivated by this fact and by the observation that real-time market can be modeled as a sort of discrete AGC, we make use of a dynamic electricity market model proposed in [3] and discretize it to represent different trading intervals. Then, we discuss the impact of the MAGC on the frequency stability when using different trading intervals and different shares of RES.

The concept of "sample-data system" has been widely discussed in the literature, and so has the effect of the sampling period on stability. For example, reference [4] proposes a delay-dependent stability criterion, based on the linear matrix inequality method, that allows calculating the delay margins of multi-area LFC schemes and choosing the upper bound of the sampling period. More recently, [5] presents a study on the stability of time-delayed cyber-physical power systems under fully distributed frequency control using a critical eigenvalue tracking algorithm. However, references [4, 5] consider only time delays, not discrete controllers. Delay-dependent controllers are, in effect, quite different compared to the discrete controllers considered in this work. That is, the former act continuously but use quantities that are "old," while the latter act "now" on the current value of the measured signal and then do nothing during a given period.

Reference [6] proposes a robust proportional-integral LFC scheme that takes into account the sampling period and the transmission delay of the communication network. Reference [7] designs a robust retarded sampled-data control with constant delays and proposes a delay-dependent sufficient criterion in the form of linear matrix inequality in order to improve the dynamic performance of power systems with the inclusion of RES. References [6, 7] consider sampling periods along with communication delays. Moreover, [6, 7] consider time periods up to a maximum of 14 s, which are hardly an issue for secondary controllers.

It is also relevant to note that all four references above, namely, [4–7], utilize a linear state-space representation of the system and, in general, very simplified models such as 2nd order model of synchronous machines. On the other hand, the analysis that carried out in this chapter is based on a set of fully-fledged nonlinear DAEs for transient stability analysis.

Other relevant examples are [8, 9] that focus on short-term dynamics of linear systems. On the other hand, research works that have utilized the market model in [3] to study its impact on power system dynamics, are based on small-signal stability analysis [10]. The market model proposed in [3] is utilized in [11] to study the impact of grid-connected MGs, but do not provide a formal analogy between AGC and MAGC.

Most existing literature, thus, does not consider the impact of discrete controllers on the long-term dynamics of nonlinear hybrid systems. In the following we discuss how long sampling time periods of the AGC can give rise to the birth of limit cycles [12]. Long sampling time periods are not expected to be an issue in conventional AGC implementations, of course. However, we show how short-term electricity markets can be interpreted as "slow" AGC and that their implementation can lead to undesirable power oscillations.

9.2 Discrete AGC

We start with a simple illustrative example to describe the effect of discrete output of the AGC on frequency stability. The IEEE WSCC 9-bus system (see Appendix B.1) is utilized to show the effect of different AGC time intervals.

For the simulations presented in this example and in the remainder of the chapter, we use the deterministic part of the power system model (A.1) given in Appendix A, namely:

$$\begin{aligned} \boldsymbol{T}\,\dot{\boldsymbol{x}} &= \boldsymbol{f}(\boldsymbol{x}, \boldsymbol{y}, \boldsymbol{u}, \boldsymbol{z}) \\ \boldsymbol{0}_m &= \boldsymbol{g}(\boldsymbol{x}, \boldsymbol{y}, \boldsymbol{u}, \boldsymbol{z})\,. \end{aligned} \tag{9.1}$$

Equations (9.1) are utilized to emulate the transient behavior of power systems. These equations include the dynamic models of synchronous machines, TGs, AVRs, PSSs and the discrete model of the AGC and MAGC, to name a few. Note that the discrete variables $\boldsymbol{z}$ play a crucial role in the following examples as they are utilized to represent the signal of the AGC.

Since the original benchmark system does not include an AGC, we have included in the model the AGC described in the Sect. 2.3.4.1. In the simulation results below, the disturbance consists in the disconnection of the load at bus 6 at $t = 1$ s.

9.2.1 Impact of AGC Gain

We first consider the effect of K_o with a continuous AGC model. It is well-known that a high gain of the AGC may lead to power system instability due to the coupling between the dynamics of the PFC and AGC. Figure 9.1 shows the transient behavior of ω_{CoI} for $K_o = \{5, 20, 50\}$ without considering the AW limiter; and for $K_o = 100$

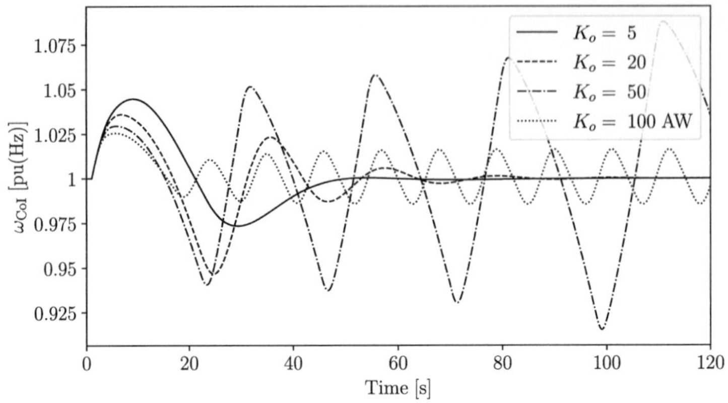

Fig. 9.1 Transient response of ω_{CoI} of the WSCC 9-bus system for different AGC gains

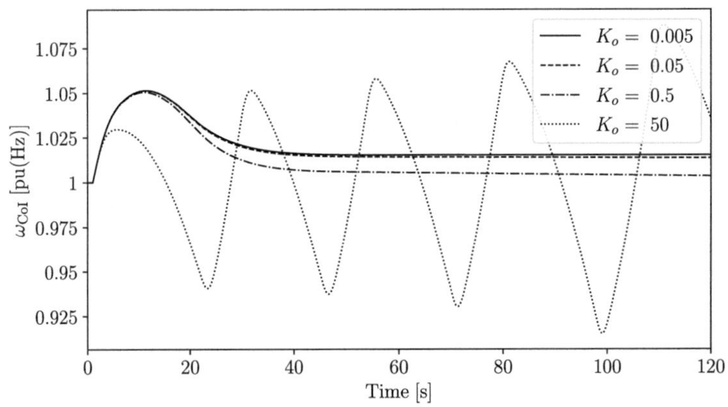

Fig. 9.2 Transient response of ω_{CoI} for different AGC gains

with inclusion of the AW limiter on Δp (40% of the total generation capacity). As expected, the system is unstable for high values of K_o, in this case, for $K_o > 20$. On the other hand, low gain values keep the system stable (see Fig. 9.2). However, the dynamic response with this small value of K_o is very slow, and consequently the controller is not effective. Finally, imposing a limiter on Δp leads to a limit cycle.

9.2.2 Impact of AGC Time Interval

First we show qualitatively that the impact of different values of K_o of continuous AGC is equivalent to different time intervals of a discrete AGC. Let us discretize the first-order differential equation (2.38) using the forward Euler method. The k-th integration step is:

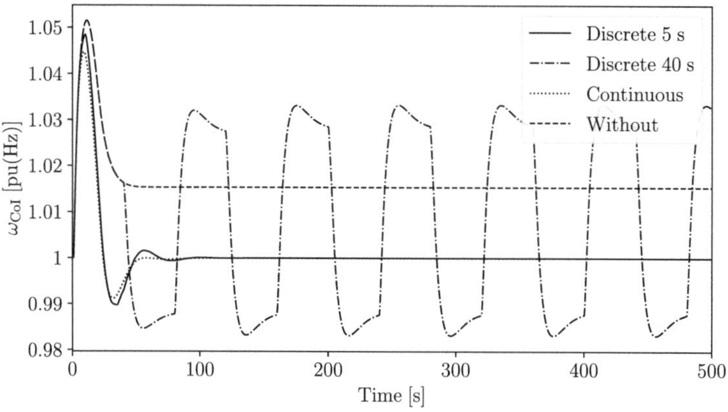

Fig. 9.3 Transient response of ω_{CoI} for different AGC time intervals

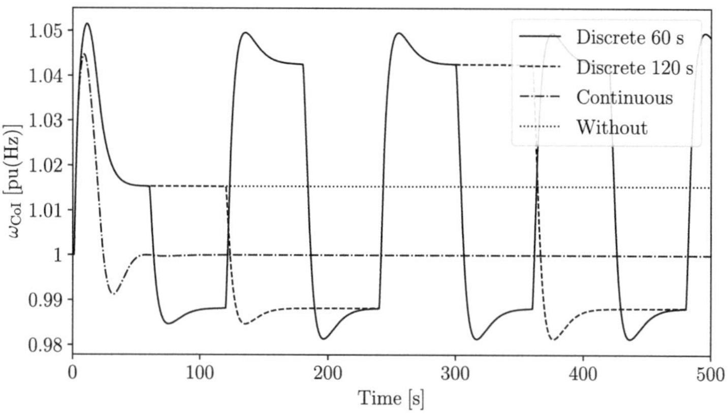

Fig. 9.4 Transient response of ω_{CoI} for different AGC time intervals

$$\Delta p^{(k+1)} = K_o \Delta t (\omega^{\mathrm{ref}} - \omega_{\mathrm{CoI}}(t)) + \Delta p^{(k)} , \tag{9.2}$$

where $\Delta p^{(k+1)}$ and $\Delta p^{(k)}$ are the next and present value of Δp, respectively; Δt is the integration time interval. From (9.2), it is apparent that increasing Δt is equivalent to increasing K_o.

Figures 9.3 and 9.4 depict the transient response of the ω_{CoI} for different AGC time intervals. With a continuous AGC or a discrete AGC with a time interval of 5, the system is stable as the controller is fast enough to bring back the frequency to the nominal value. On the other hand, the results without an AGC show a stable transient response with a steady state-frequency error. For AGC time intervals equal or greater than 40 s, the system enters into a limit cycle. As expected from the discussion above, large discretization time intervals lead to unstable system dynamics. For this reason,

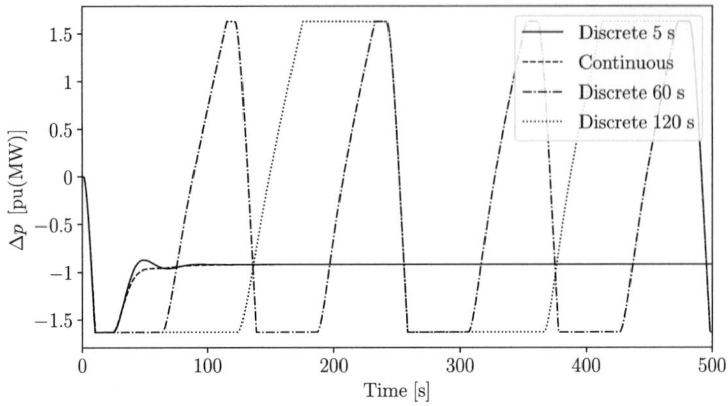

Fig. 9.5 Transient response of Δp for different AGC time intervals

in practice, the AGC time interval is in the range from 2 to 6 s, which is small enough not to originate any limit cycle.

Figure 9.5 shows the transient behavior of the state variable of the AGC (Δp) for some relevant AGC time intervals. The limit cycle occurs whenever the AGC time interval is big enough, say 120 s, to allow Δp to go to the other extreme value (limit) and saturate. Other factors that determine the period of the limit cycle are related to how fast is the AGC controller as well as the type of the contingency. Note that for very long-time intervals, say > 10 minutes, the AGC controller is not effective anymore and other mechanisms such as short-term optimal power flow are in place to compensate the power unbalance.

9.2.3 RoCoF Dead-Band

A way to remove the bang-bang phenomena shown in the previous section is to consider a dead-band on the RoCoF. In the following, the RoCoF is calculated using the following expression:

$$\text{RoCoF} = \frac{\omega_{\text{CoI}}^{(k-1)} - \omega_{\text{CoI}}^{(k)}}{\Delta t}\,, \tag{9.3}$$

where $\omega_{\text{CoI}}^{(k-1)}$ and $\omega_{\text{CoI}}^{(k)}$ correspond to the values of ω_{CoI} at the previous interval and the current time, respectively. While Δt represents the AGC time interval. Note that controlling the frequency of the CoI is not viable in practice. One can, however, implement RoCoF dead-bands in the primary frequency controllers of the power plants. These will affect the variation of the rotor speeds of the synchronous machines

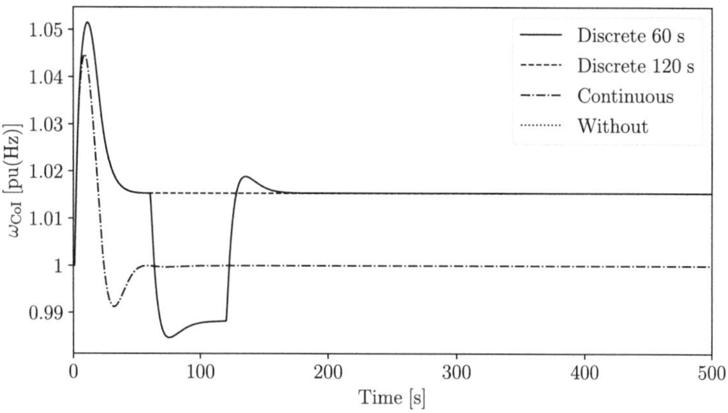

Fig. 9.6 Transient response of $\omega_{\rm CoI}$ for different AGC time intervals and with RoCoF dead-band

and, in turn, of the frequency of the CoI. The model considered in the chapter is thus an approximation but it is sufficient to show the effect of this dead-band.

Figure 9.6 depicts the transient behavior of the $\omega_{\rm CoI}$ with RoCoF dead-band of 0.0002 pu(Hz/s). The RoCoF dead-band successfully removes the limit cycle, however, it leads to a frequency steady-state error. Clearly, this is not desirable because the AGC cannot serve its main purpose.

9.2.4 Sensitivity Analysis with Respect to AGC Signal

Another solution to prevent the bang-bang phenomena is to limit the variation of the internal state variable of the AGC, namely Δp. The limit case, of course, is when the limit of $\Delta p^{\rm max} = \Delta p^{\rm min} = 0$, which basically opens the control loop. Figure 9.7 depicts the transient behavior of the $\omega_{\rm CoI}$ for a variation of Δp from 40% up to 1% and for an AGC time interval of 60 s. As expected, limiting Δp makes the AGC controller less effective. For example, allowing Δp to vary just 1% of the total generation capacity, is almost equivalent to disabling the AGC.

Based on the simulation results above, we conclude that the only effective solution to avoid the limit cycle phenomena caused by the discretization of the AGC is to impose a sufficiently small Δt.

9.3 Market-Based AGC

The previous section shows that there is a range of values, from few tens of seconds to few minutes, of the execution cycles of conventional AGC that leads to a limit cycle. Such a result serves as a motivation to study the same phenomenon when considering

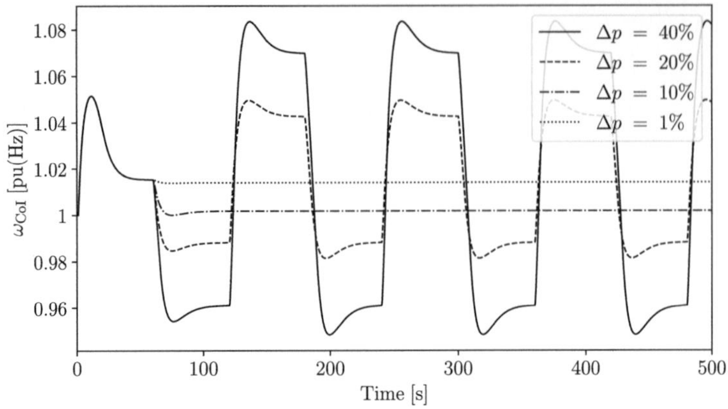

Fig. 9.7 Transient response of ω_{CoI} for different Δp limits

real-time electricity markets. These markets are currently moving to 5 minutes or even faster trading intervals to facilitate the integration of weather-dependent RES into power systems [2]. These time scales are comparable with those of the AGC. Motivated by this fact and by the observation that real-time market can be modeled as a sort of discrete AGC, we make use of a dynamic electricity market model proposed in [3] and discretize it to represent different trading intervals. Then, we discuss the impact of the MAGC on the frequency stability when using different trading intervals and different shares of renewable generation.

9.3.1 Dynamic Short-Term Electricity Market Model

Reference [3] proposes a dynamic market model to study the couplings between the dynamics of the power network and the short-term electricity market. The main idea is to model the balance between power demand and supply through as the equilibrium of a set of differential equations. The electricity price is linked to the power imbalance and represented with a state variable shared by both supply and demand.

The model is formulated as:

$$T_\lambda \, \lambda'(t) = K_E(\omega^{\mathrm{ref}} - \omega_{\mathrm{CoI}}(t)) - \mathcal{I}_d \, \lambda(t) \,, \tag{9.4}$$

$$T_{g,i} \, \Delta p'_{g,i}(t) = \lambda(t) - c_{g,i} \, \Delta p_{g,i}(t) - b_{g,i} \,, \quad i = 1, \ldots, n_g \,, \tag{9.5}$$

$$T_{d,i} \, \Delta p'_{d,i}(t) = b_{d,i} + c_{d,i} \, \Delta p_{d,i}(t) - \lambda(t) \,, \quad i = 1, \ldots, n_d \,, \tag{9.6}$$

where λ is the electricity price; ω^{ref} represents the reference frequency; ω_{CoI} represents the frequency of the Center-of-Inertia (CoI); $\Delta p_{g,i}$ and $\Delta p_{d,i}$ are the generator and load active power, respectively; $c_{g,i}$, $b_{g,i}$, $c_{d,i}$, $b_{d,i}$ are the parameters of the marginal cost and benefit of the generator and load, respectively; T_λ, $T_{g,i}$, and $T_{d,i}$

are the time constants; $\mathcal{I}_d$ is the deviation with respect to a perfect tracking integrator and for a Low-Pass Filter (LPF) it is $\mathcal{I}_d = 1$; and K_E is the feedback gain. Equation (9.4) accounts for the system power imbalance through the deviation frequency of the CoI with respect to the reference frequency. Finally, equations (9.5)–(9.6) assume that a generator/load will increase/decrease its power production/consumption if the electricity price λ is higher than its marginal cost/benefit.

Enforcing a limit on λ is desirable in order to limit the electricity price spikes. This happens, for example, in most of the European electricity markets [13]. The AW limiter on λ is as follows:

$$\begin{aligned} &\text{if } \lambda(t) \geq \lambda^{\max} \;\&\; \lambda'(t) \geq 0 : \lambda(t) = \lambda^{\max} \;\&\; \lambda'(t) = 0 \,, \\ &\text{if } \lambda(t) \leq \lambda^{\min} \;\&\; \lambda'(t) \leq 0 : \lambda(t) = \lambda^{\min} \;\&\; \lambda'(t) = 0 \,, \\ &\qquad\qquad \text{otherwise : equation 9.4} \,. \end{aligned} \tag{9.7}$$

In [3], loads responds similarly to generators with respect to the electricity price λ but with opposite strategy, that is, load bids increase as the electricity price decreases. In this chapter, however, loads are assumed to be inelastic and do not participate in the MAGC. Such an assumption is consistent with the current situation in most of the electricity markets where loads (effectively) do not respond to changes in the wholesale electricity prices. In fact, it has been shown in the literature that making loads price responsive as well will further deteriorate the stability of the system [14].

The electricity market model (9.4)–(9.6) is adequate only for real-time markets with short dispatch periods [3]. As mentioned above, some systems already use these markets and they are expected to become more common in future power systems.

If one assumes loads to be inelastic and, hence, (9.6) is not considered, then the market model (9.4)–(9.5) has a very similar structure to that of a conventional SFC as it can be seen by observing the AGC scheme shown in Fig. 2.19.

To better illustrate similarities, the control diagram of the market model (9.4)–(9.5) (or MAGC) is depicted in Fig. 9.8. The input of both controllers is the same. The AGC includes an integrator with gain K_o that has a similar function with the LPF block of the market, namely, to reduce the frequency oscillations. Finally, the outputs of the AGC and MAGC are distributed to the TGs of the synchronous generators proportionally to their droops (R_i) and bids, respectively.

A substantial difference between the two controllers concerns the meaning of the state variables, namely Δp and λ. In the MAGC, λ is a price, which does not carry any information on the energy involved in the system, as opposed to Δp, which is a power reserve. Nevertheless, the market model (9.4)–(9.5) makes λ a "physical quantity" of the system.

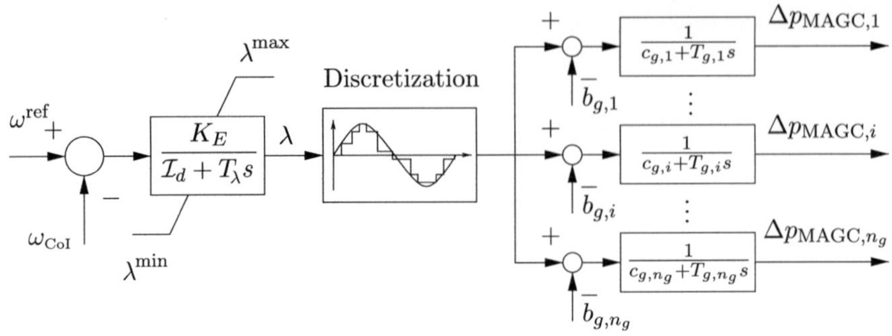

Fig. 9.8 MAGC control diagram

9.3.2 Example

The analogy discussed above allows us simulating real-time electricity markets as an equivalent AGC. In this case, however, the time interval matters because in practice electricity markets use dispatch periods of the order of minutes, not seconds as in the conventional AGC.

For a fair comparison between the AGC and MAGC, we use again the IEEE WSCC 9-bus system and we assume that the three generators have the same market data as the first three generators in the IEEE 39-bus system [3]. Unless stated otherwise, for simulation purposes, the maximum and minimum limit of λ is set to 4,000 and 0 \$/MWh, respectively. The contingency is the outage of the load at bus 6 at $t = 1$ s.

System With MAGC *Only*
In this first scenario, it is assumed that the system includes only the MAGC controller. This scenario corresponds to power systems that do not include an AGC, but instead use offline "manual" generation control [15]. The goal is to investigate whether the idea of controlling power systems exclusively based on market mechanism is feasible.

Figures 9.9 and 9.10 show the transient behavior of ω_{CoI} and λ, respectively. The discrete MAGC leads to limit cycles similar to the case of the AGC. Specifically, the limit cycle appears for a MAGC time interval equal or above 120 s. Compared to the AGC, this phenomenon appears for longer time intervals. This is due to the parameters of the MAGC control scheme. It is important to note, however, that the limit cycles always appear for a certain range of Δt, independently from the parameters of the secondary controller.

There are two ways to avoid the limit cycles. One is to use "short" time periods, for example, 5 s in Fig. 9.9. The other one is to use "long" time periods, for example, 10 minutes [16]. In this case, the MAGC approximation does not hold and is effectively decoupled from the dynamics of the system, and thus it does not cause the occurrence of instability. Figure 9.9 also suggests that real-time electricity markets with very short periods can, in principle, substitute the AGC provided that they have the same features.

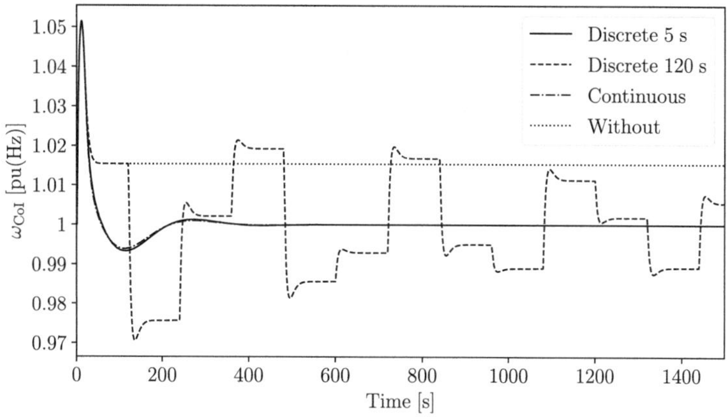

Fig. 9.9 Transient response of ω_{CoI} for different MAGC time intervals

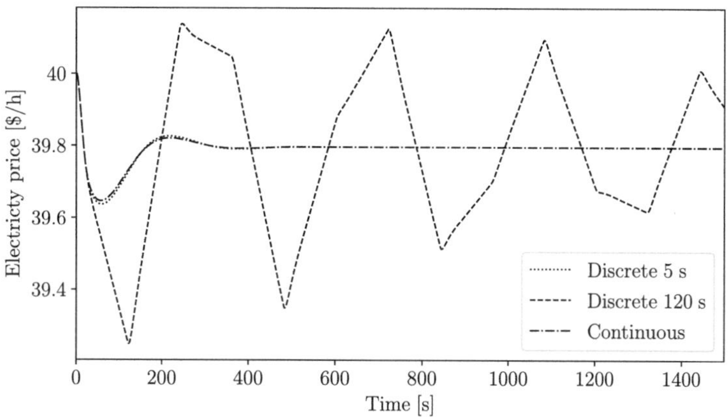

Fig. 9.10 Transient response of λ for different MAGC time intervals

Sensitivity Analysis With Respect to Electricity Price

As discussed earlier in the chapter, a solution to remove the limit cycle is to limit the variation of the state variable of the AGC controller, in this case, that of MAGC controller. In this context, Fig. 9.11 shows the transient response of the system for different λ limits. As expected, limiting the variation of the state variable removes the limit cycle. For example, allowing λ to vary just 0.75% from its nominal value (40 $/MWh) appears to be able to remove the instability, however, it leads to a steady state frequency error. Further limiting the variation of λ makes the controller even less effective [17]. As for the case of the AGC, the best solution appears to be that of using short dispatch intervals.

Systems With AGC and MAGC

This scenario considers the common scenario of power systems equipped with both an AGC and a real-time electricity market. The AGC considered here has a time

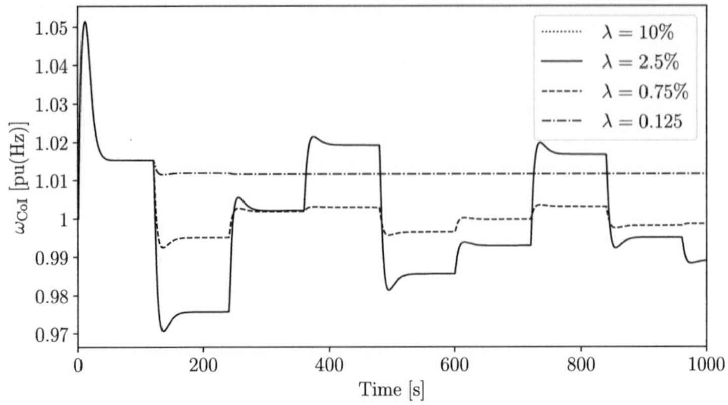

Fig. 9.11 Transient response of ω_{CoI} for different λ limits

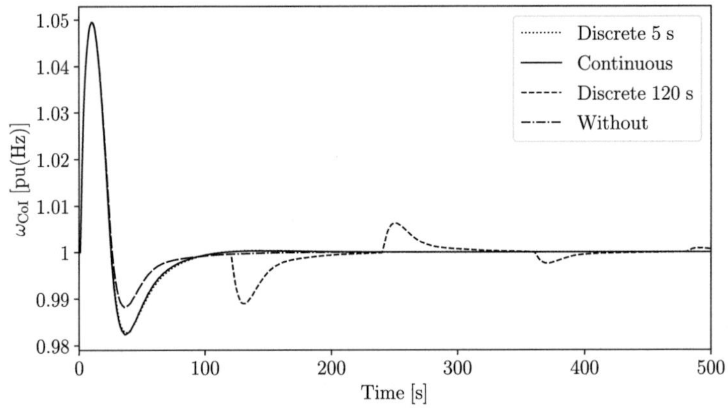

Fig. 9.12 Transient response of ω_{CoI} using the AGC and MAGC

interval of 4 s. Simulations are solved considering the same load outage as in the previous section.

The transient response of the system following the contingency is depicted in Fig. 9.12. The inclusion of the AGC removes the limit cycle. It appears that, if the system includes an AGC, then the discretization of the real-time market does not cause instability issues. It is also interesting to note that using a continuous MAGC leads to a lower frequency nadir. This is due to the dynamic coupling between the dynamics of the MAGC and the AGC. Depending on the severity of the frequency variations, the system operator may have to implement corrective actions such as load shedding. However, a discussion on corrective actions is outside the scope of this chapter.

Economic Impact of Systems With and Without an AGC

This scenario discusses the differences with respect to the profit of the generating units that are part of systems that include only MAGC controller, and both AGC

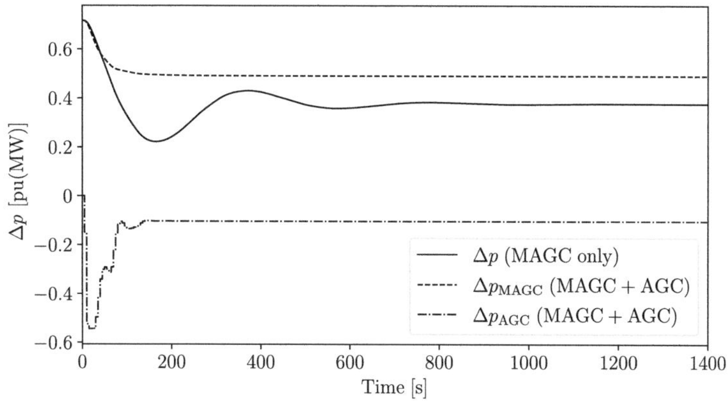

Fig. 9.13 SFC signals for the scenarios with MAGC only and with both MAGC and AGC

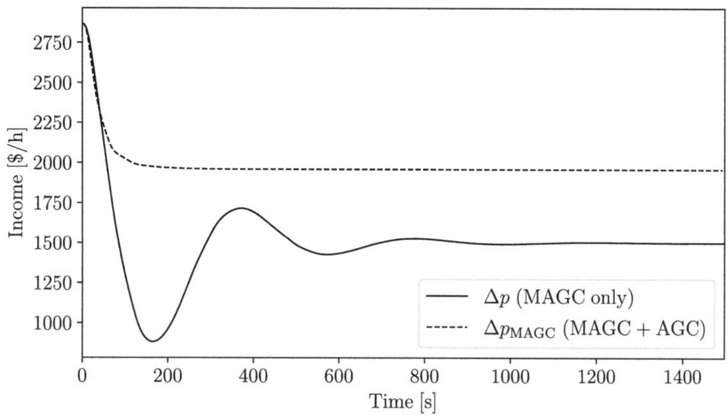

Fig. 9.14 Profit of generator 1 with MAGC only and with both MAGC and AGC

and MAGC. We assume that both MAGC controllers utilizes a time interval of 60 s, whereas the AGC utilizes a time interval of 4 s. Figure 9.13 shows the outputs of the controllers for the generator at bus 1, for the scenarios with MAGC only, and with both AGC and MAGC.

The MAGC-only controller contributes more by decreasing more its output in response to the contingency compared to the same controller when the system includes both MAGC and AGC. This is due to the fact that in the scenario that includes both controllers, the AGC is the dominant controller and takes care of power mismatches. For this reason, the profit of generating units that participate in the MAGC only is less compared to the profit of the same generating units that are part of power systems with MAGC and AGC. The other way round is true if a generator is lost. This is better shown in Fig. 9.14 where the income of the first generator is depicted for both scenarios considered in this section.

9.4 Wind Power in Real-Time Electricity Markets

So far, we have studied the impact of discrete AGC and MAGC using conventional power systems. However, it is also relevant to study such an impact on power systems with high shares of IBRs. With this aim, we consider the impact of wind power penetration.

In most electricity markets worldwide, wind power producers bid in the same way and follow same rules as conventional power plants. For instance, wind power plants are responsible for power deviations with respect to the values scheduled in day-ahead market. In this context, following the structure of the original dynamic market model in (9.4)–(9.5), one can write the equation that models wind power plant behavior with respect to their marginal cost and the price λ, as follows:

$$T_{w,k}\, p'_{w,k}(t) = \lambda(t) - c_{w,k}\, p_{w,k}(t) - b_{w,k}\,, \quad k = 1, \dots, n_w\,, \tag{9.8}$$

where $p_{w,k}$, $T_{w,k}$, $c_{w,k}$, $b_{w,k}$ have same meaning as in (9.5).

Marginal cost of wind power equal to zero
In general, the marginal cost of wind is assumed to be zero, as follows:

$$c_{w,k}\, p_{w,k}(t) - b_{w,k} = 0\,,$$

and, from (9.8), we have:

$$T_{w,k}\, p'_{w,k}(t) = \lambda(t)\,. \tag{9.9}$$

Equation (9.9) is a pure integrator. This means that the wind power plants will try to dynamically integrate and set the price $\lambda(t) = 0$. As shown above in the chapter, a pure integrator tends to be unstable. Thus, the wind power plants will integrate until their output reaches a maximum power limit. In the same vein, if the plants bid negative prices, namely:

$$c_{w,k}\, p_{w,k}(t) + b_{w,k} < 0\,,$$

then from the dynamic point of view their market secondary control is unstable. However, this does not mean that wind power plants drive the system to instability, but just that they are going to generate their maximum power all the time.

9.4.1 Example

In the following, the New England IEEE 39-bus system (see Appendix B.2) is utilized along with the market data taken from [3]. The contingency is the outage at $t = 1$ s of the load located at bus 3. For simulation purposes a MAGC with a time interval of 120 s (chosen on purpose as it is the worse case scenario) and gain $K_E = 15$ are

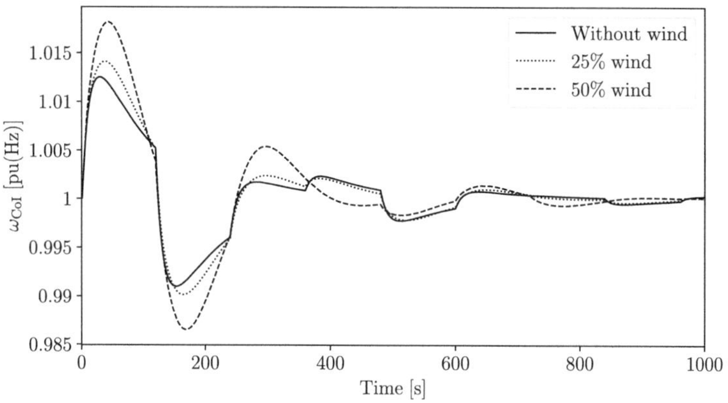

Fig. 9.15 Transient response of ω_{CoI} for different wind power shares

used. Whereas for the AGC a time interval of 4 s and gain $K_o = 2$ is used. The focus is on the transient response of the CoI following the contingency. Three scenarios are considered: (i) base case with conventional generation; (ii) 25% penetration of wind generation; and (iii) 50% penetration of wind generation. For a fair comparison, all scenarios have same loading level, control and network topology.

We assume that the system includes both AGC and MAGC and show the transient behavior of ω_{CoI} in Fig. 9.15. The scenarios with inclusion of wind generation, worsen the performance of both AGC/MAGC controllers, and consequently the dynamic performance of the system. These results indicate that future power systems with high shares of wind power will require much shorter dispatch periods compared to, for example, conventional power systems to avoid possible instabilities.

If the share of wind power generation in the electricity market increases, wind power plants may consider acting strategically as a price maker in order to increase their own profits through intentionally altering the market clearing price. This strategy, called "market power" has been well discussed in the literature. In order to become a price maker, wind power plants will have to be sufficiently big (otherwise they will be limited), and with a marginal cost that is in the same range as that of other conventional generators. In this scenario, wind power plants will participate in the real-time electricity market according to (9.8) and be part of the MAGC shown in Fig. 9.8.

Note that, if they are not coupled with BESS or include some mechanism to provide power reserve, the wind power plant can only provide down regulating service, that is, will only decrease their power production. To simulate this scenario, we consider a 25% wind power penetration by replacing 3 conventional generators with wind power plants. Furthermore, we assume that the wind generators have the same bids and market data as the conventional power plants. The Spanish electricity market is a real-world example where RES and conventional power plants bid together [18]. We also assume that the systems includes both AGC and MAGC.

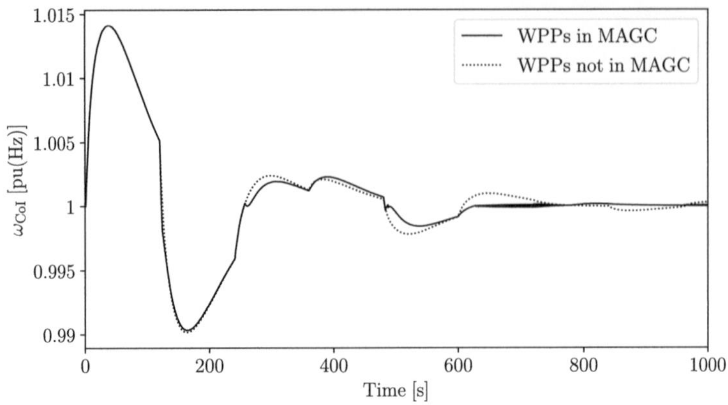

Fig. 9.16 Transient response of ω_{CoI} with wind generation

Figure 9.16 shows the results for this scenario. It appears that the participation of a wind power plant in real-time electricity markets does not make a huge difference with respect to the transient response of the system. This is due to the fact that the AGC is the dominant secondary controller, as it is faster than the MAGC and thus contributes more after the occurrence of the contingency (see Fig. 9.13). We can conclude that, in this scenario, the participation of wind power plants in real-time electricity markets does not make a significant contribution to long-term power system dynamics.

9.5 VPP Frequency Control

As briefly discussed in Chap. 3, the VPP concept can play a significant role in providing frequency support during times of high penetration of DERs. In recent years, many DER units have been operated as a VPP and participate in the ancillary service market to provide frequency control to the TSOs [19]. In this section, we study the effectiveness of VPPs in providing SFC.

European TSOs require that the power output of a VPP increases linearly during the ramp-up time [19]. If the VPP rate of change of power is higher than the agreed linear ramp, then the TSO does not pay for the excess power. If the VPP provides less power than the scheduled linear ramp, then the TSO fines the VPP. However, providing a linear ramping response is a challenge for the VPP as different generators have different characteristics, e.g. different capacity, response and ramping time. This is illustrated in Figs. 9.17 and 9.18.

The points in Fig. 9.17 have the following meaning:

- $(0, 0)$, represents the time when the TSO tells the VPP to go to the maximum production.

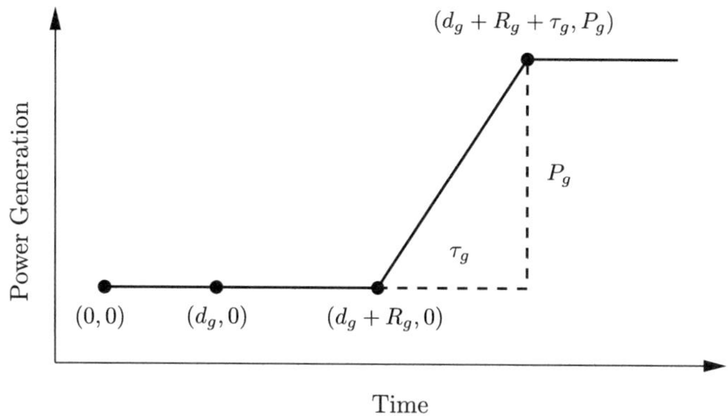

Fig. 9.17 Power production of a single small generator [19]

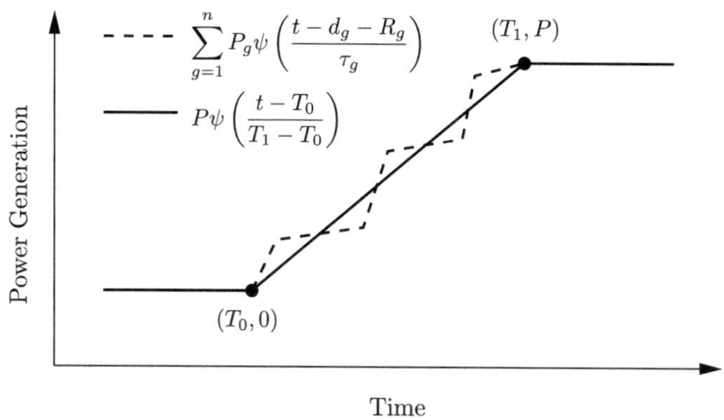

Fig. 9.18 Power production of a single large power plant (continuous line) and that of a collection of small generators (dashed line) [19]

- $(d_g, 0)$, represents the delay time of the VPP, i.e., the time to send a signal to the g-th generator to start the production.
- $(d_g + R_g, 0)$, represents the time at which the g-th generator transitions from the minimum to the ramping production, where R_g is the response time of the g-th generator of the VPP.
- $(d_g + R_g + \tau_g, P_g)$, represents the time at which the g-th generator of the VPP transitions from the ramping to the maximum production, where τ_g is the ramping time and P_g is the maximum capacity of the g-th generator.

The points in Fig. 9.18 have the following meaning:

- $(T_0, 0)$, represents the start of the ramping of the VPP.
- (T_1, P), represents the stop of the ramping of the VPP, with P being the total capacity of the small generators of the VPP.

Furthermore the functions

$$P\psi\left(\frac{t-T_0}{T_1-T_0}\right), \tag{9.10}$$

$$\sum_{g=1}^{n} P_g\psi\left(\frac{t-d_g-R_g}{\tau_g}\right), \tag{9.11}$$

represent the active power generated of a single large power plant (linear), and the total active power output of small generators of the VPP (piecewise linear), respectively.

Two approaches for the provision of SFC through VPPs are discussed below. The first is based on a MILP problem that optimally schedules the small generators of the VPP. The second is based on an AGC approach.

9.5.1 AGC-Based VPP

We consider a VPP where all its units are connected to an AGC, as illustrated in Fig. 9.19. The AGC regulates the total active power produced by the VPP rather than the frequency. This setup allows for a direct comparison with the MILP-based scheduling approach described in the next section.

The control scheme of the AGC is shown in Fig. 9.20. The signal $p_{\rm VPP}^{\rm ref}$ represents the reference power signal sent by the TSO to the VPP, whereas $p_{\rm VPP}$ represents the sum of the measured active power of the DERs included in the VPP. Then the AGC includes an integrator with gain K_0 in order to reduce the active power steady-state error to zero. Finally, the output of the integrator is sent to the TGs of the machines proportionally to their droops $\mathcal{R}$.

9.5.2 MILP-Based VPP

The MILP problem proposed in [19] is considered in this section as an example of optimal scheduling approach to coordinate the units that compose the VPP and obtain a ramping rate that is as close to linear as possible. The mathematical formulation of the MILP problem is shown below.

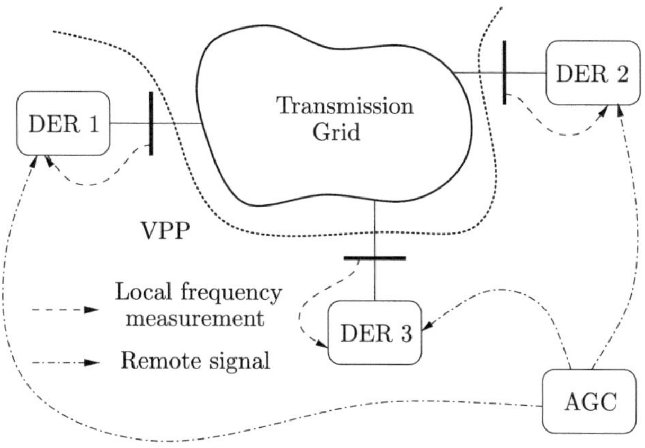

Fig. 9.19 AGC coordination scheme

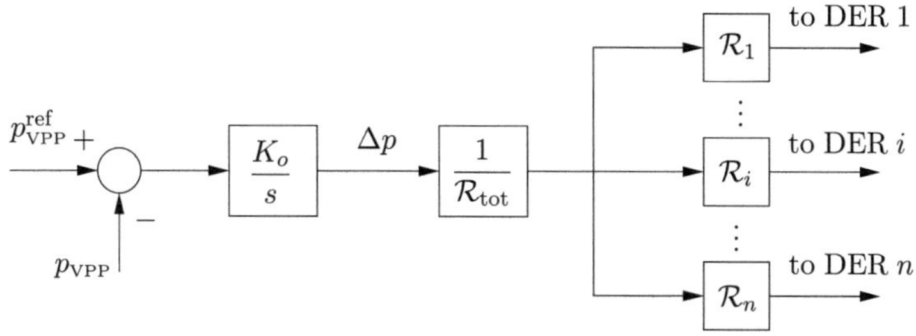

Fig. 9.20 Basic AGC control scheme for active power regulation of VPPs

$$\min \sum_t \left(p_{a,t} + K p_{b,t}\right), \tag{9.12}$$

such that

$$p_{g,t} \leq P_g, \ \ \forall g, t, \tag{9.13}$$

$$p_{g,t} = p_{g,t-1} + R_g\left(z_{g,t} - \overline{z_{g,t}}\right), \ \ \forall g, t, \tag{9.14}$$

$$z_{g,t} \geq z_{g,t-1}, \ \ \forall g, t, \tag{9.15}$$

$$\overline{z_{g,t}} \geq \overline{z_{g,t-1}}, \ \ \forall g, t, \tag{9.16}$$

$$\sum_t (z_{g,t} - \overline{z_{g,t}}) = \tau_g, \ \ \forall g, \tag{9.17}$$

$$\sum_g p_{g,t} + p_{b,t} - p_{a,t} = \frac{t \sum_g P_g}{N}, \ \ \forall t, \tag{9.18}$$

$$z_{g,t}, \overline{z_{g,t}} \in \{0, 1\}, \ \ \forall g, t, \tag{9.19}$$

$$p_{g,t}, p_{a,t}, p_{b,t} \geq 0, \ \ \forall g, t, \tag{9.20}$$

where $p_{a,t}$ and $p_{b,t}$ represent continuous variables that model the distances above and below the target linear characteristic at time t, respectively (see Fig. 9.18). K is a penalty multiplier for when the actual ramping rate is below the target line. This is needed as the VPP is penalized if it provides less power but is not if it provides more than the scheduled power. A value of $K = 10$ is considered. Equation (9.13) models the capacity limits of single small generators, where $p_{g,t}$ represents the active power generation of the g-th generator at time period t. Equalities (9.14) model the ramping limits of generating units, where the binary variables $z_{g,t}$ model the status of generating units when they are generating (1 if producing and 0 otherwise), while the binary variables $\overline{z_{g,t}}$ model the status of generating units when they are generating at maximum capacity (1 if true and 0 otherwise). Equations (9.15) and (9.16) model the logic of the binary variables. Equation (9.17) models the generators ramp time (τ_g), i.e. the sum of the differences $z_{g,t} - \overline{z_{g,t}}$ must equal τ_g. Equation (9.18) models the target ramping line, i.e., $t(\sum_g C_g)/N$, with N representing the total number of time periods. Finally, equations (9.19) and (9.20) represent variable declarations.

9.5.3 Example

The dynamic performance of MILP- and AGC-based approaches is evaluated by means of a modified New England 39-bus system (see Appendix B.2). The VPP is assumed to be composed of 10 small gas power plants, which are connected to buses 10-19. These DERs are equipped with conventional synchronous generators and PFC and voltage regulators. The data of the VPP can be found in Table B.14. In particular, a total capacity of VPP equal to 20% of the total load is considered. In order to simulate the performance of the VPP ramping response, it is assumed that at $t = 1$ s there is a 20% instantaneous load increase. Because of this contingency, it is assumed that the TSO sends a signal to the VPP to start the production and cover the increase of the load within 15 minutes.

Figure 9.21 shows the frequency response of the system as obtained using the MILP-based VPP and the AGC-based VPP. It can be seen that the AGC-based approach leads to a better frequency performance and, in particular, to a better frequency recovery. Observe that neither of the VPP approaches impact frequency nadir due to their slow response. As the MILP-based control starts up the generators one by one, starting from the cheapest one, until they reach the set-point sent by the TSO, is not able to track perfectly the reference frequency. Moreover, the MILP-based approach causes several "ripples" in the frequency, corresponding to the start-up of the units included in the VPP. The AGC-based VPP, on the other hand, starts-up the DERs all at the same time and then smoothly increase their power output. In the long-term, the AGC-based VPP leads to a better dynamic behavior than the MILP-based approach and achieves a null steady-state frequency error. These results suggest that the continuous control provided by the AGC-based approach is to be preferred from a dynamic performance point of view. As all units are started at the

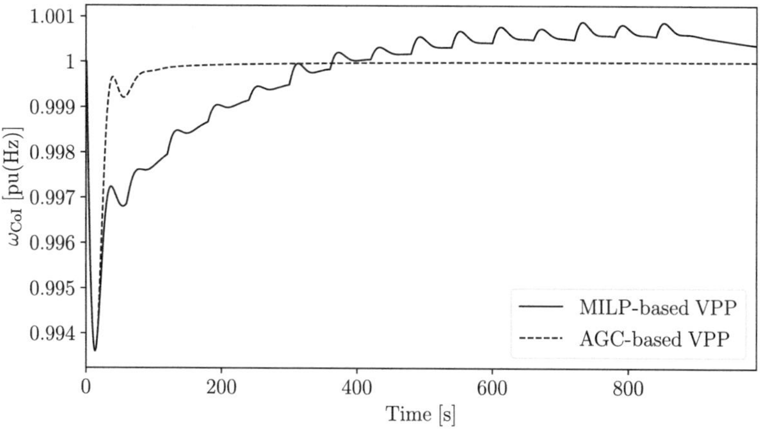

Fig. 9.21 Comparison of the frequency response of the AGC-based and MILP-based VPP

same time, it is possible that the AGC-based solution is more expensive than the MILP-based approach.

9.6 BESS and Long-Term Frequency Performance

The performance of the converter-interfaced BESS in mitigating the long-term oscillatory behavior of the frequency caused by the electricity market (see Fig. 9.9) is illustrated in this section. Such an application of the BESS is motivated by the fact that they can be used for many power system applications, including the safe integration of intermittent RES, PFC and SFC, voltage support, and RoCoF control, to name a few [20]. With this aim, in Fig. 9.8, we assume that BESS do not participate in the electricity markets, and are thus not part of the MAGC, but instead are paid for the frequency containment support that they provide. This is the situation in most of the jurisdictions worldwide where short-duration BESS are mainly used for ancillary/system services rather than for providing energy.

Here, we utilize the simplified BESS model presented in [21]. The BESS frequency control diagram is depicted in Fig. 8.6. In this case, the controller's input is the difference between the measured (ω) and reference frequency (ω^{ref}). The frequency is measured through a PLL device. Next, the BESS control scheme includes a dead-band and a LPF to make the controller less sensitive to small frequency deviations and filter out noises, respectively. Then, it includes a PI regulator with K_p being the proportional gain and K_i being the integral gain. In the control scheme shown in Fig. 8.6, the droop coefficient is obtained indirectly through the integral deviation coefficient $\mathcal{I}_d$. For $\mathcal{I}_d = 0$ the controller is a PI and hence the droop coefficient is $\mathcal{R} = 0$, whereas for $\mathcal{I}_d \neq 0$, the controller becomes a lead-lag filter, with droop coefficient $\mathcal{R} = \mathcal{I}_d/(K_i + K_p\mathcal{I}_d)$. Finally, the BESS is represented by an anti-windup first-order lag filter with time constant T_{ESS}.

9.6.1 *Example*

We discuss the effectiveness of BESS with an example based on a modified WSCC 9-bus system where a MAGC time interval of 120 s is considered. Furthermore, the contingency is a 10% instantaneous load increase (approximately 32 MW) at $t = 1$ s.

BESS devices are connected to different buses of the system, and their impact on the long-term frequency deviations is duly discussed. Three scenarios are considered, as follows:

- Without BESS limits;
- With BESS limits; and
- Comparison of the impact of droop and dead-band parameters.

All scenarios above are compared with the base case scenarion that does not include BESS (see Fig. 9.9).

9.6.1.1 BESS Without Capacity Limits

This first scenario discusses the effect of BESS on long-term power system dynamics without considering the capacity limits of the BESS. This scenario is relevant for MGs where the storage capacity can be relatively large with respect to the total installed power capacity. For simulation purposes, $\mathcal{I}_d = 0.7$ is used, and the dead-band is set to zero. Figure 9.22 shows the transient response of ω_{CoI} for three scenarios, namely with one BESS connected at bus 9, three BESS connected at buses 7, 8, and 9, and when no storage is included in the system.

Simulation results indicate that the BESS help reduce the frequency variations caused by large MAGC time intervals and, hence, considerably improve the system's dynamic performance. Moreover, the case with three BESS leads to significantly lower frequency variations than the case with one BESS. This result indicates that, from the dynamic point of view of the system, it is better to install BESS in different locations of the network rather than installing them in one location.

9.6.1.2 BESS with Capacity Limits

This section considers the same system and contingency as in the previous scenario but taking into account the BESS capacity limits. In particular, each BESS device is now assumed to have a 10 MW capacity. Thus, the case with three BESS represents almost the 10% instantaneous load increase (32 MW). Figure 9.23 depicts

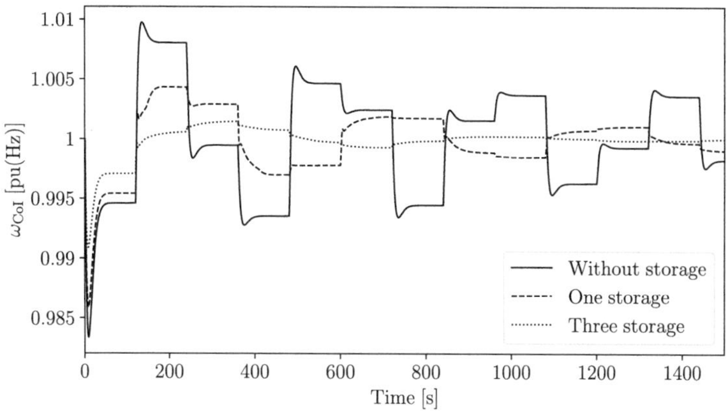

Fig. 9.22 Transient response of ω_{CoI} following a 10% instantaneous load increase without BESS limits

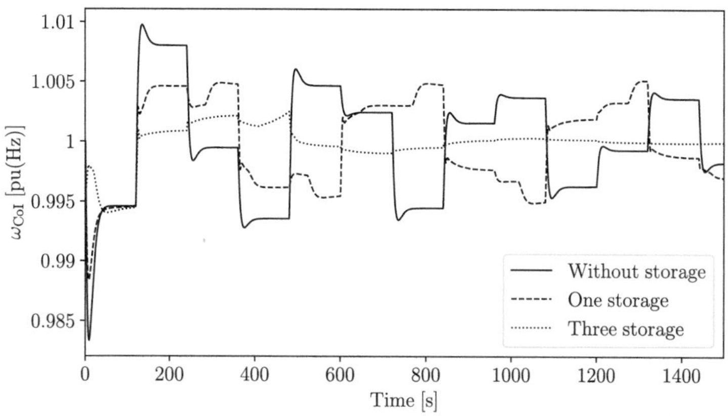

Fig. 9.23 Transient response of ω_{CoI} following a 10% instantaneous load increase with BESS limits

the transient response of ω_{CoI}. The case with three BESS leads to a similar dynamic behavior where no limits are considered (Fig. 9.22). In contrast, the case when one BESS is connected in the system is unable to solve the problem caused by the MAGC discretization. These results indicate that, as expected, the performance of the BESS is greatly impacted by its capacity.

9.6.1.3 Impact of Droop and Dead-Band

As discussed so far in the book, dead-band and droop coefficient are two critical parameters that can reduce the sensitivity of a controller to small frequency variations and noise. In the context of this chapter, we are interested to understand what is the

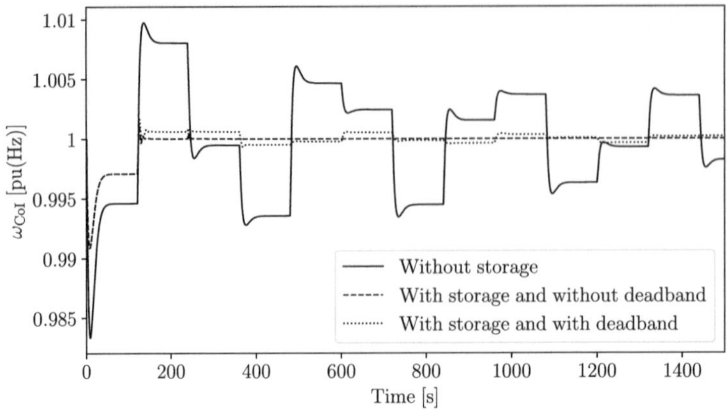

Fig. 9.24 Transient response of ω_{CoI} following a 10% instantaneous load increase with zero droop and without BESS limits

impact of these control parameters on the overall dynamic performance of the system. With this aim, four cases are considered, as follows:

1. Zero droop and without BESS limits.
2. Zero droop and with BESS limits.
3. Non-zero droop and without BESS limits.
4. Non-zero droop and with BESS limits.

When non-zero values are utilized, a dead-band value of $\mathrm{db} = 0.0006$ pu(Hz) and $\mathcal{I}_d = 0.7$ are used. Furthermore, only the case when three BESS connected at buses 7, 8, and 9 is considered below.

Zero Droop Without BESS *limits*
Figure 9.24 shows the trajectories of the ω_{CoI} for three scenarios, namely without storage, with storage and without dead-band, and with storage and with dead-band. The aim is to evaluate the impact of the dead-band on the frequency error when BESS capacity limits are not considered. Interestingly, the case with a non-zero dead-band leads to a worse dynamic behavior compared to the case without. This result leads to conclude that it is better not to include a dead-band on the frequency error.

Zero Droop With BESS Capacity Limits
The results of this scenario are shown in Fig. 9.25. If the BESS limits are taken into account and the droop is set to zero ($\mathcal{I}_d = 0$), the inclusion of the BESS does not necessarily improve the system's dynamic behavior. Moreover, introducing or not a dead-band on the frequency error does not improve the overall dynamic response of the system. These results indicate that a PI controller is not the most adequate set-up to mitigate the long-term frequency variations caused by the market.

Non-zero Droop Without BESS Capacity Limits
This scenario considers a system setup similar to the case with zero droop without BESS limits except for the value of the integral deviation coefficient assumed to be

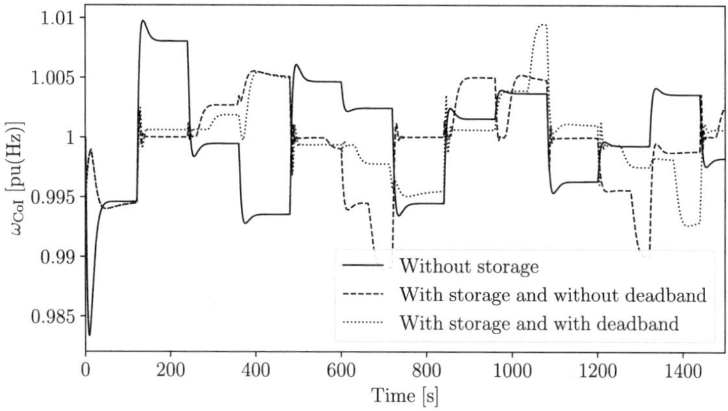

Fig. 9.25 Transient response of ω_{CoI} following a 10% instantaneous load increase with zero droop and with BESS limits

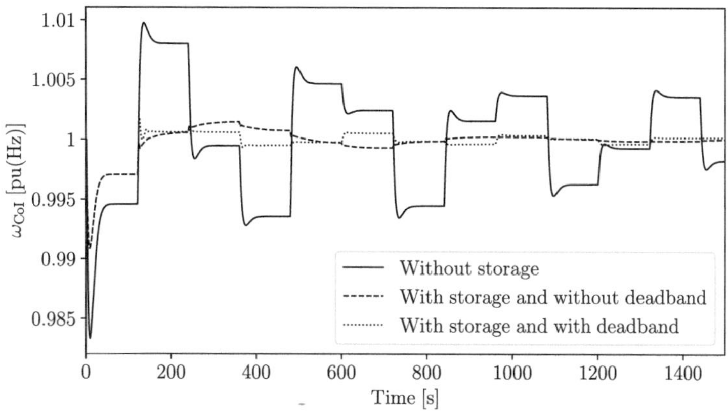

Fig. 9.26 Transient response of ω_{CoI} following a 10% instantaneous load increase with non-zero droop and without BESS limits

$\mathcal{I}_d = 0.7$. Figure 9.26 shows that the frequency deviations for this scenario are similar to those obtained in Fig. 9.24. Also, in this case, the inclusion of the dead-band on the frequency error leads to a worse dynamic behavior of the system as compared to the case without dead-band.

Non-zero Droop With BESS Capacity Limits

This last scenario considers BESS controllers with a non-zero droop and with and without a dead-band in a scenario where BESS capacity limits are binding.

Results shown in Fig. 9.27 allow concluding that, also in this scenario, the inclusion of the dead-band deteriorates the dynamic behavior of the system. Figure 9.28 shows the effect of the dead-band on the active power of an BESS. When dead-band is not zero, the BESS reaches its active power limit, which gives raise to a sort of

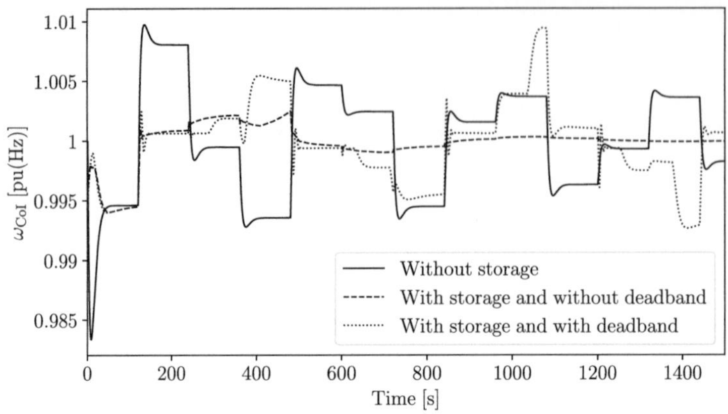

Fig. 9.27 Transient response of $\omega_{\rm CoI}$ following a 10% instantaneous load increase with non-zero droop and with BESS limits

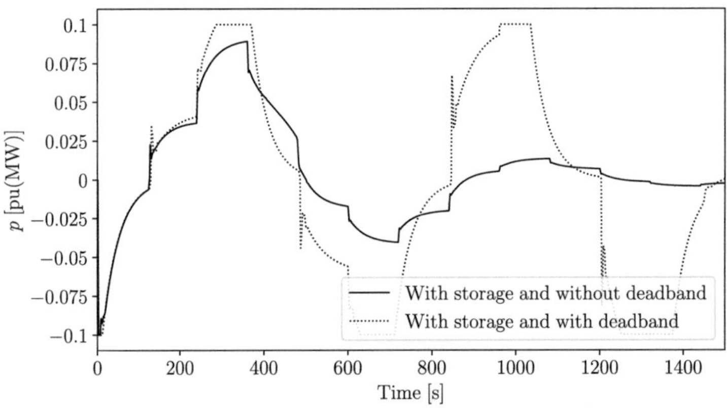

Fig. 9.28 Transient response of BESS active power following a 10% instantaneous load increase with non-zero droop and with BESS limits

bang-bang phenomenon. These results indicate that depending on the set-up of the real-time electricity market, the values of the droop and dead-band of the BESS frequency controller have to be carefully chosen.

9.7 Memory Effect of Market Participants

This section presents a short-term dynamic electricity market model with "memory effect." The model takes into account how the memory of market participants, namely suppliers and consumers, impacts on their bids. The memory effect is represented by employing proper mathematical tools from the theory of fractional calculus. The impact of the fractional-order model on system dynamics and, in particular, its interaction with a SFC, are studied by means of the WSCC 9-bus system.

9.7.1 Short-Term Electricity Markets

Short-term electricity markets such as balancing markets are becoming increasingly important as the integration of variable RES like wind and solar continues [22]. These markets offer flexibility to the power system by maintaining the power balance until physical generation and consumption. The time scale of these markets is comparable with those of the long-term power system dynamics [23]. Because of the similar time scales, there is a concern on the coupling between the dynamic response of the system and such markets [17]. The first studies that have looked at this problem date back to around two decades ago when Alvarado used first-order differential equations to describe the dynamics of a short-term electricity market model [24]. This model is an abstraction but allows studying the stability of the markets as well as interactions with the power grid [3]. Numerous works have used this model to date. For example, we cite [10, 25].

In this section, we discuss a particular dynamic model of the short-term electricity market that includes the memory effect in the behavior of the market participants such as suppliers and consumers [26]. Conventional models are all based on conventional integer-order time derivatives. These consider an infinitely small neighborhood of the "present" time, that is, the time at which the derivatives are computed [27]. As a consequence, the processes that are modeled through conventional differential equations have infinitely fast "amnesia." However, taking into account the memory of market participants is of utmost importance in economic processes as they can keep records of past economic indicators and events [28]. This information can be used to modify their behavior and take decisions.

An effective and powerful tool to model the memory effects of a dynamic system is fractional calculus, which deals with the analysis of non-integer order differentials and integrals [29]. The ability of fractional derivatives to capture physical processes better than traditional integer-order derivatives has been shown for a number of systems in physics and engineering [30]. For example, Ref. [31] provides a review of the application of fractional calculus in science and engineering. The recent work in [32] proposes a new economic model of the price dynamics of goods that takes into account the memory of the market agents. Using the work in [32] as a source of inspiration, we take into account for the first time the memory of the participants in power system markets.

9.7.2 Economic Processes with Memory

A mathematical tool that allows modeling the memory effects of market agents is fractional-order differential equations. The recent work [32] extends the well-known Evans' model, which describes price dynamics of goods [33], to take into account the memory of market participants. Based on this theoretical background, this section shows the mathematical steps that lead to the fractional market model that we discuss in the examples.

The Evans' model of price dynamics is as follows:

$$\lambda'(t) = -\gamma Q'(t)\,, \tag{9.21}$$

where λ represents the price of goods; γ is a proportional coefficient that represents the speed of response; and Q represents the stocks. The change in stocks is defined by the following differential equation:

$$Q'(t) = S(t) - D(t)\,, \tag{9.22}$$

where S and D represent the supply and demand for goods, respectively. Using Eqs. (9.22), (9.21) can be rewritten as:

$$\lambda'(t) = -\gamma(S(t) - D(t))\,. \tag{9.23}$$

The supply and demand are described by the following equations:

$$\begin{aligned} S(t) &= c_s + b_s\lambda(t)\,, \\ D(t) &= d_d + a_d\lambda(t)\,, \end{aligned} \tag{9.24}$$

where a_d, b_s, c_s, d_d are constant parameters. In particular, c_s and d_d represent the supply and demand, respectively, and do not depend on the price λ. In general, it is assumed that $a_d < 0$ and $b_s > 0$. Using equation (9.24), equation (9.23) of price dynamics can be rewritten, as follows:

$$\lambda'(t) + \gamma(b_s - a_d)\lambda(t) = \gamma(d_d - c_s)\,. \tag{9.25}$$

Equation (9.25) is a first-order differential equation. It cannot account for the memory of the market participants. With this aim, one has to consider not only the difference $S(t) - D$, but also the "history" of changes of the differences $S(\tau) - D(\tau)$ on a finite time interval $\tau \in [0, t]$. Such a dependence of λ can be described by the following equation:

$$\begin{aligned} \lambda'(t) &= -\int_0^t \gamma(t-\tau)Q'(\tau)d\tau \\ &= -\int_0^t \gamma(t-\tau)(S(\tau) - D(\tau))d\tau\,, \end{aligned} \tag{9.26}$$

where γ represents the memory function. Assuming a power-law fading memory, the function γ can be written as follows:

$$\gamma(t-\tau)=\frac{\gamma}{\Gamma(\alpha)}(t-\tau)^{\alpha-1}\,, \tag{9.27}$$

where $\Gamma(\alpha)$ is the gamma function; $0<\alpha<1$ is the fractional-order; and $t>\tau$. Using (9.27), (9.26) can be rewritten as:

$$\begin{aligned}\lambda'(t)&=-\gamma\left(I^{\alpha}_{RL;0+}Q^{(1)}\right)(t)\\&=-\gamma\left(I^{\alpha}_{RL;0+}(S-D)\right)(t)\,,\end{aligned} \tag{9.28}$$

where $I_{RL;0+}$ is the Riemann-Liouville fractional integral defined as:

$$I^{\alpha}_{RL;0+}f(t)=\frac{1}{\Gamma(\alpha)}\int_0^t(t-\tau)^{\alpha-1}f(\tau)d\tau\,, \tag{9.29}$$

where the function $f(\tau)$ is measurable on the interval $(0,t)$ and has the property $\int_0^t|f(\tau)|d\tau<\infty$. Since we are interested to obtain a differential equation for the price dynamics, we act on Eq. (9.28) by the left-sided Caputo derivative of order $\alpha>0$ defined as follows:

$$\frac{d^{\alpha}}{dt^{\alpha}}_{C;0+}f(t)=\frac{1}{\Gamma(n-\alpha)}\int_0^t(t-\tau)^{n-\alpha-1}f^{(n)}(\tau)d\tau\,, \tag{9.30}$$

and obtain the following fractional differential equation:

$$\frac{d^{\alpha+1}}{dt^{\alpha+1}}_{C;0+}\lambda(t)=-\gamma(S(t)-D(t))\,, \tag{9.31}$$

where $n-1<\alpha+1\le 1$ and $n\in\mathbb{N}$. It can be seen that for $\alpha=0$, Eq. (9.31) takes the form of the equation (9.25), while for $\alpha=1$, Eq. (9.31) takes the form of the second-order differential equation of the Evans model [32].

9.7.3 Dynamic Electricity Market with Memory Effect

The short-term electricity market model (9.4)–(9.6) neglects the memory of market participants as it uses first-order integer derivatives. On the other hand, the previous section shows that the memory of market participants can be taken into account using the mathematical tool of fractional calculus. Motivated by this theoretical background, we consider the following fractional-order version of the dynamic electricity market model (9.4)–(9.6):

$$T_\lambda\, \lambda^{(\alpha)}(t) = K_E(\omega^{\text{ref}} - \omega_{\text{CoI}}(t)) - \lambda(t)\,, \tag{9.32}$$

$$T_{g,i}\, \Delta p'_{g,i}(t) = \lambda(t) - c_{g,i}\, \Delta p_{g,i}(t) - b_{g,i}\,, \quad i = 1, \ldots, n_g\,, \tag{9.33}$$

$$T_{d,i}\, \Delta p'_{d,i}(t) = b_{d,i} + c_{d,i}\, \Delta p_{d,i}(t) - \lambda(t)\,, \quad i = 1, \ldots, n_d\,, \tag{9.34}$$

where $0 < \alpha < 1$ is the fractional order and we have used the notation:

$$\lambda^{(\alpha)}(t) = \frac{d^\alpha}{dt^\alpha}\lambda(t)\,. \tag{9.35}$$

The new market model (9.32)–(9.34) accounts for the overall memory of generators and loads through the fractional equation (9.32) of the price dynamics.

9.7.3.1 Oustaloup's Recursive Approximation

In order to implement or simulate in practice the fractional market model (9.32)–(9.34), one needs to approximate the fractional dynamics, in this case equation (9.32), with rational order transfer functions [34]. In this work, we select the Oustaloup's Recursive Approximation (ORA) method to approximate the fractional electricity price dynamics. The generalized ORA of a fractional derivative of order α is defined as [35]:

$$s^\alpha \approx \omega_h^\alpha \prod_{k=1}^{N} \frac{s + \tilde{\omega}_k}{s + \omega_k}\,, \tag{9.36}$$

where $\tilde{\omega}_k = \omega_b \omega_v^{(2k-1-\alpha)/N}$, $\omega_k = \omega_b \omega_v^{(2k-1+\alpha)/N}$, $\omega_v = \sqrt{\omega_h/\omega_b}$. In the above expressions, $[\omega_b, \omega_h]$ is the frequency range for which the approximation is designed to be valid; N is the order of the polynomial approximation. The term "generalized" means that, in (9.36), N can be either even or odd [35], while the term "recursive" implies that the values of $\tilde{\omega}_k$, ω_k result from a set of recursive equations [36]. The block diagram of ORA is shown in Fig. 9.29. Further details on the ORA method and its accuracy can be found in [34] and references therein .

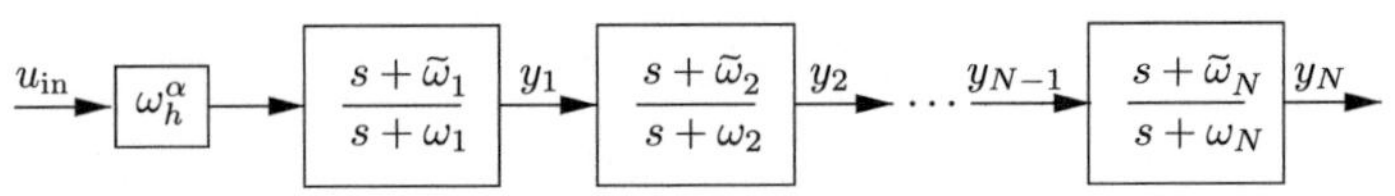

Fig. 9.29 Oustaloup's recursive approximation block diagram

9.7.4 Examples

In this section, we study the dynamic behavior of the two dynamic market models introduced in Sects. 9.3.1 and 9.7.3, namely, conventional integer-order MAGC (I-MAGC) (9.4)–(9.5) and fractional-order MAGC (F-MAGC) (9.32)–(9.33). The objective is to evaluate the impact of these models on the behavior of market participants, and on the overall dynamic response of the power system [12]. With this aim, we first discuss the impact of I-MAGC and illustrate the dynamic coupling between market and AGC in Sect. 9.7.4.1. Section 9.7.4.2 discusses a sensitivity analysis with respect to the fractional-order α of the F-MAGC and compares results with the I-MAGC. Both Sects. 9.7.4.1 and 9.7.4.2 consider a 10% sudden load increase. The dynamic response of both I-MAGC and F-MAGC following a 10% sudden load decrease is presented in Sect. 9.7.4.3.

9.7.4.1 Impact of the Frequency of Price Updates

Some long-term power system dynamics, in particular the dynamics of AGC, evolve with a time scale similar to today's short-term market dynamics [23]. For this reason, it is important to understand how the frequency with which the market price is updated impacts on the decision-making process of market participants and on power system dynamics. In the continuous market models considered in this chapter, the information on how often the price is updated is contained in the value of the gain K_E in (9.4). Hence, this section presents a sensitivity analysis with respect to the variations of the K_E.

Figure 9.30 shows that the value of K_E has a negligible impact on the overall dynamic of the system. The frequency nadir, in fact, is the same in all cases. This was expected as the MAGC is slow with respect to the PFC. Figure 9.31, on the other hand, shows that the schedule of generator active power are by the value of K_E. Specifically, the faster the price updates and the higher K_E, the faster the generator response and consequently the higher the generator schedules. This phenomenon is called *price chasing* [37]. These results indicate that how often the market updates the price (which in this continuous model is modeled by means of K_E) impacts the schedule of the suppliers or generators.

The trajectories of the AGC set-point Δp_1 of generator 1 are shown in Fig. 9.32. Higher gain values—and hence faster price updates—lead to faster AGC response and lower AGC set-points. This has to be expected as the AGC has to compensate the difference in the market schedules since at the end the total power output of the generator has to be the same. These results imply that, depending on the market design and rewards of the ancillary services, generators may prefer to compensate power unbalances through the short-term market or through the SFC.

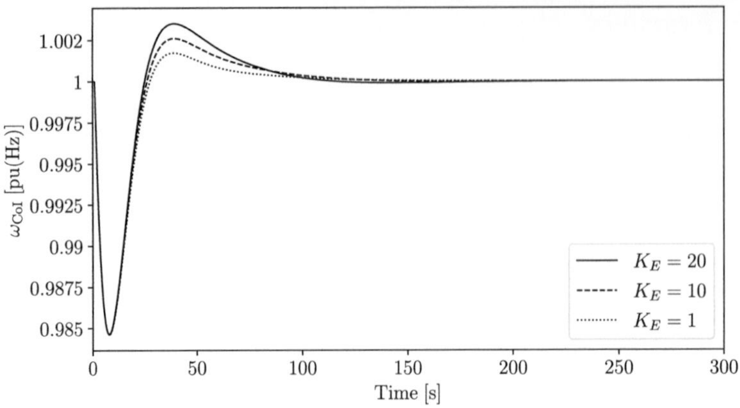

Fig. 9.30 Trajectories of the frequency of the CoI

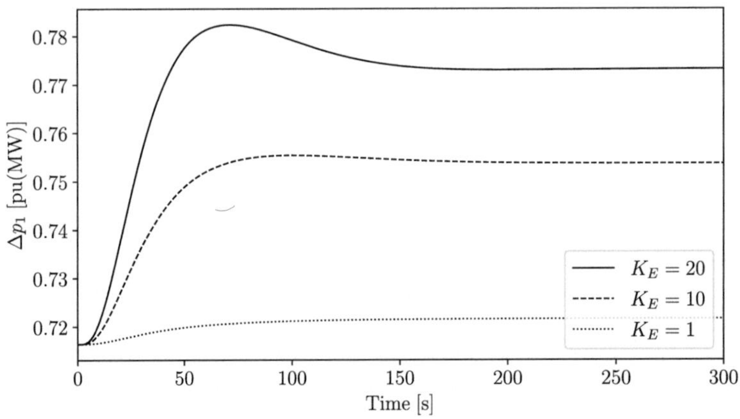

Fig. 9.31 Trajectories of the MAGC active power schedules of generator 1

9.7.4.2 Sensitivity Analysis

This section presents a sensitivity analysis with respect to the fractional-order α of the F-MAGC and compares the results with the ones obtained using the conventional integer-order market model. For fair comparison, all other parameters, namely time constants and gains, of both market models, (9.4)–(9.5) and (9.32)–(9.33), are kept the same.

First we compare the impact of I-MAGC and F-MAGC on power system dynamics. Figure 9.33 shows the trajectories of ω_{CoI} for both models. Both the I-MAGC and F-MAGC lead to the same frequency nadir and very similar frequency overshoots. The memory of market participants, thus, does not have a relevant impact on the overall power system dynamics. These results are consistent with those shown in Fig. 9.30.

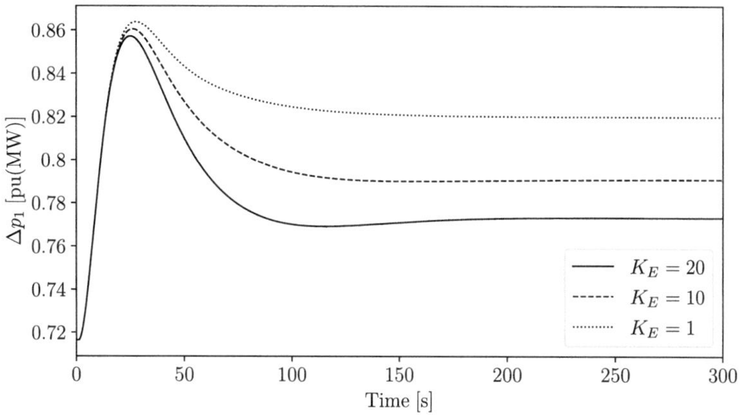

Fig. 9.32 Trajectories of the AGC active power set-point of generator 1

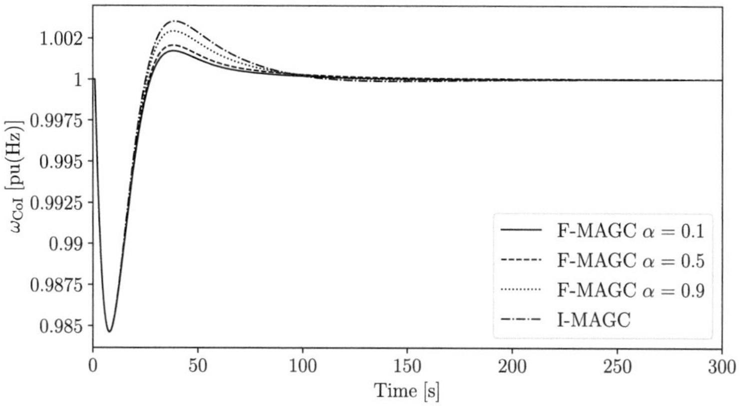

Fig. 9.33 Comparison of the trajectories of $\omega_{\rm CoI}$ as obtained with I-MAGC and F-MAGC

Next we compare the impact that different values of α have on the behavior of the generators. Figure 9.34 shows that the F-MAGC leads to different (in this case, lower) market schedules as compared to that of the I-MAGC. This result suggests that the F-MAGC is less prone to the price changes. In other words, taking into account the memory of market participants makes them more conservative. This conclusion is supported by Fig. 9.35. This figure shows that the AGC set-point for the fractional market is less prone to changes compared to the conventional market. Furthermore, the higher the fractional order α, the faster the generator response, and consequently the higher the generator market schedules.

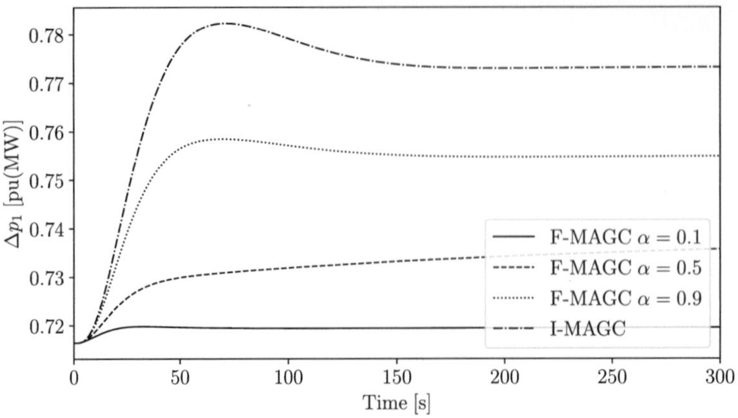

Fig. 9.34 Trajectories of the MAGC active power schedules of generator 1

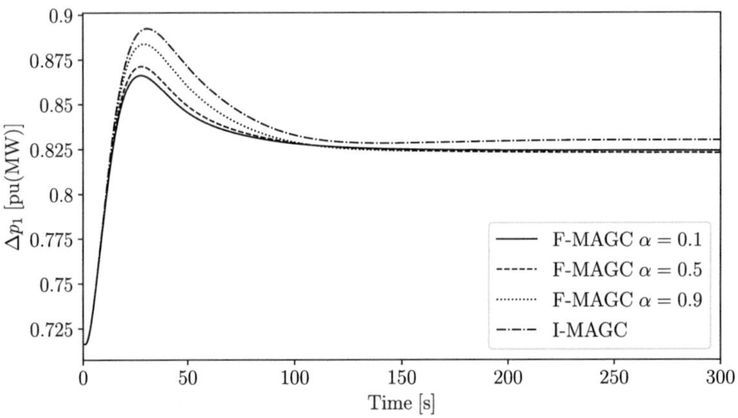

Fig. 9.35 Trajectories of the AGC active power set-point of generator 1

9.7.4.3 Impact of a Sudden Load Decrease

In this final example, we compare the impact on the performance of I-MAGC and F-MAGC of a 10% sudden load decrease occurring at $t = 1$ s.

Figure 9.36 shows that the F-MAGC is again less prone to price changes compared to the I-MAGC. For the considered contingency, such a behavior leads the market to schedule higher generator powers.

Figure 9.37 shows the AGC power output and indicates that the I-MAGC case responds faster than the F-MAGC to the contingency. This result is consistent with that obtained in the previous section. The memory effect makes the market participants less sensitive to changes in the operating point of the grid. This conservativeness, however, has to be compensated, at least in the short term, by the SFR.

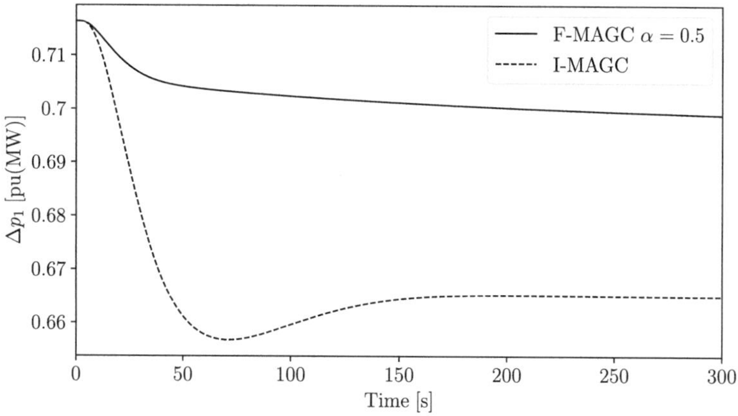

Fig. 9.36 Trajectories of the MAGC active power schedules of generator 1

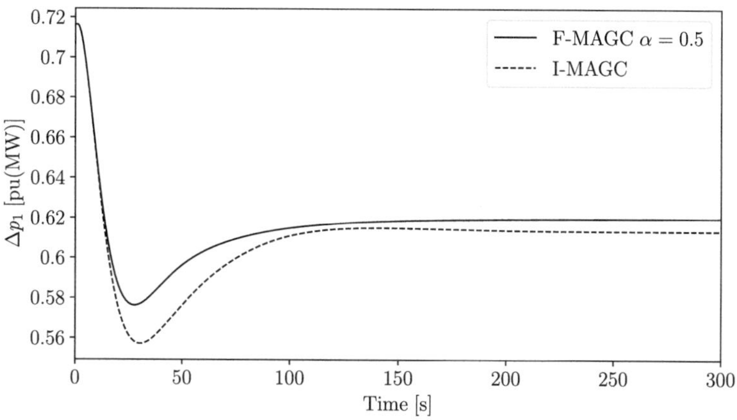

Fig. 9.37 Trajectories of the AGC active power set-point of generator 1

In Brief

- Increasing too much the execution cycles of the AGC leads to a limit cycle or bang-bang phenomena. The most effective solution is to keep as short as possible the AGC execution cycles.
- If future real-time electricity markets are based on the power imbalance to update the electricity price and use dispatch periods that range from some tens of seconds up to few minutes, then these markets might lead to limit cycles and instability.
- Integrating wind power generation into real-time electricity markets may deteriorate the performance of the AGC and, consequently, the overall power system dynamic performance.

- BESS play an essential role in reducing the frequency deviations caused by the market. However, BESS frequency control parameters, namely the droop coefficient and the dead-band on the frequency error, have to be carefully chosen not to deteriorate the system's overall dynamic performance.
- Fractional-order derivatives can be utilized to model the "memory effect" of controllers.
- Fractional-order market models lead to a conservative behavior of market participants, and can improve the overall dynamic performance of the grid.

Open Questions

The focus of this chapter and of the whole book, in fact, is on frequency. But it should not be forgotten that voltage requires regulation. Secondary voltage regulation is way less common than SFC, but it has been implemented with some success in some countries, such as, Italy and France. Moreover, secondary controllers for DERs and MGs are an emerging topic. However, the impact of these discrete secondary controllers on the dynamic performance of the system has not been fully studied yet.

A related aspect that requires more study concerns the utilization of energy storage systems, such as BESS to achieve overall best short- and long-term frequency performance as well as for load leveling and shifting.

Finally, we believe that the concept of controllers with "memory," is intriguing, may have useful practical applications in power systems, and is definitively worth more research.

References

1. Bevrani H (2009) Robust power system frequency control. Springer
2. AEMO (2018) AEMO annual report 2018—leading the transformation. www.aemo.com.au
3. Alvarado FL, Meng J, DeMarco CL et al (2001) Stability analysis of interconnected power systems coupled with market dynamics. IEEE Trans Power Syst 16(4):695–701
4. Zhang C, Jiang L, Wu QH et al (2013) Further results on delay-dependent stability of multi-area load frequency control. IEEE Trans Power Syst 28(4):4465–4474
5. Xu L, Guo Q, Wang Z et al (2021) Modeling of time-delayed distributed cyber-physical power systems for small-signal stability analysis. IEEE Trans Smart Grid 1
6. Shangguan X, Zhang CK, He Y et al (2020) Robust load frequency control for power system considering transmission delay and sampling period. IEEE Trans Ind Inf 1
7. Venkateswaran R, Joo YH (2021) Retarded sampled-data control design for interconnected power system with DFIG-based wind farm: LMI approach. IEEE Trans Cybern 1–11
8. Fridman E (2010) A refined input delay approach to sampled-data control. Automatica 46(2):421–427
9. Gao H, Xue S, Yin S et al (2017) Output feedback control of multirate sampled-data systems with frequency specifications. IEEE Trans Control Syst Technol 25(5):1599–1608

10. Stegink T, De Persis C, van der Schaft A (2017) A unifying energy-based approach to stability of power grids with market dynamics. IEEE Trans Autom Control 62(6):2612–2622
11. Ferraro P, Crisostomi E, Raugi M et al (2017) Analysis of the impact of microgrid penetration on power system dynamics. IEEE Trans Power Syst 32(5):4101–4109
12. Kërçi T, Murad MAA, Dassios I et al (2021) On the impact of discrete secondary controllers on power system dynamics. IEEE Trans Power Syst. In press
13. EPEX SPOT (2020) Trading on EPEX spot 2019–2020. www.epexspot.com
14. Nutaro J, Protopopescu V (2009) The impact of market clearing time and price signal delay on the stability of electric power markets. IEEE Trans Power Syst 24(3):1337–1345
15. Rebours YG, Kirschen DS, Trotignon M et al (2007) A survey of frequency and voltage control ancillary services—Part I: technical features. IEEE Trans Power Syst 22(1):350–357
16. EirGrid and SONI (2024) Balancing market principles statement. A guide to scheduling and dispatch in the single electricity market. https://www.eirgrid.ie/
17. Kërçi T, Giraldo JS, Milano F (2020) Analysis of the impact of sub-hourly unit commitment on power system dynamics. Electr Power Energy Syst 119:105819
18. Barquín J (2014) The new Spanish electricity market design. In: Enerday 2014–9th conference on energy economics and technology, pp 1–10
19. Beagon P, Bustamante MD, Devine MT et al (2018) Optimal scheduling of distributed generation to achieve linear aggregate response. In: Proceedings from the 141st European study group with industry. http://www.maths-in-industry.org/miis/758/
20. Milano F, Ortega Á (2019) Converter-Interfaced Energy Storage Systems: Context. Cambridge University Press, Modelling and Dynamic Analysis
21. Ortega Á, Milano F (2016) Generalized model of VSC-based energy storage systems for transient stability analysis. IEEE Trans Power Syst 31(5):3369–3380
22. Brijs T, De Jonghe C, Hobbs BF et al (2017) Interactions between the design of short-term electricity markets in the CWE region and power system flexibility. Appl Energy 195:36–51
23. Wang Q, Zhang C, Ding Y et al (2015) Review of real-time electricity markets for integrating distributed energy resources and demand response. Appl Energy 138:695–706
24. Alvarado FL (1997) The dynamics of power system markets. University of Wisconsin-Madison, Pserc Research Report, pp 97–01
25. Chiu W, Sun H, Poor HV (2013) Energy imbalance management using a robust pricing scheme. IEEE Trans Smart Grid 4(2):896–904
26. Dassios I, Kërçi T, Baleanu D et al (2023) Fractional-order dynamical model for electricity markets. Math Methods Appl Sci 46(7):8349–8361
27. Tarasov VE (2018) Generalized memory: fractional calculus approach. Fractal Fract 2(4):23
28. Tarasov VE (2019) On history of mathematical economics: application of fractional calculus. Mathematics 7(6):509
29. Tarasova VV, Tarasov VE (2018) Concept of dynamic memory in economics. Commun Nonlinear Sci Numer Simul 55:127–145
30. Baleanu D, Machado JAT, Luo AC (2011) Fractional dynamics and control. Springer Science and Business Media
31. Sun H, Zhang Y, Baleanu D et al (2018) A new collection of real world applications of fractional calculus in science and engineering. Commun Nonlinear Sci Numer Simul 64:213–231
32. Tarasov VE (2020) Fractional econophysics: market price dynamics with memory effects. Physica A 557:124865
33. Evans GC (1930) Mathematical introduction to economics. McGraw-Hill, New York
34. Tzounas G, Dassios I, Murad MAA et al (2020) Theory and implementation of fractional order controllers for power system applications. IEEE Trans Power Syst 35(6):4622–4631
35. Monje CA, Chen Y, Vinagre BM et al (2010) Fractional-order systems and controls. Springer, Fundamentals and Applications
36. Oustaloup A, Levron F, Mathieu B et al (2000) Frequency-band complex noninteger differentiator: characterization and synthesis. IEEE Trans Circuits Syst I 47(1):25–39
37. Watts D, Alvarado FL (2004) The influence of futures markets on real time price stabilization in electricity markets. In: Proceedings of the 37th annual Hawaii international conference on system sciences, pp 1–7

Chapter 10
Sub-Hourly Unit Commitment

This chapter focuses on market-driven imbalances and their impact on (long-term) frequency quality. The market-driven imbalances are studied by means of different real-time sub-hourly unit commitment models and strategies. As discussed in Chap. 3, these are still a major concern for TSOs. The impact is mostly quantified using the concept of DFDs and the standard deviation of the frequency of the CoI, σ_{CoI}.

Section 10.1 provides a comparison between self and central dispatch systems. Sections 10.2 and 10.3 comprehensively study the impact of real-time sub-hourly deterministic Unit Commitment (d-UC) and s-UC on long-term frequency quality, respectively. The impact is evaluated through a software framework developed by the authors. Section 10.4 provides final remarks.

10.1 Self Versus Central Dispatch Systems

Before studying the impact of different real-time sub-hourly UC approaches on power system frequency quality, it is convenient to provide an overview of the differences between self-dispatch and central dispatch systems. While the discussion below is based on an European context, the main principles and differences between the two different dispatch approaches remain the same.

According to EU legislation on electricity balancing a self-dispatching model is defined as [1]:

A scheduling and dispatching model where the generation schedules and consumption schedules as well as dispatching of power generating facilities and demand facilities are determined by the scheduling agents of those facilities.

While a central dispatching model is defined as:

A scheduling and dispatching model where the generation schedules and consumption schedules as well as dispatching of power generating facilities and demand

T. Kërçi and F. Milano, *Frequency Quality in Power Systems*, Power Systems,
https://doi.org/10.1007/978-3-032-04721-2_10

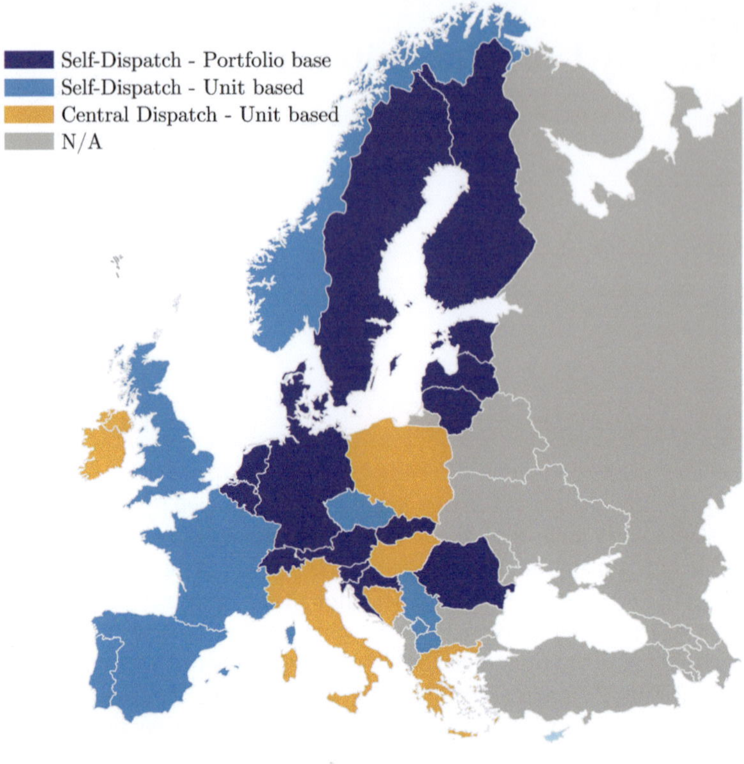

Fig. 10.1 Types of dispatch in Europe

facilities, in reference to dispatchable facilities, are determined by a TSO within the integrated scheduling process.

In this context, Fig. 10.1 shows the dispatch models implemented by different European jurisdictions. Note that GB is currently assessing the impacts of moving to a central dispatch model [2].

Figure 10.2 illustrates the differences between the two different approaches applied in different jurisdictions in Europe, for example, in the AIPS (central-dispatch) and Germany (self-dispatch).

The main difference between the two different approaches is the fact that in central dispatch systems the balancing, congestion management and reserve scheduling are performed simultaneously in an integrated process [3]. This reference also states that: (i) central dispatch is preferred in systems where the impact of market-driven imbalances is a material threat to security; and (ii) countries that implement central dispatch such as Italy, Greece, Hungary, Poland and IE and NI need to ensure system security as well as minimize the cost of energy delivery to the end consumer. However, we think that the latter aspect (minimizing cost) is something that all jurisdictions

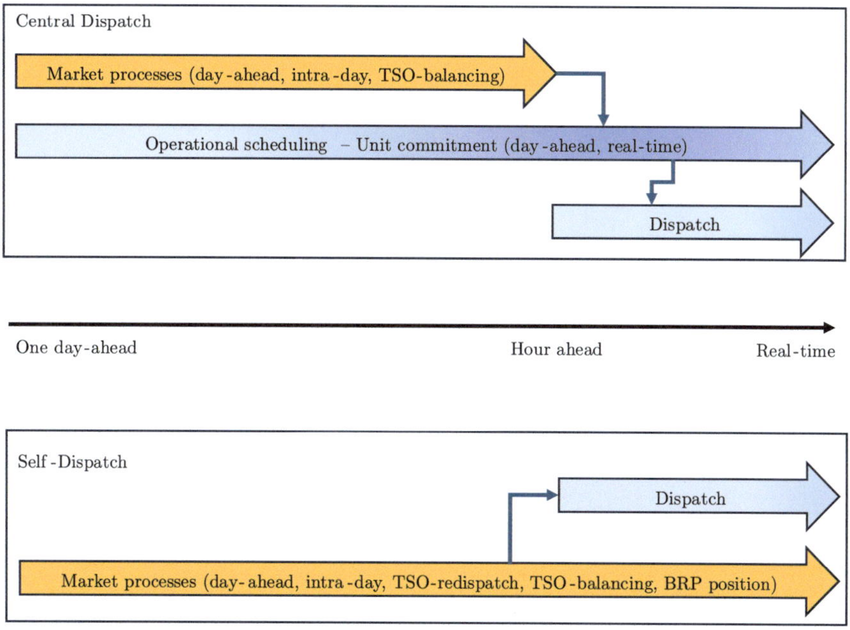

Fig. 10.2 Differences between central and self-dispatch systems

aim at achieving and, thus, the choice between self-dispatch and central dispatch seems more due to legacy reasons.

Intuitively, in a self-dispatch world market participants or, in a more technical term, BRPs, are responsible for the scheduling and dispatch of their units until real-time.[1] BRPs are responsible financially for any imbalances they cause in the system. If there are remaining imbalances in real-time, TSOs will re-actively balance the system through the balancing energy market/services, for example, by activating aFRR and Automatic Generation Control (AGC) bids, to maintain the frequency and to secure a stable operation. Some TSOs (Germany) also utilize a redispatch market if there are system constraints or congestion in their grids. Note that this is needed due to the fact that current energy markets are designed to match demand and supply but not to address technical requirements of a power system. In other words, current wholesale energy markets are unconstrained. Therefore, it can be said that self-dispatch systems heavily rely on market participants to solve the power imbalances.

On the other hand, in central-dispatch systems, relevant market outputs such as market schedules, namely, Physical Notifications (PNs), serve as an input to the TSOs operational scheduling (UC) and dispatch processes which have a time-horizon from day-ahead to real-time. The latter means that TSOs have full control of all market

[1] TSOs monitor BRP positions and can ask for clarification and corrective actions from BRPs if there will be large systematic deviations in realized production compared with the plans [4].

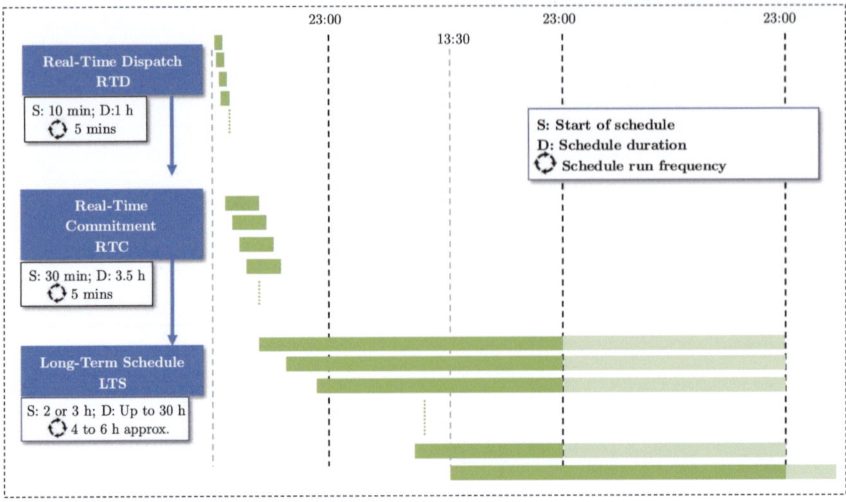

Fig. 10.3 Scheduling and dispatch process in the AIPS [5]

units from day-ahead until real-time, for example, see Fig. 10.3 which shows an outline of the scheduling and dispatch process implemented in the AIPS. We provide below a short description of each process [5]:

- *Long-Term Schedule (LTS).* LTS uses a security-constrained UC (SCUC) to produce UC advice up to 30 h ahead. The horizon reduces as the time between the run and the end of the trading day reduces. LTS is run manually by a control room operator, and the exact timing of each run can vary depending on the availability of key input data.
- *Real-Time Commitment (RTC).* Similar to LTS, RTC uses a SCUC to produce UC advice from close to real-time (from 30 min) to four hours ahead (producing a schedule which covers 3.5 h).
- *Real-Time Dispatch (RTD).* As the name suggests, RTD is a real-time market dispatch tool that uses a security-constrained economic dispatch to produce MW dispatch advice for control room operators based on real-time system conditions and forecasts for an hour from close to real-time (10 min).

The objective of TSOs is match demand and supply through the above UC and dispatch processes and, at the same time, minimize deviations from the market schedules and meet any technical requirements including network and system constraints. TSOs are thus regularly solving active power imbalances in a central-dispatch system. Moreover, there does not exist a redispatch market in central-dispatch systems as the TSOs scheduling and dispatch processes include both network and system constraints. In other words, TSOs (re)-dispatch the system in real-time for both energy and congestion reasons. The interested reader is referred to [2] for a detailed comparison between self and central-dispatch systems.

In this chapter, we focus on central-dispatch systems and, in particular, study the impact interactions between real-time sub-hourly UC and frequency variations.

10.2 Real-Time Sub-Hourly d-UC

10.2.1 Need for Sub-Hourly Modeling

The UC problem plays a crucial role in the secure operation of power systems. Most of the electric utilities in the US clear the markets using a d-UC [6]. The level of system security in these models is mainly defined by network constraints equations and the amount of scheduled reserves. Traditionally, these models are solved with an hourly time period [7]. This time scale is well decoupled from relevant power system dynamics. However, a sub-hourly UC is to be preferred in systems with high penetration of variable RES as these increase variability and uncertainty and at the same time decrease the overall inertia of the system [8].

Furthermore, sub-hourly scheduling is a way to increase flexibility without investing in physical assets [9]. Reference [10] shows that sub-hourly modeling allows to better capture the costs and the ramping capability of generators. A more recent work demonstrates that sub-hourly dispatch results in lower costs and lower reserves [11].

A sub-hourly UC with, say, 5-min resolution, can overlap with long-term dynamics. There exist several attempts to include simplified dynamic constraints into the UC problem, for example [12, 13]. However, the main limitation of these approaches is that the dynamics of the system are oversimplified and linearized. On the other hand, embedding the UC problem into a fully-fledged transient stability analysis software tool has not been thoroughly discussed so far.

10.2.2 Deterministic Unit Commitment

The main objective of the UC problem is to minimize the total operating cost and determine for a given planning horizon the ON/OFF status of the generating units needed to match the demand. It is common practice to formulate the UC as a MILP problem thanks to the good performance of the available commercial solvers [14]. In this section, a conventional MILP d-UC problem is used [15]. The mathematical formulation is given below.

Objective Function
The objective function of the d-UC to be minimized consists of the fixed, variable, start-up and shut-down costs of the generating units, as follows:

$$\sum_{t\in\Omega_{\mathrm{T}}}\sum_{g\in\Omega_{\mathrm{G}}}(C_g^F z_{g,t}^F + C_g^V p_{g,t} + C_g^{SU} z_{g,t}^{SU} + C_g^{SD} z_{g,t}^{SD}) \tag{10.1}$$

where t is the index for the time period; g is the index for the generating units; Ω_{T} is the set of time periods, $\{1, \ldots, 24\}$ hours; Ω_{G} is the set of generating units; $z^F_{g,t}$ is the binary variable that represents the status of the units in time period t (1 if ON and 0 if OFF); $z^{SU}_{g,t}$ and $z^{SD}_{g,t}$ are the binary variables that represent the status of the units at the beginning of time period t, i.e., $z^{SU}_{g,t} = 1$, $z^{SD}_{g,t} = 0$ if the generator is up and $z^{SU}_{g,t} = 0$, $z^{SD}_{g,t} = 1$ is the generator is down; and $p_{g,t}$ is the continuous variable representing the active power production during time period t.

Binary Variable Constraints
These constraints are needed to ensure the consistency of the logic of binary variables. For example, if a unit is ON in a given period t, then it can only be switched OFF but not started-up in the following period. The constraints are:

$$z^{SU}_{g,t} - z^{SD}_{g,t} = z^F_{g,t} - z^F_{g,t-1}, \quad \forall g \in \Omega_{\mathrm{G}}, \forall t \in \Omega_{\mathrm{T}}, \tag{10.2}$$

$$z^{SU}_{g,t} + z^{SD}_{g,t} \leq 1, \quad \forall g \in \Omega_{\mathrm{G}}, \forall t \in \Omega_{\mathrm{T}}, \tag{10.3}$$

$$z^F_{g,t}, z^{SU}_{g,t}, z^{SD}_{g,t} \in \{0, 1\} \quad \forall g \in \Omega_{\mathrm{G}}, \forall t \in \Omega_{\mathrm{T}} \tag{10.4}$$

Note that the sum of the start-up and shut-down binary variables cannot be greater than one. Note also that for the first time period, the initial status of the unit, namely $z^F_{g,0}$ has to be known in (10.2) and is thus an input datum. When the model steps forward, the status of the units at the end of the horizon serve as an initial status for the next planning horizon, and so on.

Power Bounds
Typical technical constraints of the generating units include their upper and lower limits:

$$P^{\min}_g z^F_{g,t} \leq p_{g,t} \leq P^{\max}_g z^F_{g,t}, \quad \forall g \in \Omega_{\mathrm{G}}, \forall t \in \Omega_{\mathrm{T}}, \tag{10.5}$$

where, $P^{\min}_g$ and $P^{\max}_g$ are the minimum and maximum active power limits, respectively.

Ramping Limits
Other important constraints of generation units are the so-called ramping limits. For instance, between two successive time periods, a unit output is bounded by a maximum value called the ramping-up limit. Also, a start-up limit is applied to the unit power output during its start-up time. Similar considerations apply to the ramping-down and shut-down ramping limits. Ramping limit constraints are:

$$p_{g,t} - p_{g,t-1} \leq R^U_g z^F_{g,t-1} + R^{SU}_g z^{SU}_{g,t}, \ \forall g \in \Omega_{\mathrm{G}}, \forall t \in \Omega_{\mathrm{T}}, \tag{10.6}$$

$$p_{g,t-1} - p_{g,t} \leq R^D_g z^F_{g,t} + R^{SD}_g z^{SD}_{g,t}, \quad \forall g \in \Omega_{\mathrm{G}}, \forall t \in \Omega_{\mathrm{T}}, \tag{10.7}$$

where, R^U_g, R^{SU}_g, R^D_g, R^{SD}_g are the ramping-up, start-up ramping, ramping-down and shut-down ramping limits, respectively. Since ramping limits are generally given in per hour, we divide the hourly data by the relevant sub-hourly scheduling interval, for example, by 4 in the 15-min case. Again, $z^F_{g,0}$ has to be assigned for the first time period.

Power Balance
The following constraints ensures the active power balance at every node of the network:

$$\sum_{g \in \Omega^n_{\mathrm{G}}} p_{g,t} - \sum_{l \in \Omega^n_{\mathrm{D}}} d_{l,t} = \sum_{m \in \Omega^n_{\mathrm{F}}} B_{nm}(\theta_{n,t} - \theta_{m,t}), \quad \forall n \in \Omega_{\mathrm{F}}, \quad \forall t \in \Omega_{\mathrm{T}}, \tag{10.8}$$

where $d_{l,t}$ is the forecast demand located at node n; Ω_{F} is the set of all branches; B_{nm} is the susceptance of transmission line $n - m$; and $\theta_{n,t}$ and $\theta_{m,t}$ are the voltage phase angles at nodes n and m, respectively. The set Ω^n_{G} indicates the generators connected to bus n. Similarly, Ω^n_{D} and Ω^n_{F} are the demands and lines, respectively, connected to bus n.

Transmission Line Limits
The power through a transmission line is limited by its thermal limit:

$$-P^{\max}_{nm} \leq B_{nm}(\theta_{nt} - \theta_{mt}) \leq P^{\max}_{nm}, \quad \forall n, \ \forall m \in \Omega^{:}_{\mathrm{F}} \ \forall t, \tag{10.9}$$

where, $P^{\max}_{nm}$ is the capacity limit of the line.

Security Constraints
System operators usually schedule some spinning reserves in order to cope with unforeseen events. So, for all time periods, the total generation available online has to be greater than the actual demand:

$$\sum_{g \in \Omega^i_{\mathrm{G}}} P^{\max}_g z^F_{g,t} \geq \sum_{l \in \Omega_{\mathrm{D}}} (d_{l,t} + r_{j,t}), \quad \forall i, \forall t \in \Omega_{\mathrm{T}}, \tag{10.10}$$

where $d_{l,t}$ is the system total forecast demand; $r_{j,t}$ accounts for reserves; and Ω^i_{G} is the set of generators that provides reserve (in the following, we assume $\Omega^i_{\mathrm{G}} \equiv \Omega_{\mathrm{G}}$). For simplicity, the amount of reserve is assumed to be a percentage of the total demand. The reserve percentage value is lower for shorter scheduling time scales assuming that better forecast is available.

Reference Angle
Finally, the voltage phase angle at some node of the network has to be assigned:

$$\theta_{n,t} = 0, \quad \forall t \in \Omega_{\mathrm{T}}, \tag{10.11}$$

where n is the node chosen to be the reference angle.

Remarks on the UC Model

Equations (10.1)–(10.11) form a conventional model of d-UC. The aim, in fact, is not to propose a novel formulation of the UC but rather to show how the UC can be embedded into the dynamic model of power systems. In the following, the d-UC is modeled using different sub-hourly time periods, ranging from 15 to 3.75-min. Moreover, a rolling (moving window) approach with a planning horizon of 24 h is considered to account for better forecast. In other words, the d-UC problem (10.1)–(10.11) is solved at every time period t for the next 24 h.

The average demand of each load is assumed to vary as a piece-wise linear function according to a predefined profile. To simulate uncertainty, the values $d_{l,t}$ utilized to solve the d-UC problem at each period differ from the actual demand of the loads by a given percentage. A normal distribution function with different standard deviations per period, say, σ_t, is used to generate the forecast error. The value of the standard deviation increases linearly as a function of t. Specifically, σ_t is null for current loading condition at $t = 0$, and is maximum for the last period of the planning horizon Ω_{T}.

10.2.3 Interaction Between d-UC and System Dynamics

We are now ready to embed the d-UC Eqs. (10.1)–(10.11) into Eqs. (A.1). One has two ways to do so: embed d-UC into an existing power system dynamic model, or the other way round. In the examples below we have adopted the former approach, namely, we have embedded the d-UC problem (10.1)–(10.11) into the TDS routine of the power system software tool Dome [16].

Figure 10.4 shows the connection and the interactions between d-UC, Turbine Governors (TGs), AGC, generators, demands and the rest of the grid. In particular, the solutions of the d-UC ($p_{g,t}$, $\forall g \in \Omega_{\mathrm{G}}$) and the output of the AGC (Δp), are utilized to change the power order set-point (p_{ord}) of the TGs of the power plants which in turn changes the mechanical (p_{m}) and electrical (p_{e}) power of synchronous generators, respectively.

10.2.4 Scheduling Time Periods and DFDs

This case study investigates the impact of different scheduling periods of the d-UC on DFDs based on a modified version of the IEEE 39-bus system (see Appendix B.2). While the rolling UC is solved for 24 h at every period, in the discussions below, we show only the first hour of the planning horizon, which has a demand forecast of 700 MW [17].

Several scenarios are considered to study the impact of the uncertainty of the net-load, on the long-term dynamic behavior of the system. The first four scenarios compare the effects of the scheduling period of the UC problem on frequency

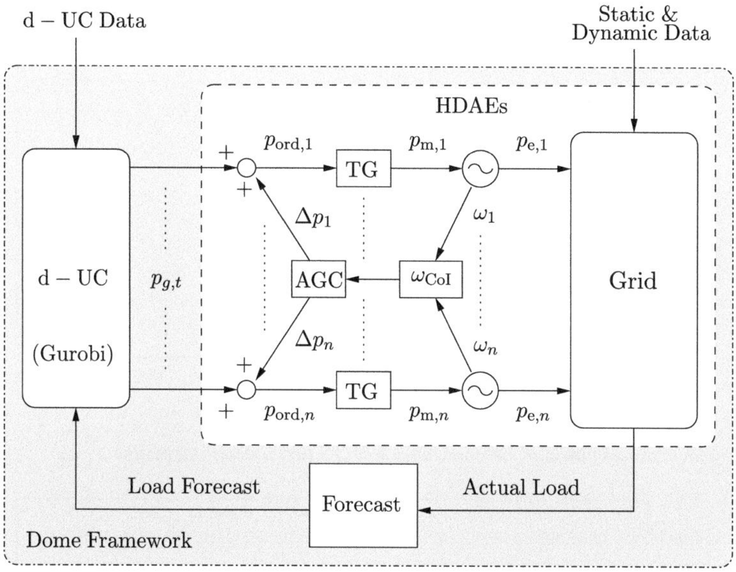

Fig. 10.4 Interaction between d-UC, AGC, TGs, synchronous machines and the grid

variations. Periods of 15, 10, 7.5 and 3.75 min are considered. Then, the impact of net-load volatility (short-term noise) on the system's frequency is discussed. Finally, the combined effect of uncertainty and a large contingency on frequency deviations is shown.

15-min Scheduling
In this scenario, d-UC is solved four times and a maximum of 30% of uncertainty is applied to the end of the first hour. Figure 10.5 shows the mechanical power of two relevant synchronous machines, whereas Fig. 10.6 depicts the frequency of the CoI of the system.

DFDs occur every 15-min due to a change in the operating points of the synchronous machines enforced by the d-UC. The simulated frequency peaks are similar to those observed in real-world power systems [18]. In the periods between two scheduling events of the d-UC, machine powers vary due to load ramps, but the frequency is almost steady-state thanks to the action of PFR and SFC. The average value of the objective function for these four periods is found to be $553,346.4.

10-min Scheduling
In this scenario, the hour is divided into six scheduling time intervals and uncertainty is added to the load, but it is proportionally lower compared to the 15-min scenario as a more reliable forecast is assumed to be available. Figure 10.7 shows that the amplitude and duration of DFDs of the ω_{CoI} decrease with respect to the 15-min scenario. This is due to both lower uncertainty and lower load variations in the shorter period.

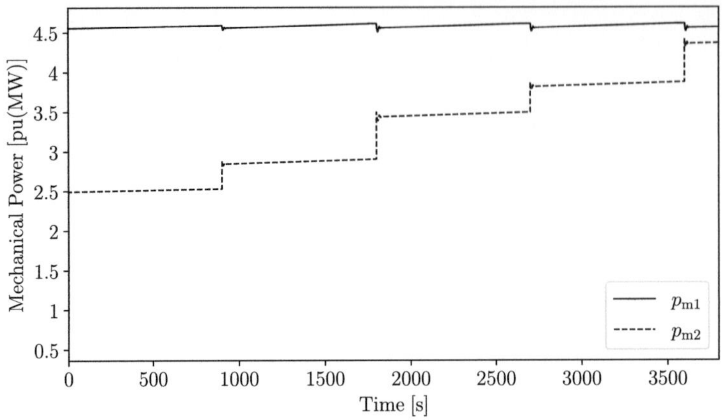

Fig. 10.5 Mechanical power of two machines with 15-min scheduling period

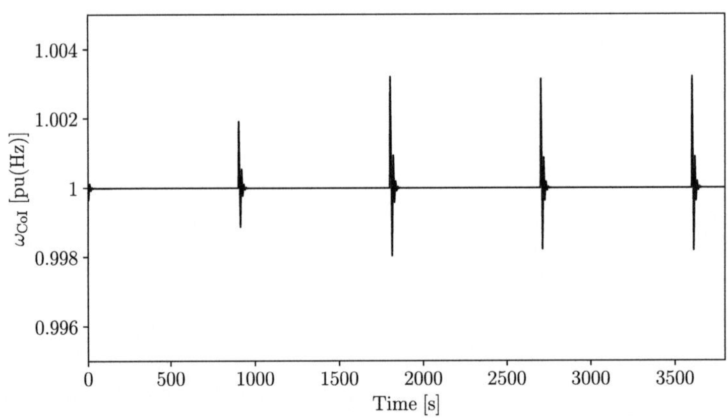

Fig. 10.6 Frequency of the CoI for the 15-min scheduling period

The average value of the objective function per period is $551,575.1, hence lower than for the 15-min scenario. The decrease of the objective function is a consequence of the lower reserves and uncertainty and, in turn, of the shorter time period. Notably, the net-load consumption is assumed to vary linearly between two consecutive periods. If different paths are assumed, the solution of the sub-hourly d-UC problem can become more expensive. A proper modeling of the behavior and the control of the system between consecutive solutions of the UC problem is thus crucial to avoid unnecessarily increasing the price of electricity.

7.5 and 3.75-min Scheduling

The results of these scenarios are shown in Figs. 10.8 and 10.9. The trend shown by the 10 min scheduling is confirmed: shorter periods leads to lower frequency variations or DFDs to the lower uncertainty. These results support the move towards

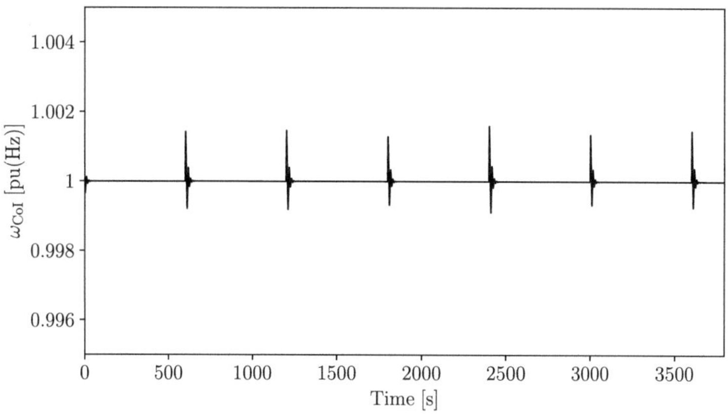

Fig. 10.7 Frequency of the CoI for the 10-min scheduling period

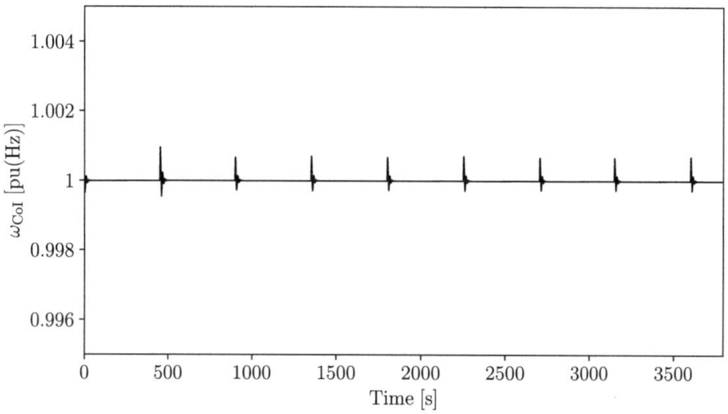

Fig. 10.8 Frequency of the CoI for the 7.5-min scheduling period

higher granularity in ISPs and market time units such as switching from an 1 h ISP to a 15-min ISP in Europe.

Also the cost of electricity decreases: the average values of the objective function for these two scenarios are $550,938.35 and $550,033.97, respectively. These results suggest that a smoother operation is not just better from a dynamic point of view, but it is also more economical.

Effect of noise
In this scenario, volatility is added to the load net consumption as proposed in [19]. This noise models both load and DER short-term fluctuations. Simulation results indicate that noise can significantly increase frequency variations. Figures 10.10 and 10.11 show the worst case scenario, namely the 15-min scheduling period with noise. In addition, the average value of the objective function is $556,689.32 and so slightly higher compared to the 15-min case without noise.

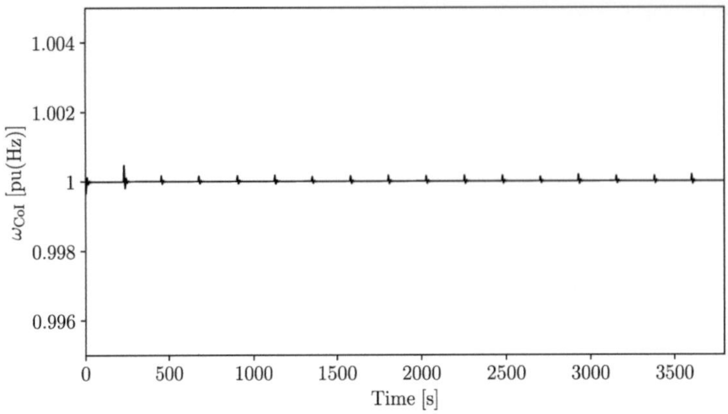

Fig. 10.9 Frequency of the CoI for the 3.75-min scheduling period

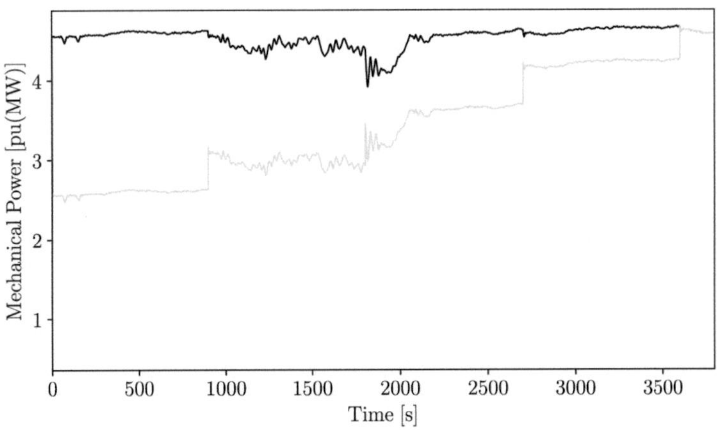

Fig. 10.10 Mechanical power of relevant machines for 15-min scheduling period and noise

Effect of Contingencies

The outage of line 1, occurring at 1,802 s and cleared after 200 ms for the scenarios of 15 and 3.75-min scheduling scenarios illustrates the impact of the d-UC on the stability of the system. The time at which the contingency occurs is chosen on purpose to be in the seconds after the solution of the d-UC problem. Figures 10.12 and 10.13 show the rotor speeds of the machines following the d-UC scheduling and the contingency. As expected, the highest impact of the contingency on the frequency occurs if a 15-min scheduling period is used.

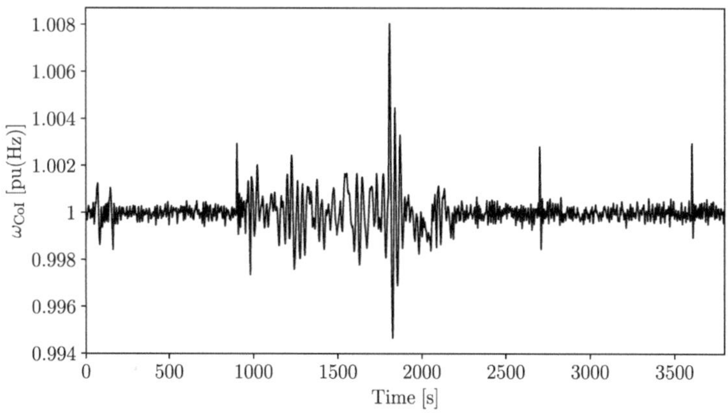

Fig. 10.11 Frequency of the CoI for 15-min scheduling period and noise

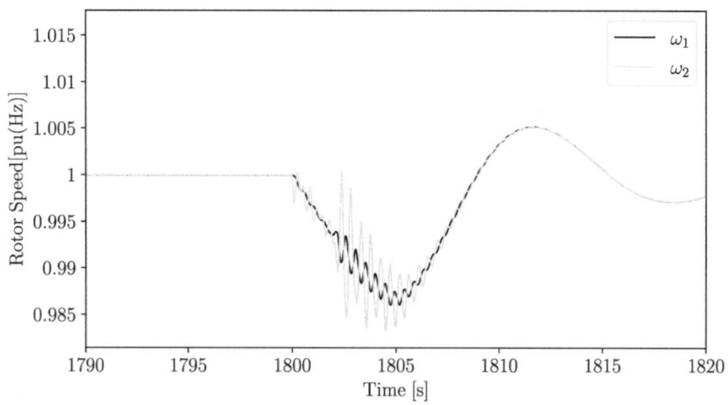

Fig. 10.12 Rotor speed of two synchronous machines following a contingency for the 15-min scheduling period

10.2.5 Net-Load Volatility and Generator Parameters

This case study carries out a sensitivity analysis with respect to net-load volatility and control/machine parameters and their impact on σ_{CoI}. Four case studies are considered on the modified IEEE 39-bus system. The first two scenarios use a 15-min scheduling period while the last two use a 5-min scheduling period. Each scenario is characterized by a different amount of load stochastic variations.

The base case (reference) machine/control parameters are set as follows: gain of the AGC is $K_o = 50$, droop of the TGs is $\mathcal{R} = 0.05$, and the inertia of the synchronous machines is taken equal to the original values, say H_o (see Table B.9 in Appendix B.2). For the sensitivity analysis, these parameters are varied, one at a time, and their impact on σ_{CoI} is evaluated. Next, to simulate uncertainty, the load is assumed to

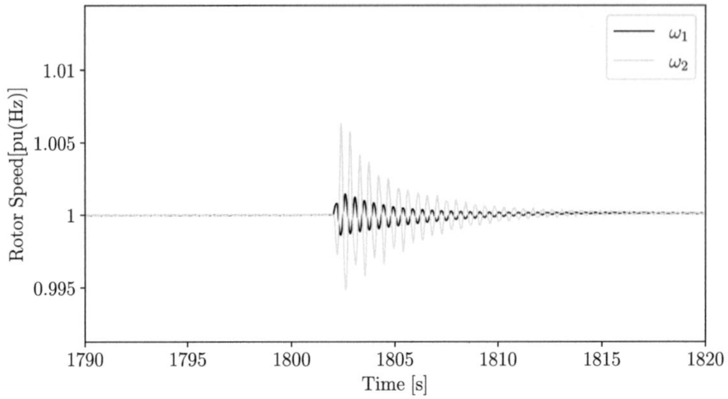

Fig. 10.13 Rotor speed of two synchronous machines following a contingency for the 3.75-min scheduling period

Table 10.1 15-min scheduling period with high noise—σ_{CoI} of the frequency for different generator parameters

K_o	σ_{CoI} $[10^{-4}$ pu(Hz)]	$\mathcal{R}$	σ_{CoI} $[10^{-4}$ pu(Hz)]	H	σ_{CoI} $[10^{-4}$ pu(Hz)]
25	9.73	0.02	4.39	$2H_o$	6.85
50	7.5	0.05	7.5	H_o	7.5
100	5.09	0.08	9.26	$0.5H_o$	7.43

differ from the forecast by a maximum of 30% in the first hour in all scenarios. In other words, the load in the last period of the first hour is chosen to be 30% higher than the forecast, and it is proportionally lower for other periods of this hour.

Scenario 1–15-min Scheduling Period with High Noise

The standard deviation of the white noise for this scenario is 1% of the base case load. During the first hour, the d-UC is solved four times and the average value of the objective function for these 4 periods is found to be \$561,784.79. It is relevant to note that this cost is lower compared to the one found in some other works [20], [14]. This is mainly due to the fact that since we are using a 15-min scheduling interval, the value of the total reserves is taken 2.5% of the total demand and so proportionally lower compared to the above works that use a value of 10%.

Simulation results of this scenario are shown in Table 10.1. These indicate that, as the gain K_o of the AGC increases, σ_{CoI} decreases. Also, as expected, if the droop of the TGs increases, σ_{CoI} increases. Interestingly enough though, decreasing the inertia of the system leads to slightly lower standard deviations of the frequency. This is due to the fact that the distribution of the frequency is not exactly a normal distribution because of the jumps of the power imposed by the d-UC [21].

Scenario 2–15-min Scheduling Period with Low Noise

Table 10.2 15-min scheduling period with low noise—σ_{CoI} for different generator parameters

K_o	σ_{CoI} [10^{-4} pu(Hz)]	$\mathcal{R}$	σ_{CoI} [10^{-4} pu(Hz)]	H	σ_{CoI} [10^{-4} pu(Hz)]
25	3.74	0.02	1.57	$2H_o$	2.84
50	2.98	0.05	2.98	H_o	2.98
100	2.31	0.08	3.99	$0.5H_o$	3.01

Table 10.3 5-min scheduling period with high noise—σ_{CoI} for different generator parameters

K_o	σ_{CoI} [10^{-4} pu(Hz)]	$\mathcal{R}$	σ_{CoI} [10^{-4} pu(Hz)]	H	σ_{CoI} [10^{-4} pu(Hz)]
25	6.68	0.02	3.71	$2H_o$	3.84
50	5.29	0.05	5.29	H_o	5.29
100	4.71	0.08	6.97	$0.5H_o$	6.33

The standard deviation of the load noise in this scenario is chosen to be five times lower compared to Scenario 1. The value of the objective function is found to be \$559,191.47 and hence lower than Scenario 1. This is because the net-load volatility is lower. The simulation results of the sensitivity analysis are shown in Table 10.2. As expected, increasing the gain K_o of the AGC decreases σ_{CoI}; increasing the droop $\mathcal{R}$ of TGs leads to higher values of σ_{CoI}; and decreasing the inertia H of synchronous machines increases σ_{CoI}.

Scenario 3–5-min Scheduling Period with High Noise
In this scenario, load noise is proportionally lower to scenario one. The value of the objective function is found to be \$557,373.86 and hence lower than the other two scenarios. This is the result of lower reserves and lower noise and uncertainty of the net-load. Therefore, it can be said that shorter scheduling periods of the d-UC leads to lower operating costs. On the other hand, the sensitivity analysis with respect to different parameters is shown in Table 10.3. Similar to scenario one, increasing the gain K_o of the AGC decreases σ; increasing $\mathcal{R}$ leads to higher values of σ_{CoI}; and decreasing inertia H increases the value of σ_{CoI}.

Scenario 4–5-min Scheduling Period with Low Noise
The white noise of the load is reduced 5 times with respect to Scenario 3. The objective function is found to be \$556,421.26 and, hence, lower than all other scenarios. It can be concluded that the shorter the scheduling period of the UC, the lower the total operating costs. Moreover, the sensitivity analysis is shown in Table 10.4. Again, similar to Scenario 1, increasing the inertia leads to a higher σ_{CoI}. Following the same rationale discussed in Scenario 1, a higher value of σ_{CoI} with high inertia is due to the fact that, after each d-UC scheduling, the frequency takes a longer time to recover to its nominal value.

Table 10.4 5-min scheduling period with low noise—σ_{CoI} of the frequency for different control/machine parameters

K_o	σ_{CoI} [10^{-4} pu(Hz)]	$\mathcal{R}$	σ_{CoI} [10^{-4} pu(Hz)]	H	σ_{CoI} [10^{-4} pu(Hz)]
25	3.8	0.02	1.73	$2H_o$	2.99
50	2.79	0.05	2.79	H_o	2.79
100	2.16	0.08	3.62	$0.5H_o$	2.87

10.3 Real-Time Sub-Hourly s-UC

Large amounts of stochastic RES can significantly impact the performance of the system [8, 22]. Stochastic programming has been introduced in recent years to properly account for uncertainty when scheduling the system. For example, [23] shows that a stochastic market-clearing procedure allows greater wind penetration than a deterministic approach. Moreover, the stochastic real-time market with 15-min dispatch intervals proposed in [24] appears to outperform the deterministic approach. In [25], a stochastic economic dispatch (15-min dispatch interval) leads to lower operating costs compared to those obtained with deterministic approaches. Nevertheless, the computational burden of the stochastic approach is the main limitation for practical applications.

This section focuses on the impact of the sub-hourly s-UC on long-term frequency quality. If sub-hourly scheduling time scales are used, say 15 or 5-min, in fact, then these time scales can overlap with long-term dynamics [26]. Therefore, it appears worthwhile and timely to embed the sub-hourly s-UC problem into a fully-fledged transient stability analysis software tool [27, 28] and study the dynamic behavior of power systems.

The remainder of the section is organized as follows. Section 10.3.1 presents the stochastic long-term power system model. Section 10.3.2, 10.3.3 and 10.3.4 describe the formulation of the complete, simplified and alternative s-UC problem, respectively. Section 10.3.5 provides a modified version of the proposed software framework able to embed the various s-UC formulations in the TDS routine of Dome. Sect. 10.3.6 discusses the impact of different scheduling periods of the complete s-UC problem on long-term frequency deviations. Section 10.3.7 focuses on the effects of different s-UC and frequency controllers/machine parameters on σ_{CoI}. Finally, conclusions are drawn in Sect. 10.4.

10.3.1 Stochastic Long-Term Power System Model

The large-scale deployment of stochastic RES significantly increases the random perturbations and inevitably affects power systems' safe and stable operation [29].

As a result, the adoption of stochastic DAEs to model this randomness in power systems has gained significant interest in recent years. However, most of the literature focus on the impact of wind power variability on transient and small-signal stability, respectively [30]. On the other hand, stochastic long-term stability analysis of power systems has received little attention. This is of particular importance considering the expected trends regarding the integration of RES into modern power systems.

We consider the stochastic long-term dynamic model (A.1) [19]. In particular, it is necessary to consider both electromechanical and long-term dynamic models, when one performs stochastic long-term dynamic simulations [31]. With this regard, (A.1) includes the dynamic models of conventional machines (4th order models) and their primary controllers; AGC; wind power plants (5th order Doubly-Fed Induction Generator)[32]; and the model of sub-hourly s-UC. It is worth noting that we have not included in the system a fast frequency control in wind power plants. This controller, in fact, while can help improve the dynamic response of the system does not alter the conclusions that are drawn in the rest of this chapter.

Modeling the stochastic nature of wind power is critical in power system dynamic studies [33]. In this context, we include in the DAE model (A.1) wind power variations with respect to the forecast wind generation as a stochastic perturbations. An Ornstein-Uhlenbeck process is used to model the stochastic nature of the wind speed ν that enters into the wind turbine, as follows:

$$\begin{aligned} \nu(t) &= \nu_o + \eta_w(t)\,, \\ \eta_w'(t) &= a_w(\mu_w - \eta_w(t)) + b_w\zeta_w(t)\,, \end{aligned} \tag{10.12}$$

where ν_o is the wind speed initial value; η_w is the stochastic variable that is dependent on the drift $a_w(\mu_w - \eta_w)$, and diffusion term b_w of the stochastic differential equation; a_w is the mean reversion speed that indicates the rate at which η_w tends to the mean value μ_w; and ζ_w represents the white noise.

10.3.2 Complete Stochastic Unit Commitment

As the penetration of highly variable RES increases, so does the uncertainty in power systems. This complicates the real-time balance between generation and demand. Therefore it is of particular importance to model the uncertainty when scheduling the system. There are different methodologies and techniques proposed for optimization under uncertainty, with one of the most popular being the two-stage stochastic programming. In the context of UC, the two-stage s-UC makes use of a probabilistic model for the uncertain input parameters and is usually approximated by a set of scenarios representing the plausible realizations of these random parameters [34].

In this chapter, a standard MILP s-UC problem is implemented based on [35], in which wind power production is considered as an uncertain parameter of the system, as follows:

$$\min_{\Omega_{\mathrm{H}},\Omega_{\mathrm{H}}} \sum_{t\in\Omega_{\mathrm{T}}} \sum_{g\in\Omega_{\mathrm{G}}} (C_g^F z_{g,t}^F + C_g^{SU} z_{g,t}^{SU} + C_g^{SD} z_{g,t}^{SD}) \tag{10.13}$$

$$+ \sum_{s\in\Omega_{\mathrm{S}}} \varpi_s \left[\sum_{t\in\Omega_{\mathrm{T}}} \sum_{g\in\Omega_{\mathrm{G}}} C_g^V p_{g,t,s} + \sum_{t\in\Omega_{\mathrm{T}}} \sum_{l\in\Omega_{\mathrm{D}}} C^L d_{l,t,s}^{SH} \right]$$

such that

$$z_{g,t}^{SU} - z_{g,t}^{SD} = z_{g,t}^F - z_{g,t-1}^F \tag{10.14}$$

$$(\forall g \in \Omega_{\mathrm{G}}, \forall t \in \{2, ..., T\})$$

$$z_{g,t}^{SU} - z_{g,t}^{SD} = z_{g,t}^F - IS_g \tag{10.15}$$

$$(\forall g, \forall t \in \{1\})$$

$$z_{g,t}^{SU} + z_{g,t}^{SD} \leq 1 \tag{10.16}$$

$$(\forall g, \forall t \in \{1, ..., T\})$$

$$z_{g,t}^F = IS_g \tag{10.17}$$

$$(L_g^{UP} + L_g^{DW} > 0, \forall g, \forall t \leq L_g^{UP} + L_g^{DW})$$

$$\sum_{\tau=t-UT_g+1}^{t} z_{g,\tau}^{SU} \leq z_{g,t}^F \tag{10.18}$$

$$(\forall g, \forall t > L_g^{UP} + L_g^{DW})$$

$$\sum_{\tau=t-DT_g+1}^{t} z_{g,\tau}^{SD} \leq 1 - z_{g,t}^F \tag{10.19}$$

$$(\forall g, \forall t > L_g^{UP} + L_g^{DW})$$

$$\sum_{g\in\Omega_{\mathrm{G}}^n} p_{g,t,s} - \sum_{l\in\Omega_{\mathrm{D}}^n} d_{l,t} + \sum_{l\in\Omega_{\mathrm{D}}^n} d_{l,t,s}^{SH} + \sum_{k\in\Omega_{\mathrm{K}}^n} w_{k,t,s} \tag{10.20}$$

$$- \sum_{k\in\Omega_{\mathrm{K}}^n} w_{k,t,s}^{SP} = \sum_{m\in\Omega_{\mathrm{M}}^n} \frac{(\theta_{n,t,s} - \theta_{m,t,s})}{X_{n,m}}$$

$$(\forall n, \forall t, \forall s \in \Omega_{\mathrm{S}})$$

$$p_{g,t,s} \leq P_g^{\max} z_{g,t}^F \tag{10.21}$$

$$(\forall g, \forall t, \forall s \in \Omega_{\mathrm{S}})$$

$$p_{g,t,s} \geq P_g^{\min} z_{g,t}^F \tag{10.22}$$

$$(\forall g, \forall t, \forall s \in \Omega_{\mathrm{S}})$$

$$p_{g,t,s} \leq (P_g^{IS} + RU_g) z_{g,t}^F \tag{10.23}$$

$$(\forall g, \forall t \in \{1\}, \forall s \in \Omega_{\mathrm{S}})$$

$$p_{g,t,s} \geq (P_g^{IS} - RD_g) z_{g,t}^F \tag{10.24}$$

$$(\forall g, \forall t \in \{1\}, \forall s \in \Omega_S)$$

$$p_{g,t,s} - p_{g,t-1,s} \leq (2 - z^F_{g,t-1} - z^F_{g,t})P^{SU}_g + (1 + z^F_{g,t-1} - z^F_{g,t})RU_g) \tag{10.25}$$

$$(\forall g, \forall t \in \{2, ..., T\}, \forall s \in \Omega_S)$$

$$p_{g,t-1,s} - p_{g,t,s} \leq (2 - z^F_{g,t-1} - z^F_{g,t})P^{SD}_g + (1 - z^F_{g,t-1} + z^F_{g,t})RD_g) \tag{10.26}$$

$$(\forall g, \forall t \in \{2, ..., T\}, \forall s \in \Omega_S)$$

$$d^{SH}_{l,t,s} \leq d_{l,t} \tag{10.27}$$

$$(\forall l, \forall t, \forall s \in \Omega_S)$$

$$w^{SP}_{k,t,s} \leq w_{k,t,s} \tag{10.28}$$

$$(\forall k, \forall t, \forall s \in \Omega_S)$$

$$-P^{\max}_{n,m} \leq \frac{(\theta_{n,t,s} - \theta_{m,t,s})}{X_{n,m}} \leq P^{\max}_{n,m} \tag{10.29}$$

$$(\forall n, m \in \Omega^n_M, \forall t, \forall s \in \Omega_S)$$

$$p_{g,t,s}, d^{SH}_{l,t,s}, w^{SP}_{k,t,s} \geq 0 \tag{10.30}$$

$$(\forall g, \forall l, \forall k, \forall t, \forall s \in \Omega_S)$$

$$z^F_{g,t}, z^{SU}_{g,t}, z^{SD}_{g,t} \in \{0, 1\} \tag{10.31}$$

$$(\forall g, \forall t)$$

where $\mathcal{H} = \{z^F_{g,t}, z^{SU}_{g,t}, z^{SD}_{g,t}\}$ and $\mathcal{W} = \{p_{g,t,s}, d^{SH}_{l,t,s}, w^{SP}_{k,t,s}\}$; and the initial state conditions are as follows:

$$IS_g = \begin{cases} 1 & \text{if } ON_g > 0 \\ 0 & \text{if } ON_g = 0 \end{cases}$$

$$L^{UP}_g = \min\{T, (UT_g - ON_g)IS_g\}$$

$$L^{DW}_g = \min\{T, (DT_g - OFF_g)(1 - IS_g)\}$$

Equations (10.13) represent the total cost to be minimized which includes the fixed, start-up, shut-down and variable cost of the generating units, as well as the cost of involuntarily demand curtailment. Equations (10.14)–(10.16) model the logical expression between the binary variables, namely, start-up and shut-down of generating units. Equations (10.17)–(10.19) model the minimum and maximum up- and down-time constraints. The power balance constraint is modeled through equations (10.20). While the capacity limits of generating units are modeled through equations (10.21)–(10.22) and their respective ramping limits through (10.23)–(10.26). Equations (10.27)–(10.28) model the limits of the involuntary demand curtailment and wind power spillage, respectively. Transmission capacity limits are enforced by equations (10.29). Finally, Eqs. (10.30)–(10.31) refer to the variable declarations.

The model shown in (10.13)–(10.31) is the deterministic equivalent of the original two-stage stochastic programming problem. It is called a two-stage problem since there are first-stage and second-stage variables, also known as here-and-now and wait-and-see variables, respectively [34]. In particular, $z^F_{g,t}$, $z^{SU}_{g,t}$, $z^{SD}_{g,t}$ are first-stage decision variables that represent the status of generating unit g in time period t and represent ON/OFF status, start-up and shut-down, respectively. These decisions do not depend on uncertainty realization s, and are generally made one day in advance. Similarly, $p_{g,t,s}$, $d^{SH}_{l,t,s}$, $w^{SP}_{k,t,s}$, $\theta_{n,t,s}$ are second-stage decision variables that represent the active power dispatch of generating units g in time period t and scenario s, the involuntary power curtailment from load j in time period t and scenario s, wind power spillage from wind production unit k in time period t and scenario s, and voltage angle at node n in time period t and scenario s, respectively. All second-stage decision variables depend on uncertainty realization s. Further details of the s-UC can be found in [35] and references therein.

Scenarios and Rolling Horizon within the s-UC

To illustrate the modeling of s-UC wind uncertainty and volatility, and rolling planning horizon used in this chapter, we show below the power balance equations of the complete s-UC, as follows:

$$\sum_{g\in\Omega^n_{\mathrm{G}}} p_{g,t,s} - \sum_{l\in\Omega^n_{\mathrm{D}}} d_{l,t} + \sum_{l\in\Omega^n_{\mathrm{D}}} d^{SH}_{l,t,s} + \sum_{k\in\Omega^n_{\mathrm{K}}} w_{k,t,s} - \sum_{k\in\Omega^n_{\mathrm{K}}} w^{SP}_{k,t,s} = \sum_{m\in\Omega^n_{\mathrm{M}}} \frac{(\theta_{n,t,s} - \theta_{m,t,s})}{X_{n,m}}, \ (\forall n, \forall t, \forall s \in \Omega_{\mathrm{S}}) \tag{10.32}$$

where $p_{g,t,s}$ is the active power of conventional generating units g, at time period t, and scenario s, which is equivalent to the second-stage variable $u_{f,t,s}$ in Sect. 10.3.2; $d_{l,t}$ is the demand for load l at time period t; $d^{SH}_{l,t,s}$ is the power curtailment from load l, at time period t, and in scenario s; $w_{k,t,s}$ and $w^{SP}_{k,t,s}$ represent the power generation and curtailment, respectively, from wind unit k, in time period t, and scenario s; $X_{n,m}$ is the reactance of line $n-m$; $\theta_{n,t,s}$ represent the voltage angle at node n, time period t, and scenario s; and $\mathcal{K}_n$, $\mathcal{M}_n$ are the sets of stochastic power generation located at node n, and nodes $m \in \Omega_{\mathrm{N}}$ connected to node n by transmission line, respectively.

Modeling Wind Uncertainty

Similar to [36], a wind power penetration level $w_{k,t,s}$ is defined as a percentage of the demand $d_{l,t}$, and named the medium scenario, $w^m_{k,t,s}$. Then, high and low wind power scenarios ($w^h_{k,t,s}$, $w^l_{k,t,s}$) are built as percentages of the medium scenario, as follows:

$$\begin{aligned} w^l_{k,t,s} &= w^m_{k,t,s} \times (1 - j/100) \\ w^h_{k,t,s} &= w^m_{k,t,s} \times (1 + j/100) \end{aligned} \tag{10.33}$$

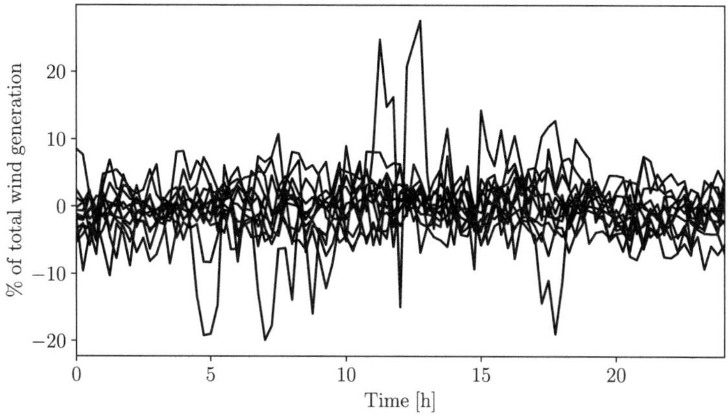

Fig. 10.14 Wind power change in 15 min for 12 typical days in the AIPS in 2018

where j is the percentage of deviation of the the high and low scenarios with respect to the medium one.

The consistency of the wind power scenarios with real-world information is compared using wind power data of the Irish system [37]. Specifically, the 15-min change of wind power is determined based on one typical day per month of 2018 (see Fig. 10.14). Based on these data, wind power does not appear to change more than 10% in 15 min. For this reason, in the case study, $j = 10\%$ is assumed.

Modeling wind Volatility

Wind power volatility is modeled as small fluctuations with respect to the average value for a given period. Hence, uncertainty is related to wind power forecast, while volatility is considered as a percentage, on top of the wind power forecast [38]. A normal distribution $\mathcal{N}(0, \sigma^2)$ with zero mean and given standard deviation is attached to each wind power scenario, as follows:

$$
\begin{aligned}
w^{l1}_{k,t,s} &= w^{l}_{k,t,s} + \mathcal{N}(0, \sigma^2) \\
w^{m1}_{k,t,s} &= w^{m}_{k,t,s} + \mathcal{N}(0, \sigma^2) \\
w^{h1}_{k,t,s} &= w^{h}_{k,t,s} + \mathcal{N}(0, \sigma^2)
\end{aligned}
\qquad (10.34)
$$

where $w^{l1}_{k,t,s}$, $w^{m1}_{k,t,s}$, $w^{h1}_{k,t,s}$ are the new low, medium and high wind power scenarios, respectively, after adding the volatility.

An important aspect to keep in mind when building the scenarios is the relationship between the wind power level and its standard deviation σ. With this aim, two typical days are analyzed for two months, namely January (high wind) and July (low wind). The wind power profile for these typical days is shown in Fig. 10.15. It appears that wind varies more in January than in July. More specifically, the standard deviation of wind power generation is found to be 234.78 MW and 68.46 MW, for January and July, respectively, and that high wind leads to higher σ.

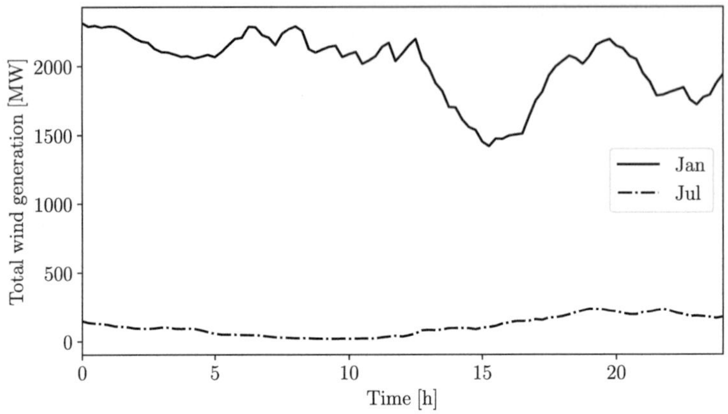

Fig. 10.15 Wind power profile for two typical days in January and July in the AIPS in 2018

Note that the goal is not to propose new s-UC models to deal with wind power uncertainty and volatility but rather to study the impact of a well assessed s-UC formulation on long-term frequency quality. For this reason, it is not necessary to consider more sophisticated s-UC models. As a matter of fact, the first case study shows that, depending on the wind penetration and planning horizon, a s-UC might not be needed at all.

Modeling the Rolling Planning Horizon
Scheduling the system more than once a day allows having better wind and load forecasts. As a result, fewer reserves are required. This chapter uses a rolling planning approach for updating the wind power forecast ($w_{k,t,s}$) with a planning horizon of 24 h. During the first hour of the planning horizon, the high/low wind scenarios increase/decrease as a linear function from the medium scenario and after that, they have a fixed error. The s-UC model is solved at every period t during the first 12 h with a planning horizon of 24 h. When rolling forward, the status of the units of the previous horizon serves as an initial status for the next horizon. Between two scheduling events of the s-UC, wind and load profiles are modeled as linear ramps.

10.3.3 Simplified Stochastic Unit Commitment

The level of complexity of s-UC formulations proposed so far in the literature varies significantly [39]. For example, a well-assessed MILP s-UC formulation is provided in [35]. Such a model considers several technical constraints such as ramping limits of generators and capacity limits of transmission lines, to name a few. These constraints improve the performance of the schedules but are not crucial in this chapter, which focuses on the impact on long-term power system dynamics. Hence, a simplified model of s-UC is considered, as follows:

$$\underset{z^F_{g,t},p_{g,t,s}}{\text{minimize}} \sum_{t\in\Omega_T}\sum_{g\in\Omega_G} C^F_g z^F_{g,t} + \sum_{s\in\Omega_S} \varpi_s \left[\sum_{t\in\Omega_T}\sum_{g\in\Omega_G} C^V_g p_{g,t,s}\right] \tag{10.35}$$

such that

$$p_{g,t,s} \leq P^{\max}_g z^F_{g,t}, \qquad \forall g, \forall t, \forall s \in \Omega_S \tag{10.36}$$

$$p_{g,t,s} \geq P^{\min}_g z^F_{g,t}, \qquad \forall g, \forall t, \forall s \in \Omega_S \tag{10.37}$$

$$\sum_{g\in\Omega_G} p_{g,t,s} - \sum_{l\in\mathcal{D}_n} d_{l,t} + \sum_{k\in\mathcal{K}_n} w_{k,t,s} = \sum_{m\in\mathcal{M}_n} \frac{(\theta_{n,t,s}-\theta_{m,t,s})}{X_{nm}}, \quad \forall n, \forall t, \forall s \in \Omega_S \tag{10.38}$$

$$z^F_{g,t} \in \{0,1\}, \qquad \forall g \in \Omega_G, \forall t \in \mathcal{T} \tag{10.39}$$

where $z^F_{g,t}$ is a first-stage decision variable that models the status (ON/OFF) of the conventional machines in time period t; C^F_g and C^V_g represents the fixed and variable production cost of generation unit g, respectively; ϖ_s is the probability of wind power scenario s; $p_{g,t,s}$ is a variable that represents the active power of the machine g in scenario s and time period t; $P^{\max}_g$ and $P^{\min}_g$ represents the maximum and minimum active power limits of generation unit g, respectively; $d_{l,t}$ represents the demand for load l at time period t; $w_{k,t,s}$ is the wind power production of wind generation unit k, $\theta_{n,t,s}$ are the voltage angles at node n and at time period t in scenario s, respectively; $X_{n,m}$ is the reactance of the transmission line $n-m$; and $\mathcal{K}_n$, $\mathcal{M}_n$ represents the sets of wind power generation connected at node n, and nodes $m \in \Omega_N$ connected to node n through a line, respectively. Finally, constraints (10.35) to (10.39) model the objective function, maximum and minimum power output of generators, nodal power balance equation (dc power flow), and variable declarations, respectively.

10.3.4 Alternative Stochastic Unit Commitment

The literature provides several different formulations of the s-UC problem [40]. Therefore, it is relevant to compare and study the impact that these models have on the power system dynamic behavior. In this context, an alternative sub-hourly s-UC model with respect to the one discussed in the previous section has been adapted based on [41], as follows:

$$\underset{z^F_{g,t}, p_{g,t}, r^{\mathrm{U}}_{g,t,s}, r^{\mathrm{D}}_{g,t,s}, w_{k,t}}{\text{minimize}} \sum_{t\in\Omega_{\mathrm{T}}} \sum_{g\in\Omega_{\mathrm{G}}} C^{SU}_{g,t} + C^V_g p_{g,t} \tag{10.40}$$

$$+ \sum_{s\in\Omega_{\mathrm{S}}} \varpi_s \left[\sum_{t\in\Omega_{\mathrm{T}}} \sum_{g\in\Omega_{\mathrm{G}}} C^V_g (r^{\mathrm{U}}_{g,t,s} - r^{\mathrm{D}}_{g,t,s}) \right]$$

such that

$$C^{SU}_{g,t} \geq C^{SU}_g (z^F_{g,t} - z^F_{g,t-1}), \quad \forall g, \forall t \tag{10.41}$$

$$C^{SU}_{g,t} \geq 0, \quad \forall g, \forall t \tag{10.42}$$

$$z^F_{g,t} P^{\min}_g \leq p_{g,t} \leq z^F_{g,t} P^{\max}_g, \quad \forall g, \forall t \tag{10.43}$$

$$\sum_{g\in\Omega_{\mathrm{G}}} p_{g,t} + \sum_{k\in\mathcal{K}} w_{k,t} = l_t, \quad \forall t \tag{10.44}$$

$$z^F_{g,t} P^{\min}_g \leq (p_{g,t} + r^{\mathrm{U}}_{g,t,s} - r^{\mathrm{D}}_{g,t,s}) \leq z^F_{g,t} P^{\max}_g, \quad \forall g, \forall t, \forall s \tag{10.45}$$

$$\sum_{g\in\Omega_{\mathrm{G}}} (p_{g,t} + r^{\mathrm{U}}_{g,t,s} - r^{\mathrm{D}}_{g,t,s}) + \sum_{k\in\mathcal{K}} w_{k,t,w} = l_t, \quad \forall t, \forall s \tag{10.46}$$

$$0 \leq r^{\mathrm{U}}_{g,t,s} \leq R^{\mathrm{U}\max}_{g,t}, \quad \forall g, \forall t, \forall s \tag{10.47}$$

$$0 \leq r^{\mathrm{D}}_{g,t,s} \leq R^{\mathrm{D}\max}_{g,t}, \quad \forall g, \forall t, \forall s \tag{10.48}$$

$$z^F_{g,t} \in \{0, 1\}, \quad \forall g, \forall t \tag{10.49}$$

where, apart from the variables and parameters already defined above, $C^{SU}_{g,t}$ models the start-up cost of generation unit g incurred at the beginning of time period t; $r^{\mathrm{U}}_{g,t,s}$ and $r^{\mathrm{D}}_{g,t,s}$ are decision variables and represents the increase and decrease in the active power output of generation unit g at time period t, respectively, say, during real-time operation; while all other parameters and variables have analogous meanings as in the previous s-UC formulation.

The main difference between problem (10.40)–(10.49) and problem (10.35)–(10.39) is that in the former the active power of generation units, $p_{g,t}$, is a first-stage variable and does not adapt to the uncertainty realization. Also, note that model (10.35)–(10.39) has a slightly different objective function, namely it does not include the start-up cost ($C^{SU}_{g,t}$) as compared to that of model (10.40)–(10.49).

10.3.5 Interaction Between s-UC and System Dynamics

It is time to merge all of the above in a single framework. Generally speaking, the goal is to embed s-UC Eqs. (10.13)–(10.31), (10.33)–(10.34), (10.35)–(10.39), and (10.40)–(10.49) into Eqs. (A.1).

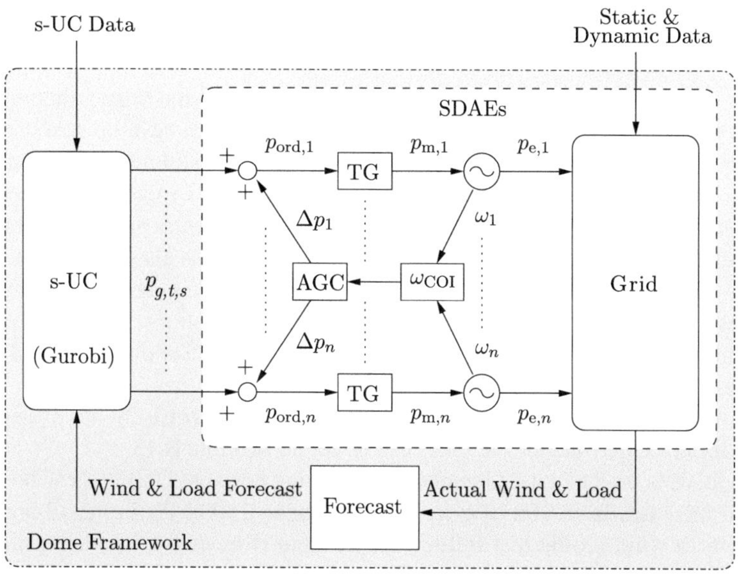

Fig. 10.16 Interaction between the s-UC problem, AGC, TGs, synchronous machines and the grid

Figure 10.16 shows the overall structure of the recently proposed co-simulation framework. The tool is composed of two parts, namely the dynamic model of power systems (A.1) and the discrete model of s-UC. Dome coordinates the exchange of information between the s-UC and the system dynamic model. In particular, the output of the s-UC model, namely the active power of generating units ($p_{g,t,s}$), serves as an input to the DAEs by changing the power set-point (p_{ord} in Figure B.1) of the TGs of the synchronous generators. Finally, a Monte Carlo Method (MC) is utilized to simulate large sets of realizations of the stochastic processes of wind and loads. Each realization defines the "reality" that needs to be updated to solve the next s-UC problem. Such feedback is required to update the forecast of wind ($w_{k,t,s}$) and loads as utilized in the s-UC problem.

The s-UC model (10.13)–(10.31) uses the active power of the generators, namely $p_{g,t,s}$, as a second-stage decision variable. In other words, $p_{g,t,s}$ adapts to the uncertainty realization s. Since we are interested in having a single power dispatch for each generator and for each time period, a reasonable tradeoff consists in taking a weighted-sum of the scenarios s. In the literature, one may find different formulations of s-UC for the active power of generators. For example, in [41], the active power is utilized as a first-stage decision variable and up/down reserve deployment as a second-stage decision variable to accommodate wind variability in real-time.

10.3.6 Impact of Uncertainty and Volatility

From a system operator point of view, it is useful to study the impact that different levels of uncertainty and volatility, within the s-UC model have on power system frequency performance. At the time of this writing, TSOs still rely on d-UC formulations. Therefore, a comparison between d-UC and s-UC approaches is relevant. Also, since different TSOs use different scheduling time scales and different rolling approaches, a comparison of the effect of these strategies is also carried out in this section. Moreover, the impact of contingency and renewable penetration on the transient response of the system and long-term frequency deviations, respectively, using different s-UC strategies is shown as well. Finally, the impact of different wind power scenarios of the s-UC on σ_{CoI} is discussed.

All simulations are based on a modified version of the IEEE 39-bus system (see Appendix B.2). The data of the s-UC can be found in Table B.13.

The focus is on the first 12 hours of the planning horizon. During these hours the demand increases from 700 MW in the first hour to 1500 MW in the 12 hour. For simplicity, a wind profile that follows the demand is modeled. In other words, we assume the same wind penetration level for the medium scenario during these hours, namely 25%, and based on this we build the low and high scenario accordingly. Such a relationship between demand and wind power corresponds to a typical day in 2018 in the Irish system. It should be noted here that one may choose any other profile for the demand and wind but according to our studies that does not change the relevant conclusions. Wind generation is given by three wind power plants connected at bus 20, 21 and 23, respectively, with a nominal capacity of 300 MW each.

The total number of state and algebraic variables of the stochastic DAE model for all scenarios are 173 and 277, respectively. Regarding the s-UC variables, the model includes three first-stage variables, namely the ON/OFF, start-up and shut-down status of generating units, and four second-stage variables, namely the active power of conventional units (set-points), the demand and wind power curtailment, and the bus voltage angle, respectively. The total numbers of the first-stage and second-stage variables for the 15-min model are 2,880 and 20,448, respectively. While the total numbers of these variables for the 5-min model are 8,640 and 61,344, respectively. Therefore even considering only three wind power scenarios, shortening the time period of s-UC, lead to a huge increase in the size of the s-UC model. In fact, this is one of the main limitations of s-UC approaches, especially when considering their use for real-time operations of power systems.

Finally, a MC is used in all scenarios (100 simulations are solved for each scenario). The standard deviation of the frequency of the CoI, σ_{CoI}, (computed as the average of the standard deviation obtained for each trajectory) is utilized as an index to quantify the impact of s-UC on long-term frequency quality. The software tool Dome used in the simulations includes a set of dynamic models similar to the ones provided by commercial software tools but with the additional feature of being able to model and properly integrate stochastic processes. Note that Electro-Magnetic Transient (EMT) models are not considered because they are not suitable for the simulation of several minutes of an interconnected power system due to their computational burden.

15-min Scheduling
In this section, a 15-min scheduling time period is used. The average value of the objective function is found to be approximately \$412,000, hence, lower than the value found in, for example, [14]. This is due to the fact that wind generation is explicitly accounted in the objective function, and since its marginal production cost is considered zero, it leads to lower operational costs. Each scenario is characterized by an appropriate amount of wind stochastic variations. For the sensitivity analysis, the s-UC probabilities are varied, and their impact on σ_{CoI} is evaluated.

In order to compare results, a base-case scenario is considered with the following properties: s-UC probabilities for the low, medium and high wind power scenarios are set to 20%, 60%, 20%, respectively. Similarly, when we run the MC-TDS, the system is assumed to be with the following probabilities: 20%, 60%, 20% for the low, medium and high wind power scenario, respectively. As mentioned in the rolling planning section, the low and high scenario differs from the medium scenario by 10%.

This base case is shown in Fig. 10.17, 10.18, while Fig. 10.17 shows the trajectories of $\omega_{\text{CoI}}(t)$, and Fig. 10.18 shows the trajectories of the wind speed scenarios. frequency jumps due to a change in the operating point of the machines forced by the s-UC. These jumps, also known as DFDs, are very similar to real-world power systems behavior observed in, for example, the CE grid [42, 43]. Finally, note that the wind speed profile in Fig. 10.18 is obtained by adding some stochastic noise on each of the three wind scenarios.

Impact of Different s-UC Strategies
Table 10.5 shows some of the most relevant results of the sensitivity analysis. Specifically, scenario 1 assumes a s-UC with wind probabilities 20%, 60%, 20% and a MC-TDS with the same probabilities. Thus, it is assumed that what was forecast by the s-UC will actually happen in reality. σ_{CoI} for this scenario is $8.47 \cdot 10^{-4}$ pu(Hz). In Scenario 2 the probabilities of the s-UC differ from that of the system by a relevant value. The value of σ_{CoI} is $8.59 \cdot 10^{-4}$ pu(Hz), and thus higher than Scenario 1 due to the error in the s-UC probabilities.

Scenario 3 assumes a s-UC with 100% low wind (one scenario, equivalent to d-UC). The value of σ_{CoI} for this scenario is $8.67 \cdot 10^{-4}$ pu(Hz) and so higher than scenario 1 for the same reason above. Similarly, Scenario 4 assumes a s-UC with 100% medium wind. This leads to higher frequency variations compared to scenario 1, with $\sigma_{\text{CoI}} = 8.72 \cdot 10^{-4}$ pu(Hz). Next, Scenario 5 assumes a s-UC with 100% high wind. Quite surprisingly, this scenario appears to be the best from the frequency point of view with $\sigma_{\text{CoI}} = 8.02 \cdot 10^{-4}$ pu(Hz).

To analyze this relevant case, more scenarios are considered. In Scenarios 6 to 9 in Table 10.5, the probabilities of s-UC are varied from a s-UC with 100% high wind to a s-UC with 100% medium, and it can be seen that σ_{CoI} increases almost linearly. Therefore, even though Scenario 5 assumes an error in the s-UC probabilities, synchronous machines and the respective controls (primary and secondary) can regulate it very fast.

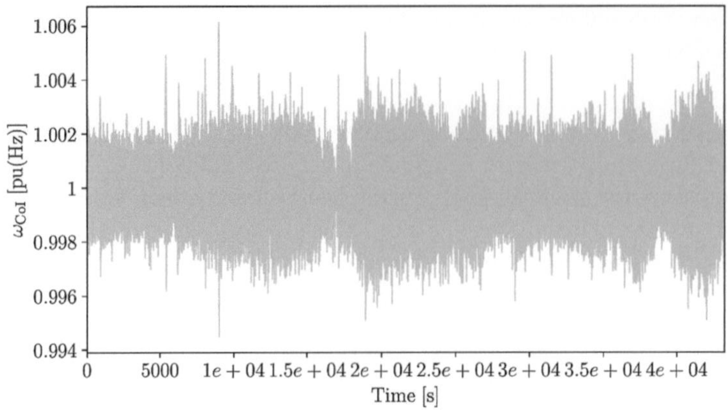

Fig. 10.17 15-min scheduling period—Trajectories of ω_{CoI} for 12 h

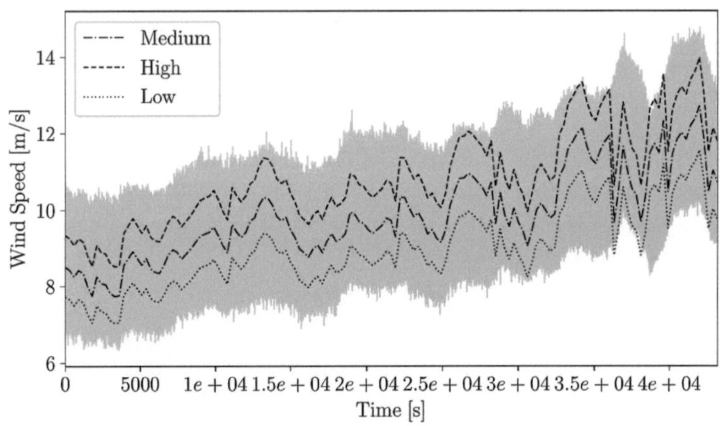

Fig. 10.18 15-min scheduling period—Trajectories of wind speeds for 12 h. Dashed lines indicate the high, medium and low scenarios, respectively, while black line indicate the same scenarios with inclusion of volatility

To further analyze this, in Table 10.6 the wind power uncertainty level is increased from $j = 10\%$ to $j = 30\%$ with a step of 10% for both, Scenario 1 and Scenario 5, respectively. For $j \geq 30\%$, Scenario 1 gives the better dynamic behavior. It appears that, if the wind forecast error is small, then from the dynamic performance viewpoint of the system, it is better to solve a d-UC with high wind power.

It is worth observing that the differences in the system's long-term frequency deviations between scenarios are marginal (maximum of 3.5 mHz). This is mainly because, since the scheduling is repeated with a short period, it reduces the forecast error, which leads different s-UC strategies to produce similar schedules for the generators. This indicates that system operators may prefer to use deterministic approaches when scheduling the system as the complexity of the stochastic one

Table 10.5 15-min scheduling period—σ_{CoI} for different s-UC probabilities with $j = 10\%$

Scenario	s-UC [%]	MC-TDS [%]	$\sigma_{\mathrm{CoI}}[10^{-4}\,\mathrm{pu(Hz)}]$
1	20 60 20	20 60 20	8.47
2	40 50 10	20 60 20	8.59
3	100 0 0	20 60 20	8.67
4	0 100 0	20 60 20	8.72
5	0 0 100	20 60 20	8.02
6	0 20 80	20 60 20	8.17
7	0 40 60	20 60 20	8.32
8	0 60 40	20 60 20	8.41
9	0 80 20	20 60 20	8.52

Table 10.6 15-min scheduling period—σ_{CoI} for different s-UC probabilities and different j

Scenario j	s-UC [%]	MC-TDS [%]	σ_{CoI} $[10^{-4}\,\mathrm{pu(Hz)}]$	
1	10	20 60 20	20 60 20	8.47
1	20	20 60 20	20 60 20	12.58
1	30	20 60 20	20 60 20	17.20
5	10	0 0 100	20 60 20	8.02
5	20	0 0 100	20 60 20	12.41
5	30	0 0 100	20 60 20	17.37

Table 10.7 15-min scheduling period—σ_{CoI} for different s-UC probabilities with $j = 10\%$

Scenario	s-UC [%]	MC-TDS [%]	σ_{CoI} $[10^{-4}\,\mathrm{pu(Hz)}]$
1	0 0 100	0 0 100	8.01
2	0 100 0	0 100 0	2.90
3	100 0 0	100 0 0	1.46

does not provide a solution with a significant added value for the operation of the system.

Finally, Table 10.7 compares three deterministic cases, namely low, medium and high s-UC wind power scenarios. The deterministic low wind power scenario leads to better dynamic behavior, that is, lower σ_{CoI}.

Impact of s-UC Wind Uncertainty

To simulate the impact of the s-UC wind power uncertainty, the uncertainty level is increased from $j = 10\%$ to $j = 40\%$ with a step of 10% (volatility is kept constant). Four σ_{CoI} are calculated and Fig. 10.19 shows σ_{CoI} as a function of wind power uncertainty. The higher wind uncertainty, the higher σ_{CoI}. This relationship is almost linear within the used range. This suggests that as power systems accommodate more

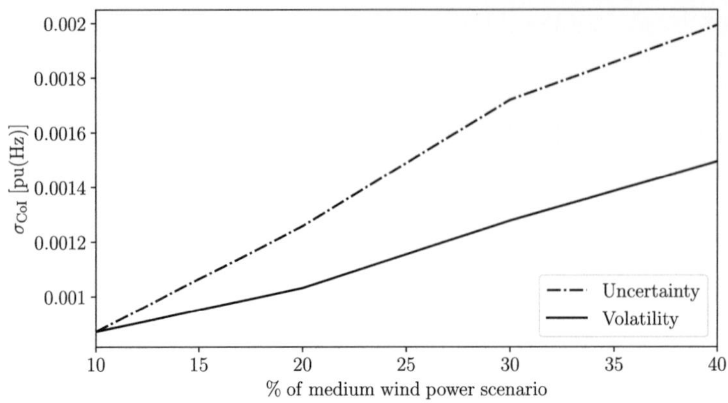

Fig. 10.19 15-min scheduling period—σ_{CoI} as a function of wind uncertainty and volatility

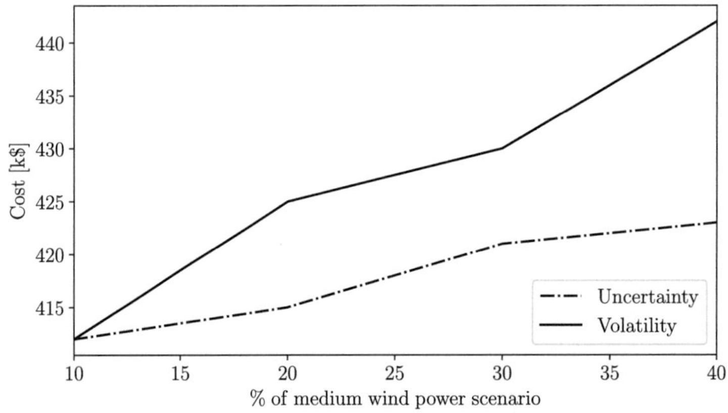

Fig. 10.20 15-min scheduling period—Expected cost as a function of wind uncertainty and volatility

RES, TSOs will have to ensure they have the necessary sources (power reserves) to cope with this uncertainty.

Figure 10.20 shows the total cost as a function of wind uncertainty. The higher the wind uncertainty, the higher the cost due to more ramp-up and ramp-down of generators. Hence, despite RES being a cheap source of energy, their intermittent nature, requires more reserves to cope and so there will be an increase of the cost of ancillary service.

Impact of s-UC Wind Volatility
To assess the impact of s-UC wind power volatility on power system dynamics, we consider different volatility levels. In particular, the standard deviation of wind scenarios is increased from 10% to 40% with a step of 10% while uncertainty is kept constant.

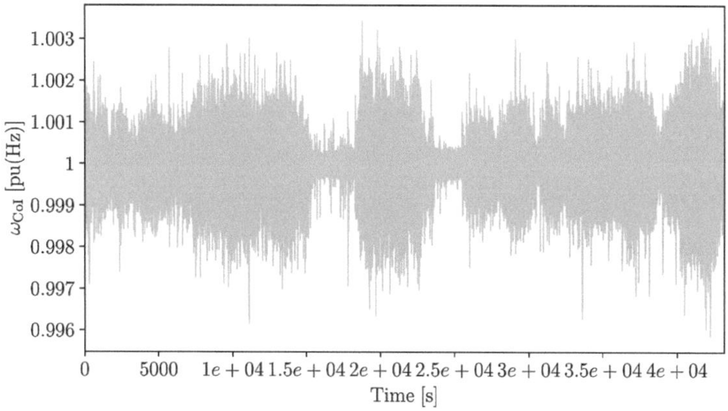

Fig. 10.21 5-min scheduling period—Trajectories of ω_{CoI} for 12 h

Figure 10.19 shows σ_{CoI} as a function of wind power standard deviation. The higher the wind power volatility, the higher the frequency variations. Similarly to the results shown in Fig. 10.19, this relationship appears to be almost linear within the considered range. As mentioned above, this supports the idea for increased ancillary services by TSOs in the future.

Figure 10.20 shows the total cost as a function of wind volatility. It can be seen that higher wind power volatility leads to higher costs due to higher ramping of generating units. Wind power volatility has thus a more significant effect than uncertainty on costs.

5-min Scheduling

In this scenario, the wind power uncertainty level is assumed proportional lower compared to the 15-min scheduling period ($j = 3.33\%$). This assumption is made based on the 15-min rate of change of wind power shown above (Fig. 10.14). Next, the base case scenario is depicted in Figs. 10.21, 10.22. Compared to the base-case scenario in the 15-min case study (Fig. 10.17), frequency variations are lower (Fig. 10.21). Note also that high frequency variations correspond to the wind ramp-up. Regarding the total operating cost, its average value is approximately \$408,000, and hence, lower than in the 15-min case study due to lower uncertainty.

Impact of Different s-UC Strategies

Furthermore, a sensitivity analysis similar to the 15-min scheduling is carried out and Table 10.8 shows relevant results. A reduction of the value of σ_{CoI} is common to all scenarios. This is due to the lower level of wind power uncertainty. In particular, if we compare scenarios 1, 3 and 5, respectively, we can see that the differences are less significant. It appears that, if the uncertainty level is low and the system is scheduled more frequently, then differences between d-UC and s-UC becomes less evident. This result can be explained by the fact that, even if there is an error in the forecast, the machines and the relevant controls will easily account for it. Therefore, if shorter interval of s-UC are used and the system is scheduled more frequently, like,

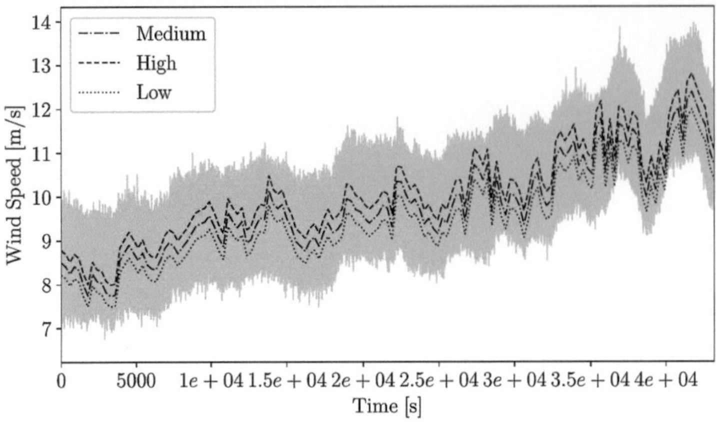

Fig. 10.22 5-min scheduling period—Trajectories of wind speeds for 12 h. Dashed lines indicate the high, medium and low scenarios, respectively, while black line indicate the same scenarios with inclusion of volatility

Table 10.8 5-min scheduling period—σ_{CoI} for different s-UC probabilities with $j = 3.33\%$

Scenario	s-UC [%]	MC-TDS [%]	σ_{CoI} [10^{-4} pu(Hz)]
1	20 60 20	20 60 20	5.45
2	40 50 10	20 60 20	5.51
3	100 0 0	20 60 20	5.44
4	0 100 0	20 60 20	5.70
5	0 0 100	20 60 20	5.48
6	0 20 80	20 60 20	5.53
7	0 40 60	20 60 20	5.55
8	0 60 40	20 60 20	5.55
9	0 80 20	20 60 20	5.59

Table 10.9 5-min scheduling period—σ_{CoI} for different s-UC probabilities with $j = 3.333\%$

Scenario	s-UC [%]	MC-TDS [%]	σ_{CoI} [10^{-4} pu(Hz)]
1	0 0 100	0 0 100	5.46
2	0 100 0	0 100 0	3.38
3	100 0 0	100 0 0	2.43

for example, in Australia [44], then system operators can still rely on deterministic approaches without compromising the dynamic performance of the system.

The sensitivity analysis in the case of the perfect forecast is shown in Table 10.9. The deterministic case with low wind gives better long-term frequency behavior, thus confirming the conclusions drawn for the 15-min s-UC.

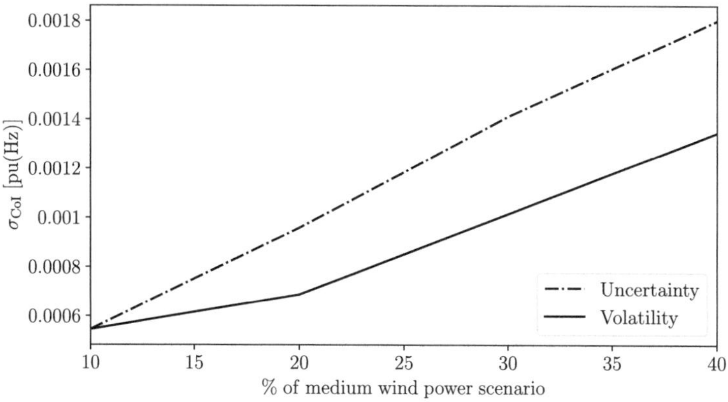

Fig. 10.23 5-min scheduling period—σ_{CoI} as a function of wind uncertainty and volatility

Impact of s-UC Wind Uncertainty

Moreover, similar to the 15-min case, the effect of wind power uncertainty on power system dynamics using 5-min scheduling is discussed. With this aim, the wind power uncertainty level is increased from $j = 10\%$ to $j = 40\%$ with a step of 10%.

Figure 10.23 shows σ_{CoI} as a function of wind power uncertainty. Again, we can see that such a relationship is almost linear within the used range. This suggests that, even using shorter s-UC time scales, higher shares of RES will likely affect the system's dynamic performance.

Impact of s-UC Wind Volatility

Following the same procedure as in the 15-min case study, the standard deviation of wind power scenarios is increased from 10% to 40% with a step of 10%. Then, Fig. 10.23 shows σ_{CoI} as a function of wind power volatility. This relationship is almost linear within the considered range, and these findings support the conclusions made above.

Finally, the impact of wind power volatility on costs is shown in Figure 10.24. Results indicate that the higher the wind power volatility, the higher the cost due to more ramping of generating units.

Impact of Contingency on the Transient Response of the System Using s-UC and d-UC

This section discusses whether a contingency leads to different dynamic behavior of the system if using a s-UC or d-UC. With this aim, the comparison is performed using scenario 1 (stochastic) and scenario 5 (deterministic) from Table 10.1.

A three-phase fault is applied at $t = 900$ s and cleared after 200 ms by disconnecting line 1. The impact of the contingency is shown in Figs. 10.25 and 10.26. Specifically, Fig. 10.25 depicts the trajectories of the rotor speed of the relevant machines during the contingency when using a s-UC. Similarly, Fig. 10.26 depicts the trajectories of the rotor speed of the relevant machines during the contingency when using a d-UC. Results indicate that for the considered case study the impact

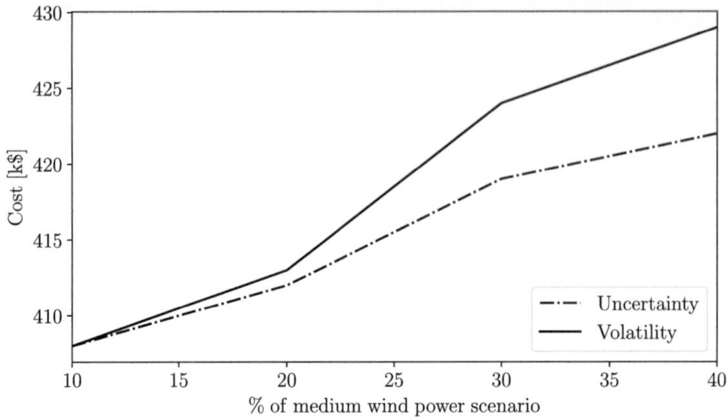

Fig. 10.24 5-min scheduling—Expected cost as a function of wind uncertainty and volatility

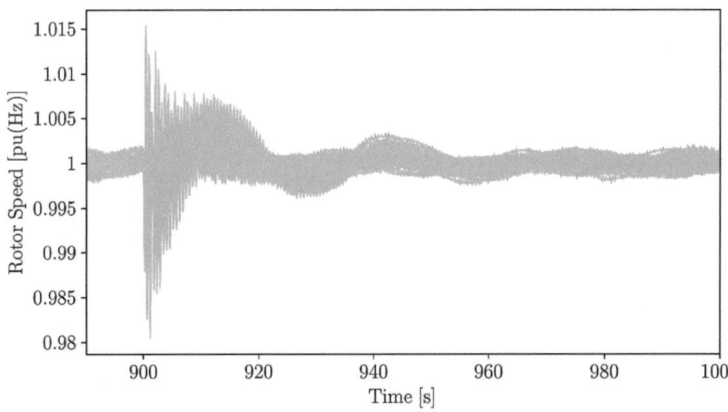

Fig. 10.25 s-UC and 25% wind penetration—Trajectories of the rotor speed of relevant machines following a contingency

of contingency is almost identical. This is because the generator schedules obtained with the s-UC and d-UC do not differ significantly.

The same contingency is applied for the case of 50% wind penetration. Figures 10.27 and 10.28 show the trajectories of the rotor speed of the relevant machines during the contingency when using a s-UC and d-UC, respectively. In this case, the s-UC leads to a better transient response of the system following a contingency. It appears that one cannot know a priori which strategies of s-UC are better from the dynamic viewpoint of the system before solving both s-UC and d-UC problems.

Impact of Renewable Penetration on Long-Term Frequency Deviations Using Different s-UC Strategies

Increasing the penetration levels of RES changes the system's stability and dynamic performance and makes real-time system operation more difficult for TSOs [22]. In

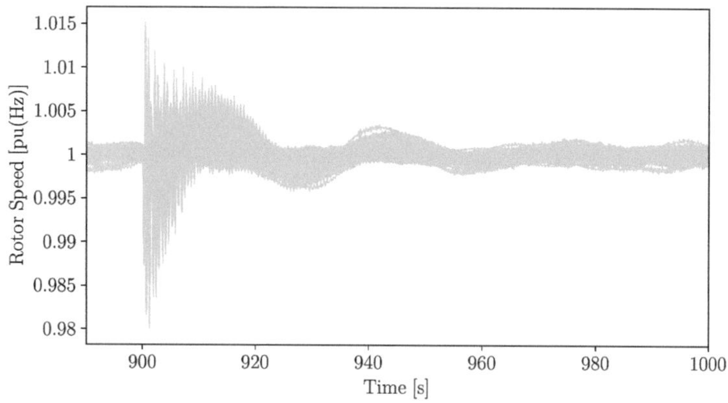

Fig. 10.26 d-UC and 25% wind penetration—Trajectories of the rotor speed of relevant machines following a contingency

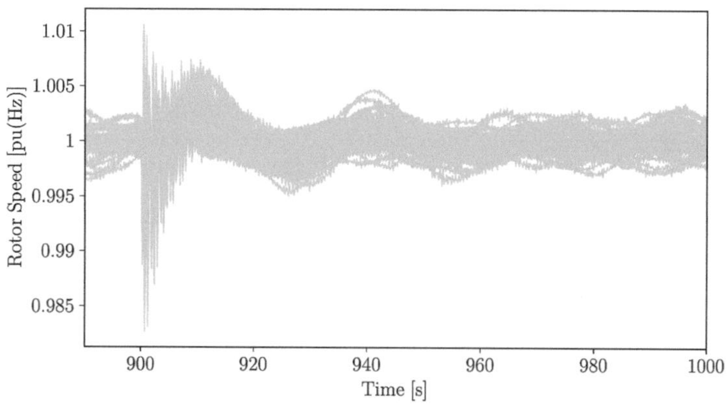

Fig. 10.27 s-UC and 50% wind penetration—Trajectories of the rotor speed of relevant machines following a contingency

this context, this section focuses on the impact of high penetration levels of RES, namely 50%, on the long-term frequency deviations using different s-UC strategies. A sensitivity analysis with respect to different s-UC probabilities for the low, medium and high wind power scenario is carried out with this aim.

Table 10.10 shows some relevant results of the analysis. There is a significant increase in the value of σ_{CoI} in all scenarios compared to 25% of wind penetration. This is to be expected as fewer synchronous generators that provide frequency regulation (primary and secondary) are now scheduled to be online and more power is being produced by stochastic sources.

Interestingly, the deterministic scenario with high wind power (scenario 5) is the worst with $\sigma_{\mathrm{CoI}} = 2.61 \cdot 10^{-3}$. Scenario 5, in fact, schedules fewer synchronous generators to be online compared to the same scenario in the previous Section. In

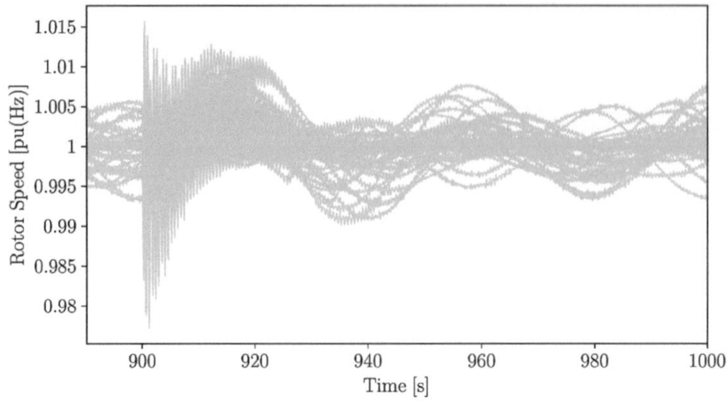

Fig. 10.28 d-UC and 50% wind penetration—Trajectories of the rotor speed of relevant machines following a contingency

Table 10.10 15-min scheduling period—σ_{CoI} for different s-UC probabilities with $j = 10$ and 50% wind penetration

Scenario	s-UC [%]	MC-TDS [%]	σ_{CoI} [10^{-4} pu(Hz)]
1	20 60 20	20 60 20	19.03
2	40 50 10	20 60 20	18.95
3	100 0 0	20 60 20	19.09
4	0 100 0	20 60 20	19.09
5	0 0 100	20 60 20	26.18

other words, there is less frequency regulation available online to cope with wind power uncertainty. Therefore, depending on the level of wind power uncertainty and wind penetration level, TSOs can solve a s-UC or d-UC with high wind. Specifically, according to our results, it is better to solve a d-UC with high wind power for low wind power uncertainty ($j < 30\%$) and 25% wind penetration level. On the other hand, if the wind penetration level is 50%, TSOs can solve a unit-commitment with medium and low wind power.

Impact of s-UC Wind Scenarios on the Dynamic Response of the System

In this section, we compare the impact of the number of s-UC wind scenarios on long-term frequency deviations. The comparison is performed using a 15-min scheduling time period and 25% wind penetration. With this aim, we consider 10 wind power scenarios and perform a sensitivity analysis. Wind scenarios are built according to the wind maximum variation width j (see Fig. 10.14). For comparison, the base-case scenario is assumed to have the following features: s-UC probabilities for the 10 scenarios (starting from low to high wind) are set as follows: 5, 5, 10, 10, 30, 10, 10, 10, 5, 5%. Similarly, when we run the MC-TDS, the system is assumed to be

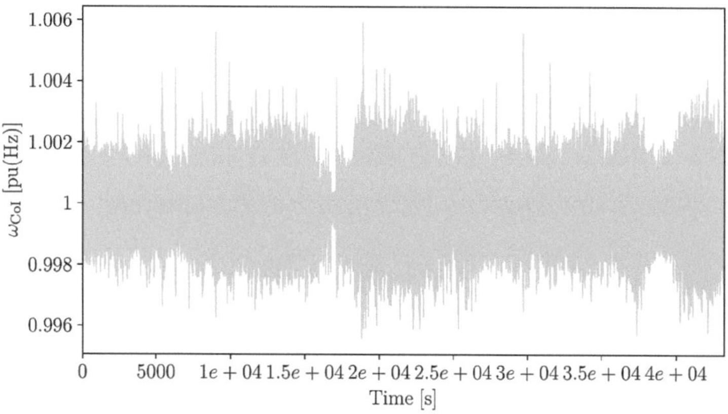

Fig. 10.29 15-min scheduling period—Trajectories of $\omega_{\rm CoI}$ for 10 s-UC wind power scenarios

Table 10.11 Different s-UC probabilities with $j = 10\%$ and 10 wind scenarios

Scenario	s-UC [%]
1	5, 5, 10, 10, 30, 10, 10, 10, 5, 5
2	20, 5, 5, 10, 10, 5, 10, 10, 5, 20
3	100, 0, 0, 0, 0, 0, 0, 0, 0, 0
4	0, 0, 0, 0, 100, 0, 0, 0, 0, 0
5	0, 0, 0, 0, 0, 0, 0, 0, 0, 100

Table 10.12 15-min scheduling period—$\sigma_{\rm CoI}$ for different s-UC probabilities with $j = 10\%$ and 10 wind scenarios

Scenario	MC-TDS [%]	$\sigma_{\rm CoI}$ [10^{-4} pu(Hz)]
1	5, 5, 10, 10, 30, 10, 10, 10, 5, 5	8.74
2	5, 5, 10, 10, 30, 10, 10, 10, 5, 5	8.77
3	5, 5, 10, 10, 30, 10, 10, 10, 5, 5	8.82
4	5, 5, 10, 10, 30, 10, 10, 10, 5, 5	8.99
5	5, 5, 10, 10, 30, 10, 10, 10, 5, 5	8.90

with the following probabilities: 5%, 5%, 10%, 10%, 30%, 10%, 10%, 10%, 5%, 5%, starting from low to high wind scenario, respectively.

Figure 10.29 shows the trajectories of $\omega_{\rm CoI}(t)$ for this base case scenario. Compared to Fig. 10.17 (3 s-UC wind power scenarios), there is no significant difference in the system's dynamic behavior. To further support this, Tables 10.11 and 10.12 show some of the relevant results of the sensitivity analysis. As it can be seen, the long-term frequency deviations are similar to those in Table 10.1 and do not differ significantly between scenarios. It appears that, for the considered case, increasing the number of s-UC wind power scenarios from 3 to 10 does not have a significant impact on the dynamic response of the system. Furthermore, these results support the above conclusion that a highly sophisticated s-UC might not be necessary if the scheduling is repeated with a short period.

10.3.7 Sensitivity Analysis W.r.t. Control Parameters

This example provides a sensitivity analysis of the impact of different frequency controllers/machine parameters using a s-UC, a d-UC and different scheduling time periods [45]. This analysis allows comparing and drawing conclusions on the effect of these parameters on the dynamic performance of the system. The case study also compares the impact that different formulations of s-UC have on the dynamic performance of power systems.

The focus of this case study is on the first 4 h of the planning horizon. Wind power uncertainty, volatility and rolling planning horizon within the s-UC are modeled as in [46]. All other data are the same as in case study 1. The case study results are based on a MC, where 50 simulations are considered for each scenario.

Sensitivity Analysis with Respect to Control and Machine Parameters
In this section, we vary (one at a time) three relevant frequency control/machine parameters and evaluate their impact on $\sigma_{\rm CoI}$.

The following base-case scenario is considered: the value of the gain of the AGC is taken equal to $K_o = 50$, the value of the droop of the TGs is taken equal to $\mathcal{R} = 0.05$, and the value of inertia of the machines is taken equal to the original values. In the scenarios below, the total inertia of the system and the gain of AGC is decreased up to 45% from the base case. Similarly, the aggregated system droop is increased up to 45% from the base case. Finally, since system operators still rely on d-UC and different sub-hourly scheduling periods [47], this sensitivity is performed using 15- and 5-min periods and different s-UC formulations.

In the following, the sub-indexes S and D indicate control parameters for stochastic and deterministic scenarios, respectively.

s-UC with 15-min Scheduling Period
In this scenario, we use a 15-min scheduling time period and set the s-UC probabilities for the low, medium and high wind power scenario equal to 20%, 60%, 20%, respectively. Same probabilities are used when solving the MC-TDS. Next, the relevant frequency control/machine parameters are varied accordingly up to 45% of their base case value.

Figure 10.30 shows the effect of the variation of these parameters on $\sigma_{\rm CoI}$. The gain of AGC, K_S, has the highest impact on $\sigma_{\rm CoI}$. The relationship between the gain K_S, the droop $\mathcal{R}_S$, and $\sigma_{\rm CoI}$ is almost linear within the used range. This indicates larger frequency deviations as synchronous generators are replaced with RES (assuming that RES will not provide frequency regulation). On the other hand, the inertia H_S appears to have a small impact on long-term frequency deviations. This result indicates that while inertia is the main parameter impacting on the frequency dynamics

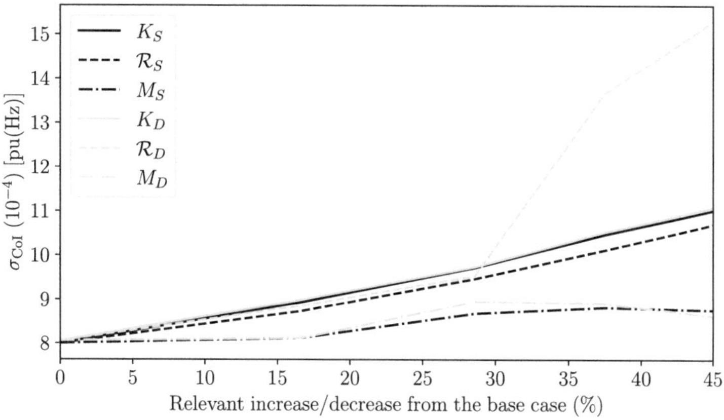

Fig. 10.30 15-min scheduling period—σ_{CoI} as a function of generator parameters using the complete s-UC and d-UC models

following a major contingency (see (1.67)), this is not the case on its impact on the σ_{CoI} (see also [48], which draws a similar conclusion).

d-UC with 15-min Scheduling Period

As discussed above, system operators still rely on a d-UC formulation when scheduling the system. Therefore, it is essential to compare the impact of the variation of the relevant control/machine parameters on σ_{CoI}, using a s-UC and a d-UC. With this aim, we set the d-UC probabilities for the low, medium and high wind power scenarios equal to 0%, 100%, 0%, respectively. Thus, perfect forecast, which corresponds to the medium scenario, is assumed. However, when solving the MC-TDS, 20%, 60%, 20% probabilities are used to generate the three wind power scenarios. This creates a mismatch between forecast and actual wind variations and allows evaluating the robustness of the s-UC and d-UC formulations.

Figure 10.30 shows the impact on σ_{CoI} of varying control/machine parameters. A relevant difference concerning the scenario above is that the droop $\mathcal{R}_D$ has the highest impact on σ_{CoI} when its value is $\geq 30\%$ with respect to the base case. On the other hand, the gain of the AGC, K_D and the inertia H_D have similar impact on σ_{CoI} as in the previous scenario. It appears that, the impact of different control parameters on σ_{CoI} depends on the UC formulation (deterministic or stochastic). Thus, it is not obvious which parameter, either $\mathcal{R}$ or K, has the highest impact on σ_{CoI}. Figure 10.30 compares the results of scenarios 1 and 2 and shows that using a s-UC leads to lower frequency variations. Furthermore, it should be noted that the differences between scenarios in Fig. 10.30, for example, $\mathcal{R}_S$ and $\mathcal{R}_D$ after 28%, are due to the fact that both models, namely s-UC and d-UC, produce different schedules for generators.

Sensitivity Analysis Using the Simplified and Alternative s-UC, and 15-min Scheduling Period

Here, we perform the same sensitivity analysis as above, but this time using the simplified and alternative s-UC models, respectively. Such an analysis allows

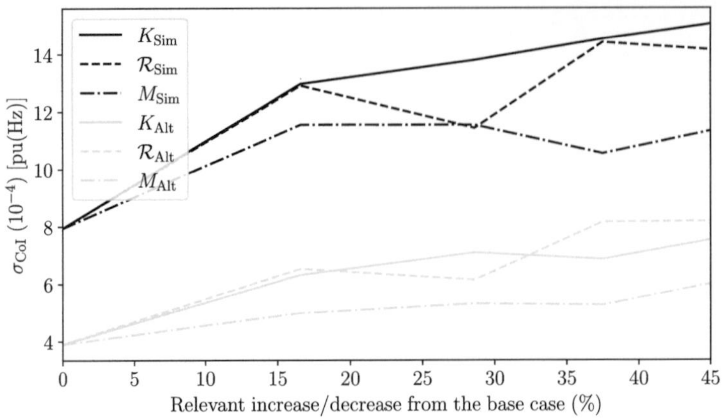

Fig. 10.31 15-min scheduling period—σ_{CoI} as a function of generator parameters using the simplified and alternative s-UC models

comparing and drawing conclusions on the impact of different frequency control/machine parameters on σ_{CoI}, using different sub-hourly s-UC models. With this aim, Fig. 10.31 shows the relevant results of the sensitivity analysis. Note that the sub-indexes Sim and Alt indicate control parameters for simplified and alternative s-UC scenarios, respectively. As expected, results are, in general, very similar to the ones discussed in previous sections. The gain of the AGC, in this case, K_{Sim} and K_{Alt}, has most of the time the highest impact on σ_{CoI}, as well as the inertia of the machines appears to have a small impact. Moreover, using an alternative s-UC model and varying its relevant parameters leads to lower frequency variations. This is because the alternative s-UC model schedules more generators to be online compared to the other three UC models. Thus, depending on the s-UC model, control parameters such as K and $\mathcal{R}$ can have a different impact on σ_{CoI}.

s-UC with 5-min Scheduling Period

This scenario investigates whether a shorter scheduling time period of s-UC, namely 5-min, changes the results and conclusions drawn for the 15-min scheduling period. Due to the shorter time period, the wind power uncertainty level within s-UC is lower compared to the previous section. Both s-UC and MC-TDS probabilities for the low, medium and high scenarios are set equal to 20%, 60% and 20%, respectively.

Figure 10.32 shows the effect of the variation of the relevant control parameters on σ_{CoI}. The most visible effect of reducing the time period to 5-min is that the value of σ_{CoI} decreases in all scenarios due to a lower wind power uncertainty. Also in this case, the gain of the AGC, K_S, has the highest impact on σ_{CoI}. The relationship is linear in the considered range, which indicates the need to increase the frequency regulation to keep long-term frequency deviations within certain limits. Finally, the inertia has little effect on σ_{CoI}, supporting the conclusions above.

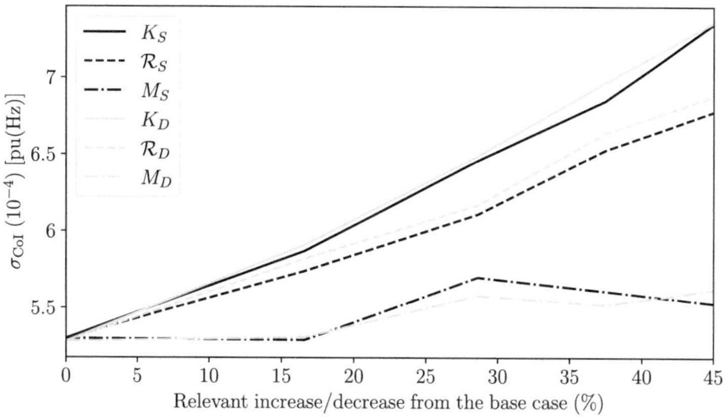

Fig. 10.32 5-min scheduling period—σ_{CoI} as a function of generator parameters

d-UC with 5-min Scheduling Period

Figure 10.32 shows the effect of the variation of control/machine parameters using a 5-min scheduling and d-UC. Results are very similar to the 5-min scheduling s-UC. To better show the differences, Fig. 10.32 compares the results of both scenarios, which shows that the s-UC leads, in general, to lower frequency variations. Also, in both scenarios, the gain of the AGC has the highest impact on σ_{CoI}, whereas the inertia has a small impact.

Comparison of Different s-UC Models

Even though system operators are skeptical regarding s-UC approaches due to their complexity, they acknowledge the need to better represent uncertainty when scheduling the system [40].

For this reason, the objective of this section is to compare the impact that these models have on long-term frequency quality. Such a comparison is made using a 15-min scheduling period.

We first compare the impact on the dynamic response of the power system of the s-UC formulations (10.13)–(10.31) (complete s-UC) and (10.35)–(10.39) (simplified s-UC).

Figures 10.33 shows the ω_{CoI} obtained sing the complete s-UC model. The trajectory of ω_{CoI} for the basic s-UC model are substantially identical to that of the complete one. In fact, $\sigma_{\mathrm{CoI}} = 8.00 \cdot 10^{-4}$ pu(Hz) for the complete s-UC and that of the simplified s-UC is $\sigma_{\mathrm{CoI}} = 7.94 \cdot 10^{-4}$.

Figures 10.35 and 10.36 show the mechanical power of the conventional synchronous generators 1, 2 and 4, of the complete and simplified s-UC model. The two models produce similar schedules. In fact, at the beginning of the planning horizon, the simplified s-UC model alternates the schedules for generators 2 and 4, and after 6,000 s it produces the same schedules as the complete s-UC model. It appears, thus, that for this particular system and for normal operation conditions, the differences between using a complete s-UC model and a simplified one are negligible. These

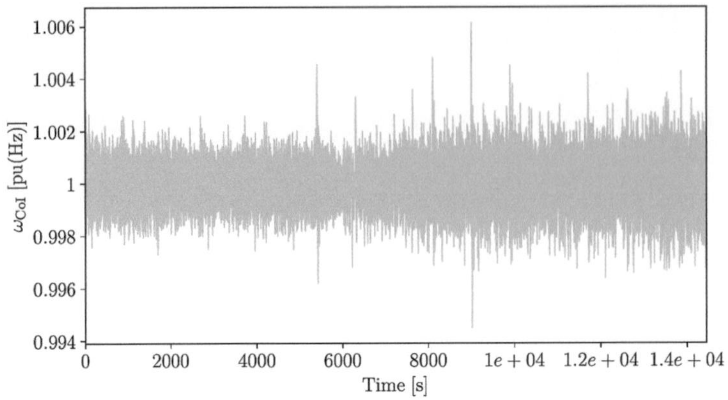

Fig. 10.33 Trajectories of ω_{CoI} for 15-min scheduling period and complete s-UC model

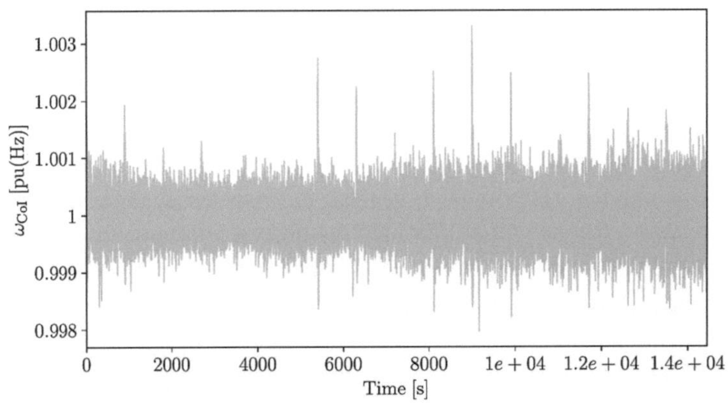

Fig. 10.34 Trajectories of ω_{CoI} for 15-min scheduling period and alternative s-UC model

results suggest that an involved UC formulation is not necessarily the best in normal operating conditions of the system.

Next, we compare the impact on long-term frequency deviations of the problem (10.13)–(10.31) (complete s-UC) and the problem (10.40)–(10.49) (alternative s-UC). With this aim, a MC-TDS per each s-UC formulation is carried out.

Figure 10.34 shows ω_{CoI} for the alternative s-UC model. Compared to the complete s-UC model (see Fig. 10.33), the alternative formulation leads to lower frequency variations, namely, $\sigma_{\text{CoI}} = 3.90 \cdot 10^{-4}$. In other words, the differences between the models is about 48%. This is due to fact that the two s-UC formulations produce slightly different schedule for generators. Specifically, the alternative s-UC model schedules a few more generators (Fig. 10.37) compared to the complete s-UC model (Fig. 10.35). This, in turn, implies more regulation, which helps better manage wind uncertainty. This is shown in Fig. 10.34 where, in the time between two scheduling

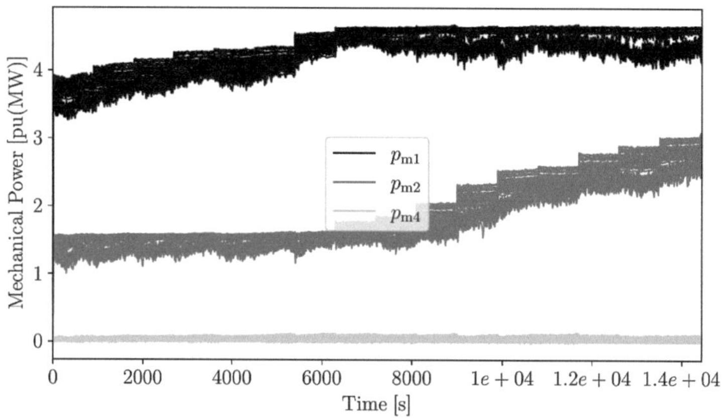

Fig. 10.35 Mechanical power of synchronous generators 1, 2 and 4, for 15-min scheduling period and complete s-UC model

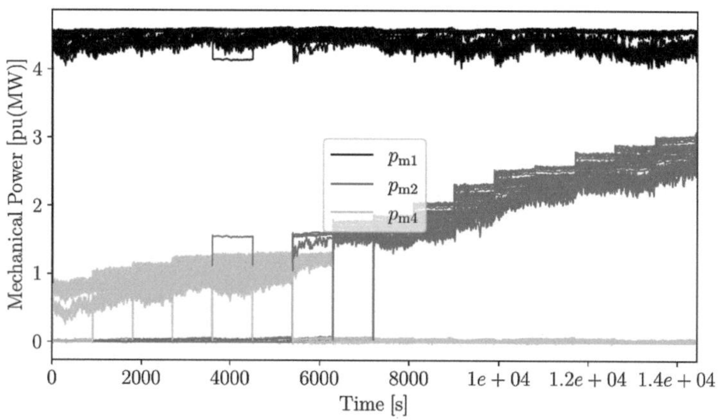

Fig. 10.36 Mechanical power of synchronous generators 1, 2 and 4, for 15-min scheduling period and simplified s-UC model

events, frequency variations are lower due to the increased frequency regulation available in the system.

This result had to be expected. From the dynamic point of view, in fact, it is better to schedule more conventional synchronous generators, which provide both PFR and SFC, rather than wind generation.

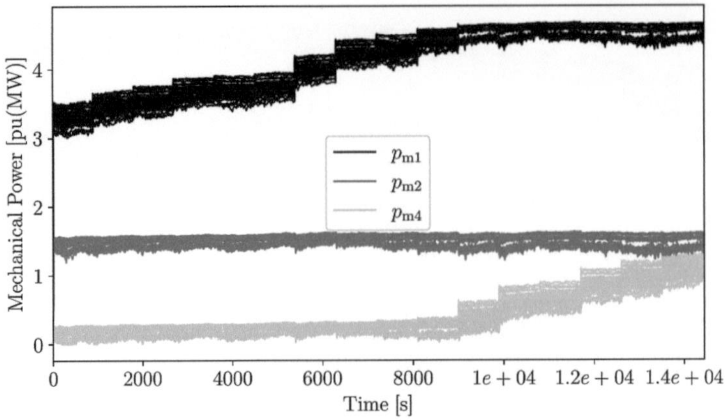

Fig. 10.37 Mechanical power synchronous generators 1, 2 and 4, for 15-min scheduling period and alternative s-UC model

10.4 Final Remarks

This chapter presents a framework to embed the sub-hourly d-UC and s-UC problems into TDS. The framework allows studying the impact of different scheduling periods of the d-UC/s-UC and the effect of the net load variability and uncertainty on long-term frequency quality, particularly on DFDs and σ_{CoI}. In the proposed approach, the d-UC/s-UC problems are modeled as a slow "discrete controllers" that responds to the time-varying loading condition of the grid by changing the power set-points of the TGs. In particular, sub-hourly MILPs d-UC/s-UC are used to accommodate the variable net-load.

Deterministic UC

The results of the first part of the chapter indicate that reducing the time intervals at which the d-UC problem is solved help: (i) reduce frequency variations in terms of DFDs; (ii) reduce electricity price; and (iii) mitigate the impact of volatility and contingencies. These findings suggest that power system operators should switch to the utilization of the sub-hourly modeling instead of conventional hourly resolution. This is particularly important given the need for higher flexibility in modern power systems characterized by high shares of RES.

In addition, the results show that: (i) increasing the gain K_o of the AGC decreases σ_{CoI}; (ii) increasing $\mathcal{R}$ leads to higher values of σ_{CoI}; and (iii) decreasing inertia H increases the value of σ_{CoI}. However, depending on the scheduling period used and the amount of noise in the system, increasing the inertia H may lead to a higher σ_{CoI}. As aforementioned, this is because the frequency takes a longer time to recover to its nominal value after each d-UC scheduling.

Stochastic UC

The second part of the chapter performs a similar analysis but now focusing on the impact of the sub-hourly s-UC problem on long-term frequency performance. More specifically, the chapter focuses on the effect of different s-UC strategies and different wind uncertainty and volatility scenarios included in the s-UC on frequency variations. Also, the chapter performs a thorough sensitivity analysis to assess the impact on long-term frequency quality of different frequency control/machine parameters and different sub-hourly s-UC formulations. With this aim, various standard and sophisticated s-UC are used to capture wind uncertainty, while the variability is captured through a sub-hourly UC. Then, the sub-hourly s-UC are embedded into a TDS, and a rolling approach is used to account for wind and load forecast updates.

Simulation results based on MC-TDS show no significant difference on long-term frequency deviations of the system when using different s-UC strategies. Results also suggest that for low wind uncertainty and 25% wind penetration level, system operators may want to solve a d-UC with high wind. However, as the s-UC wind uncertainty increases, then a s-UC approach is to be preferred. In addition, if the system is scheduled more frequently, then differences between stochastic and deterministic approaches become less evident. From the dynamic performance point of view, thus, the computational complexity of the stochastic methods does not provide an added value compared to that of deterministic ones.

Furthermore, results show that s-UC wind power uncertainty has a more substantial impact than volatility on σ_{CoI}. They also indicate an almost linear relationship between s-UC wind uncertainty and volatility and frequency variations. This means that as the share of RES deepen into modern power systems, it will be difficult for TSOs to manage the real-time balance between generation and demand. Therefore, there is a clear need for a linear increase of the active power reserves, especially if wind and solar Photovoltaic (PV) are not geographically well-distributed.

Regarding the impact of s-UC wind scenarios, the case study shows that increasing the number of scenarios does not lead to any significant difference in the long-term frequency deviations. This result supports the conclusion that a highly sophisticated s-UC might not be needed (from a frequency quality perspective) if the scheduling is repeated frequently and with a short time period.

Moreover, results demonstrate that s-UC and d-UC for 25% wind penetration lead to an almost identical transient response of the system following a contingency. On the one hand, s-UC leads to a better transient response of the system in case of 50% wind penetration. On the other hand, when the wind penetration level reaches 50%, solving a s-UC or d-UC with medium and low wind power leads to lower long-term frequency deviations.

Simulation results reveal that the gain of the AGC is the main parameter impacting the long-term frequency deviations. On the other hand, synchronous machine inertia has a negligible effect on σ_{CoI}. The results also highlight that the power system's frequency behavior differences are marginal when using a complete or a simplified s-UC model. Furthermore, a s-UC leads to lower variations of the frequency compared to a d-UC. An interesting result is that different formulations of the s-UC problem can schedule conventional power plants. These differences can significantly impact long-term power system dynamics such as long-term frequency quality, and hence the formulation of the s-UC has to be carefully chosen.

In Brief

- The simulation framework discussed in this chapter appears as an useful tool for operating power systems with growing volumes of RES.
- Shorter scheduling periods of the d-UC and s-UC are to be preferred.
- TSOs can rely on deterministic approaches when scheduling the system for wind power penetration levels below 25%.
- For wind power penetration levels higher than 25%, stochastic approaches should be employed and preferred by TSOs.
- There is a need for a linear increase of active power reserves as the share of RES increases, especially if wind and solar PV are not geographically well-distributed.
- The gain of the AGC and the formulation of the s-UC have to be carefully chosen as they impact significantly the dynamic performance of the system.

Open Questions

Despite the hundreds, if not thousands, of papers on its potential application in power systems, no real-world application of s-UC has been implemented yet. Thus, it remains unclear whether s-UC brings the benefits that are described in the literature. Moreover, dispatch of BESS is still an issue as TSOs need to consider state of charge management. It is important to understand how this will work in practice as it could have a significant impact on both short- and long-term frequency quality.

References

1. European Commission (2017) Commission Regulation (EU) 2017/2195 of 23 Nov 2017 establishing a guideline on electricity balancing. https://www.eur-lex.europa.eu
2. Economics F (2024) Assessing the impacts of implementing central dispatch in GB. www.frontier-economics.com
3. European Union Emissions Trading System (2023) Central dispatch model. https://www.emissions-euets.com
4. ENTSO-E (2024) Nordic balancing philosophy . https://www.eepublicdownloads.entsoe.eu
5. EirGrid and SONI (2024) Balancing market principles statement. A guide to scheduling and dispatch in the single electricity market. https://www.eirgrid.ie/
6. Fu Y, Li Z, Wu L (2013) Modeling and solution of the large-scale security-constrained unit commitment. IEEE Trans Power Syst 28(4):3524–3533
7. Fu Y, Shahidehpour M, Li Z (2005) Security-constrained unit commitment with ac constraints. IEEE Trans Power Syst 20(3):1538–1550

8. Milano F, Drfler F, Hug G et al (2018) Foundations and challenges of low-inertia systems (invited paper). In: Power systems computation conference (PSCC), pp 1–25
9. Hillberg E, Zegers A, Herndler B et al (2019) Flexibility needs in the future power system, ISGAN, Technical report. Discussion paper
10. Deane JP, Drayton G, Gallachóir BÓ (2014) The impact of sub-hourly modelling in power systems with significant levels of renewable generation. Appl Energy 113:152–158
11. Gangammanavar H, Sen S, Zavala VM (2016) Stochastic optimization of sub-hourly economic dispatch with wind energy. IEEE Trans Power Syst 31(2):949–959
12. Luo J, Teng F, Bu S (2020) Stability-constrained power system scheduling: a review. IEEE Access 8:219 331–219 343
13. Paturet M, Markovic U, Delikaraoglou S et al (2020) Stochastic unit commitment in low-inertia grids. IEEE Trans Power Syst 35(5):3448–3458
14. Carrión M, Arroyo JM (2006) A computationally efficient mixed-integer linear formulation for the thermal unit commitment problem. IEEE Trans Power Syst 21(3):1371–1378
15. Conejo A, Baringo L (2017) Power system operations. Springer International Publishing (2017)
16. Milano F (2013) A Python-based software tool for power system analysis. IEEE PES Gen Meet 1–5
17. Chen Y, Liu F, Wei W et al (2016) Robust unit commitment for large-scale wind generation and run-off-river hydropower. CSEE J Power Energy Syst 2(4):66–75
18. ENTSO-E (2020) Report on deterministic frequency deviations. https://www.eepublicdownloads.entsoe.eu
19. Milano F, Zárate-Miñano R (2013) A systematic method to model power systems as stochastic differential algebraic equations. IEEE Trans Power Syst 28(4):4537–4544
20. Kazarlis SA, Bakirtzis AG, Petridis V (1996) A genetic algorithm solution to the unit commitment problem. IEEE Trans Power Syst 11(1):83–92
21. Mele FM, Zárate-Miñano R, Milano F (2019) Modeling load stochastic jumps for power systems dynamic analysis. IEEE Trans Power Syst 34(6):5087–5090
22. Yamashita K, Villanueva SM, Custem TV et al (2018) Modelling of inverter-based generation for power system dynamic studies. CIGRE Tech Brochure C **4**
23. Bouffard F, Galiana FD (2008) Stochastic security for operations planning with significant wind power generation. IEEE Trans Power Syst 23(2):306–316
24. Wang B, Hobbs BF (2016) Real-time markets for flexiramp: a stochastic unit commitment-based analysis. IEEE Trans Power Syst 31(2):846–860
25. Gu Y, Xie L (2017) Stochastic look-ahead economic dispatch with variable generation resources. IEEE Trans Power Syst 32(1):17–29
26. Kundur P, Paserba J, Ajjarapu V et al (2004) Definition and classification of power system stability IEEE/CIGRE joint task force on stability terms and definitions. IEEE Trans Power Syst 19(3):1387–1401
27. Kërçi T, Milano F (2019) A framework to embed the unit commitment problem into time domain simulations. In: IEEE international conference on environment and electrical engineering, 1–5 June 2019
28. Kërçi T, Milano F (2019) Sensitivity analysis of the interaction between power system dynamics and unit commitment. IEEE Milan PowerTech 2019:1–6
29. Yuan B, Zhou M, Li G et al (2015) Stochastic small-signal stability of power systems with wind power generation. IEEE Trans Power Syst 30(4):1680–1689
30. Wang X, Chiang H, Wang J et al (2015) Long-term stability analysis of power systems with wind power based on stochastic differential equations: model development and foundations. IEEE Trans Sustain Energy 6(4):1534–1542
31. Kundur P (1994) Power system stability and control. McGraw-Hill, New York
32. Milano F (2010) Power system modelling and scripting. Springer, London
33. IEA-WIND (2019) Design and operation of power systems with large amounts of wind power. www.vttresearch.com/sites/default/files/pdf/technology/2019/T350.pdf
34. Conejo A, Carrión M, Morales J (2010) Decision making under uncertainty in electricity markets. Springer

35. Blanco I, Morales JM (2017) An efficient robust solution to the two-stage stochastic unit commitment problem. IEEE Trans Power Syst 32(6):4477–4488
36. Morales JM, Conejo AJ, Pérez-Ruíz J (2009) Economic valuation of reserves in power systems with high penetration of wind power. In: IEEE PES general meeting, 1 July 2009
37. EirGrid, System and renewable data. https://www.eirgrid.ie//how-the-grid-works/renewables
38. Wang J, Shahidehpour M, Li Z (2008) Security-constrained unit commitment with volatile wind power generation. IEEE Trans Power Syst 23(3):1319–1327
39. Hberg M (2019) Fundamentals and recent developments in stochastic unit commitment. Electr Power Energy Syst 109:38–48 (2019)
40. Zheng QP, Wang J, Liu AL (2015) Stochastic optimization for unit commitment—a review. IEEE Trans Power Syst 30(4):1913–1924
41. Gómez Expósito A, Conejo AJ, Cañizares C (2018) Electric energy systems: analysis and operation. CRC Press (2018)
42. ENTSO-E (2019) Continental Europe significant frequency deviations—Jan 2019. https://www.eepublicdownloads.entsoe.eu
43. Schäfer B, Timme M, Witthaut D (2018) Isolating the impact of trading on grid frequency fluctuations. In: 2018 IEEE PES innovative smart grid technologies conference Europe (ISGT-Europe), pp 1–5
44. MacGill I (2010) Electricity market design for facilitating the integration of wind energy: experience and prospects with the Australian national electricity market. Energy Policy 38(7):3180–3191
45. Kërçi T, Giraldo JS, Milano F (2020) Sensitivity analysis of the impact of the sub- hourly stochastic unit commitment on power system dynamics. Energies 13(6)
46. Kërçi T, Giraldo JS, Milano F (2020) Analysis of the impact of sub-hourly unit commitment on power system dynamics. Electr Power Energy Syst 119:105819
47. Costley M, Feizollahi MJ, Ahmed S et al (2017) A rolling-horizon unit commitment framework with flexible periodicity. I J Electr Power Energy Syst 90:280–291
48. Vorobev P, Greenwood DM, Bell JH et al (2019) Deadbands, droop, and inertia impact on power system frequency distribution. IEEE Trans Power Syst 34(4):3098–3108

Chapter 11
Wind and Solar Generation

This chapter studies the impact of wind and solar generation on frequency quality, particularly long-term frequency quality. It does so based on operational data from the AIPS and through a correlation analysis to determine whether the penetration of wind and solar generation and the magnitude of system frequency fluctuations show any statistical correlation. This analysis is important for using TSOs as it is not yet clear, especially in large interconnected power systems, if wind and solar PV negatively impact the quality of system frequency. To answer these questions, we quantify, using Pearson correlation coefficient and p-value, the correlation between wind (Sect. 11.1) and solar PV (Sect. 11.2) penetration and frequency fluctuations.

11.1 Correlation Between Wind Penetration and Grid Frequency Variations

The power generated by wind and photovoltaic power plants, often referred to as variable RES, is affected by uncertainty and volatility. Moreover, RES, unlike conventional generation, is connected to the grid through power electronic converters and is thus *non-synchronous* in the sense that it does not respond to grid power unbalances by varying its frequency. For this reason, the high penetration of RES makes frequency control more complex as fewer synchronous generators are available to provide the system with inertia and power reserve.

The penetration of RES is expected to consistently increase in the future. For example, the Irish government has set a target to achieve 80% consumption of electricity from the RES by 2030. This is a significant step with respect to the current situation, see Fig. 11.1.

The reduction of conventional generation and, thus, of the total inertia available in the system, as well as the increase of uncertainty and volatility due to RES can lead to systems with potentially high frequency deviations, which, in turn, can lead

T. Kërçi and F. Milano, *Frequency Quality in Power Systems*, Power Systems,
https://doi.org/10.1007/978-3-032-04721-2_11

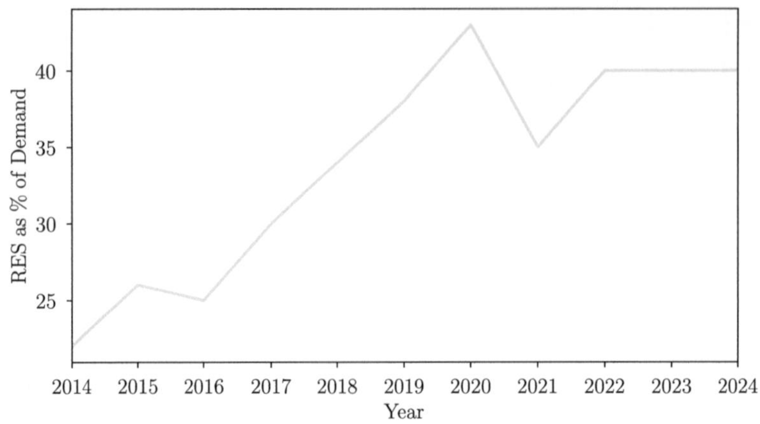

Fig. 11.1 Evolution of RES as % of AIPS demand

to higher risks of instability. With this regard, the first part of this chapter focuses on the impact of wind penetration on the frequency fluctuations in the grid based on historical data. The goal is to quantify the statistical correlation between wind generation and σ_f.

Several studies have been carried out to define the minimum level of the percent share of instantaneous demand to be supplied by synchronous generators and ensure a proper inertial response of the system. In [1], the impact of high penetration of RES on the frequency stability in the CE system is analyzed. In [2], the impact of higher penetration of RES on the frequency stability of the US Eastern Interconnection is discussed. A study on the Spanish grid, carried out in [3], considers RoCoF to calculate the limit of maximum penetration for wind energy in the grid. Given a certain power unbalance, in fact, the RoCoF can be used as an indirect measure of the total inertia present in the system.

The capacity of an electrical grid has a great impact on the definition of a maximum penetration of RES. The size of CE network makes it more robust as far as system frequency dynamics are concerned [1] (see also Chap. 5). On the other hand, the isolated systems, such as the AIPS, are potentially more fragile from the dynamic point of view [4]. A study based on simulation analyzes the impact of high RES, in isolated electrical power system, on frequency stability and system security [5]. The maximum SNSP and minimum inertia requirements in the AIPS based on simulation results is discussed in [6].

The most critical periods for frequency control are the periods with high RES output and low demand. This means fewer synchronous generators are available to fulfill the system inertia requirements. Such periods have a significant effect on frequency dynamics and thus on power system security [3]. In IE, the Delivering a Secure, Sustainable Electricity System (DS3) is aimed at increasing the SNSP level up to 75% [7]. Previously, the DS3 project has demonstrated that providing system-wide "synthetic inertia" FFR can be a solution to alleviate high RoCoF rates [8].

The aforementioned studies consider software simulation and predetermined scenarios. A different approach, based on actual measurements, is used in [9] to determine the correlation between IBR-based generation and RoCoF for the German and Austrian grids. This chapter makes a similar study as in [9] but considers σ_f instead of RoCoF.

11.1.1 Operational Constraints in the AIPS

Wind speed variations include *uncertainty* (variation with respect to a forecast average value in a given time period) and *volatility* (fast variations around the average value). Volatility is a low amplitude noise that averages out if the number of wind power plants is high. Volatility has thus a small impact on frequency deviations and, ultimately, cannot be distinguished from load variations and other noises present in the grid. The analysis carried out here focuses exclusively on the impact of the uncertainty of wind generation.

Another important aspect that has to be taken into account is the fact that, in the AIPS, wind generation is often not fully dispatched (this operation is called *wind dispatch down*) p_{WD}:

$$p_{\mathrm{WD}} = p_{\mathrm{wind-avail}} - p_{\mathrm{wind-gen}} , \tag{11.1}$$

where $p_{\mathrm{wind-gen}}$ is the actual wind power generation and $p_{\mathrm{wind-avail}}$ is the total power available from the wind. If $p_{\mathrm{WD}} > 0$, there is a wind power reserve and thus the stochastic variations of the wind do not affect the power unbalance of the network, and are consequently not responsible of frequency variations. Equation (11.1) is also illustrated in Fig. 11.2 below:

To properly decide the correlation between wind generation and frequency deviations, some precautions have to be taken into account. In particular, we have to exclude from the analysis the periods during which the load demand varies significantly (known as *demand ramping*). The load variation, in fact, leads to generator rescheduling that causes fast variations of the frequency. These variations are clearly independent from the wind generation.

The remainder of this section outlines the wind dispatch-down procedure and demand ramping up and down in the AIPS.

11.1.1.1 Wind Dispatch-Down

Renewable-rich power systems face several challenges such as ensuring dynamic stability [11], economically maintaining a balance of supply and demand, and managing local congestion/constraint [12]. For example, EirGrid and SONI are, at times, forced to dispatch down RES when there is more power available than can be accommodated by the grid. This may be necessary for various reasons, among others, local congestion (known as "Constraint") and stability driven limits on IBR-based

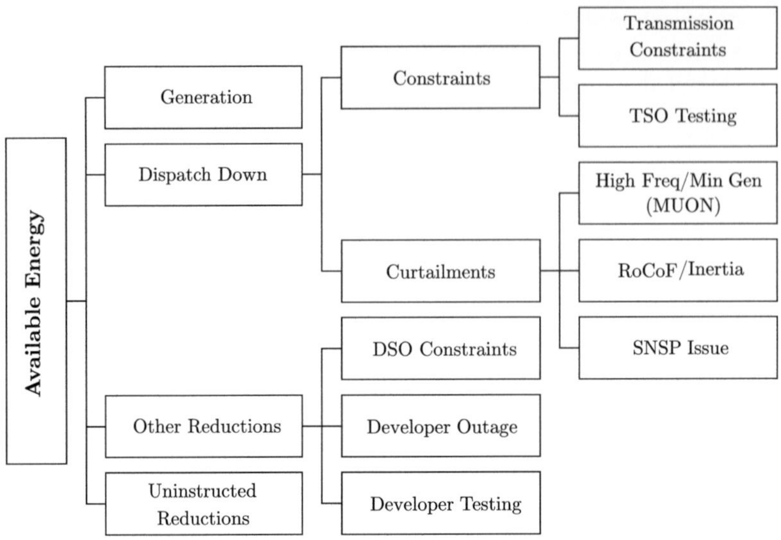

Fig. 11.2 Overview of dispatch down of wind in the AIPS [10]

generation (known as "Curtailment"). The term *dispatch down*, thus, is the result of the combined effect of constraint and curtailment (see Fig. 11.2).

Research in renewable and, in particular, wind energy curtailment in power systems with very high renewable penetration levels is vast. One of the main well-established conclusions in the literature is that curtailment will be the "new normal" for everyday operations [13]. This is because curtailment is seen both as a source of flexibility (maintain the supply-demand balance) and operating reserves [14]. For instance, references [15–17] propose to use the curtailed wind or solar power output to provide ancillary services such as AGC, ramping and spinning reserve to the TSO. In the same vein, reference [18] discusses the use of emerging technologies such as BESS to effectively balance the excess renewable energy output.

With respect to current and expected curtailed volumes of renewable output, reference [19] presents an analysis of solar power curtailment for four key countries namely Chile, China, Germany, and the United States. The paper finds that about 6.5 million MWh of solar power output was curtailed in these countries in 2018 and one of the main reasons is attributable to limited transmission capacity. Reference [20], on the other hand, uses a 2025 model of the Nordic power system to estimate the amount of future curtailment. The study considers two wind power scenarios namely 26 and 33 GW and finds curtailment to be below 0.3% and 1.7%, respectively. The study suggests that the most effective measures to decrease curtailment is increased transmission capacity, as well as flexibility of nuclear generation.

Regarding research on wind curtailment in AIPS, reference [21] discusses the effectiveness of CAES for mitigating wind curtailment on a 2020 scenario and shows that a 270 MW CAES plant in conjunction with a 75% SNSP limit can reduce wind

Table 11.1 Wind dispatch-down for the AIPS in 2014–2017

Year	2014	2015	2016	2017
Jan	–	4.3	3.5	–
Feb	–	4.2	3.1	1.7
Mar	–	8.8	–	3.3
Apr	–	2.0	1.3	3.6
May	–	4.3	1.2	3.5
Jun	–	4.8	–	4.1
Jul	3.4	3.7	–	3.2
Aug	3.6	5.6	–	2.9
Sep	1.8	2.5	–	5.1
Oct	–	3.9	1.8	10.6
Nov	–	–	1.3	2.6
Dec	4.9	6.3	3.3	–

curtailment levels to 2.6%. Using again a 2020 scenario of AIPS, reference [22] shows that when the SNSP limit is increased from 60 to 75% there is a reduction in wind curtailment from 14 to 7%, while for higher SNSP values wind curtailment is influenced primarily by the inclusion of transmission constraints. These conclusions only in part align with the analysis presented in this chapter.

Table 11.1 shows the volume of monthly wind dispatch-down as percentage of the total available wind energy per year under study [23].

Table 11.2 shows a breakdown of wind dispatch down on the AIPS over the last decade, including the part of constraint. Overall, curtailment is higher than constraint. These results suggest that as the AIPS accommodates higher and higher levels of wind, significant levels of wind curtailment are expected if no actions are taken to address them. As discussed in Sect. 11.1.1.2, EirGrid and SONI are working toward relaxing the operational constraints in order to make more room for wind and, in general, for more renewable generation in the future.

The technical procedures and constraints implemented by EirGrid are outlined below.

11.1.1.2 Operational Policy Constraints in AIPS

As discussed in Chap. 6 and shown in Table 6.2, EirGrid and SONI have developed various system-wide operational constraints to ensure system security and stability is maintained at all times. In particular, in order to ensure system stability, the AIPS has in place four operational constraints namely: (i) SNSP limit; (ii) MUON limit; (iii) RoCoF limit; and (iv) minimum inertia floor. These constraints are explicitly modeled in the scheduling and dispatch process of the TSOs. The evolution and relaxation of these constraints is crucial to achieve the IE and Northern Ireland (NI)

Table 11.2 Wind dispatch-down for the AIPS in 2011–2021

Year	Constraint [%]	Curtailment [%]	Total Dispatch Down Levels [%]
2011	0.4	1.8	2.2
2012	0.8	1.3	2.1
2013	0.9	2.3	3.2
2014	1.4	2.6	4.1
2015	1.8	3.3	5.1
2016	1.4	1.5	2.9
2017	1.2	2.7	4.0
2018	2.2	3.8	6.0
2019	4.0	3.7	7.7
2020	6.2	5.9	12.1
2021	4.4	3.0	7.4

targets of 80% of annual electricity demand coming from renewable generation by 2030 [24, 25]. Note that renewable generation in AIPS are mostly IBR-based unlike, for example, countries with large hydro (Norway) or geothermal resources (Iceland). A brief description of the above constraints is provided below.

System Non-synchronous Penetration

With increasing level of installed IBR-based generation capacity, and further HVDC interconnection to the AIPS, it is necessary to measure and limit the SNSP to ensure secure and prudent operation of the system [26]. Its formulation is shown in the equation below:

$$\text{SNSP } (\%) = \frac{\text{Wind} + \text{Solar} + \text{HVDC Imports}}{\text{System Demand} + \text{HVDC Exports}} 100\,. \tag{11.2}$$

This calculation is currently being modified to incorporate BESS and split the net import and export to sum of exports and imports only, respectively. The SNSP metric was developed more than a decade ago based on extensive studies to provide a single constraint (modeled in the scheduling and dispatch process of TSOs) that captures the range of issues. Its evolution over the years is shown in Fig. 11.3. SNSP limit is currently set at 75%. TSOs plan to relax and eventually remove SNSP as a constraint but maintain it as a key operational reporting metric.

Minimum Number of Conventional Units Online

The MUON constraint is a system-wide ("catch-all") constraint. Currently, it is set to 7 (4 in IE and 3 in NI) large conventional generation units (from a selected subset of the available generators). It serves to provide stability for a range of potential system phenomena, such as frequency stability, voltage, and system strength. However, there is a need to relax and eventually remove this constraint to make room for more

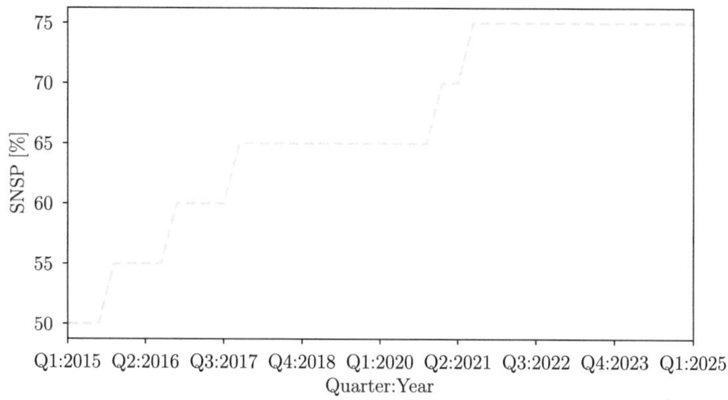

Fig. 11.3 SNSP evolution over the years

renewable generation in the system and thus achieve both governments' targets. In this context, the plan is to achieve secure system operation with ≤ 3 conventional generation units by 2030 and the ability to operate with no conventional generation units online by 2035 [27].

Rate of Change of Frequency

RoCoF is applied as a constraint in the scheduling and dispatch process of the TSOs. Its mathematical formulation is given by (1.67). This constraint is put in place to account for the largest single infeed/outfeed. Around a decade ago, the TSOs performed extensive studies and found out that it would be difficult to ensure stable system operation with a RoCoF of ± 0.5 Hz/s at high SNSP levels, for example, $\geq 50\%$. As a result, a program of work was followed by the TSOs, the generator market participants and the DSOs to work toward changing the RoCoF setting on all assets to ± 1 Hz/s. The TSOs are currently operating the AIPS with a RoCoF limit of ± 14 Hz/s.

Inertia Floor

According to (1.67), inertia is inversely proportional to RoCoF, hence, higher inertia means lower RoCoF. Currently, this limit is set to 23 GWs and it is planned to monitor this limit in the future. The inertia floor can be met by a combination of the units that are constrained on by MUON. Deployment of other low carbon inertia sources and potentially IBRs would contribute to providing inertia in the future.

System Strength

The development of system strength metrics and strategy for operational system strength and monitoring capability is ongoing. The introduction of a system strength policy in the planning and operations time frames is expected to replace the SNSP and MUON metrics [27].

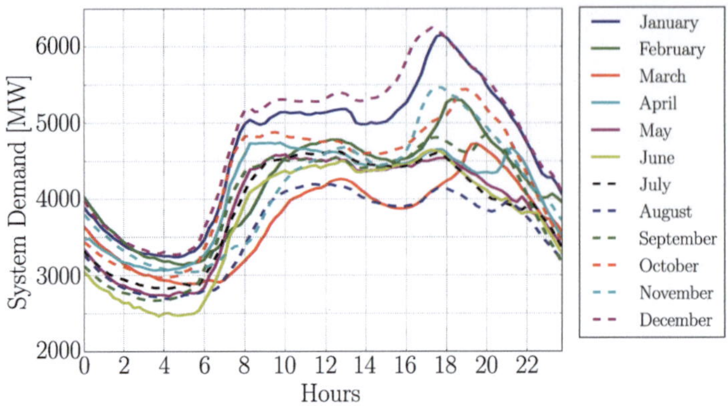

Fig. 11.4 Demand for a particular day for all the months in 2016

Constraints

The dispatch-down of wind due to technical constraints imposed by the network are known as constraints. Firstly, constraints can be understood as localized power carrying capacity of the network at the region of wind production. Secondly, outages in the network that may occur due to maintenance, upgrade works or faults. The dispatch-down of wind in the AIPS remains almost the same throughout the day irrespective of demand levels [23].

Demand Ramps

Figure 11.4 shows the load profile of the AIPS during a typical day, for different months. Conventionally, the period from 10:00 to 16:00 is called *day hours* and the period from 16:00 to 10:00 is called *night hours*. The system demand generally ramps down between 18:00 and 04:00. Then system demand ramps up from 04:00 to 10:00 and from 16:00 to 18:00 h. Load ramping leads to greater variations of the grid frequency during night hours. As discussed above, the effect of load ramping has to be separated as much as possible from the frequency deviations in order to be able to identify the impact of wind generation on the system. In the case study, thus, only day hours are considered.

11.1.2 Correlation Indices

We consider two statistical indices to evaluate the correlation between wind generation and frequency deviations, namely, the Pearson correlation coefficient and the p-value.

11.1.2.1 Pearson Correlation Coefficient

The Pearson correlation coefficient is a measurement of the linear correlation between two variables [28], as follows:

$$r = \frac{\sum_i^N (X_i - \overline{X})(Y_i - \overline{Y})}{\sqrt{\sum_i^N (X_i - \overline{X})^2 \sum_i^N (Y_i - \overline{Y})^2}}, \tag{11.3}$$

where N is the number of observations; X_i and Y_i are the values of the two time series, with length N, whose correlation is to be calculated; $\overline{X}$ and $\overline{Y}$ are the mean values of the time series X_i and Y_i, respectively; r can take values from -1 to 1. If $r = \pm 1$ exactly, then the relationship between, for example, the wind/solar PV penetration (% of demand) $p_{\text{wind}\%}$ and, say, the standard deviation of the frequency, σ_f, can be described by means of a linear equation. On the other hand, $r = 0$ indicates that there is no linear relationship between $p_{\text{wind}}\%$ and σ_f. Further, if $r > 0$, means that if $p_{\text{wind}}\%$ increases then also σ_f increases and vice-versa if $r < 0$. Only positive correlation coefficients are observed in the wind generation case study discussed below.

11.1.2.2 Pearson Correlation Coefficient

The Pearson correlation coefficient reflects the degree of correlation between two variables but does not provide any information weather such a correlation is significant or not. The index used to express the statistical significance of a correlation is known as p-value [29].

Given the student t-distribution:

$$t(r) = \frac{r\sqrt{N-2}}{\sqrt{1-r^2}}, \tag{11.4}$$

the p-value is defined as:

$$\text{p-value} = 2\Pr(T > t), \tag{11.5}$$

where T follows a t-distribution with $N - 2$ degrees of freedom. Hence the p-value is twice the probability (for double tail events) to obtain the current value of r if the correlation were actually zero (*null hypothesis*). The null hypothesis for this study is defined as the lack of correlation between wind generation and the hourly standard deviation of the frequency.

Being a probability, the p-value range is [0, 1]. A small p-value implies the rejection of the null hypothesis and imposes that the correlation r is significant. The conventional threshold p-value $= 0.05$ is chosen in the case study to validate statistical significance of a correlation between the variables [29]. So, if p-value < 0.05, we assume that frequency fluctuations are statistically correlated with the penetration of wind generation in the system.

11.1.3 Example

TSOs generally keep a record of the power generation along the years but very rarely frequency measurements obtained with PMUs or other instrumentation are stored for a long time. Typically, only major events that lead to high frequency deviations are recorded. During the development of the AMPSAS project, funded by Science Foundation Ireland and carried out at University College Dublin, the frequency within the university campus in Belfield was recorded for a period of four years, from 2014 to 2017, using a sampling rate of 0.1 s. The measurements were obtained with a Frequency Disturbance Recorder (FDR), courtesy of Prof. Yilu Liu, University of Tennessee [30].[1]

Archival wind generation data have been provided by EirGrid for the same period of four years (2014–2017). The data-set acquired consists of instantaneous power in MW for wind production, system demand and total generation in 15-min time series records. These values have been averaged using minutely measurements over a period of 15 minutes from the Supervisory Control and Data Acquisition (SCADA) system of the AIPS.

The Pearson correlation coefficients and p-values are calculated taking $\mathrm{X} = p_{\mathrm{wind\%}}$, that is, the 15-minute values averaged over 1 h of wind energy produced as percentage share of system demand:

$$p_{\mathrm{wind\%}} = \frac{\text{Hourly Averaged Wind Production}}{\text{Hourly Averaged System Demand}} \cdot 100 \,, \tag{11.6}$$

and $\mathrm{Y} = \sigma_f$, that is, the standard deviation of the system frequency over the same period for which $p_{\mathrm{wind\%}}$ is calculated.

Table 11.3 shows the correlation of $p_{\mathrm{wind\%}}$ with σ_f per month in the period from 2014 to 2017.[2] The wind penetration and frequency fluctuation show a relatively large correlation ($r > 0.4$) in most of the months.

Table 11.4 shows the p-values for the same months considered in Table 11.3. All the values are well below 0.01 except for three months (January 2015, April 2016 and June 2017), which, consistently, are the same months that show the lowest values of the Pearson correlation coefficients. Interestingly, these three months are all in different years.

---

[1] The FDR is a FNET/GridEye device, developed at Virginia Tech, that measures the frequency, phase angle and voltage of the power signal found at ordinary electrical outlets. We also utilized the measurement obtained with this device to generate Fig. 2.17.

[2] Frequency data were not available for some months.

Table 11.3 Pearson coefficients for $p_{\text{wind}\%}$ for the AIPS in 2014–2017

Year	2014	2015	2016	2017
Jan	–	0.2400	0.4939	–
Feb	–	0.5919	0.4233	0.4595
Mar	–	0.3923	–	0.3599
Apr	–	0.4756	0.2075	0.4971
May	–	0.5009	0.4127	0.5374
Jun	–	0.4198	–	0.1424
Jul	0.3692	0.5791	–	0.3987
Aug	0.5033	0.5514	–	0.4029
Sep	0.4513	0.3615	–	0.3063
Oct	–	0.5759	0.5793	0.3580
Nov	–	–	0.5997	0.4053
Dec	0.4619	0.3660	0.3374	–

Table 11.4 Frequency standard deviation for the AIPS in 2014–2017

Year	2014	2015	2016	2017
Jan	–	$1.51 \cdot 10^{-2}$	$< 10^{-6}$	–
Feb	–	$< 10^{-6}$	$6.80 \cdot 10^{-6}$	$< 10^{-6}$
Mar	–	$1.21 \cdot 10^{-7}$	–	$5.84 \cdot 10^{-5}$
Apr	–	$< 10^{-6}$	$1.92 \cdot 10^{-2}$	$< 10^{-6}$
May	–	$< 10^{-6}$	$5.84 \cdot 10^{-5}$	$< 10^{-6}$
Jun	–	$< 10^{-6}$	–	$9.55 \cdot 10^{-2}$
Jul	$9.82 \cdot 10^{-5}$	$< 10^{-6}$	–	$3.44 \cdot 10^{-6}$
Aug	$< 10^{-6}$	$< 10^{-6}$	–	$< 10^{-6}$
Sep	$4.15 \cdot 10^{-7}$	$1.33 \cdot 10^{-5}$	–	$1.31 \cdot 10^{-4}$
Oct	–	$< 10^{-6}$	$< 10^{-6}$	$3.36 \cdot 10^{-6}$
Nov	–	–	$< 10^{-6}$	$7.71 \cdot 10^{-6}$
Dec	$< 10^{-6}$	$6.97 \cdot 10^{-6}$	$4.65 \cdot 10^{-5}$	–

The least correlated month is June 2017, while the maximum correlated month is November 2016. Figures 11.5 and 11.6 present the scatter plot where x-axis represents $p_{\text{wind}\%}$ and y-axis is σ_f for the months of June 2017 and November 2016, respectively. In June 2017, the wind penetration has been greater than 50% for a significant number of hours, whereas, in November 2016, the wind penetration remained below 50% all the time. Still wind penetration and frequency fluctuations are more correlated in November 2016 than in June 2017. Moreover, in June 2017, there are several hours with a high σ_f but these hours are mostly characterized by low value of $p_{\text{wind}\%}$. In November 2016, the hours with higher σ_f are mostly characterized by high $p_{\text{wind}\%}$.

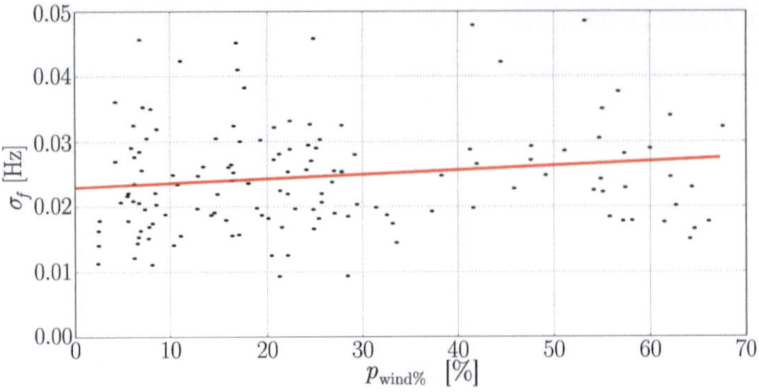

Fig. 11.5 Scatter plot of σ_f versus $p_{\text{wind}\%}$ in June 2017

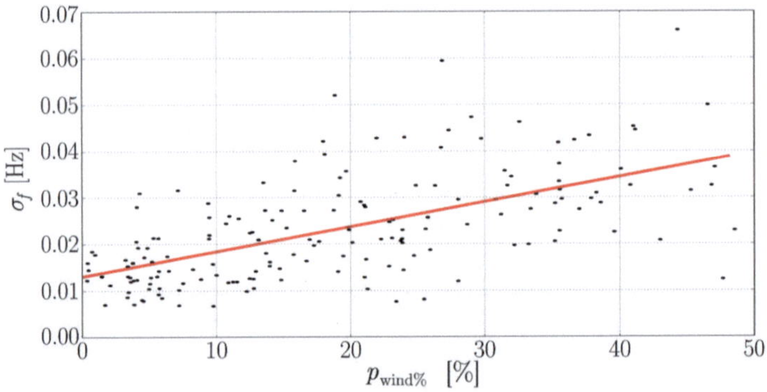

Fig. 11.6 Scatter plot of σ_f versus $p_{\text{wind}\%}$ in November 2016

These apparently mixed results can be explained by comparing the values of p_{WD} in different periods. Figure 11.7 shows the histogram of p_{WD} for four relevant months, where x-axis represents p_{WD} and y-axis shows the number of hours for which the wind dispatch-down happened. Comparing the histograms and looking at the values in Table 11.3, it is evident that the month with greater number of hours during which p_{WD} is high shows a relatively low correlation between wind generation and frequency variations. This supports the argument made in Sect. 11.1.1 that the higher the amount of wind rejected, the lower the correlation in a given month.

April 2016 is an exception to this rule. This month shows a low correlation between wind and frequency variations despite having a lower p_{WD} and fewer hours of wind curtailment, compared to January 2015. However, we note that, in 2016, the AIPS faced a significant number of the transmission outages, mainly due to maintenance and refurbishment of the transmission system [23]. These outages led to significant changes in the transmission network topology, which could be the cause for such a low correlation in April 2016.

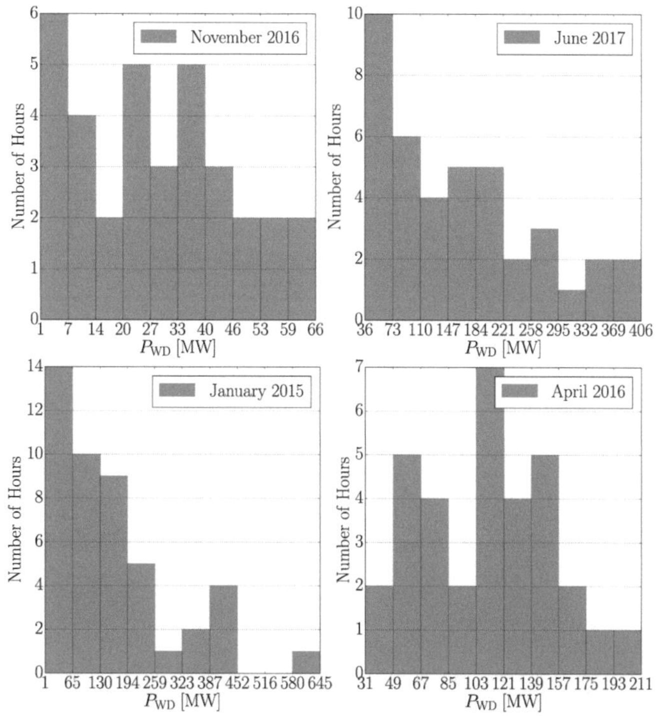

Fig. 11.7 Histogram of p_{WD} in November 2016, June 2017, January 2015 and April 2016

11.1.4 Remarks

The above results allow concluding that a statistically significant correlation exists between wind penetration and frequency deviations. This suggest that there is a need to deploy more and faster reserve resources to deal with the ever increasing penetration of stochastic and intermittent RES. This correlation is also illustrated in Fig. 11.8 where σ_f is plotted against the total wind generation of the AIPS in 2022. Figure 11.8, in fact, shows that frequency quality has a time-varying nature, in this case, it depends on wind generation.

The relationship between the total wind generation and σ_f is almost perfect. However, poor frequency quality is not *per se* linked to high amounts of RES. As we have discussed in previous chapters, IBR-based RES are excellent providers of frequency control and frequency regulation. Figure 11.8 and the correlation results only confirms that if one replaces conventional generators that provide frequency control and regulation with RES that *do not* provide similar services, then normally one should expect a deterioration of frequency quality in the grid. In turn, these results reaffirm the importance of utilizing RES to actively participate in the provision of frequency and voltage support services at all times in RES-dominated grids. This is

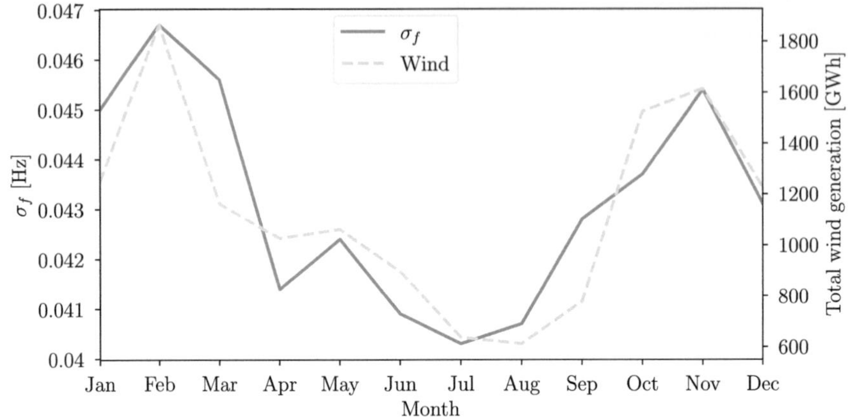

Fig. 11.8 Correlation between σ_f and wind generation based on 2022 real-world data from the AIPS

the case, for example, in Spain where RES are actively participating in balancing markets and providing ancillary services.

Load variations have a strong impact on frequency fluctuations and thus, the correlation between wind and frequency appears only if the load demand does not vary too much. As explained in Sect. 11.1.1, forecast error is not the cause of the frequency variations in the AIPS because wind is subject to curtailments. The results of the statistical correlation analysis confirm that if in a month the wind dispatch-down is high for long periods, the correlation between wind generation and frequency variations is relatively low. These results support the argument that wind production is no longer a stochastic variable of the system if subjected to curtailments and hence the wind penetration is not crucial, as one would expect, for the fluctuation of the frequency. It is rather the non-synchronous nature of wind generation that determines the quality of the system dynamic behavior.

A limitation of the study discussed above is that available wind generation data have a resolution of only 4 values per hour. On the other hand, frequency data obtained with the FDR have a resolution of 10 values per second. It appears desirable to obtain a higher resolution data for instantaneous wind penetration. This will help to perform a multivariate correlation analysis that includes the information of standard deviation of instantaneous penetration of the wind in the network. Further information about the ramp rates of the instantaneous wind penetration can be included. A higher resolution of wind data can allow developing a precise regression model which can be utilized for model predictive control.

11.2 Solar PV Penetration and Frequency Variations

To achieve global decarbonization goals the direction in Europe and worldwide has been, and still is, clear and ambitious regarding RES targets [31]. For instance, European Network of TSOs for Electricity (ENTSO-E) reports that in 2023 solar PV accounted for almost 10% of European Union (EU) electricity generated, with an installed capacity of approximately 150 GW (strong growth in solar PV generation seen in recent years) [32]. IE and NI are no different to this. More specifically, IE and NI have set ambitious RES targets for 2030 including a target of 80% of electricity met by RES in IE, while NI has a target of at least 70% [33]. In IE and NI, renewable energy is predominantly sourced from wind (approximately 6 GW installed as of 2022), although solar energy has grown in size and significance in recent years driven by an increasingly favorable legislative and policy framework. Regarding SNSP, the TSOs accommodate up to 75% at any point in time and plan to raise this limit to 95% by 2030 and to 100% by 2035 [27].

With regard to solar PV, the latest IE climate action plan foresees up to 5 GW and 8 GW solar PV capacity by 2025 and 2030, respectively [33]. The DSO in IE, in fact, expected to have over 1 GW of distributed-connected solar in IE by the end of 2023, making it the fastest-growing renewable industry [34]. For 2024, the DSO in IE states that there are around 1.6 GW of solar PV connected in 2024 including from domestic roof-top to large utility scale solar projects [35]. From the 1.6 GW figure, 554 MW come from utility-scale solar, 26 MW from mini-generation (with capacities between 17 kW and 50 kW) and 373 from micro-generation, among others. In fact, there are more than 94,000 homes in IE equipped with solar panels in 2024, with an with an average of 65-100 installations per day. While in NI, there is approximately 182 MW of large-scale solar PV as of the end of 2022 [36], and a couple of hundreds MW of distributed PV (roof-top PV) installed [37]. In terms of future outlook, it remains "sunny" and promising as it is expected that by 2030 solar PV will be a fundamental predictable part of the AIPS with around 8 GW of it being installed.

These solar PV levels have started introducing challenges such as maintaining frequency, voltage and operational demand within limits and managing its dispatch down levels [38]. This section aims at analyzing the frequency regulation challenge using actual data and provide insights into how those can be addressed.

11.2.1 Solar PV Characteristics

Transitioning to a renewable-based future means TSOs need to be prepared to deal with their inherent characteristics such as variability and uncertainty. For instance, real-world experiences show that due to cloud passing, solar PV output can drop as high as 63% of the rated capacity/minute [39]. This variability is significantly higher compared to, for example, wind power. Hence, controlling the frequency in power systems with high shares of solar PV is a challenging task.

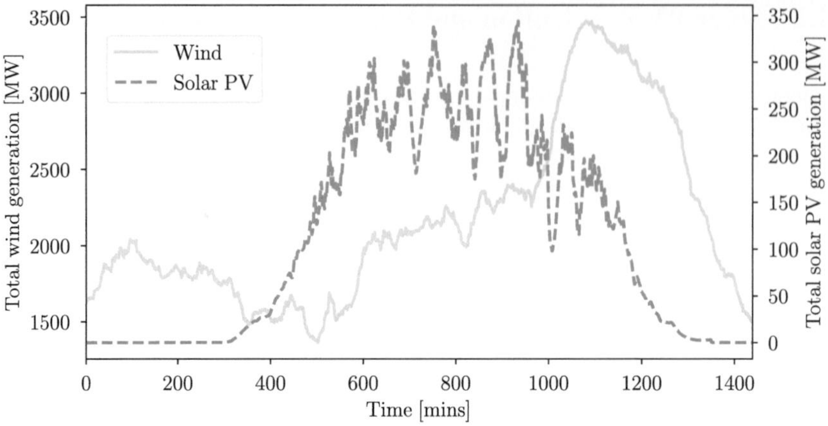

Fig. 11.9 Typical daily solar PV and wind power profiles in the AIPS

To illustrate these differences, we plot in Fig. 11.9 typical solar PV and wind power profiles in the AIPS. It can be seen that solar PV generation is much more volatile throughout the day making it hard to predict with good accuracy. As it will be shown later in the chapter, such volatility, despite still low in magnitude, introduces challenges in terms of frequency and voltage management, among others.

Figure 11.9 also suggest that TSOs are already operating a "different" power system during the daylight and another one during the night. In fact, as the penetration of wind and solar PV and converter-based load, among others, continuous, it is to be expected that TSOs will operate multiple different power systems during a day.

While the literature is rich in studying the impact of solar PV on the transient frequency response of power systems [40], that is not the case for its impact on long-term frequency deviations. An exception is reference [41], which shows that a value of 3% of the regulation reserve might not be enough to keep frequency within operational limits during a cloudy day. Similarly, reference [42] uses a 2040 transmission system model of IE and shows that while frequency remains within normal operating range (49.80 Hz to 50.2 Hz), its standard deviation increases to 105% when moving from 60% and 80% RES penetration levels.

Another challenge associated with high shares of PV penetration is managing its curtailment. For example, significant levels of PV curtailment (greater than 1% of potential output) have been recorded in Chile, China, Germany, and specific markets in the US [19]. The issue of distributed PV curtailment in low voltage networks is discussed in [43] using real-world data from Australia. Indeed, the curtailment may be the last action for TSOs to manage the minimum operational demand problem ("duck curve") [44]. On the other hand, reference [45] stress the need to convert solar PV to a dispatchable source to address its intermittency nature. Other solutions to reduce solar PV curtailment proposed in the literature are demand-side management and economic dispatch [46].

Unlike for wind generation, the degree of utilization for solar PV exhibits a distinctive daily and annual/seasonal pattern [47]. Dealing with seasonal changes represents a challenge for TSOs. For instance, ENTSO-E reports that TSOs have taken proactive steps in dealing with significant surplus of available energy, particularly solar PV, during the months of April-May 2025 where demand is low due to long weekends and holiday period [32]. These steps include:

- Establish dedicated TSOs operational teams to prepare and manage critical demand-supply situations.
- Support from regional coordination centers in terms of assessment of available flexibility updated up to near real-time.
- Enhance information sharing between the TSOs to ensure coordinated cross-border actions.
- Monitor continuously the situation by TSOs dispatching teams and apply necessary measures to keep system secure.

ENTSO-E remarks that market participants have also a critical role to play in managing the above situation, when and if needed, by keeping portfolios balanced and reducing feed-in.

11.2.2 Example

Although current IE and NI solar PV capacity is still significantly lower than wind power, it is still introducing challenges such as: (i) dispatch down levels; (ii) long-term frequency deviations; (iii) voltage magnitude variations; and (iv) operational demand variations. If certain operational constraints are binding, such as SNSP, then the TSOs instruct solar PV units to dispatch down their generation, similar to wind generation described above. The AIPS has in place four operational constraints/limits that impact solar PV dispatch down levels (see Sect. 11.1.1.2 and the evolution of constraints in Table 6.2).

In this section, we discuss the frequency regulation challenge based on actual data obtained from the TSO SCADA. Unless stated otherwise, frequency data sampling rate is seconds whereas solar PV generation sampling rate is minutes.

11.2.2.1 Impact on Minutes Outside ±100 mHz Range

Until now, Irish TSOs have mainly been concerned with the impact of wind power penetration on the AIPS. For example, more than a decade ago, the Irish TSOs carried our various wind integration studies and identified different potential issues with high values of SNSP limit, for example, 75%. Two of the main issues identified were RoCoF (±0.5 Hz/s limit back at the time) and large wind ramps. To address these issues, the Irish TSOs introduced relevant system services such as fast frequency response and ramping margin products, among others [48].

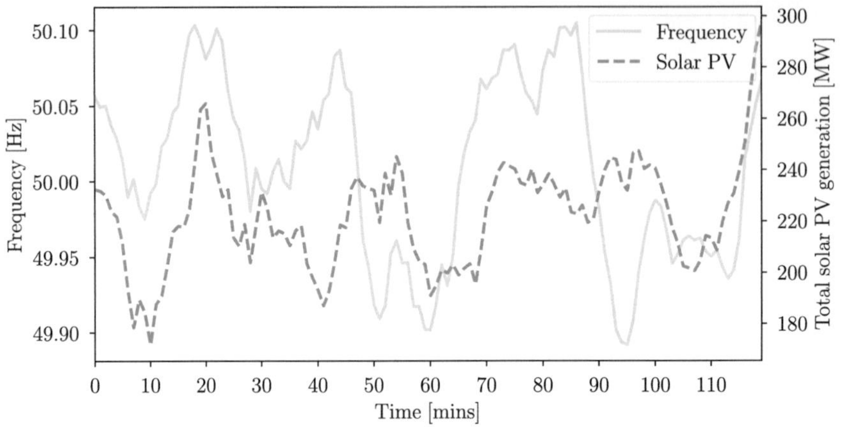

Fig. 11.10 Frequency and solar PV traces for two hours in the AIPS

An emerging issue is higher frequency deviations due to increased PV penetration. PV power profiles like that in Fig. 11.9 create additional challenges for control room operators. In particular, it is becoming increasingly difficult to manage frequency within operational limits (±200 mHz) when PV output power changes quickly due to cloud passes. We illustrate this issue in Fig. 11.10, where solar PV variations and frequency are plotted against each other for two hours of a relevant day. In the first and last 30 minutes, there is an almost perfect linear relationship between solar and frequency variations. However, when the frequency is about to drift outside the ±100 mHz range, then the control room operators implement manual operations, including conventional generation redispatch, to bring back system frequency within limits.

Figure 11.11 illustrates another recent frequency fluctuation in the AIPS caused by variations in solar generation. More than 100 MW fluctuation in two solar units, due to passing clouds, resulted in a similar pattern in frequency, reaching a zenith of 50.2 Hz. Note that the standard frequency range in the AIPS is set between 49.8 and 50.2 Hz (see Table 3.1), and the fluctuations brought the frequency to the maximum of this limit. Currently, these issues can be managed with existing operational capabilities, given that the total installed capacity of utility-scale solar generation in IE is approximately 500-600 MW. However, by 2030, the capacity of solar generation (both utility-scale and embedded) is expected to increase dramatically to around 8 GW, with 2.5 GW being embedded. This significant increase can exacerbate frequency fluctuations due to the concentration of an increasing number of solar farms between 100 and 200 MW in specific regions, necessitating enhancements in TSOs operational capabilities.

Table 11.5 compares the violation minutes for April to August 2022 and 2023. During this period, the solar PV share in the system rose from 0.5% to 2.5% on average but, of course, it is still much lower than wind (around 20-30% on average). However, due to increased solar PV variations, frequency is spending more time outside the ±100 mHz range. It is counter-intuitive that low solar penetration impacts more on

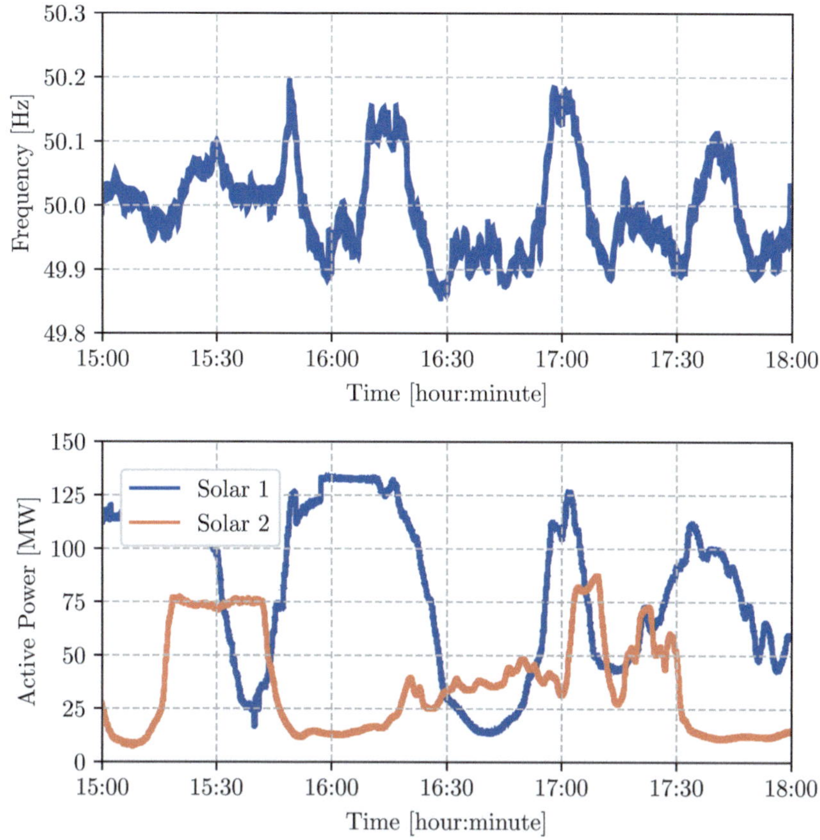

Fig. 11.11 Frequency fluctuations due to solar generation variations in the AIPS, April 16, 2024

Table 11.5 Solar and wind share, PV dispatch down and violation minutes (±100 mHz criteria ≥98% of time) for the period April–August of 2022 and 2023

Month	April	May	June	July	August
$p_{\text{solar\%}}$ 2022	0.5	0.6	0.6	0.5	0.6
$p_{\text{wind\%}}$ 2022	32	34	30	21	20
Dispatch Down 2022	6.4%	3.4%	6.3%	4.1%	4.2%
Violation Minutes 2022	411	433	333	230	316
$p_{\text{solar\%}}$ 2023	1.4	2.6	2.8	2.4	2.2
$p_{\text{wind\%}}$ 2023	35	21	21	33	33
Dispatch Down 2023	2.9%	3.9%	10.6%	11.3%	11.4%
Violation Minutes 2023	486	348	494	626	624

frequency deviations than high wind generation. However, what matters is not the average power, but the variance of this power. The low level of solar PV penetration, in fact, comes with a high volatility as PV power plants in the AIPS are currently located in few specific areas of the network. Thus, the averaging effect that reduces the variance of wind-generated power cannot take place for solar generation. This also suggests a way to smooth and reduce solar PV system output variability, that is, through geographical aggregation, namely, an even geographical installation of solar power plants in the whole network [49].

The increase in solar PV penetration not accompanied by geographical aggregation is one of the main factors leading to almost doubling the violation minutes in 2023. A crucial frequency service that can address this problem is to turn on Active Power Control (APC) (see Chap. 4) for wind and solar plants [50]. Another important operational policy change that has impacted the increased number of violation minutes is the MUON limit change from 8 to 7 (end of May 2023). Operating with one conventional unit less in the system means there is less regulation reserve provided (by means of ± 15 mHz dead-band) on average and, in turn, a deterioration of frequency quality [50].

11.2.2.2 Pearson Correlation Coefficient

We utilize again the Pearson correlation coefficient to provide a more quantitative analysis of the impact of solar PV penetration on long-term frequency deviations in the AIPS. The Pearson correlation coefficients are calculated taking $\mathrm{X} = p_{\mathrm{solar}\%}$, namely, the instantaneous value of solar energy as percentage share of system demand:

$$p_{\mathrm{solar}\%} = \frac{\text{Averaged Solar}}{\text{Averaged Demand}} 100, \tag{11.7}$$

and $\mathrm{Y} = \sigma_f$, that is, the standard deviation of the system frequency over the same period for which $p_{\mathrm{solar}\%}$ is calculated. For comparison, we calculate the Pearson correlation coefficients for wind and SNSP as well ($\mathrm{X} = p_{\mathrm{wind}\%}$ and $\mathrm{X} = \mathrm{SNSP}$, and $\mathrm{Y} = \sigma_f$) with wind energy share equation as follows:

$$p_{\mathrm{wind}\%} = \frac{\text{Averaged Wind}}{\text{Averaged Demand}} 100. \tag{11.8}$$

Since large-scale PV installations in IE went live mostly around April-May 2023, we perform the analysis for months April-August 2023 [36]. The focus is on the day hours, that is, the period from 10:00 to 16:00, in order to minimize the effect of load ramping.

The results of the analysis are shown in Tables 11.6, 11.7 and 11.8 using three different resolutions of data, that is, 15, 5 and 1 min, respectively. For the 5-min case we have also calculated the Pearson coefficients for standard deviations of

Table 11.6 Pearson coefficients for $p_{solar\%}$, $p_{wind\%}$ and SNSP%, respectively, and σ_f for the AIPS in the period April–August 2023 and using 15-min resolution

Month	Solar	Wind	SNSP
April	0.0573	0.1006	0.1125
May	−0.1483	0.3323	0.2857
June	−0.1766	0.3257	0.3074
July	0.1472	−0.0852	−0.0631
August	0.1056	0.1825	0.1935

solar generation (σ_{solar}) and wind (σ_{wind}). The Pearson coefficients for solar are, in general, lower than those for wind and SNSP. This was to be expected considering the penetration of solar PV at the time of writing (significantly lower than wind). On the other hand, the Pearson coefficients for SNSP is a combination of the solar and wind coefficients. This was also expected considering the SNSP calculation [51]. However, interestingly when using a 1-min resolution of data, all the Pearson coefficients are reduced and take a negative value. Using a higher sampling rate, in fact, increases the effect of uncorrelated noise and decreases that of solar variations. However, if we refer to the Pearson coefficients for σ_{Solar} and σ_{Wind}, and σ_f in Table 11.7, then solar generation shows a comparable correlation to wind and coefficients take a positive value. These results support the conclusion above that solar PV impacts on long-term frequency deviations. The TSOs are addressing the frequency regulation challenge by reviewing all frequency reserve products. This is the argument of the next chapter.

In Brief

- Wind and solar PV generation deteriorate long-term frequency quality, in particular, if they do not provide Primary Frequency Regulation (PFR).
- There is an almost perfect linear relationship between increased wind and solar PV penetration and increased frequency deviations.
- Wind and solar PV generation should provide frequency support at all times under normal and abnormal system conditions similar to conventional synchronous generators.
- Despite its (still) relatively low penetration compared to wind power penetration, solar PV is challenging the real-time operation of the AIPS, such as, for example, maintaining frequency within operational limits.

Table 11.7 Pearson coefficients for $p_{\text{solar}\%}$, $p_{\text{wind}\%}$, SNSP%, σ_{solar}, and σ_{wind}, respectively, and σ_f for the AIPS in the period April–August 2023 with 5-min resolution

Month	Solar	Wind	SNSP	$(\sigma_{\text{solar}}, \sigma_f)$	$(\sigma_{\text{wind}}, \sigma_f)$
April	0.0436	0.0974	0.1008	0.1460	0.3975
May	−0.1104	0.3240	0.3058	0.2170	0.2650
June	−0.1152	0.3212	0.3324	0.2088	0.2894
July	0.1225	−0.0819	−0.0630	0.2385	0.2799
August	0.0582	0.1571	0.1551	0.2444	0.4097

Table 11.8 Pearson coefficients for $p_{\text{solar}\%}$, $p_{\text{wind}\%}$, and SNSP%, respectively, and σ_f for the AIPS in the period April–August 2023 and using 1-min resolution

Month	Solar	Wind	SNSP
April	−0.0274	0.0739	0.0505
May	−0.1343	0.2640	0.2353
June	−0.0569	0.3073	0.3537
July	−0.0079	−0.0660	−0.0607
August	−0.0346	0.1399	0.1114

Open Questions

The impact of wind and solar PV on long-term frequency quality is an emerging issue that has not been fully discussed. How to quantify, through indices, for example, this impact is currently an open question. The lack of publicly available high-resolution wind and solar PV generation data is part of the problem along with the need for more precise short-term solar PV forecasts. Then, the coordination between IBRs and synchronous generators still in operation to provide frequency regulation appears needed. But this will likely require properly incentivizing, scheduling and eventually dispatching wind and solar PV resources.

References

1. Wang Y, Silva V, López-Botet-Zulueta M (2016) Impact of high penetration of variable renewable generation on frequency dynamics in the Continental Europe interconnected system. IET Renew Power Gener 10(1):10–16
2. Miller NW, Shao M, D'aquila R et al (2015) Frequency response of the US Eastern interconnection under conditions of high wind and solar generation. In: 2015 seventh annual IEEE green technologies conference, Apr 2015, pp 21–28
3. Fernández-Bernal F, Egido I, Lobato E (2014) Maximum wind power generation in a power system imposed by system inertia and primary reserve requirements. Wind Energy 18(8):1501–1514

4. Wang Y, Delille G, Bayem H et al (2013) High wind power penetration in isolated power systems – assesment of wind inertial and primary frequency responses. IEEE Trans Power Syst 28(3):2412–2420
5. Vasconcelos H, Moreira C, Madureira A et al (2015) Advanced control solutions for operating isolated power systems: examining the Portuguese islands. IEEE Electrif Mag 3(1):25–35
6. Martin Almenta M, Morrow DJ, Best RJ et al (2017) An analysis of wind curtailment and constraint at a nodal level. IEEE Trans Sustain Energy 8(2):488–495 (2017)
7. Eirgrid and SONI (2016) DS3 programme operational capability outlook 2016. https://www.eirgrid.ie/
8. Eirgrid and SONI, RoCoF alternative and complementary solutions project. https://www.eirgrid.ie/
9. Xypolytou E, Gawlik W, Zseby T et al (2018) Impact of asynchronous renewable generation infeed on grid frequency: analysis based on synchrophasor measurements. Sustainability 10:(5)
10. EirGrid (2016) Wind dispatch down reports user guide. https://www.eirgrid.ie/
11. Milano F, Drfler F, Hug G et al (2018) Foundations and challenges of low-inertia systems (invited paper). In: Power systems computation conference (PSCC), pp 1–25
12. Mai T, Denholm P, Brown P et al (2022) Getting to 100%: six strategies for the challenging last 10%. Joule 6(9):1981–1994
13. Frew B, Sergi B, Denholm P et al (2021) The curtailment paradox in the transition to high solar power systems. Joule 5(5):1143–1167
14. Olson A, Jones RA, Hart E et al (2014) Renewable curtailment as a power system flexibility resource. Electr J 27(9):49–61
15. Rebello E, Watson D, Rodgers M (2020) Ancillary services from wind turbines: automatic generation control (AGC) from a single type 4 turbine. Wind Energy Sci 5(1):225–236
16. Ge X, Zhu X, Fu Y et al (2022) Optimization of reserve with different time scales for wind-thermal power optimal scheduling considering dynamic deloading of wind turbines. IEEE Trans Sust Energy 13(4):2041–2050
17. Kroposki B, Johnson B, Zhang Y et al (2017) Achieving a 100% renewable grid: operating electric power systems with extremely high levels of variable renewable energy. IEEE Power Energ Mag 15(2):61–73
18. Root C, Presume H, Proudfoot D et al (2017) Using battery energy storage to reduce renewable resource curtailment. IEEE PES ISGT, 1–5
19. O'Shaughnessy E, Cruce JR, Xu K (2020) Too much of a good thing? Global trends in the curtailment of solar PV. Sol Energy 208:1068–1077
20. Nycander E, Söder L, Olauson J et al (2020) Curtailment analysis for the Nordic power system considering transmission capacity, inertia limits and generation flexibility. Renew Energy 152:942–960
21. Cleary B, Duffy A, O'Connor A et al (2015) Assessing the economic benefits of compressed air energy storage for mitigating wind curtailment. IEEE Trans Sustain Energy 6(3):1021–1028
22. Mc Garrigle EV, Deane JP, Leahy PG (2013) How much wind energy will be curtailed on the, 2020 Irish power system?. Renew Energy 55(544–553):2013
23. Eirgrid and SONI, Annual renewable energy constraint and curtailment reports. https://www.eirgrid.ie/
24. Government of Ireland, Climate action plan 2021 securing our future. assets.gov.ie/224574/be2fecb2-2fb7-450e-9f5f-24204c9c9fbf.pdf
25. Government of Northern Ireland (2022) Climate change act (Northern Ireland). www.legislation.gov.uk/nia/2022/31/enacted
26. Eirgrid and SONI, System non-synchronous penetration definition and formulation. https://www.eirgrid.ie/
27. Eirgrid and SONI (2025) Operational policy roadmap 2025–2035. cms.eirgrid.ie
28. Kendall M, Stuart A (1979) The advanced theory of statistics. Inference and relationship, vol 2, 4th ed. Griffin, London, UK
29. Fisher RA (1990) Statistical methods, experimental design, and scientific inference. Oxford Science Publications, Oxford, UK

30. Wang L, Burgett J, Zuo J et al (2007) Frequency disturbance recorder design and developments. IEEE PES Gen Meet, 1–7
31. Kërçi T, Tzounas G, Milano F (2022) A dynamic behavioral model of the long-term development of solar photovoltaic generation driven by feed-in tariffs. Energy 256:124506
32. ENTSO-E (2025) Sunny Spring and risks of electricity surplus: TSOs coordinate. www.entsoe.eu/news
33. Government of Ireland (2022) Climate action plan 2023 CAP23. assets.gov.ie
34. Irish Solar Energy Association (2023) Ireland's solar revolution. www.irishsolarenergy.org
35. Irish Solar Energy Association (2024) Scale of solar. www.irishsolarenergy.org
36. EirGrid (2023) Renewable energy. https://www.eirgrid.ie/
37. Statista (2023) Cumulative installed capacity of solar photovoltaic power in Northern Ireland from 2010 to 2022. www.statista.com
38. Kërçi T, Hurtado M, Tweed S et al (2024) Emerging challenges of integrating solar PV in the Ireland and Northern Ireland power systems. IEEE PES Gen Meet, 1–5
39. Alam MJE, Muttaqi KM, Sutanto D (2014) A novel approach for ramp-rate control of solar PV using energy storage to mitigate output fluctuations caused by cloud passing. IEEE Trans Energy Convers 29(2):507–518
40. You S et al (2017) Impact of high PV penetration on U.S. Eastern interconnection frequency response. IEEE PES Gen Meet, 1–5
41. Yuan H, Tan J (2020) Multi-timescale integrated dynamic and scheduling model (midas-solar). Tech. Rep, NREL, Golden, CO
42. Hellmuth S, Kuschke M, Strunz K et al (2019) System frequency variations and the effect of wind power: analysis based on an Irish transmission system test model. CIGRE Sci Eng
43. Yildiz B, Stringer N, Klymenko T et al (2023) Real-world data analysis of distributed PV and battery energy storage system curtailment in low voltage networks. Renew Sustain Energy Rev 186:113696
44. Hou Q, Zhang N, Du E et al (2019) Probabilistic duck curve in high PV penetration power system: concept, modeling, and empirical analysis in China. Appl Energy 242:205–215
45. Liu Z, Du Y (2023) Evolution towards dispatchable PV using forecasting, storage, and curtailment: a review. Electric Power Syst Res 223:109554
46. Sambasivam B, Xu Y (2023) Reducing solar PV curtailment through demand-side management and economic dispatch in Karnataka, India. Energy Policy 172:113334
47. Scherer M, Andersson G (2015) How future-proof is the Continental European frequency control structure? IEEE Eindhoven PowerTech 2015:1–6
48. SEMC (2013) DS3 system services technical definitions decision paper SEM-13-098. www.semcommittee.com/news-centre/ds3-system-services-technical-definitions-decision-paper
49. Aldeman M, Jo J, Loomis D et al (2023) Reduction of solar photovoltaic system output variability with geographical aggregation. Renew Sustain Energy Trans 3:100052
50. Kërçi T, Hurtado M, Gjergji M et al (2023) Frequency quality in low-inertia power systems. IEEE PES Gen Meet 1–5
51. Hurtado M, Kërçi T, Tweed S et al (2023) Analysis of wind energy curtailment in the Ireland and Northern Ireland power systems. IEEE PES Gen Meet 1–5

Chapter 12
New Reserves and Volume Forecasting

This last chapter summarizes a review conducted by EirGrid and SONI of the reserve services for the AIPS. The chapter also presents the Volume Forecasting Methodology (VFM) the TSOs have developed to forecast reserve volumes over different time frames namely long-term (10 years ahead), medium-term (1–2 years ahead), and short-term (week to day-ahead). The review and VFM aim at ensuring reserve services can help to achieve 2030 RES targets and, at the same time, addressing several trends and evolving risks such as OF events, faster-happening frequency nadir/zenith, and higher RoCoF.

Section 12.1 describes the current frequency control employed in the AIPS to ensure frequency quality. Section 12.2 shows the current procured reserve volumes and technology mix. An analysis of various frequency quality parameters of the AIPS is then presented based on actual data in Sect. 12.3. Section 12.4 provides an illustrative example of the effectiveness of AGC as a potential approach in reducing σ_f in the AIPS. Section 12.5 discusses the recent reserves review and its outcomes. Finally, Sect. 12.6 provides the proposed VFM to forecast reserve needs over different time frames.

12.1 Frequency Control

Ancillary services such as frequency control reserves are indispensable to the safe and secure operation of power systems. Recently, they are evolving to successfully integrate large and small-scale RES such as wind and solar energy [1, 2]. For example, EirGrid and SONI have integrated world-leading levels of RES, particularly wind and, recently, solar energy as well [3, 4]. DS3 reserve services have been key to this successful integration of RES in the AIPS [5].

These services, most of which were mainly designed based on a predominately conventional generation fleet, are:

T. Kërçi and F. Milano, *Frequency Quality in Power Systems*, Power Systems,
https://doi.org/10.1007/978-3-032-04721-2_12

- Fast Frequency Response (FFR)
- Primary Operating Reserve (POR)
- Secondary Operating Reserve (SOR)
- Tertiary Operating Reserve 1 (TOR1)
- Tertiary Operating Reserve 2 (TOR2)
- Replacement Reserve Synchronized (RRS)
- Replacement Reserve Desynchronized (RRD)

In 2013, the DS3 program developed in conjunction with the RAs the FFR service to address frequency stability, and various long-term ramping margin products to manage variable RES and changes in IC flows while maintaining system security [5]. Similar FFR products have also been defined by other TSOs operating renewable-dominated power grids such as ERCOT and AEMO [6–8].

The aim of the DS3 services was to meet 2020 policy objectives namely 40% of total electricity consumption to be provided by RES [9]. Current DS3 reserves have the following main objectives (see Table 3.1 for more details):

- Maintaining frequency within the standard frequency range (49.8–50.2 Hz) and, in particular, within 49.9–50.1 Hz range for 98% of time, as monitored and reported annually in the TSOs performance report [10].
- Mitigating large disturbances to avoid a maximum instantaneous frequency deviation larger than ± 1000 mHz and a RoCoF larger than ± 1 Hz/s (as measured over a rolling 500 ms period in Eq. (1.66)).

Table 12.1 shows current DS3 reserve services. It should be stressed that, historically, for the AIPS, but also for the vast majority of power systems worldwide, the main risk has been from an UF event. For this reason, the reserve products procured

Table 12.1 Reserve definitions and minimum real-time requirements

Reserve service	Direction (Upward/Downward)	Min. volume requirement (% LSI or MW)	FAT (s)	Response duration (max. time after the event in s)
Fast frequency response	Upward	N/A	2	10
Primary operating reserve	Upward	75% LSI	5	15
Secondary operating reserve	Upward	75% LSI	15	90
Tertiary operating reserve 1	Upward	100% LSI	90	300
Tertiary operating reserve 2	Upward	100% LSI	300	1200
Replacement reserve desynchronized	Upward	325 MW	1200	3600
Replacement reserve synchronized	Upward	125 MW	1200	3600

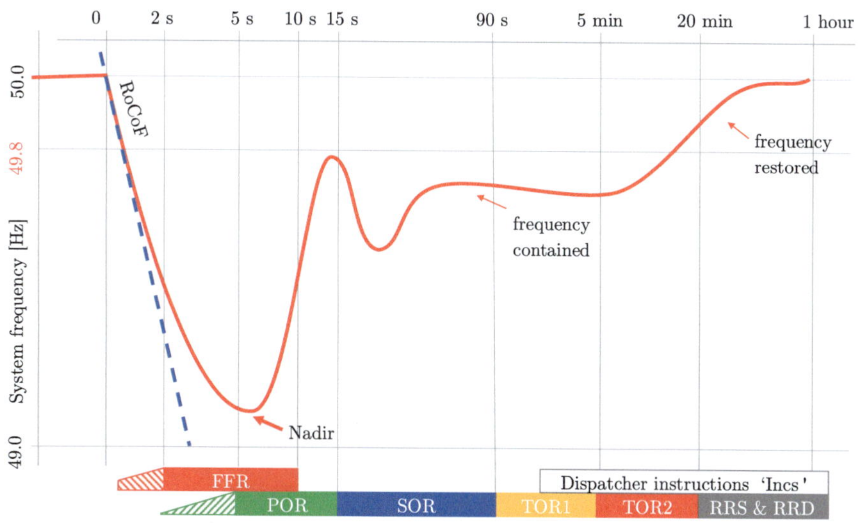

Fig. 12.1 UF event and response from reserves

by the TSOs require an upward response, that is, generation increase or demand reduction. Then, in contrast to other reserves such as minimum POR and SOR volumes, there is currently no operational policy with respect to the dimensioning of FFR requirements such as as a % of the LSI/Largest Single Outfeed (LSO). However, FFR has been a key service in reaching 75% SNSP.

Figure 12.1 depicts the sudden loss of a large conventional generator or of an importing IC results in a rapid and significant frequency drop. First, initial RoCoF, computed with (1.67), is mainly maintained within the ± 1 Hz/s standard by ensuring sufficient levels of inertia. Next, FFR and POR prevent reaching the minimum frequency nadir of 49 Hz. Then, POR and SOR recover the frequency to a steady state value. The frequency is further restored toward 50 Hz by automatic response and manual instructions during the TOR1 and TOR2 time frames. Whereas RRS and RRD are manually activated to replace the required levels of reserves.

The evolution of frequency nadir and RoCoF over the years is shown in Figs. 12.2, 12.3 and 12.4. These figures show that AIPS is experiencing faster nadir and higher RoCoF compared to the past as more inertialess units namely wind and solar PV is added to the grid. Despite EirGrid and SONI not yet explicitly contracting OF response products, similar frequency evolution and characteristic exist for OF events. This is due to the fact that conventional units' TGs work symmetrically and, given there is sufficient foot and head room available, will respond similarly in UF and OF situations.

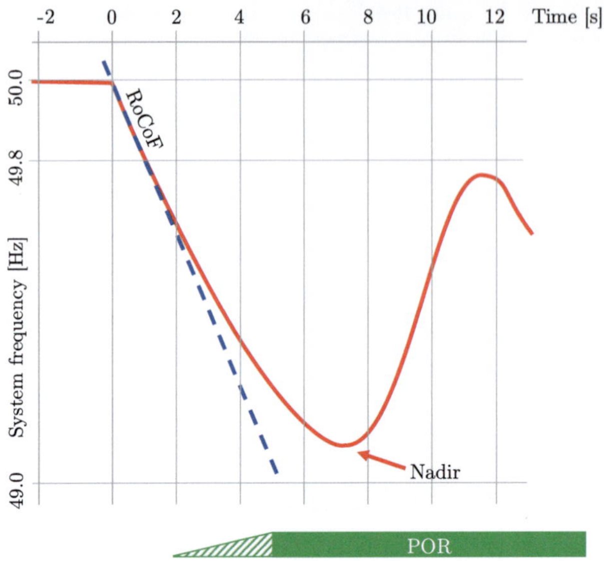

Fig. 12.2 UF event in 2010

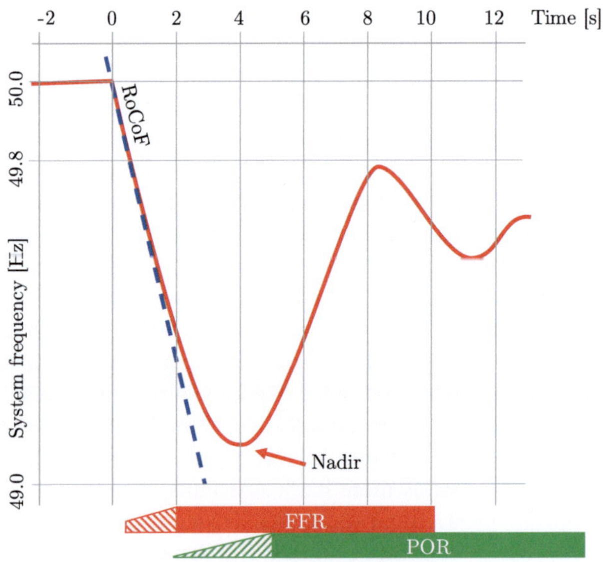

Fig. 12.3 UF event in 2022

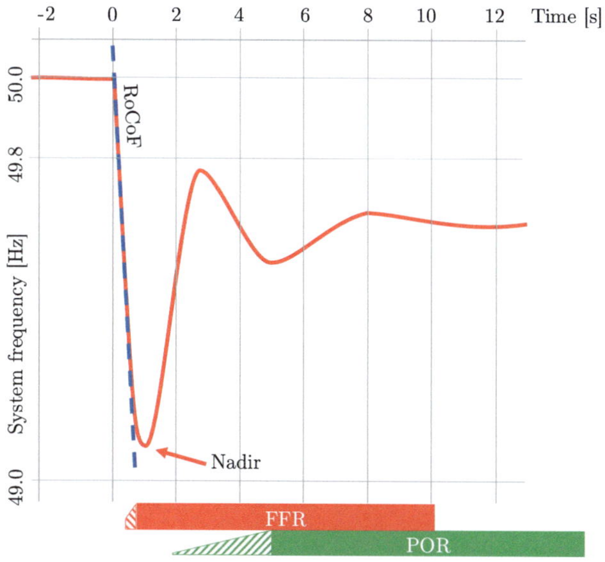

Fig. 12.4 UF event in 2026

The Irish TSOs employ several other mitigation strategies and operational constraints to deal with contingency events. These include:

- The operational constraint for MUON which is currently set at 7.
- Minimum inertia floor which is currently set at 23 GWs. Future inertia requirements will increasingly be met by LCIS, which will reduce the dependency on running conventional generation.
- A restriction on the scheduling of peaking conventional generators to ensure sufficient RR provision.
- IC run-back schemes to help mitigate OF events.

12.1.1 Frequency Regulation

Frequency regulation from an AIPS perspective is achieved by utilizing manual actions of balancing bids, and PFC with narrow dead-band or PFR. TSOs maintain a minimum volume of primary regulating reserve provision in the POR, SOR, TOR1 and TOR2 time frames through operational constraints in their central dispatch optimization. These minimum volumes are 75-87 MW and 50 MW in IE and NI, respectively. Currently these operational constraints are applicable only to conventional generators with ±15 mHz dead-band. As there does not exist an automatic SFC in the AIPS, manual activations of conventional units through the EDIL and

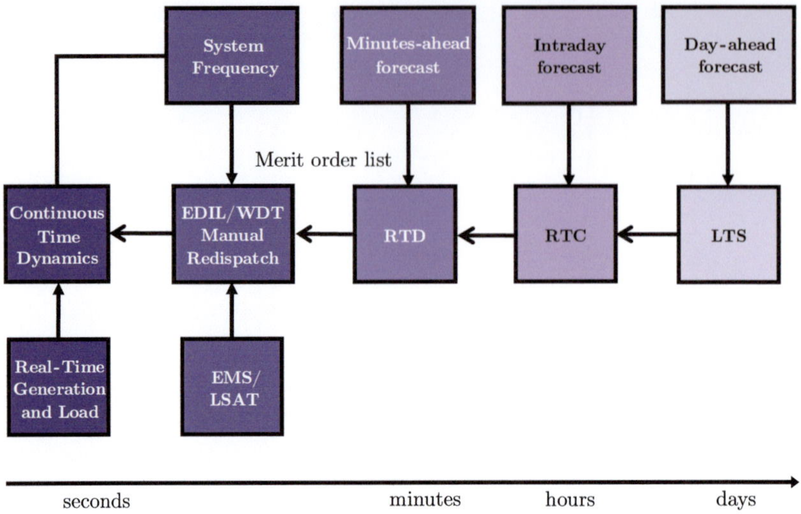

Fig. 12.5 Frequency control and regulation in the AIPS

of wind and solar units through the WDT are utilized to regulate system frequency proactively, as shown in Fig. 12.5.

APC from wind and solar units is another crucial frequency regulation service impacting long-term frequency quality and currently in place in the AIPS. As discussed in Chapters 2 and 4, APC is a droop-based PFC service and is mandatory for all dispatchable wind farms in IE [11]. It involves a selectable dead-band setting of ±200 or ±15 mHz. The default value of the dead-band is ±200 mHz but, when the frequency control is challenging, EirGrid changes remotely this dead-band to ±15 mHz. In this way, wind farms adjust their output much more dynamically and contribute to the control of system frequency under normal, pre-contingency conditions. In the near future, it is expected that other technologies such as solar power will have APC functionality enabled in order to help maintain the frequency quality parameters.

12.2 Procured Reserve Volumes and Technology Mix

As part of the DS3 program, TSOs run a procurement process every six months to procure the current upward reserve products. Table 12.2 shows contracted volumes by technology type as of October 2024.

These reserves can currently be provided by a range of technologies including short-duration BESS, conventional generators, HVDC ICs, RES, aggregated generator units and DSUs. BESS is currently mostly used for system services rather than providing energy to the grid.

Table 12.2 Contracted volumes of upward reserves

Unit	FFR [MW]	POR [MW]	SOR [MW]	TOR1 [MW]	TOR2 [MW]	RRS [MW]	RRD [MW]
Conventional	375	845	1250	1473	2289	4878	2878
DSUs	182	212	225	290	299	0	414
Aggregated gen. units	0	0	16	61	74	10	88
Wind	145	298	313	307	0	0	70
HVDC IC	200	200	200	200	200	0	0
BESSBESS	934	949	949	949	929	0	382
Hybrid	2	2	2	2	2	0	0
Solar	24	24	23	23	0	0	0
Total	1867	2529	2978	3306	3793	4888	3832

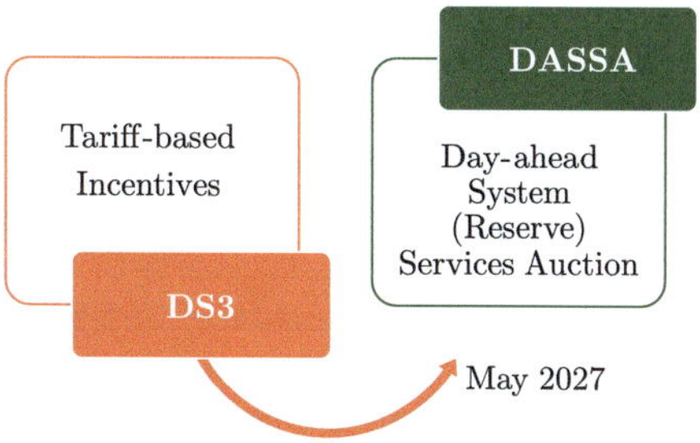

Fig. 12.6 Transition from DS3 to DASSA arrangements for the procurement of system services in the AIPS

Current reserve volumes are more than sufficient with respect to minimum volumes requirements shown in Table 12.1. Indeed, these high reserve volumes have significantly increased DS3 expenditure. Because of this, and to meet EU regulation requirements for day-ahead balancing capacity procurement,[1] there is the need to move from the existing DS3 regulated arrangements (tariff-based procurement) to a set of future competitive arrangements known as the future arrangements for system services [12]. New arrangements will include DASSA procurement of a subset of the system services from 2027 namely reserve services [13]. The transition is illustrated in Fig. 12.6.

[1] *"Balancing capacity" means a volume of reserve capacity that a balancing service provider has agreed to hold and in respect to which the balancing service provider has agreed to submit bids for a corresponding volume of balancing energy to the TSO for the duration of the contract. "Balancing energy" means energy used by TSOs to perform balancing and provided by a balancing service provider* [12].

12.3 Evolution of Frequency Quality

The AIPS which is a real-world large-scale low-inertia synchronous island power system currently accommodates up to 75% of IBR-based generation. For this reason the AIPS is an ideal benchmark to analyze the evolution of frequency quality in low-inertia power systems. This evolution involves dealing with different technical challenges such as ensuring power system stability and security. An emerging challenge is to maintaining frequency quality parameters within acceptable limits. In this section, we discuss through actual data, recent trends of some frequency quality parameters of the AIPS such as σ_f and the phenomenon of slow frequency restoration.

12.3.1 Maximum Frequency Deviations

Figure 12.7 shows the trend of the maximum (zenith) and minimum (nadir) frequency deviations in the AIPS over the last years. Interestingly, both parameters are within limits and improving in their performance. Note also that allowable limits are currently set to ±1000 mHz according to Table 3.1. The performance of these parameters is strongly related to the reduced number of large generator tripping, wind generation response, HVDC ICs, demand response, and BESS (see FFR volumes in Figure 5.8). In other words, there are large volumes of contracted reserve service providers.

The yearly frequency nadir in 2024 has been 49.591 Hz. This value was reached after the tripping of two ICs (N-2) something that is beyond the normal operating standard (N-1). Despite this, frequency still remained well within the limit. The real-world observations shown in Fig. 12.7 mean that low-inertia may not necessarily

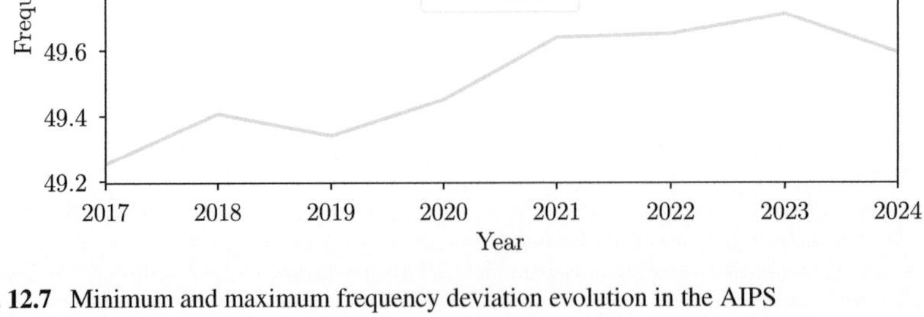

Fig. 12.7 Minimum and maximum frequency deviation evolution in the AIPS

lead to worse transient frequency performance. The performance depends on several other factors as discussed above.

12.3.2 Frequency Standard Deviation

Figure 12.8 shows the evolution of three relevant parameters, namely, minutes above and below the standard frequency range of ±200 mHz and σ_f. There are a number of factors that have led to a dramatic improvement in the quality of these parameters during 2009-2018:

- Change from verbal dispatch to electronic logged dispatch of generation units, leading to improved unit operator response;
- Newer generating units in the latter years having the latest electronic governor controls;
- Retrofit of the electro-mechanical control systems on older generating units with modern electronic controls;
- Increase of system inertia as a result of more generating units to meet increases in system demand; and
- Response of HVDC ICs and BESS and reduction in the number of large generator tripping.

Similar improvement in the minutes outside the standard frequency range has been observed in other synchronous areas such as in the Nordic synchronous area in recent

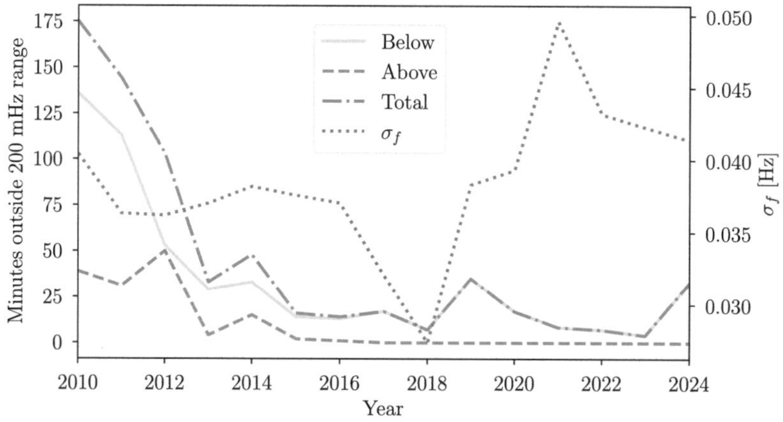

Fig. 12.8 Evolution of minutes outside the standard frequency range (±200 mHz) and standard deviation of the frequency over the last decade in the AIPS

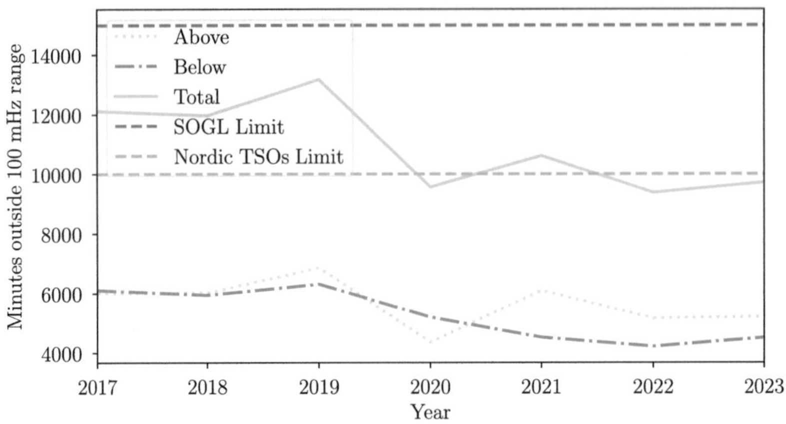

Fig. 12.9 Evolution of minutes outside the standard frequency range (±100 mHz) in the Nordic synchronous area in recent years

years. This is shown in Fig. 12.9.[2] Nevertheless, OF issues seem to be increasing in the Nordic power system. Nordic TSOs believe that the reason behind increasing over-frequencies is the high penetration of RES. For this reason, they are actively reviewing their mFRR capacity for downward activation, and are establishing a common day-ahead reserve capacity market as an additional mitigation measure [14]. The need for increased downward reserves is, in general, a worldwide power system trend. We discuss this trend below from an AIPS perspective.

On the other hand, there has been an increase in σ_f in the AIPS during 2019-2021. This is mainly due to: (i) reduction in regulating resources or PFR; (ii) increasing proportion of the PFC reserves from IBRs that are not configured to regulate frequency; and (iii) aging of conventional generating portfolio. This trend will likely continue as more wind and solar power are integrated into the system and the operational policy evolves by reducing the number of conventional units online and, consequently, also reducing the inertia.

While the EU legislation and the synchronous area operational agreement (Table 3.1) require EirGrid to keep frequency within its standard range (±200 mHz), the national RA in IE have put in place an incentive to keep frequency within an even tighter range namely ±100 mHz for $\geq$ 98% of the time [15]. This is also known as the ±100 mHz criteria. Since frequency is a robust metric of reliability and quality of supply, the incentive is appropriate and recognizes the challenge of system frequency control as RES penetration increases. Its evolution over the years is shown in Fig. 12.10. While there has been a deterioration in the last years, the percentage of time within the (49.9, 50.1) Hz range is still meeting the requirements.

---

[2] While SOGL defines a requirement of 15,000 minutes/year (see Table 3.1), the Nordic TSOs have agreed upon a goal for frequency deviations outside standard frequency range to be no more than 10,000 minutes/year.

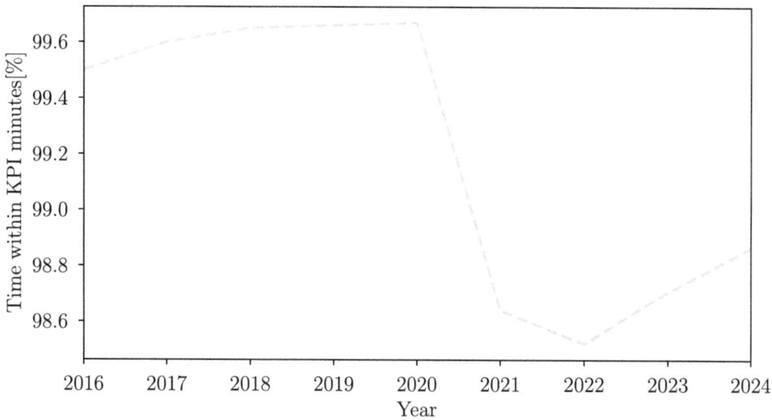

Fig. 12.10 Evolution of 100 mHz criteria over the last years in the AIPS

Table 12.3 Frequency KPI statistics for 2021-2022 ($\pm$100 mHz criteria $\geq$ 98% of time)

Month	Minutes	KPI violation minutes
August 2021	44,640	372
September 2021	43,200	420
October 2021	44,640	862
November 2021	43,200	935
December 2021	44,640	1,803
January 2022	44,640	1,135
February 2022	40,320	1,066
March 2022	44,640	1,078
April 2022	43,200	411
May 2022	44,640	433
June 2022	43,200	333
July 2022	44,640	230
Total minutes (12 months)	525,600	9,078
Time within KPI limits		98.27%

Table 12.3 shows the violation minutes in 2021 and 2022. Frequency spent more time outside $\pm$100 mHz range during winter due to more wind generation. This is interesting considering that, compared to 2020, there was less wind available in 2021.

There are a number of potential solutions to address the performance degradation of the frequency regulation in the last years (see Figs. 12.8 and 12.10), as follows:

- Increase the minimum primary regulating reserve requirement which currently comes from conventional generation with governor dead-bands of $\pm$15 mHz—this would seem a backward step as it would result in additional conventional unit commitment;

- Narrow frequency dead-bands on BESS/HVDC ICs;
- Update rules for activation of wind farm APC, including NI wind farms when feasible;
- Implement an AGC; and
- Introduce new market products to assist with frequency regulation.

In Sect. 12.4, we discuss the effectiveness of one of the options to improve frequency performance, that is, implementing an AGC.

12.3.3 Frequency Recovery

This section illustrates another frequency quality issue, that is, the slow frequency recovery following large disturbances in the AIPS. With this aim, Fig. 12.11 shows the frequency trace for an event on August 9, 2022 when the LSI tripped from 530 MW import. It took the frequency almost 15 minutes to recover to 50 Hz. Recall that 15 minutes is the AIPS limit for "time to restore frequency" in Table 3.1. In particular, frequency recovers to around 49.87 Hz in almost 3 minutes but then stays there for a long time. The fast recovery in the 3 minutes is mainly because of FFR from BESS. At the time of writing, the majority of BESS installed in AIPS have a trigger point between 49.9 Hz to 49.8 Hz.

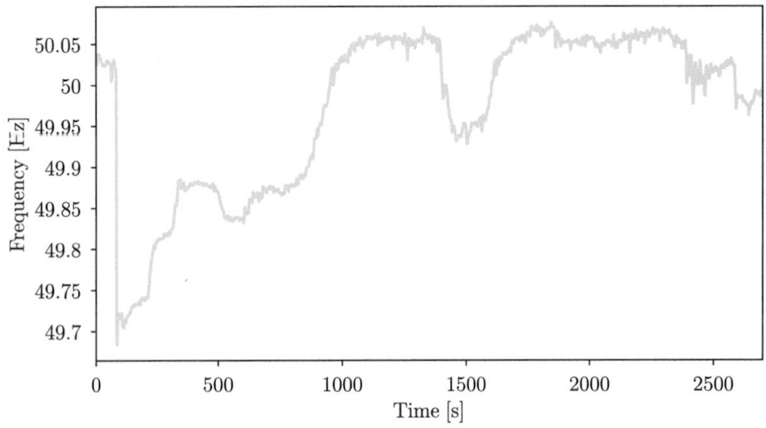

Fig. 12.11 LSI trip in August 2022

12.4 Example

EU Network Codes and the national energy regulators require EirGrid, SONI and NESO to justify the need to install or not an AGC every few years [16]. The purpose of this section is to illustrate the effectiveness of AGC to reduce frequency fluctuations and its performance in terms of long-term frequency quality enhancement.

Frequency quality is impacted by several dynamical processes starting from fast ones such as the inertial response of synchronous machines, the stochastic wind speed variations, to longer ones, such as PFC and SFC of conventional generators. To capture and model all of these dynamics, we consider a combined short- and long-term dynamic power system model represented by a set of nonlinear stochastic DAEs (see equations (A.1)). The AGC is modeled as in Figure 2.19 and updates the power order set-points of dispatchable generators at certain time intervals, for example, every 4 s.

Simulations are based on a modified IEEE 39-bus system (see Appendix B.2) where 25% of the generator has been replaced with wind power plants. Two scenarios are considered: (i) impact of stochastic noise (given by both load and wind); and (ii) same scenario 1 plus the introduction of wind and load step and ramp variations [17].

Figure 12.12 shows the results of the first scenario with and without the AGC. In this scenario, the inclusion of the AGC does not appear to have any visible impact on frequency fluctuations. This is due to the fact that the AGC controller is slow compared to the dynamics of the noise (stochastic process). On the other hand, Fig. 12.13 compares the effect of AGC under both noise and wind/load step and ramp power variations. Since wind and load ramp time scales are closer to that of the AGC, in this case, the inclusion of the AGC allows reducing frequency deviations. Specifically, σ_f of the frequency with and without AGC is 0.05395 Hz and 0.0765 Hz,

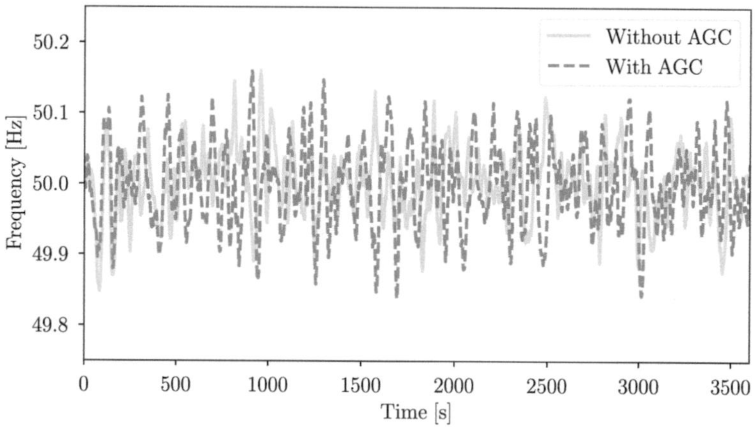

Fig. 12.12 Scenario 1: Impact of noise

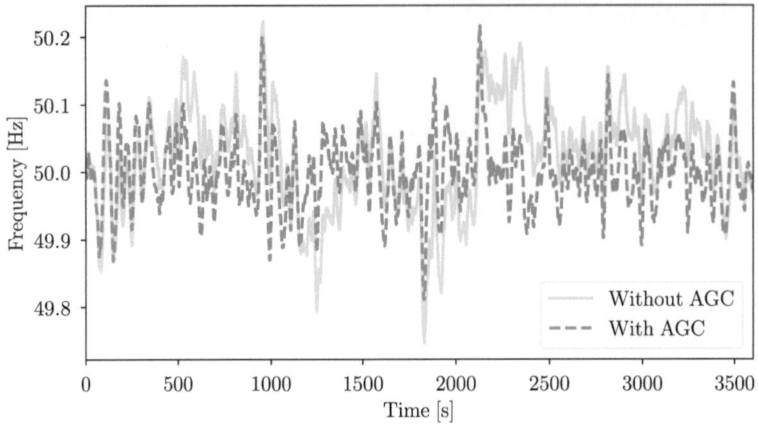

Fig. 12.13 Scenario 2: Impact of noise and wind/load step and ramp variations

respectively. These results indicate that an AGC implementation may be an option to improve frequency quality in the AIPS in the future.

12.5 Reserves Review

In this section, we present updated reserve product definitions to address several real-world trends and evolving risks in the AIPS. Most notably, increased levels of IBR-based sources namely wind, solar and HVDC imports continue to replace conventional synchronous generation leading to a reduction in system inertia. These lower inertia conditions have the potential to lead to larger frequency deviations and higher RoCoF. Based on these evolving challenges and detailed analysis and nonlinear Time Domain Simulations (TDSs), reserve services have been reviewed.

As part of this review, the Irish TSOs have modified some of the above services and defined new ones that after consultation with industry were approved by the respective RAs in October 2024. These services help achieve 2030 RES targets, namely, 80% of electricity to be provided by RES, and align with EU requirements in terms of downward reserves. They are to be procured in DASSA in 2027, see Fig. 12.6. In particular, the Irish TSOs recognize based on actual system incidents [18] and system analysis (see Sect. 12.5.1) that maintaining both short and long-term frequency quality will face challenges as the system moves toward a system mix of 80% electricity from RES by 2030. Main causes are:

- Reduced inertia levels as the running of conventional units is reduced (7 large conventional units). Lower level of inertia will lead to, in general, faster frequency dynamics and, in particular, to larger potential RoCoF occurrences [19]. In addition, the elapsed time between incident and frequency nadir/zenith is expected to be significantly shorter.

- More variability and forecast uncertainty from wind and solar generation and also more volatile demand, will lead to larger imbalances, requiring a greater volume of balancing actions.
- Increased numbers of HVDC ICs which could ramp at greater speeds than today.

12.5.1 System Needs

12.5.1.1 Need for Downward Reserves

Irish TSOs have traditionally relied on upward reserves to deal with UF events that may ultimately lead to disconnection of loads through automatic schemes such as low-frequency demand disconnection. For instance, if the system frequency is below 49.0 Hz the UF scheme is triggered to automatically disconnect a portion of the demand to prevent further system collapse. Downward reserves, on the other hand, have been less needed as large loads rarely disconnected demand following system incidents. However, this situation has changed in recent years in the AIPS. For example, demand units can reduce demand automatically following transmission faults due to lack of FRT capability [18]. In addition, contingencies such as the loss of the N-S tie-line between IE and NI (system separation) and exporting IC, can lead to OF issues [20].

Because of these potential risks, Irish TSOs introduced OF generation protection schemes for renewable (wind) generation as a tool to mitigate the severity of such incidents. These emerging risks are likely to increase with increasing levels of large converter-based loads, increased volumes of IC capacity and RES [18]. For this reason, the types of reserve products the Irish TSOs will need in the future need to evolve to manage these risks. Moreover, the reduction of conventional units online will not only impact inertia but also the total governor response provided by them both in upward and downward direction.

Figure 12.14 shows the potential frequency nadir, zenith and RoCoF issues as % of total frequency insecurities as reported by Look-ahead Security Assessment Tool (LSAT) for one year namely October 2023—September 2024.

Figure 12.14 indicates that frequency zenith issues are greater compared to frequency nadir issues. The main drivers for these zenith issues are low-inertia conditions in IE or NI following a potential system separation. Also, the significant differences between October-November 2023 and August 2024 are due to having very few total frequency insecurities. For the reasons stated above, the TSOs introduced downward reserve service products FFR, POR, SOR, TOR1, TOR2 and RR that will mirror the upward reserve products. These proposals align with requirements outlined in the EU regulation that require reserves to be dimensioned both in an upward and downward direction [16].

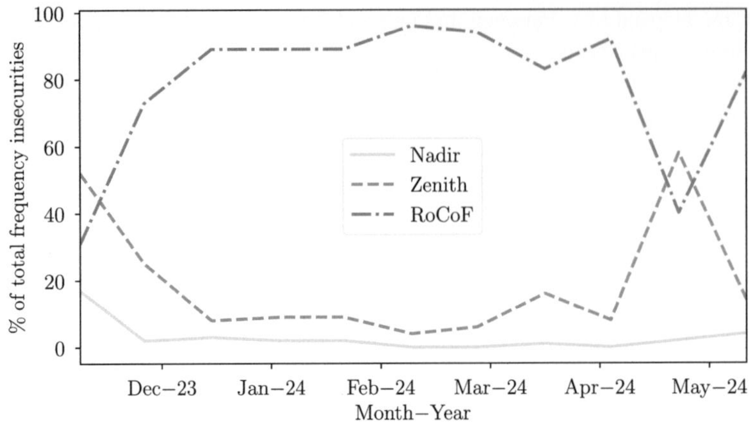

Fig. 12.14 Nadir, zenith and RoCoF as % of total frequency insecurities

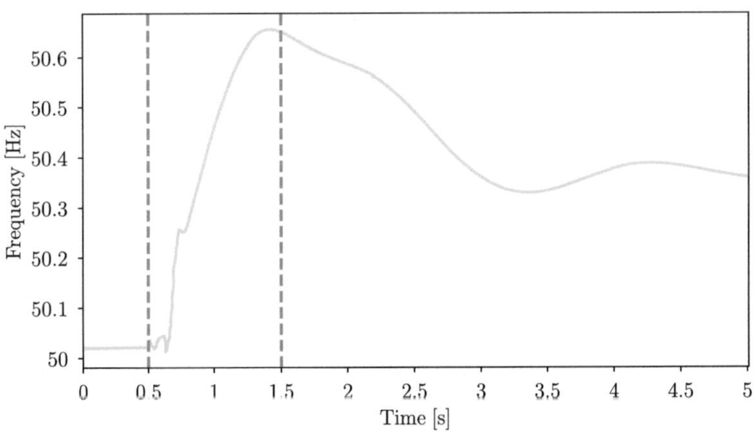

Fig. 12.15 Example of an OF event (based on TSOs studies)

12.5.1.2 Need to Reduce FFR FAT

When the AIPS is running with very few large units online and low inertia, it is likely that frequency deviations of ±1000 mHz could be reached within 1 s if there are insufficient fast responding reserves that act within 1 s. This is illustrated in Fig. 12.15 using a 2025 fully-fledged dynamic AIPS model and running frequency stability studies with LSAT.

Figure 12.15 shows that apart from the need to have downward reserves, there is also a need to reduce the FAT of FFR as frequency zenith is reached before 1 s (contingency is applied at 0.5 s). That is, the current 2 s FAT of FFR is considered too slow for future requirements. Hence, the TSOs reduced the FAT of FFR product from 2 to 1 s. In addition, to maintain current FFR capability in terms of incentives to

Table 12.4 Proposed reserve services

Reserve service	Direction [Upward/Downward]	Minimum volume requirement [% of RI]	FAT [s]	Response duration [s after the event]
FFR (Cat. 1)	Upward & Downward	60-80	0.15	≤ 10
FFR (Cat. 2)	Upward & Downward	60-80	≤ 0.3	≤ 10
FFR (Cat. 3)	Upward & Downward	60-80	≤ 1	≤ 10
POR	Upward & Downward	100	5	≤ 15
SOR	Upward & Downward	100	15	≤ 90
TOR1	Upward & Downward	100	90	≤ 300
TOR2	Upward & Downward	100	300	≤ 1200
RR	Upward & Downward	100	1200	≤ 3600

reduce FAT down to 150 ms through product scalar [5], TSOs defined FFR in three activation time sub-categories namely 150 ms, 300 ms, and 1 s (see Table 12.4 below). An UF event would be mitigated in a similar way as an OF event, that is, an upward FFR product with a FAT of less than 1 s is needed.

12.5.1.3 Need to Keep Frequency Within the Standard Range

The change in the electricity sector landscape means that the AIPS requires continued provision of frequency regulation both now and in the future to address emerging frequency quality challenges [4, 21]. For example, as the AIPS now has large volumes of RES, their inherent variability can cause increased volatility in system frequency. While the Irish TSOs have not proposed a new explicit PFR product at this time, similar to Dynamic Moderation (DM) and Dynamic Regulation (DR) regulation in Great Britain (GB) (see Figure 2.7), they have ensured that the proposed product definitions and technical requirements such as dead-band capability of ± 15 mHz could enable a similar dynamic response capability to be procured through DASSA. Note that by dynamic response the Irish TSOs mean resources such as conventional units and IBRs that can continuously adapt generation to the actual frequency. In contrast, DSUs typically provide static/discrete response by reducing load in blocks at specified frequency triggers and restoring this demand once frequency recovers to a threshold.

12.5.2 New Reserve Services

Table 12.4 summarizes the new reserve products defined by Irish TSOs, while Fig. 12.16 and Table 12.5 illustrate and provide key requirements for all upward and downward reserves, respectively. In particular, note that Irish TSOs decided to:

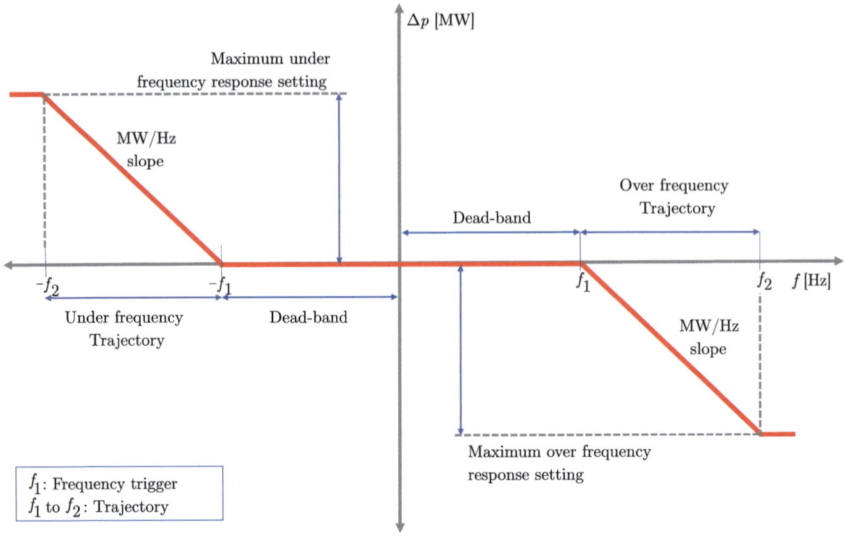

Fig. 12.16 Frequency response characteristics

Table 12.5 Proposed upward and downward reserve services

Criteria for	Trigger F_1	End of trajectory F_2	Reserve step sizes	Reserve step triggers
Static FFR, POR, SOR, TOR1 and TOR2	Configurable for each step between: $49.3 \leq F_1 \leq 49.8$ Hz (up) and $50.2 \leq F_1 \leq 50.7$ Hz (down)	N/A	1 or more steps of $\leq$ 75 MW for a single discrete step	Smallest available discrete step in response at any time must be $\geq 20\%$ of the MW value of the providing unit's largest available step at that time
Dynamic FFR, POR, SOR, TOR1 and TOR2	Configurable in range: $49.5 \leq F_1 \leq 49.985$ Hz (up), and $50.015 \leq F_1 \leq 50.5$ Hz (down)	Configurable in range: $49.3 \leq F_2 \leq 49.8$ Hz and $F_1 - F_2 \geq 200$ mHz (up), $50.2 \leq F_2 \leq 50.7$ Hz and $F_2 - F_1 \geq 200$ mHz (down)	N/A	N/A

- Introduce downward reserves (FFR, POR, SOR, TOR1, TOR2, and RR) that will mirror existing upward products.
- Reduce the current standard FAT of FFR from 2 to 1 s, to deliver faster response capability.
- Procure FFR in three activation time sub-categories, as very fast provision of FFR is essential for system operation.
- Combine RRS and RRD products into one RR product. In proposing the merging of the two products the TSOs considered EU requirements (where only one RR

product exists) and the future requirements of the power system, where access to RR from all types of generation, demand and storage technologies and in both the upward and downward directions is needed.

- Introduce minimum capability requirements on frequency dead-bands, trajectories, reserve step sizes and reserve step triggers. These will remain configurable by the TSOs as part of requirements specified in the grid services code, and also to avoid potential undesirable phenomena such as oscillations or asymmetry in the frequency distribution [22]. For instance, conventional generation, RES and BESS can provide frequency control based on specified droops. Specifically, the minimum capability requirements include:
  - A dead-band for dynamic FFR, POR, SOR, TOR1, and TOR2 that is configurable between ±15 mHz and ±500 mHz, and a trajectory capability of at least 200 mHz. Narrow dead-bands such as ±15 mHz and relatively slower dynamic response are critical to provide frequency regulation similar to conventional plants today. On the other hand, large dead-bands with a fast response are needed to contain larger event driven frequency deviations.
  - A dead-band for static FFR, POR, SOR, TOR1, and TOR2 that is configurable between ±200 mHz and ±700 mHz.

12.6 Volume Forecasting Methodology

Current methodology to determine minimum real-time reserve requirements in the AIPS is mainly based on the LSI as shown in Table 12.1. However, the TSOs have decided that reserve requirements, including for the AIPS and jurisdictional (minimum reserves in IE and NI), should be set based on the Reference Incident for Infeed (RII) and Reference Incident for Outfeed (RIO) losses plus the relevant Consequential Losss (CLs) namely Consequential Loss for Infeed (CLI) and Consequential Loss for Outfeed (CLO), as follows:

$$\mathrm{RII}^{\mathrm{AIPS}} = \mathrm{LSI}^{\mathrm{AIPS}} + \mathrm{CLI}^{\mathrm{AIPS}}, \tag{12.1}$$

$$\mathrm{RIO}^{\mathrm{AIPS}} = \mathrm{LSO}^{\mathrm{AIPS}} + \mathrm{CLO}^{\mathrm{AIPS}}, \tag{12.2}$$

$$\mathrm{RIO}^{\mathrm{NI}} = \mathrm{Flow}^{\mathrm{N}\rightarrow\mathrm{S}} + \mathrm{CLO}^{\mathrm{NI}}, \tag{12.3}$$

$$\mathrm{RII}^{\mathrm{NI}} = \mathrm{Flow}^{\mathrm{S}\rightarrow\mathrm{N}} + \mathrm{CLI}^{\mathrm{NI}}, \tag{12.4}$$

$$\mathrm{RIO}^{\mathrm{IE}} = \mathrm{Flow}^{\mathrm{S}\rightarrow\mathrm{N}} + \mathrm{CLO}^{\mathrm{IE}}, \tag{12.5}$$

$$\mathrm{RII}^{\mathrm{IE}} = \mathrm{Flow}^{\mathrm{N}\rightarrow\mathrm{S}} + \mathrm{CLI}^{\mathrm{IE}}. \tag{12.6}$$

Note that CLI and CLO are typically inadvertent and caused by, for example, lack of FRT capability of the concerned demand/generation. Additionally, reserves

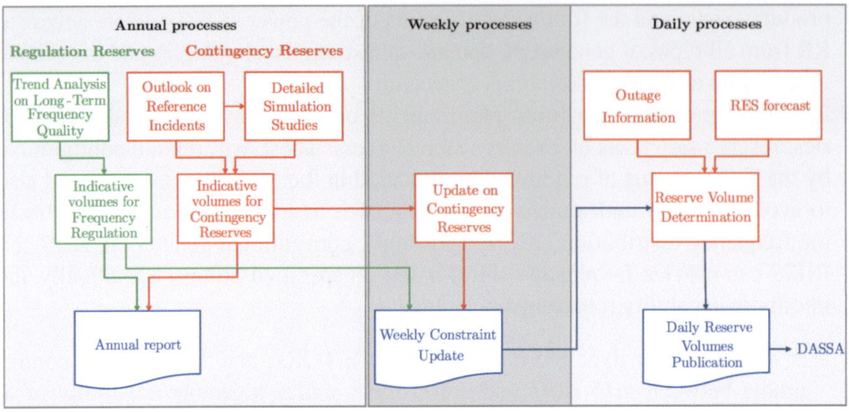

Fig. 12.17 High-level overview of proposed VFM processes

should also account for potential loss of reserve provision from the units setting the LSI/LSO and unavailability of reserve providing units in real-time, respectively [23].

Figure 12.17 provides a high-level overview of the proposed VFM for the reserve services. Irish TSOs propose a "front-loaded" process with annual, weekly and daily components. While the majority of the analysis will be performed annually, the weekly and daily processes allow for updates to reserve volume forecasts as as more up-to-date information becomes available. This methodology is considered the most appropriate by the TSOs.

Relevant reasons are as follows.

- At the day-ahead stage, that is, at 10 am, there is limited additional information available in comparison to week and year-ahead.
- Running detailed dynamic simulations such as TDSs to determine reserve requirements is a time consuming and a very specialized task. Given the fact that at the day-ahead stage (at 10 am) there is still a lot of uncertainty in terms of RES forecast and market results, additional TDS may unlikely yield better volume results/forecasts. For this reason, the TSOs will apply rule-based approaches such as those based on parameters defined in the annual process, in determining reserve volume forecasts in the daily process.
- The long-term frequency quality measured as % of time outside 49.9-50.1 Hz range during a year is generally and typically evaluated on an annual basis. Thus, the current proposal is that minimum reserve volumes will be reviewed and updated, where necessary, during the annual forecasting process. However, note that TSOs reserve the right to take relevant actions to remedy significant deteriorating frequency quality more than annually. The latter is in line with the continuous monitoring of long-term frequency by TSOs that include, for example, monthly reports on the ± 100 mHz criteria.

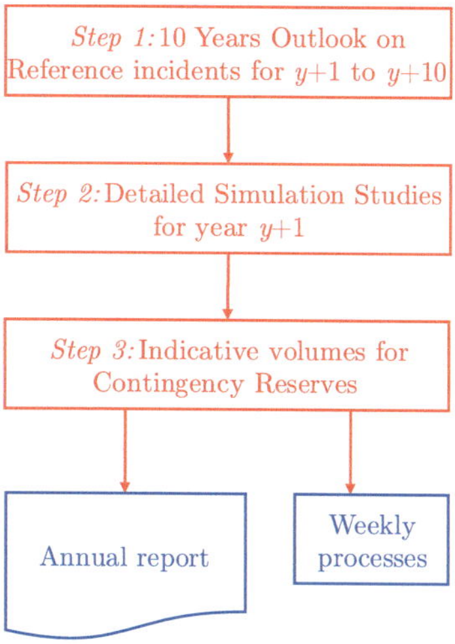

Fig. 12.18 Proposed annual process for VFM of reserves to mitigate large disturbances

In terms of the overall timing process for the VFM, the following processes are proposed by the Irish TSOs:

- *Annual Process*: Figure 12.18 shows the proposed annual process. This process includes a ten-year forecast with an indicative overview of the required reserve volumes, including different characteristics such as FFR FAT, dynamic vs static reserves and jurisdictional requirements. In doing so, it is considered that the forecast would be based on a high-level review of the system needs for the next 10 years, including consideration of anticipated changes in the system. For instance, the annual VFM could take on board the separate EirGrid and SONI long-term views of the system in terms of Transmission Development Plans, and the joint TSO Ten Year Transmission Forecast Statement. Furthermore, the process will include a more detailed assessment for the forthcoming DASSA procurement year, that is, "year+1." Such a detailed assessment of year+1 will be supported by historical trend analysis and TDSs which will also provide the necessary input to the subsequent weekly and daily volume requirements. The aim of the proposed annual process is to address two main purposes of reserves: (i) requirements for maintaining frequency quality within defined limits under normal conditions; and (ii) mitigating contingencies.
- *Weekly Process*: At this stage, the TSOs will provide a weekly update of the reserve requirements as part of the Weekly Operational Constraints Update document. This update will consider, among others, the impact of planned outages and the applicability of the results of the annual assessments.
- *Daily Process*: By 10 am each day, the TSOs will publish the required reserve volumes that will be procured in DASSA. Considering the output from the weekly process, and utilizing updated information such as outages and RES forecasts. The DASSA will specify volumes for all upward and downward reserve products namely FFR, POR, SOR, TOR1, TOR2, and RR.

Regarding large disturbance counteractions, Irish TSOs will make sure they have sufficient reserves to prevent an instantaneous frequency deviation of more than ± 1000 mHz, or a RoCoF $> \pm 1$ Hz/s, after the loss of LSI or LSO, or the loss of ICs or tie lines, including the loss of both circuits of the N-S ICs and any CLs. The following requirements will need to be assessed:

- Volumes of FFR, POR, SOR, TOR1, TOR2 and RR;
- Minimum volumes for the three FFR categories;
- Minimum volumes of dynamic POR, SOR, TOR1, and TOR2;
- Minimum volumes for IE and NI.

Step 1: 10-year outlook on potential Reference Incidents (RIs)
As mentioned above, the 10-year forecast will utilize information and assessments from other TSOs established processes such as Ten Year Transmission Forecast statements. As part of this process, the TSOs will also apply operational knowledge and expert engineering judgment. The proposal includes the preparation of an annual update of potential Reference Incidents (RIs) for the forthcoming 10 years including a list of units that could set the RIs.
Step 2: Detailed simulation studies for year $y + 1$
This step would focus on a subset of potential RIs for year $y + 1$ only. Specifically, detailed TDS studies will be performed for the following RIs:

- AIPS RI for LSI losses, considering the largest potential RI in positive direction applicable to year $y + 1$.
- AIPS RI for LSO losses, considering the largest potential RI in negative direction applicable to year $y + 1$.

And, at least until the second N-S IC is built and is in operation:

- Island split when the N-S IC is flowing at full capacity in the direction NI to IE, considering the impact in both IE and NI.
- Island split when the N-S IC is flowing at full capacity in the direction IE to NI, considering the impact in both IE and NI.

The requirements and assumptions for the power system TDS model include:

- A power system model for year $y + 1$ that includes dynamic models of all relevant generation and demand sources.
- Total POR, SOR, TOR1 and TOR2 volumes will be set equal to 100% of the RI.
- A share of dynamic PFC reserves based on last year's minimum reserves, an evaluation of historical experience, and engineering judgement.
- OF, UF, over-voltage and under-voltage protection schemes implemented in the model and enabled.

Table 12.6 Example Step 3 output: TDS studies for upward reserves

Upward	AIPS		IE		NI	
	Min Vol	Min Dyn	Min Vol	Min Dyn	Min Vol	Min Dyn
RI	IC full import		Trip of Tie-Line		Trip of Tie-Line	
FFR3	70% of RI	80% of FFR3	80% of RI	80% of FFR3	100% of RI	80% of FFR3
FFR1	60% of FFR3	80% of FFR1	70% of FFR3	80% of FFR1	80% of FFR3	80% of FFR1
FFR2	70% of FFR3	80% of FFR2	80% of RI	80% of FFR2	90% of FFR3	80% of FFR2
POR-TOR2	100% of RI	80% of POR- TOR2	100% of RI	80% of POR- TOR2	100% of RI	80% of POR- TOR2
RR	100% of RI	N/A	100% of RI	N/A	100% of RI	N/A

The output of the detailed studies will include:

- The minimum shares of FFR, FFR category 1 and FFR category 2 for which the instantaneous frequency deviation at the nadir or zenith will not exceed ±800 mHz and the RoCoF will not exceed ±0.9 Hz/s.[3]
- A confirmation of the minimum dynamic POR and SOR for which the frequency will stabilize. Other dynamic PFC reserves such as TOR1 will not be considered as current dynamic simulation capabilities are limited to POR and SOR timelines.

Table 12.6 provides an example of the potential output of the TDS studies.

Step 3: Indicative reserve volumes for mitigating large disturbances

Irish TSOs need to consider in the DASSA reserve volume determination the potential unavailability of reserve providing units between gate-closure of the DASSA and real-time. For instance, one or more units could become unavailable due to a forced outage or a transmission restriction/fault which, in turn, will limit the provision of their service. Table 12.6 will include an additional allowance for subsequent unavailability of reserve service from one or a group of reserve service providers. This additional reserve value will apply to FFR, POR, SOR, TOR1, TOR2 and RR.

The final output of this process will represent indicative reserve volume requirements for year $y + 1$ as shown in Table 12.7. The output will also include the required minimum levels of FFR, FFR categories, dynamic response and jurisdictional requirements. Table 12.7 will serve as an input to the weekly process.

The TSOs will configure the PFC dead-band settings of reserve providers that have been awarded DASSA contracts for dynamic PFC reserve provision namely for POR-TOR2. This means sufficient reserve capacity is available to the TSOs for ensuring system frequency remains within the standard frequency range 49.8–50.2 Hz. It should be noted that for the purposes of this illustration, no CLs have been

[3] A margin of ±200 mHz and ±0.1 Hz/s will be applied to the maximum frequency deviations and RoCoF, respectively.

Table 12.7 Example of Step 3 output: Indicative upward reserve volumes for mitigating large disturbances for year $y + 1$

Upward	AIPS		IE		NI	
	Min Vol	Min Dyn	Min Vol	Min Dyn	Min Vol	Min Dyn
	[MW]		[MW]		[MW]	
RI	500	–	450	–	400	–
FFR3	425	340	435	348	475	380
FFR1	255	204	305	244	380	304
FFR2	298	238	348	278	428	342
POR-TOR2	575	460	525	420	475	380
RR	575	N/A	525	N/A	475	N/A

included in Table 12.7. Additionally, a fixed allowance of 75 MW has been assumed for service providers unavailability. For years $y + 2$ and $y + 10$ the RI, the TSOs will provide an indicative overview of POR-RR volumes based on the development of the RI.

12.6.1 Consequential Losses

CLs are typically inadvertent (from a system perspective) and caused by a lack of FRT capability of the concerned demand/generation. As discussed in Chaps. 2 and 7, CLs represent a significant concern for TSOs. For instance, following the GB incident in 2019 that resulted in a series of consequential events and partial blackout, the Government there decided to [24]: *recommend that there should be a review into the reserve and response holding policy of the NESO and whether it is fit for purpose going forward.*

For this reason, EirGrid and SONI have taken a prudent approach and have proposed that the RI will also need to account for potential CLs of, for example, trips of generation or reduction of demand from demand facilities triggered by the same incident.

Figure 12.19 illustrates these steps. Considering the limited foresight at day-ahead stage, the TSOs consider including a fixed allowance for CL factors CLI and CLO. These factors will be determined in the annual process. In particular, for CLs, the steps will be carried out as follows.

Step 1: 10 years outlook on RIs *for* $y + 1$ *to* $y + 10$

A 10-year forecast on CLs, based on additional capacity of new Large Energy User (LEU) and new Distributed Energy Resource (DER), their FRT capability and the development and implementation of new (Grid Code) requirements on FRT capability.

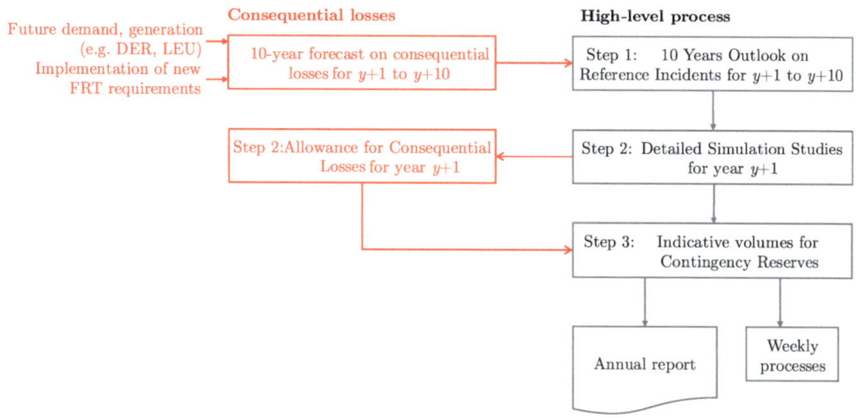

Fig. 12.19 High-level methodology for the calculation of CLs

Step 2: Detailed TDS *studies for year* $y + 1$

In performing the detailed TDS studies for $y + 1$, special attention will be paid to CLs. Hence, the output will show both components of the RIs: the LSI or LSO and the CLs. Based on the results for the RI, the allowance for CLs is determined. This allowance will consider that a share of a certain infeed or outfeed category could consequentially trip, for example, $X\%$ of LEU or $Y\%$ of DER. Where appropriate, the allowance may be time-varying, for example, solar PV generation will not be at maximum capacity for the full day, and accordingly the potential CLs may vary.

Step 3: Indicative volumes for contingency reserves

The allowance for CLs resulting from step 2 is applied to determine the indicative reserve volumes for mitigating large disturbances.

Figure 12.20 shows an example of loss of outfeed followed by CL (see Chapter 7 for real-world impact of CLs on system frequency and RoCoF). The instantaneous change of active power concerns a trip (due to a fault) of a HVDC cable exporting 500 MW before the trip. The fault causes a voltage dip that propagates to various connections of LEU. This voltage dip instantaneously triggers the LEU under voltage protections that reduce 200 MW of LEU demand (left-hand side of the figure). Accordingly, the “size of the RI shall be the largest imbalance” triggered by this incident. This “imbalance” is the sum of instantaneous change of 500 MW + the CL of 200 MW triggered by the under voltage protection = 700 MW (downward reserves).

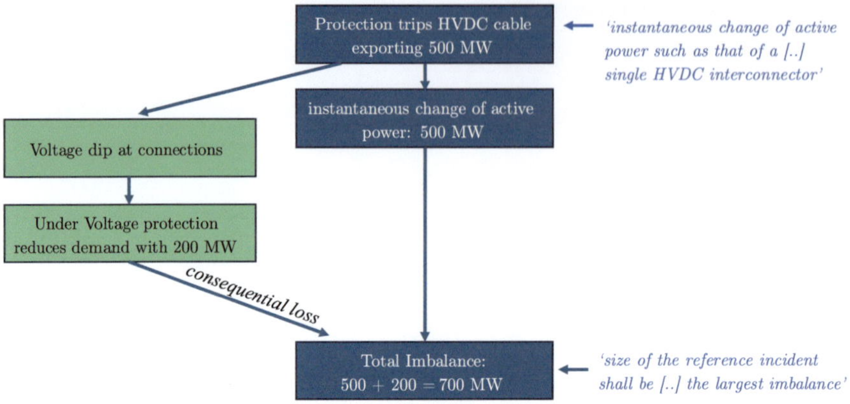

Fig. 12.20 Illustrative example of CL (demand side)

In Brief

- While some frequency quality parameters such as minutes below and above ±200 mHz have dramatically improved over the last years, others such as σ_f has increased linearly between 2018 and 2021.
- New downward reserve have been defined that mirror the existing upward reserves to deal with more and more potential OF events.
- The FAT of FFR is reduced from 2 to 1 s to deal with faster nadir/zenith and higher RoCoF.
- The RRS and RRD are combined into one RR product.
- Minimum capability requirements are set on dead-bands, trajectories, reserve step sizes and reserve step triggers, that will remain configurable by the TSOs.
- New VFM for reserve has been developed to determine volume requirements on a day-ahead forecasting basis, where uncertainty on wholesale energy market outcomes, RES forecasts and IC flows exists.
- The RI is now based on the sum of LSI/LSO and CLs.
- AGC can be a solution to address the recent deterioration in frequency quality in the AIPS.

Open Questions

While AGC is widely utilized by many jurisdictions worldwide, its effectiveness in the context of the AIPS is not yet clear. Future work need to focus on testing the effectiveness of different AGC approaches on a model of the AIPS. This work will then feed into the assessment of the range of solutions for managing frequency quality on the AIPS.

Open questions exist whether standard frequency reserve products defined by EU legislation namely FCR and FRR can be introduced in the AIPS in the future. More work is also needed to understand how to accurately predict LSI/LSO and CLI/CLO. Finally, it is still not clear how the AIPS but, in general, all RES-dominated power systems, will ensure frequency quality remains within defined standards when all demand is being met by DERs.

References

1. Rancilio G, Rossi A, Falabretti D et al (2022) Ancillary services markets in Europe: evolution and regulatory trade-offs. Renew Sustain Energy Rev 154:111850
2. Lew D, Bartlett D, Groom A et al (2019) Secrets of successful integration: operating experience with high levels of variable, inverter-based generation. IEEE Power Energ Mag 17(6):24–34
3. Hurtado M, Kërçi T, Tweed S et al (2023) Analysis of wind energy curtailment in the Ireland and Northern Ireland power systems. In: IEEE PES general meeting, pp 1–5
4. Kërçi T, Hurtado M, Tweed S et al (2024) Emerging challenges of integrating solar PV in the Ireland and Northern Ireland power systems. In: IEEE PES general meeting, pp 1–5
5. Delaney N, Deegan, Val Escudero M et al (2020) Innovative system services for facilitating the integration of high levels of renewable generation in Ireland and Northern Ireland. IET Renew Power Gen 14(19): 3954–3960
6. Du P, Mago NV, Li W et al (2020) New ancillary service market for ERCOT. IEEE Access 8(178):391–178 401
7. Cataln MN, Du P, Mago N et al (2021) Ancillary service requirements analysis with increasing solar generation in the ERCOT interconnection. In: 2021 IEEE power and energy society general meeting (PESGM), pp 1–5
8. Meng L, Zafar J, Khadem SK et al (2020) Fast frequency response from energy storage systems—a review of grid standards, projects and technical issues. IEEE Trans Smart Grid 11(2):1566–1581
9. O'Sullivan J, Coughlan Y, Rourke S et al (2012) Achieving the highest levels of wind integration—a system operator perspective. IEEE Trans Sustain Energy 3(4):819–826
10. EirGrid and SONI (2024) All-island transmission system performance report 2023. https://www.eirgrid.ie/
11. EirGrid (2024) EirGrid Grid Code. Version 14.2. https://www.eirgrid.ie/
12. European Commission (2017) Commission regulation (EU) 2017/2195 of 23 Nov 2017 establishing a guideline on electricity balancing. eur-lex.europa.eu
13. SEM Committee, System services future arrangements high level design decision SEM-22-012. www.semcommittee.com
14. ENTSO-E (2024) 2023 annual load-frequency control report. eepublicdownloads.entsoe.eu
15. Commission for Regulation of Utilities (2020) PR5 regulatory framework, incentives and reporting, 2020. www.cru.ie/wp-content/uploads/2020/12/CRU20154-PR5-Regulatory-Framework-Incentives-and-Reporting-1.pdf
16. European Commission (2017) Establishing a guideline on electricity transmission system operation. eur-lex.europa.eu
17. Kërçi T, Giraldo JS, Milano F (2020) Analysis of the impact of sub-hourly unit commitment on power system dynamics. Electr Power Energy Syst 119:105819
18. Kërçi T et al (2024) Impact of converter-based demand on frequency quality in the Ireland and Northern Ireland power systems. CIGRE Paris Session CIGRE 2024:1–11
19. Milano F, Drfler F, Hug G et al (2018) Foundations and challenges of low-inertia systems (invited paper). In: Power systems computation conference (PSCC), pp 1–25

20. Hurtado M, Jafarian M, Kërçi T et al (2024) Stability assessment of low-inertia power systems: a system operator perspective. In: IEEE PES general meeting, pp1–5
21. Kërçi T, Hurtado M, Gjergji M et al (2023) Frequency quality in low-inertia power systems. In: IEEE PES general meeting, pp 1–5
22. Kri T, Milano F (2025) Asymmetry of frequency distribution in power systems: sources, estimation, impact and control. IEEE Open Access J Power Energy 12:135–145
23. EirGrid and SONI (2024) Day-ahead system services auction (DASSA) product review and locational methodology consultation paper v1.0. consult.eirgrid.ie
24. Department for Bussines, Energy and Industrial Strategy (2019) GB power system disruption—9 Aug 2019. assets.publishing.service.gov.uk

Appendix A
Dynamic Model of Power Systems

The power system model considered in this book is the following set of hybrid stochastic Differential-Algebraic Equation (DAE) [1]:

$$\begin{aligned} \mathbf{T}\boldsymbol{x}' &= \boldsymbol{f}(\boldsymbol{x}, \boldsymbol{y}, \boldsymbol{u}, \boldsymbol{z}, \boldsymbol{\eta}) , \\ \mathbf{0}_m &= \boldsymbol{g}(\boldsymbol{x}, \boldsymbol{y}, \boldsymbol{u}, \boldsymbol{z}, \boldsymbol{\eta}) , \\ \boldsymbol{\eta}' &= \boldsymbol{a}(\boldsymbol{\eta}) + \mathcal{B}(\boldsymbol{\eta})\, \boldsymbol{\xi} . \end{aligned} \tag{A.1}$$

where $\boldsymbol{x} \in \mathbb{R}^n$ are the deterministic state variables, e.g., rotor speeds and fluxes of synchronous machines; $\boldsymbol{y} \in \mathbb{R}^m$ are the algebraic variables, e.g., bus voltage magnitudes and phase angles; $\boldsymbol{z} \in \mathbb{R}^p$ are discrete variables, e.g., the status of breakers and the tap ratio of ULTC; and $\boldsymbol{u} \in \mathbb{R}^q$ are the inputs, e.g. active power schedules and reference voltage of the automatic voltage regulators. $\boldsymbol{f}$ are deterministic differential equations that represent, e.g., the dynamic models of synchronous machine and wind power plants, as well as their controllers; $\boldsymbol{g}$ are the algebraic equations that represent the lumped models of transmission lines and transformers, as well as auxiliary constraints.

Matrix $\mathbf{T}$ is a *mass* matrix that contains the time constants of the state variables. This matrix allows to switch off the dynamic of any i-th state by simply setting to zero the i-th diagonal element. In general, $\mathbf{T}$ can include off-diagonal terms, thus leading to a semi-implicit formulation [2], and its elements can be time-varying. Unless indicated otherwise, in the models utilized in the book, we assume that $\mathbf{T}$ is diagonal, full rank and time-invariant. Whenever required to improve readability and avoid unnecessary mathematical derivations, we also assume that $\mathbf{T} = \mathbf{I}_n$, where $\mathbf{I}_n$ is identity matrix of the same order of the states, and assume that the information of the time constants is embedded in equations $\boldsymbol{f}$.

The stochastic processes $\boldsymbol{\eta} \in \mathbb{R}^r$ are expressed as set of stochastic differential equations, where $\boldsymbol{\xi} \in \mathbb{R}^s$ is the vector of white noise processes that formally represent the time derivatives of the Wiener processes. The stochastic processes are defined by a drift vector $\boldsymbol{a}$ and a diffusion term represented by matrix $\mathcal{B}$. These processes represent any noise that perturbs the system, e.g., random variations of load power

T. Kërçi and F. Milano, *Frequency Quality in Power Systems*, Power Systems,
https://doi.org/10.1007/978-3-032-04721-2

consumption or of the wind speed. Unless stated otherwise in the text, η are assumed to be normally distributed.

Mathematical tools for the stability analysis of nonlinear stochastic hybrid dynamical systems are quite limited. The nonlinearity prevents the use of methods that require a linear set of equations and this eliminates the vast majority of available techniques. The Lyapunov stability theory has been widely utilized for the analysis of nonlinear systems, including power systems. However, due to the dissipative nature of power systems, even assuming that a Lyapunov function can be found, the Lyapunov theory would not provide sufficient and necessary conditions and is thus of limited practical use. The only fully general approach to study the stability of power systems that does not requires shortcuts or simplifications is the Time Domain Simulation (TDS). This is the approach utilized in most examples discussed in the book. All simulations are performed using the power system analysis software tool Dome [3].

Appendix B
Test Systems

This appendix reports the data of the Western Systems Coordinating Council (WSCC) 9-bus and the New England 39-bus test systems, which are utilized as base case for the examples presented in the book.

The synchronous machine model is a standard 4th-order per-unit model [4]:

$$\begin{aligned} \delta' &= \omega_n(\omega - 1)\,, \\ 2H\,\omega' &= p_m - p_e - D(\omega - 1)\,, \\ T_{d10}\,e'_{q1} &= -e_{q1} - (X_d - X_{d1})\,i_d + v_f\,, \\ T_{q10}\,e'_{d1} &= -e_{d1} + (X_q - X_{q1})\,i_q\,. \end{aligned} \tag{B.1}$$

where the state variables δ, ω, e_{q1} and e_{d1} are the rotor angle, rotor angular speed, and transient dq-axis internal emfs, respectively; v_f is the field voltage; p_m is the mechanical power; p_e is the electrical power:

$$p_e = (v_q + R_a i_q)\,i_q + (v_d + R_a i_d)\,i_d\,, \tag{B.2}$$

and v_d and v_q are the stator voltage components on machine dq-axis reference frame. These are linked to the grid voltage $v\angle\theta$ through the rotation:

$$\begin{aligned} v_d &= v\sin(\delta - \theta)\,, \\ v_q &= v\cos(\delta - \theta)\,. \end{aligned} \tag{B.3}$$

Active and reactive power injected at the machine terminal bus are:

$$\begin{aligned} p &= v_d\,i_d + v_q\,i_q\,, \\ q &= v_q\,i_d - v_d\,i_q\,. \end{aligned} \tag{B.4}$$

T. Kërçi and F. Milano, *Frequency Quality in Power Systems*, Power Systems,
https://doi.org/10.1007/978-3-032-04721-2

Table B.1 Synchronous machine parameters

Parameter	Description
D	Damping coefficient
H	Inertia constant
R_a	Armature resistance
T_{d10}	d-axis transient time constant
T_{q10}	q-axis transient time constant
X_d	d-axis synchronous reactance
X_{d1}	d-axis transient reactance
X_q	q-axis synchronous reactance
X_{q1}	q-axis transient reactance
ω_n	Synchronous reference frequency in rad/s

Finally, the following relationships between voltages and currents hold:

$$\begin{aligned} 0 &= v_q + R_a\, i_q - e_{q1} + X_{d1}\, i_d\,, \\ 0 &= v_d + R_a\, i_d - e_{d1} - X_{q1}\, i_q\,. \end{aligned} \tag{B.5}$$

Machine parameters are summarized in Table B.1. Note that, for the base case scenario, the damping coefficient D is zero for all machines of both test systems.

The control diagram of turbine governor, Automatic Voltage Regulator (AVR) and Power System Stabilizer (PSS) models utilized in the simulations are shown in Figs. B.1, B.2 and B.3, respectively. The saturation function of the exciter of the AVR is:

$$\text{sat}(v_f) = A_e \exp(B_e |v_f|)\,. \tag{B.6}$$

In the remainder of this Appendix, all per-unit parameters are referred to a power base of 100 MVA.

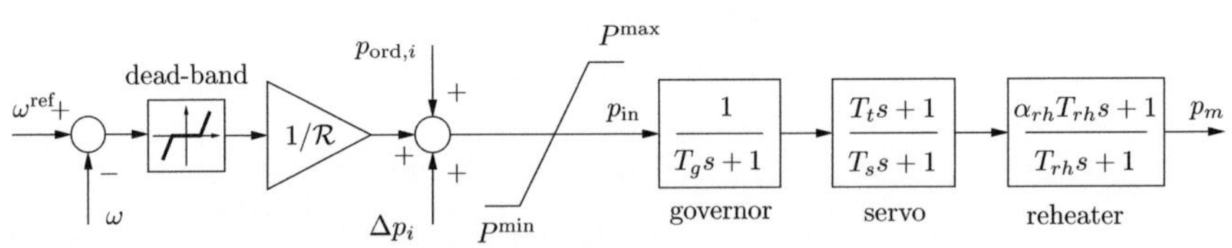

Fig. B.1 Turbine governor model. Signal Δp is the output of the AGC

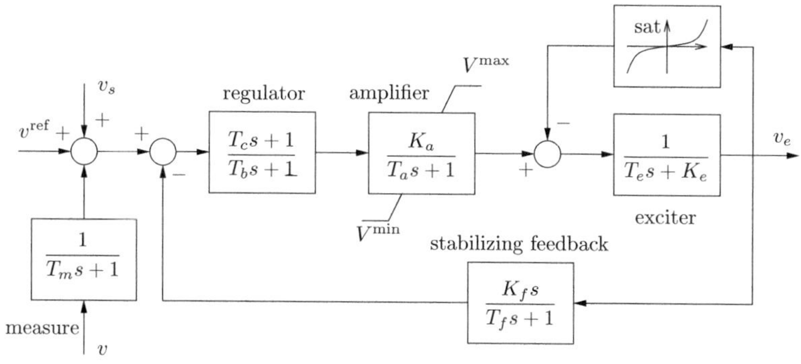

Fig. B.2 AVR model. Signal v_s is the output of the PSS

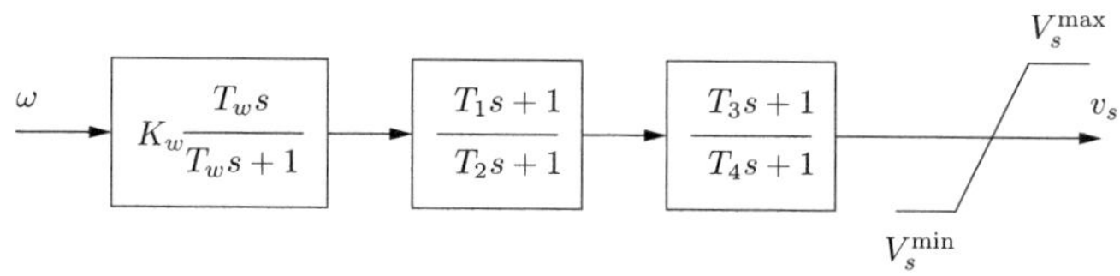

Fig. B.3 PSS model

B.1 WSCC 9-Bus System

This section provides data from the WSCC 9-bus 3-machine system, hereinafter called the *WSCC system*, utilized in several examples discussed in this book. This system is described in many references, for example, [5]. Bus 1 is the slack. Figure B.4 shows the single line diagram of the system.

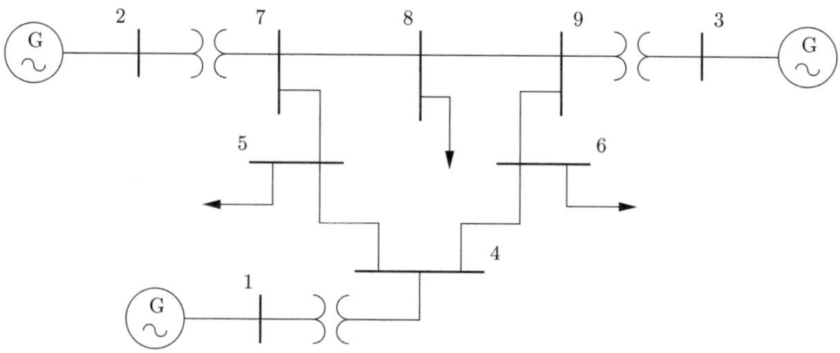

Fig. B.4 WSCC 3-machine 9-bus system

B.1.1 Network Data

The transmission system consists of four voltage levels, 13.8, 16.5, 18 and 230 kV. Lower voltage levels corresponds to three synchronous machine terminal buses before the step-up transformers, which connect buses 1–4, 2–7 and 3–9. Table B.2 shows all branch data.

B.1.2 Static Data

The power flow data and solution for the base-case operating point are shown in Table B.3.

Table B.2 Branch data of the WSCC system

Branch #	From #	To #	R_L [pu(Ω)]	X_L [pu(Ω)]	B_L [pu(Ω^{-1})]	Type
1	6	4	0.0170	0.0920	0.1580	Line
2	5	4	0.0100	0.0850	0.1760	Line
3	7	5	0.0320	0.1610	0.3060	Line
4	9	6	0.0390	0.1700	0.3580	Line
5	7	8	0.0085	0.0720	0.1490	Line
6	9	8	0.0119	0.1008	0.2090	Line
7	1	4	0	0.0576	0	Transf.
8	2	7	0	0.0625	0	Transf.
9	3	9	0	0.0586	0	Transf.

Table B.3 Base-case power flow solution of the WSCC system

Bus #	V_n [kV]	v [pu(kV)]	θ [rad]	p_G [pu(MW)]	q_G [pu(MVAr)]	p_D [pu(MW)]	q_D [pu(MVAr)]
1	16.5	1.0400	0	0.7164	0.2705	0	0
2	18.0	1.0250	0.1620	1.6300	0.0665	0	0
3	13.8	1.0250	0.0814	0.8500	–0.1086	0	0
4	230.0	1.0258	–0.0387	0	0	0	0
5	230.0	0.9956	–0.0696	0	0	1.25	0.50
6	230.0	1.0127	–0.0644	0	0	0.90	0.30
7	230.0	1.0258	0.0649	0	0	0	0
8	230.0	1.0159	0.0127	0	0	1.00	0.35
9	230.0	1.0324	0.0343	0	0	0	0

Table B.4 Synchronous machine data of the WSCC system

Machine	V_n [kV]	$2H$ [MWs/MVA]	T_{d10} [s]	T_{q10} [s]	X_d [pu(Ω)]	X_{d1} [pu(Ω)]	X_q [pu(Ω)]	X_{q1} [pu(Ω)]
1	16.5	47.28	8.96	0.310	0.1460	0.0608	0.0969	0.0969
2	18.0	12.80	6.00	0.535	0.8958	0.1198	0.8645	0.1969
3	13.8	6.02	5.89	0.600	1.3125	0.1813	1.2578	0.2500

Table B.5 Turbine governor data of the WSCC system

Machine	p^{max} [pu(MW)]	p^{min} [pu(MW)]	$\mathcal{R}$ [pu(MW)]	T_s [s]	T_{rh} [s]	T_c [s]	T_t [s]	α_{rh} –
1	1.6	0	0.05	0	50.0	0.45	0.1	0.25
2	3.2	0	0.05	0	50.0	0.45	0.1	0.25
3	1.7	0	0.05	0	50.0	0.45	0.1	0.25

Table B.6 AVR data of the WSCC system

Machine	K_a –	T_a [s]	T_b [s]	T_c [s]	K_e –	T_e [s]	K_f –	T_f [s]
All	20.0	0.2	0	0	1.0	0.314	0.063	0.35
	A_e –	B_e –	T_m [s]	V^{max} [pu(kV)]	V^{min} [pu(kV)]			
	0.0039	1.555	0.001	5.0	−5.0			

B.1.3 Dynamic Data

The parameters of the synchronous machines of the WSCC are shown in Table B.4.

The dynamic data of the turbine governors and the AVR of the WSCC system are shown in Tables B.5 and B.6, respectively.

B.2 New England 39-Bus System

The New England 39-bus 10-machine system has been widely utilized in the literature to test transient and frequency stability analysis and control techniques [6]. The slack is bus 31. Figure B.5 shows the single line diagram of the system.

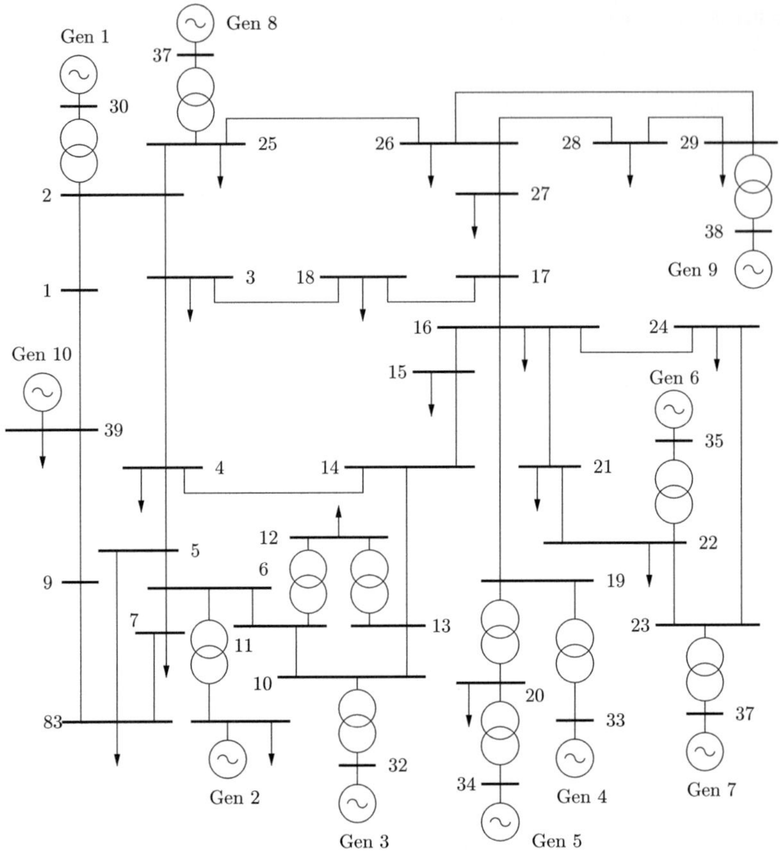

Fig. B.5 New England 10-machine 39-bus system

B.2.1 Network Data

The transmission system consists of four voltage levels, 15, 125, 230 and 380 kV. Table B.7 shows transmission line and transformer data. K_T indicates the off-nominal tap ratio of the transformers.

B.2.2 Static Data

Power flow data and solution for the base-case operating point are shown in Table B.8.

Table B.7 Branch data of the New England system

Branch #	From #	To #	R_L [pu(Ω)]	X_L [pu(Ω)]	B_L [pu(Ω^{-1})]	K_T $\left[\frac{pu(kV)}{pu(kV)}\right]$	Type
1	1	2	0.0035	0.0411	0.6987	–	Line
2	1	39	0.0010	0.0250	0.7500	–	Line
3	2	3	0.0013	0.0151	0.2572	–	Line
4	2	25	0.0070	0.0086	0.1460	–	Line
5	3	4	0.0013	0.0213	0.2214	–	Line
6	3	18	0.0011	0.0133	0.2138	–	Line
7	4	5	0.0008	0.0128	0.1342	–	Line
8	4	14	0.0008	0.0129	0.1382	–	Line
9	5	6	0.0002	0.0026	0.0434	–	Line
10	5	8	0.0008	0.0112	0.1476	–	Line
11	6	7	0.0006	0.0092	0.1130	–	Line
12	6	11	0.0007	0.0082	0.1389	–	Line
13	7	8	0.0004	0.0046	0.0780	–	Line
14	8	9	0.0023	0.0363	0.3804	–	Line
15	9	39	0.0010	0.0250	1.2000	–	Line
16	10	11	0.0004	0.0043	0.0729	–	Line
17	10	13	0.0004	0.0043	0.0729	–	Line
18	13	14	0.0009	0.0101	0.1723	–	Line
19	14	15	0.0018	0.0217	0.3660	–	Line
20	15	16	0.0009	0.0094	0.1710	–	Line
21	16	17	0.0007	0.0089	0.1342	–	Line
22	16	19	0.0016	0.0195	0.3040	–	Line
23	16	21	0.0008	0.0135	0.2548	–	Line
24	16	24	0.0003	0.0059	0.0680	–	Line
25	17	18	0.0007	0.0082	0.1319	–	Line
26	17	27	0.0013	0.0173	0.3216	–	Line
27	21	22	0.0008	0.0140	0.2565	–	Line
28	22	23	0.0006	0.0096	0.1846	–	Line
29	23	24	0.0022	0.0350	0.3610	–	Line
30	25	26	0.0032	0.0323	0.5130	–	Line
31	26	27	0.0014	0.0147	0.2396	–	Line
32	26	28	0.0043	0.0474	0.7802	–	Line
33	26	29	0.0057	0.0625	1.0290	–	Line
34	28	29	0.0014	0.0151	0.2490	–	Line
35	12	11	0.0016	0.0435	0	1.006	Transf.
36	12	13	0.0016	0.0435	0	1.006	Transf.
37	6	31	0	0.0250	0	1.070	Transf.
38	10	32	0	0.0200	0	1.070	Transf.
39	19	33	0.0007	0.0142	0	1.070	Transf.
40	20	34	0.0009	0.0180	0	1.009	Transf.
41	22	35	0	0.0143	0	1.025	Transf.
42	23	36	0.0005	0.0272	0	1.000	Transf.
43	25	37	0.0006	0.0232	0	1.025	Transf.
44	2	30	0	0.0181	0	1.025	Transf.
45	29	38	0.0008	0.0156	0	1.025	Transf.
46	19	20	0.0007	0.0138	0	1.060	Transf.

Table B.8 Base-case power flow solution of the New England system

Bus #	V_n [kV]	v [pu(kV)]	θ [rad]	p_G [pu(MW)]	q_G [pu(MVAr)]	p_D [pu(MW)]	q_D [pu(MVAr)]
1	230.0	1.0474	−0.1473	0	0	0	0
2	230.0	1.0487	−0.1004	0	0	0	0
3	230.0	1.0302	−0.1501	0	0	3.2200	0.0240
4	230.0	1.0039	−0.1677	0	0	5.0000	1.8400
5	230.0	1.0053	−0.1503	0	0	0	0
6	230.0	1.0077	−0.1387	0	0	0	0
7	230.0	0.9970	−0.1767	0	0	2.3380	0.8400
8	230.0	0.9960	−0.1853	0	0	5.2200	1.7600
9	230.0	1.0282	−0.1802	0	0	0	0
10	230.0	1.0171	−0.0947	0	0	0	0
11	230.0	1.0127	−0.1097	0	0	0	0
12	125.0	1.0001	−0.1090	0	0	0.0750	0.8800
13	230.0	1.0143	−0.1064	0	0	0	0
14	230.0	1.0117	−0.1336	0	0	0	0
15	230.0	1.0154	−0.1350	0	0	3.2000	1.5300
16	230.0	1.0318	−0.1080	0	0	3.2900	0.3230
17	230.0	1.0335	−0.1274	0	0	0	0
18	230.0	1.0309	−0.1435	0	0	1.5800	0.3000
19	230.0	1.0499	−0.0178	0	0	0	0
20	125.0	0.9912	−0.0352	0	0	6.2800	1.0300
21	230.0	1.0317	−0.0660	0	0	2.7400	1.1500
22	230.0	1.0498	0.0117	0	0	0	0
23	230.0	1.0448	0.0083	0	0	2.4750	0.8460
24	230.0	1.0373	−0.1059	0	0	3.0860	−0.9200
25	230.0	1.0576	−0.0762	0	0	2.2400	0.4720
26	230.0	1.0521	−0.0965	0	0	1.3900	0.1700
27	230.0	1.0377	−0.1308	0	0	2.8100	0.7550
28	230.0	1.0501	−0.0352	0	0	2.0600	0.2760
29	230.0	1.0499	0.0130	0	0	2.8350	0.2690
30	230.0	1.0475	−0.0582	2.5000	1.4618	0	0
31	380.0	0.9820	0	5.2081	1.9827	0.0920	0.0460
32	15.0	0.9831	0.0448	6.5000	2.0516	0	0
33	125.0	0.9972	0.0732	6.3200	1.0994	0	0
34	15.0	1.0123	0.0554	5.0800	1.6578	0	0
35	15.0	1.0493	0.0983	6.5000	2.1246	0	0
36	15.0	1.0635	0.1453	5.6000	1.0121	0	0
37	15.0	1.0278	0.0423	5.4000	0.0045	0	0
38	15.0	1.0265	0.1363	8.3000	0.2285	0	0
39	230.0	1.0300	−0.1755	10.000	0.8829	11.040	2.5000

Table B.9 Synchronous machine data of the New England system

Machine	V_n [kV]	$2H$ [MWs/MVA]	T_{d10} [s]	T_{q10} [s]	X_d [pu(Ω)]	X_{d1} [pu(Ω)]	X_q [pu(Ω)]	X_{q1} [pu(Ω)]
30	230.0	84.0	10.2	0.10	0.1000	0.0310	0.0690	0.0080
31	380.0	60.6	6.56	1.50	0.2950	0.0697	0.2820	0.0697
32	15.0	71.6	5.70	1.50	0.2495	0.0531	0.2370	0.0876
33	125.0	57.2	5.69	1.50	0.2620	0.0436	0.2580	0.1660
34	15.0	52.0	5.40	0.44	0.6700	0.1320	0.6200	0.1660
35	15.0	69.6	7.30	0.40	0.2540	0.0500	0.2410	0.0814
36	15.0	52.8	5.66	1.50	0.2950	0.0490	0.2920	0.1860
37	15.0	48.6	6.70	0.41	0.2900	0.0570	0.2800	0.0911
38	15.0	69.0	4.79	1.96	0.2106	0.0570	0.2050	0.0587
39	230.0	1,000.0	7.00	0.70	0.0200	0.0060	0.0190	0.0080

Table B.10 Turbine governor data of the New England system

Machine	p^{max} [pu(MW)]	p^{min} [pu(MW)]	$\mathcal{R}$ [pu(MW)]	T_g [s]	T_{rh} [s]	T_c [s]	T_t [s]	α_{rh} –
1	5.0	0	0.05	0	0	10.0	3.0	0
2	15.0	0	0.05	0	0	10.0	3.0	0
3	13.0	0	0.05	0	0	10.0	3.0	0
4	12.7	0	0.05	0	0	10.0	3.0	0
5	10.2	0	0.05	0	0	10.0	3.0	0
6	13.0	0	0.05	0	0	10.0	3.0	0
7	11.2	0	0.05	0	0	10.0	3.0	0
8	10.8	0	0.05	0	0	10.0	3.0	0
9	16.6	0	0.05	0	0	10.0	3.0	0
10	20.0	0	0.05	0	0	10.0	3.0	0

B.2.3 Dynamic Data

In the New England system, synchronous machines are represented by the simplified 4th order model. Machine parameters are shown in Table B.9.

The dynamic data of the turbine governors, the automatic voltage regulators and PSS of the New England system are shown in Tables B.10, B.11 and B.12, respectively. The gain of the integrator of the AGC is $K_o = 0.01$.

B.2.4 Unit Commitment Data

Table B.13 show data for deterministic Unit Commitment (d-UC) and stochastic Unit Commitment (s-UC) problems utilized in the examples of Chap. 10. The data can be also found in [7]. The value of load curtailment is taken equal to \$1000/MWh [8].

Table B.11 AVR data of the New England system

Machine	K_a	T_a	T_b	T_c	K_e	T_e	K_f	T_f
	–	[s]	[s]	[s]	–	[s]	–	[s]
All	200.0	0.015	10.0	1.0	0.01	0	0	0
	A_e	B_e	T_m	$V^{\max}$	$V^{\min}$			
	–	–	[s]	[pu(kV)]	[pu(kV)]			
	0	0	0.01	5.0	−5.0			

Table B.12 PSS data of the New England system

Machine	K_w	T_w	T_1	T_2	T_3	T_4	$V_S^{\max}$	$V_S^{\min}$
	–	[s]	[s]	[s]	[s]	[s]	[pu(kV)]	[pu(kV)]
1	1.0	10.0	5.0	0.60	3.0	0.50	0.2	−0.2
2	0.5	10.0	5.0	0.40	1.0	0.10	0.2	−0.2
3	0.5	10.0	3.0	0.20	2.0	0.20	0.2	−0.2
4	2.0	10.0	1.0	0.10	1.0	0.30	0.2	−0.2
5	1.0	10.0	1.5	0.20	1.0	0.10	0.2	−0.2
6	4.0	10.0	0.5	0.10	0.5	0.05	0.2	−0.2
7	7.5	10.0	0.2	0.02	0.5	0.10	0.2	−0.2
8	2.0	10.0	1.0	0.20	1.0	0.10	0.2	−0.2
9	2.0	10.0	1.0	0.50	2.0	0.10	0.2	−0.2
10	1.0	10.0	1.0	0.05	3.0	0.50	0.2	−0.2

Table B.13 Unit commitment data of the New England system

Unit	$P_G^{\max}$ [MW]	$P_G^{\min}$ [MW]	C_G^F [$/h]	C_G^V [$/MWh]	C_G^{SU} [$]	C_G^{SD} [$]
1	455	150	1,000	16.19	9,000	9,000
2	455	150	970	17.26	10,000	10,000
3	130	20	700	16.60	1,100	1,100
4	130	20	680	16.50	1,120	1,120
5	162	25	450	19.70	1,800	1,800
6	80	20	370	22.26	349	349
7	85	25	480	22.74	520	520
8	55	10	660	25.92	60	60
9	55	10	665	27.27	60	60
10	55	10	670	27.79	60	60

The cost of wind is assumed zero; and the value of the fixed (C_g^F) and variable (C_g^V) cost coefficients is taken equal to the fixed and proportional cost coefficients a ($/h) and b ($/MWh), respectively, in [9]. Table B.14 shows the data utilized in Sect. 9.5 for the example on Virtual Power Plant (VPP).

Table B.14 DER data of the New England system for the MILP-based VPP

Generator	Capacity [MW]	Response time [min]	Ramping time [min]
1	0.68	2.73	13.17
2	3.16	6.13	49.61
3	3.74	10.71	39.00
4	1.68	6.91	34.51
5	4.32	1.78	35.49
6	3.89	11.34	28.52
7	1.74	11.23	43.04
8	4.92	9.00	15.14
9	1.02	1.51	4.17
10	4.80	9.48	33.20

Appendix C
All-Island Power System

Figure C.1 shows the 2025 All-Island Power System (AIPS) transmission map. Due to its relatively small size and the high penetration of wind/solar Photovoltaic (PV) power, the AIPS can be considered a real-world benchmark test system for experimenting instantaneous penetration of non-synchronous generation up to 100%. This also means being one of the first power systems experiencing all potential issues of operating such grids including but not limited to, frequency stability, converter-driven interactions leading to oscillations, system strength, harmonic issues and wind and solar power curtailments. But if these issues could be fully addressed in small isolated power grids then other larger power grids could follow these real-world examples. As a matter of fact, EirGrid and System Operator for Northern Ireland (SONI) is approached on a regular basis by many Transmission System Operator (TSO) and research institutions worldwide on how they have done integrating extremely high levels of non-synchronous Renewable Energy Sources (RES) while ensuring system security and stability.

As of 2025, the AIPS has three 500 MW High Voltage Direct Current (HVDC) Interconnector (IC). New HVDC IC are expected in the future such as that with France (700 MW HVDC link, around end of 2026/early 2027) and other potential IC with Great Britain (GB). In terms of wind and solar PV geographical distribution, it can be seen that (onshore) wind is fairly distributed throughout the system while that is not (currently) the case for solar generation (located mostly in East/South-East). The distribution of these variable RES is critical from a long-term frequency quality perspective as it smooths their aggregate active power output.

With respect to voltage levels, they range from 400 kV to 100 kV. In particular, it is worth noting that the Northern Ireland (NI) system does not include the 400 kV voltage level but rather the 275 kV. It can also be said that from a very high voltage perceptive (i.e., 400 kV and 275 kV), both Ireland (IE) and NI have limited transmission lines. In this context, the TSOs recognize that the timely and efficient development of transmission networks is crucial to delivery of the renewable ambition [10]. A critical line is the North-South (N-S) 275 kV double-circuit line connecting the IE and NI power systems. This line is currently treated as a credible N-1 contingency as

T. Kërçi and F. Milano, *Frequency Quality in Power Systems*, Power Systems,
https://doi.org/10.1007/978-3-032-04721-2

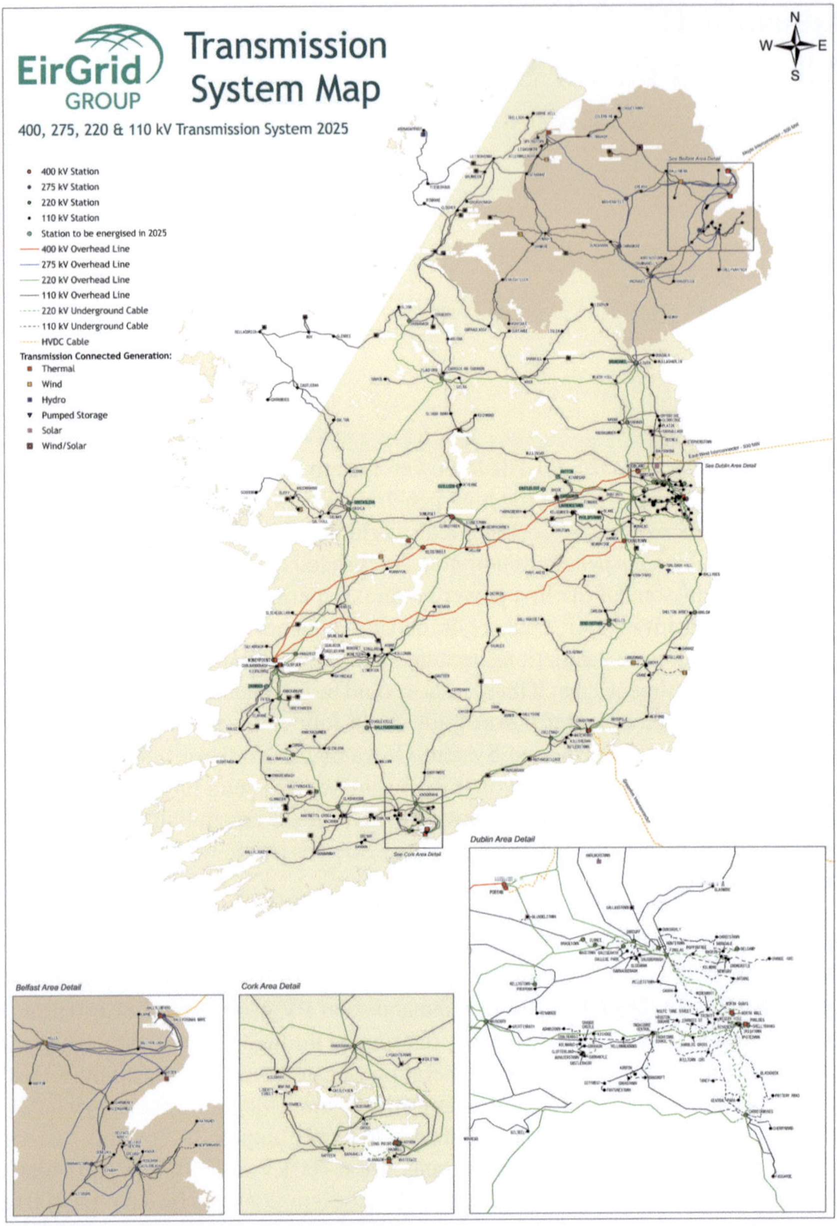

Fig. C.1 Map of the AIPS transmission system

per SONI operating security standards and a potential trip of it will lead to a system split which could be hard to deal from a frequency stability point of view. There are plans for a second N-S tie-line between the countries but this is expected to become operational beyond 2030 time frames.

Furthermore, and as expected, the major load centers are located around the capital cities of IE and NI namely around the Dublin and Belfast areas. This makes it difficult to transport energy from the rest of the AIPS. Moreover, load centers are interconnected through high voltage underground dc cables making it very difficult from a high voltage perspective, especially during night hours. These voltage issues require reactive power compensation devices such as conventional generators/condensers or Static Synchronous Compensator (STATCOM)/reactors.

Simulation results included in the book that refer to the AIPS are based on a model that includes 1479 buses, 1851 transmission lines and transformers, 22 synchronous generators, along with their appropriate control systems including 6 PSS, 169 wind power plants with a mix of Doubly-Fed Induction Generator (DFIG) and CSIG, and 245 loads. Unless otherwise specified, all wind power plants are assumed to be Grid-Following Converter (GFL) and not to provide any inertial response nor fast-frequency regulation. Generator and controller models are determined based on their technology and use commonly-accepted models. Dynamic data are guessed based on typical values for each model and technology. Thus, simulation results, while realistic, do not represent the dynamic performance of the actual AIPS.

Glossary

We collect below most relevant definitions related to frequency quality. The definitions of Fast Frequency Response (FFR), frequency control, frequency quality, frequency strength, and frequency response are original of the book.

Abnormal operating condition implies a power system state where there are faults or unit trips.

Active power control refers to the power regulation of individual units, relieving network constraints, or increasing system stability and efficiency [11].

Balancing capacity means a volume of reserve capacity that a balancingservice provider has agreed to hold and in respect to which the balancing service provider has agreed to submit bids for a corresponding volume of balancing energy to the TSO for the duration of the contract [12].

Balancing energy means energy used by TSO to perform balancing and provided by a balancing service provider [12].

Electricity ancillary service is a service necessary for the operation of a transmission or distribution system, including balancing and non-frequency ancillary services, but not including congestion management.

Fast frequency response refers to the automatic increase or decrease of power production or consumption by power system resources in response to frequency fluctuations. Power adjustments are performed through automatic Primary Frequency Control (PFC) with narrow (e.g., ± 15 mHz) or wide deadbands (e.g., ± 200 mHz) in the time frame from tens of milliseconds to tens of seconds.

T. Kërçi and F. Milano, *Frequency Quality in Power Systems*, Power Systems,
https://doi.org/10.1007/978-3-032-04721-2

Flexibility in power systems refers to the ability to manage demand and supply, solve network constraints, and avoid congestion.

Frequency control refers to the automatic or manual increase or decrease of power production or consumption by power system resources in response to frequency fluctuations and time error. Power adjustments are performed as a result of PFC, Secondary Frequency Control (SFC), Tertiary Frequency Control (TFC) and Time Error Control (TEC) in all time scales.

Frequency quality is concerned with deviations of the frequency from predefined limits/standards during normal and abnormal operating conditions. It is ensured by having adequate frequency response in place at all times.

Frequency quality (European Network of TSOs for Electricity (ENTSO-E)) is a measure of the power systems ability to maintain a stable operation during the changes of consumption and production that are creating imbalances; and to handle disturbances [13].

Frequency strength is the ability of power systems to resist and control changes in frequency during normal and abnormal operating conditions.

Frequency response refers to the natural, automatic, and manual increase or decrease of power production or consumption by power system components and resources in response to frequency fluctuations and time error. Power adjustments are performed as a result of the inherent dynamic behavior of power system components as well as any controller, in all time scales.

Frequency response dead-band means an interval used intentionally to make the frequency control unresponsive [14].

Frequency stability refers to the ability of a power system to maintain steady frequency following a severe system upset resulting in a significant imbalance between generation and load. It depends on the ability to maintain/restore equilibrium between system generation and load, with minimum unintentional loss of load. Instability that may result occurs in the form of sustained frequency swings leading to tripping of generating units and loads [15].

Normal operating condition implies a power system state where most quantities are within technical limits and there are no faults or unit trips.

Power quality power quality is concerned with deviations of the voltage from its ideal waveform (voltage quality) and deviations of the current from its ideal waveform (current quality) [16].

Power system inertia refers to the ability of a power system to oppose changes in system frequency due to resistance provided by rotating masses.

Primary frequency control refers to actions provided by the Interconnection to arrest and stabilize frequency in response to frequency deviations. Primary Control comes from automatic generator governor response, load response (typically from motors), and other devices that provide an immediate response based on local (device-level) control systems [17].

Quality is the standard of something when it is compared to other things like it; how good or bad something is [18].

Secondary frequency control refers to actions provided by an individual Balancing Authority (BA) or its Reserve Sharing Group to correct the resource–load unbalance that created the original frequency deviation, which will restore both Scheduled Frequency and Primary Frequency Response. Secondary Control comes from either manual or automated dispatch from a centralized control system [17].

Tertiary frequency control refers to actions provided by Balancing Authorities on a balanced basis that are coordinated so there is a net zero effect on area control error [19].

Time error control refers to a process for time control, where time control is a control action carried out to return the electrical time deviation between synchronous time and UTC time to zero [14].

References

1. Milano F, Zárate-Miñano R (2013) A systematic method to model power systems as stochastic differential algebraic equations. IEEE Trans Power Syst 28(4):4537–4544
2. Milano F (2016) Semi-implicit formulation of differential-algebraic equations for transient stability analysis. IEEE Trans Power Syst 31(6):4534–4543
3. Milano F (2013) A Python-based software tool for power system analysis. In: IEEE PES general meeting, pp 1–5
4. Milano F (2010) Power system modelling and scripting. Springer, London
5. Sauer PW, Pai MA (1998) Power system dynamics and stability. Prentice Hall, Upper Saddle River, NJ
6. Illinois Center for a Smarter Electric Grid (2013) IEEE 39-bus system. [Online]. http://publish.illinois.edu/smartergrid/ieee-39-bus-system
7. Chen Y, Liu F, Wei W et al (2016) Robust unit commitment for large-scale wind generation and run-off-river hydropower. CSEE J Power Energy Syst 2(4):66–75
8. Conejo A, Carrión M, Morales J (2010) Decision making under uncertainty in electricity markets. Springer
9. Carrión M, Arroyo JM (2006) A computationally efficient mixed-integer linear formulation for the thermal unit commitment problem. IEEE Trans Power Syst 21(3):1371–1378
10. EirGrid & SONI (2023) Shaping our electricity future roadmap. [Online]. https://www.eirgrid.ie/
11. Khan A, Hosseinzadehtaher M, Shadmand MB et al (2020) On the stability of the power electronics-dominated grid: a new energy paradigm. IEEE Ind Electron Mag 14(4):65–78
12. European Commission (2017) Commission regulation (EU) 2017/2195 of 23 November 2017 establishing a guideline on electricity balancing. [Online]. https://eur-lex.europa.eu
13. ENTSO-E (2015) NAG–frequency quality report. [Online]. https://www.statnett.no
14. European Commission (2017) Establishing a guideline on electricity transmission system operation. [Online]. https://eur-lex.europa.eu
15. Kundur P, Paserba J, Ajjarapu V et al (2004) Definition and classification of power system stability IEEE/CIGRE joint task force on stability terms and definitions. IEEE Trans Power Syst 19(3):1387–1401
16. Bollen MHJ (2000) Understanding power quality problems, vol 3. IEEE Press, New York
17. NERC (2012) Frequency response initiative report-the reliability role of frequency response
18. Oxford English Dictionary (2023) Flexibility. [Online]. https://www.oed.com
19. NERC Rs Committee (2021) Balancing and frequency control: reference document. Prepared by the NERC resources subcommittee, Tech, Rep, NERC Technical Report

Index

T. Kërçi and F. Milano, *Frequency Quality in Power Systems*, Power Systems,
https://doi.org/10.1007/978-3-032-04721-2

MIX
Papier aus verantwortungsvollen Quellen
Paper from responsible sources
FSC® C105338

If you have any concerns about our products,
you can contact us on
ProductSafety@springernature.com

In case Publisher is established outside the EU,
the EU authorized representative is:
Springer Nature Customer Service Center GmbH
Europaplatz 3, 69115 Heidelberg, Germany

Printed by Libri Plureos GmbH
in Hamburg, Germany